ENERGY IN TRANSITION 1985–2010

ENERGY IN TRANSITION 1985–2010

FINAL REPORT OF THE COMMITTEE ON NUCLEAR AND ALTERNATIVE ENERGY SYSTEMS

NATIONAL RESEARCH COUNCIL

National Academy of Sciences
Washington, D.C. 1979

W. H. Freeman and Company
San Francisco

The project that is the subject of this report was approved by the Governing Board of the National Research Council, whose members are drawn from the Councils of the National Academy of Sciences, the National Academy of Engineering, and the Institute of Medicine. The members of the committee responsible for the report were chosen for their special competences and with regard for appropriate balance.

This report has been reviewed by a group other than the authors according to procedures approved by a Report Review Committee consisting of members of the National Academy of Sciences, the National Academy of Engineering, and the Institute of Medicine.

This study and report were supported under Contract EX-76-C-10-3784 between the Energy Research and Development Administration and the National Academy of Sciences.

Library of Congress Cataloging in Publication Data

National Research Council. Committee on Nuclear and
Alternative Energy Systems.
Energy in transition, 1985–2010.

Includes bibliographical references and index.
1. Power resources—United States. 2. Energy
policy—United States. I. Title.
TJ163.25.U6N382 1980 333.79 79-27389
ISBN 0-7167-1227-X
ISBN 0-7167-1228-8 pbk.

Printed in the United States of America

9 8 7 6 5 4 3 2

25 December 1979

The Honorable Charles W. Duncan, Jr.
Secretary of Energy
Washington, D.C.

Dear Mr. Secretary:

I have the honor to transmit a report entitled *Energy in Transition, 1985–2010* prepared by the Committee on Nuclear and Alternative Energy Systems (CONAES) of the National Research council (NRC) and supported by Contract EX-76-C-10-3784 with the Energy Research and Development Administration (ERDA).

On April 1, 1975, Dr. Robert C. Seamans, then Administrator of ERDA, wrote to me to request that the NRC undertake "a detailed and objective analysis of the risks and benefits associated with alternative conventional and breeder reactors as sources of power." After due deliberation, the Governing Board of the NRC indicated that it would prefer "a comprehensive and objective study of the role of nuclear power in the context of alternative energy systems." These expanded terms of reference proved acceptable to ERDA, and the resultant contract between ERDA and the National Academy of Sciences so specified. Administrative management of the study within the NRC was assigned to the Assembly of Engineering.

The charge to our committee was nothing less than a detailed analysis of all aspects of the nation's energy situation. The dimensions of this charge were without precedent in the NRC. Our committees, consisting of highly qualified, public-spirited experts who serve without fee, have generally been called on to address much more narrowly circumscribed questions. The breadth of compass in this instance constituted a staggering challenge.

Harvey Brooks, then Dean of Engineering and Applied Physics at Harvard University, and Edward L. Ginzton, Chairman of the Board of Varian Associates, accepted our invitations to serve as co-chairmen of the study. The balance of the committee was then appointed after wide consultation with appropriate individuals and organizations. It was evident that the ultimate credibility of their report would rest upon public perception of the committee as balanced in composition and, in that sense, impartial. In discussing the NRC committee appointment process, my introduction to the Annual Report of the NRC for 1978 described CONAES as follows:

> An illustration of this art is afforded by the Committee on Nuclear and Alternative Energy Systems, engaged in the most complex task ever

attempted by the National Research Council. It is co-chaired by an applied physicist who is a university professor and an industrial engineer whose company manufactures scientific instruments, both of whom had previously chaired major NRC committees with great success. In all, 10 members are from academic institutions, 1 from a government laboratory, 1 from the research arm of an oil company, 1 from an instrument manufacturer, 1 from a utility company, 1 from a bank, and 1 from a law firm. From a disciplinary standpoint, there are 5 engineers, 3 physicists, 1 geophysicist, 2 economists, 1 sociologist, 1 banker, 1 physician-radiobiologist, 1 biological ecologist, and 1 "public interest" lawyer. . . . In a general way, by my appraisal when the study began, about one-third were negative, perhaps 3 were positive, and the others were genuinely open-minded concerning nuclear energy. At this writing, it is clear that the ideas that have come to be uppermost in the committee's collective thinking were central to the views of few if any of the committee members when they first met.*

The routine procedures of the NRC demand, as a condition of appointment, that each committee member file with us a disclosure of "Potential Sources of Bias" and that, at the first committee meeting, each member reveal to his colleagues the substance of that disclosure as well as the sense of his current views of the subject to be considered by the committee. That first meeting of CONAES was remarkable; the tension seemed almost physical; profound suspicion was evident; first names were rarely used; the polarization of views concerning nuclear energy was explicit. Four years later, that polarization persists, and many of the same positions are still regularly defended. But the committee has developed its own dynamic, the antagonists are personally friendly, and a very substantial measure of consensus has been achieved.

Patently, no single committee such as CONAES could embrace full competence and knowledge of all the many technical matters that would demand consideration. To provide that competence, CONAES, as described in the preface, brought into being a set of 4 major panels supported by 22 resource groups and a number of consultants, thereby acquiring the knowledge and insights of about 300 additional individuals of highly diverse backgrounds. (See Appendix C.) During January and February 1976, CONAES conducted public hearings in five major cities across the nation to test its plans for conduct of the study and to listen to approximately 100 witnesses who asked to testify. No complete summary of those hearings is available, nor did they prove particularly fruitful, but this process began the education of the CONAES members in attendance at these hearings. On 1 August 1976, CONAES adopted a Work Plan and on 12 January 1977 transmitted an Interim Report to ERDA, a planning document that remains a landmark statement of the kinds of

*In the time since, two of the original members have found it necessary to withdraw from the committee.

understandings that must be obtained if the nation is to formulate a successful energy policy.

Conduct of the study over this four-year period has been complicated by numerous developments in the nation's turbulent energy situation.

There were gasoline shortages and price rises, electricity blackouts, natural gas shortages, public debate over power plant sitings, large negative balances of payments for petroleum and for technology. Growing environmental concern was paralleled by concern that regulation is inhibiting industrial innovation and productivity. Rising prices and the debate over decontrol were accompanied by growing public distrust of the energy industries and of statements concerning the magnitude of hydrocarbon reserves. Political instability in nations on which we depend for petroleum imports made all too obvious the precariousness of the flow of imported oil. Three Mile Island revealed both the resilience designed into nuclear plants and the significance of the human factor in the operation of such plants. Established energy companies began to develop capabilities in new energy technologies, and a host of new, smaller companies entered the market for such technologies as solar heating, windmills, biomass utilization, insulation, etc.

President Carter, particularly concerned that nuclear weapons should not proliferate, took action to defer reprocessing of spent nuclear materials and to delay commercialization of a breeder reactor, while the pace of the much debated Clinch River breeder project was deliberately slowed. The President also presented to the nation energy messages emphasizing conservation, decontrol of petroleum and natural gas prices, vigorous exploration for new domestic sources, as well as a substantial synthetic fuels program to be financed from a windfall profits tax.

During this period, CONAES resource groups and panels were variously reporting that domestic uranium will be less plentifully available than had earlier been suggested, and that the linkage between growth of the energy supply and real growth of the GNP is more flexible than many had previously considered. A panel of the NRC Geophysics Research Board flagged attention to the fact that continuing buildup of atmospheric CO_2, thought to be largely due to fossil fuel combustion, would drastically alter climate, although the timing and manner of change are not yet reliably predictable. The CONAES Risk and Impact Panel reported its comparison of risks associated with various energy technologies. The work of the NRC Committee on Biological Effects of Ionizing Radiation (BEIR III) revealed the controversy concerning the biological effects of low level ionizing radiation, although, as a guide to policy makers, the differences between contending factions would appear to be rather small. The problem of planning for disposal of radioactive wastes assumed greater urgency and increasingly claimed public attention. An *ad hoc* committee under the aegis of our Committee on Science and Public Policy presented an independent analysis of the risks inherent in the nuclear fuel cycle, an analysis that highlighted, *inter alia,* the fact that uranium mining and the mine tailings are, day by day, the most hazardous elements of the system, rather than accidents at power plants or the disposal of high level waste. Numerous analyses of various aspects of our energy situation were reported by diverse groups and individuals under several auspices. And, since CONAES finished its work, an *ad hoc* conference convened by the NRC in early October concluded that use of western oil shales must be a major contributor if the President's goals for a synthetic fuels program are to be met.

ERDA was phased out and the Department of Energy was created. The new Department, not quite responsible for initiation of this effort and concerned about the lengthy time that had already elapsed, placed a ceiling on its financial support of the CONAES endeavor. During September 1978 the funds provided by ERDA and the Department were exhausted. Since then, this effort has been supported by the private funds of NAS, in a total amount of about $300,000.

Through all of these events, CONAES labored on through draft after draft. Preparation of chapter 1, in effect a short version of the report, took on the character of negotiation of a treaty; individual words and phrases were debated at wearying length. The penultimate draft of this report was sent to our Report Review Committee during the summer of 1979. A specially appointed review panel of 22 highly qualified individuals, largely members of NAS and NAE, read it with utmost care and returned to CONAES a lengthy, extremely detailed critique. CONAES responded equally carefully, accepting much of the criticism and amending the report accordingly in many cases, preferring its own position or language in others.

Most reports of this length offer a brief, explicitly designated "summary." Determined to complete its task and nearing exhaustion, CONAES eschewed preparation of such a statement. However, an equivalent of such a summary will be found in the attached letter of transmittal, to me, by the two co-chairmen, a statement which closely coincides with that which concludes chapter 1. Readers will find it helpful to study that statement before addressing the body of the report.

Most importantly, the report is addressed to a great challenge, management of the medium-term future of our energy economy, viz., the turbulent period of transition from major dependence on fossil hydrocarbons, domestic and imported, to a more stable era of utilization of energy sources that are either renewable or available on a scale sufficient for centuries. While most current public and governmental concern is necessarily focussed on the energy difficulties of the day, it is the period of this transition that must be the principal subject of major energy policy. The present report offers no prescription for such policy but does provide an analytical base and a description of alternate future scenarios that should be of considerable assistance to those who must formulate such policy.

One aspect of the CONAES exercise was the development by various panels and resource groups of a series of models of conceivable national energy and economic futures. Whereas much of the report would retain its validity in the absence of these models, their implications significantly affected the committee's thinking as it engaged in the numerous evaluations to be found in the report. Since the validity of these models rests on the validity, completeness, and consistency of their underlying assumptions, some of them quite dramatic, and since, patently, the energy futures so described flow from these premises, the reader will be well advised to examine those assumptions carefully. The variety of alternate energy futures here contemplated and their consequences for the national economy and life-style are impressive features of this report.

The report stresses the necessity to reduce national dependence on imported petroleum, to be accomplished by both conservation and switching to alternate technologies. The opportunities for conservation, and their scale and timing, are presented in some detail. Public decision concerning the major opportunities for non-petroleum-based energy production is constrained by concern for their attendant risks and environmental impact. A major feature of this report is its analysis of the state-of-the-art of these alternate technologies and a comparative assessment of their associated risks and impacts.

An unusual aspect of this report is its conclusion that future decisions concerning nuclear energy will be determined by public perceptions of risks and benefits at least as much as by rigorous conclusions drawn by scientists on the basis of scientific analysis. That circumstance places an unusually heavy burden of objectivity on those whose statements help to fashion public opinion. Excessive attention to either the risk or the benefit side of the equation, or failure to consider the alternatives, could seem to lead, on the one hand, to denial to the nation of all major energy sources or, on the other, to a false sense of security.

By design, the composition of CONAES reflected a wide spectrum of opinion concerning most aspects of the nation's energy problems, although, to be sure, none were advocates of the most extreme positions. Members frequently offered the special viewpoints expected from their places in society, as utility company executive, environmental advocate, investment banker, regulator, ecologist, physician, economist, etc., speaking on behalf of their own constituencies, as it were. Hence, the present report is unique in the growing literature concerning energy. It is particularly noteworthy precisely because it emerges from a reasonably representative microcosm of the conflicting relevant interests and viewpoints abroad in the land, rather than from a more homogeneous group with a unifying ideology.

To the extent possible, CONAES sought genuine consensus. But where the committee was significantly divided, both points of view are presented in the text. In addition, all members were invited to offer personal comments when they wished to clarify or to take exception to statements in the text that otherwise reflect the preponderance of CONAES opinion. These statements, some quite eloquent, will be found in footnotes and in Appendix A. The divisions of opinion indicated in the text and the disagreements noted in footnotes and in Appendix A, while by no means trivial, should not be permitted to lessen appreciation of the force of the analysis here presented or of the general agreement achieved on some of the most critical questions considered.

Despite the long time required to complete this effort (in large measure a consequence of the initial polarized composition of CONAES) the report could not have been more timely than it is today. Some readers may find themselves disappointed by the absence of a set of crisp recommendations for federal policy and programs. But such was not our purpose. It is the thorough analysis of almost all aspects of our energy circumstances and the detailed consideration of the possible alternatives available to the nation that constitute the principal contribution of this report. The major decisions yet to be taken must occur in the political arena and in the marketplace. It is our hope that, by illuminating our circumstances and future prospects, this report will increase the likelihood that those future decisions will be rational and based on the longer-term national interest rather than on the painful exigencies of any given moment.

Much of the material earlier available to CONAES, i.e., the reports of several of its panels and resource groups, has already been published. Several more remain to be published. Appendix D is a compilation of these titles. Each has been carefully considered and used by CONAES, but they have not been put through the normal review procedures of the NRC.

In all, about 350 individuals have contributed to various aspects of this exercise. There may well be no participant who agrees with the entirety of the CONAES report, but most participants will find themselves in substantial agreement with most of this report. An unanticipated value of this endeavor may well prove to be the educations that all participants received; the insights and understandings so gained have already found their way into the national debate as these now even more knowledgeable scientists have also participated in a multiplicity of other committees, Congressional hearings, reports, classroom teaching, and boardroom discussions. Thus, by this avenue, also, the CONAES exercise will have contributed constructively to future national energy policy.

One intrinsically political aspect of our national energy circumstance is not fully discussed by CONAES, the fact that the great uncertainty concerning our energy future has, in turn, generated innumerable other public uncertainties. These uncertainties constrain decisions by energy-producing and energy-utilizing industry; they affect personal decisions concerning housing and transportation; they inhibit foreign policy formulation and, in general, cast a pall on life in these United States. The challenge to the nation is to avoid taking, prematurely, those decisions that CONAES suggests be deferred until they can be taken with greater understanding and wisdom while, as soon as possible, enunciating and beginning to follow a stated course that will hold open as many options as possible. It is our hope that *Energy in Transition, 1985–2010* will be of assistance in that regard.

Allow me to take this opportunity to make public acknowledgment of our great debt to Harvey Brooks, who, more than any other, fashioned this report through endless hours of devoted effort and attention to all of its facets. His co-chairman, Edward L. Ginzton, earned our gratitude both by his considerable substantive contributions and by his determined drive to push the task to completion. And I am pleased to acknowledge the huge contribution of all the members of CONAES, who attended several dozen meetings and read reams of reports and drafts, who individually wrote innumerable drafts of paragraphs, pages, and chapters, and who maintained their goodwill and good humor during this prolonged exercise. Finally, let me express our profound appreciation to the panels, resource groups, consultants, and dedicated staff, without whom this report would not have been possible.

Mr. Secretary, the National Research Council is pleased, proud, and considerably relieved, to make this report available to the Department of Energy and to all Americans seriously concerned for the health of our nation's future energy economy.

Sincerely yours,

P Handler

PHILIP HANDLER
Chairman, National Research Council
President, National Academy of Sciences

Enclosure

November 6, 1979

Dr. Philip Handler
Chairman
National Research Council
2101 Constitution Avenue, N.W.
Washington, D.C. 20418

Dear Dr. Handler:

It is our pleasure to submit to you for transmittal to the Department of Energy the final report of the National Research Council Committee on Nuclear and Alternative Energy Systems (CONAES).

The purpose of the CONAES study is indicated by its title: to assess the appropriate roles of nuclear and alternative energy systems in the nation's energy future, with a particular focus on the period between 1985 and 2010. The study is intended to assist the executive and legislative branches of the government, as well as the American people as a whole, in formulating energy policy by illuminating the kinds of options the nation may wish to keep open in the future, by considering the attendant problems, and by describing the actions that may be required to do so.

Because it was central to the study's charter to assess the need and direction for nuclear power developments, the various nuclear options are considered in considerable detail. However, the decisions regarding the proper role of nuclear energy and of the several alternatives cannot be made in a contextual vacuum. We found that neither the prospective growth of our population nor other social and economic factors rigidly determine the needs of the nation for energy in the future. The study, therefore, tried to describe and relate the many economic, social, and technical factors that bear on the country's energy development and the options that must remain open to our society until ultimate decisions need to be made. Many of these decisions are not yet timely and could well be strategically in error if made too soon and based on insufficient knowledge.

This committee has studied at length the many factors and relationships involved in our nation's energy future and offers in chapter 1 some technical and economic observations that decision makers may find useful as they develop energy policy in the larger context of the future of our society. Because of their significance it seems appropriate to bring them to the reader's attention at this point, while noting that chapter 1 records also, in footnotes, the comments and reservations of individual members of CONAES concerning these major conclusions.

Our observations focus on (1) the prime importance of energy conservation, (2) the critical near-term problem of fluid fuel supply, (3) the desirability of a balanced combination of coal and nuclear fission as the only large-scale intermediate-term options for electricity generation, (4) the need to keep the breeder option open, and (5) the importance of investing now in research and development to ensure the availability of a strong range of new energy options sustainable over the long term.

Policy changes both to improve energy efficiency and to enhance the supply of alternatives to imported oil will be necessary. The continuation of artificially low prices would inevitably widen the gap between domestic supply and demand, and this could only be made up of increased imports, a policy that would be increasingly hazardous and difficult to sustain.

The most vital of these observations is the importance of energy demand considerations in planning future energy supplies. There is great flexibility in the technical efficiency of energy use, and there is correspondingly great scope for reducing the growth of energy consumption without appreciable sacrifices in the growth of GNP or in nonenergy consumption patterns. Indeed, as energy prices rise, the nation will face important losses in economic growth if we do not significantly increase the economy's energy efficiency. Reducing the growth of energy demand should be accorded the highest priority in national energy policy.

In the very near future, substantial savings can be made by relatively simple changes in the ways we manage energy use, and by making investments in retrofits of existing capital stock and consumer durables to render them more energy efficient.

The most substantial conservation opportunities, however, will be fully achievable only over the course of two or more decades, as the existing capital stock and consumer durables are replaced. There are economically attractive opportunities for such improvements in appliances, automobiles, buildings, and industrial processes at today's prices for energy, and as prices rise, these opportunities will multiply.

This underscores the importance of clear signals from the economy about trends in the price of energy. New investments in energy-consuming equipment should be made with an eye to energy prices some years in the future. Without clear ideas of the replacement cost of energy and its impact on operating costs, consumers will be unlikely to choose appropriately efficient capital goods. These projected cost signals should be given prominence and clarity through a carefully enunciated governmental pricing policy. They can be amplified where desirable by regulation; performance standards, for example, are useful in cases (such as the automobile) where fuel prices are not strongly reflected in operating costs.

Although there is some uncertainty in these conclusions because of possible feedback effects of energy consumption on labor productivity, labor-force participation, and the propensity for leisure, calculations indicate that, with sufficiently high energy prices, an energy/GNP ratio one half of today's could be reached, over several decades, without significant adverse effects on economic growth. Of course, so large a change in this ratio implies large price increases and consequent structural changes in the economy. This would entail major adjustments in some sectors, particularly those directly related to the production of energy and of some energy-intensive products and materials. However, given the slow introduction of these changes, paced by the rate of turnover in capital stock and consumer durables, we believe neither their magnitude nor their rate will exceed those experienced in the past owing to changes in technology and in the conditions of economic competition among nations. The possibility of reducing the nation's energy/GNP ratio should serve as a stimulus to strong conservation efforts. It should not, however, be taken as a dependable basis for foregoing simultaneous and vigorous efforts on the supply programs discussed in this report.

The most critical near-term problem in energy supply for this country is fluid fuels. World supplies of petroleum will be severely strained beginning in the 1980s, owing both to the expectation of peaking in world production about a decade later and to new world demands. Severe problems are likely to occur earlier because of political disruptions or cartel actions. Next to demand-growth reduction, therefore, highest priority should be given to the development of a domestic synthetic fuels industry, for both liquids and gas, and to vigorous exploration for conventional oil and gas, enhanced recovery, and development of unconventional sources (particularly of natural gas).

As fluid fuels are phased out of use for electricity generation, coal and nuclear power are the only economic alternatives for large-scale application in the remainder of this century. A balanced mix of coal- and nuclear-generated electricity is preferable to the predominance of either. After 1990, for example, coal will be increasingly required for the production of synthetic fuels. The requirements for nuclear capacity depend on the growth rate of electricity demand; this study's projections of electricity growth between 1975 and 2010 (for up to 3 percent annual average GNP growth) are considerably below industry and government projections, and in the highest conservation cases actually level off or decline after 1990. Such projections are sensitive also to assumptions about end-use efficiency, technological progress in electricity generation and use, and the assumed behavior of electricity prices in relation to those of primary fuels. They are therefore subject to some uncertainty.

At relatively high growth rates in the demand for electricity, the attractiveness of a breeder or other fuel-efficient reactor is greatest, all other things being equal. At the highest growth rates considered in this study, the breeder can be considered a probable necessity. For this reason, this committee recommends continued development of the LMFBR breeder, so that it can be deployed early in the next century if necessary. Any decision on deployment, however, should be deferred until the future courses of electricity demand growth, fluid fuel supplies, and other factors become clearer.

In terms of public risks from routine operation of electric power plants (including fuel production and delivery), coal-fired generation presents the highest overall level of risk, with oil-fired and nuclear generation considerably safer, and natural gas the safest. With respect to accidents, the generation of electricity from fossil fuels presents a very low risk of catastrophic accidents. The projected mean number of fatalities associated with nuclear accidents is probably less than the risk from routine operation of the nuclear fuel cycle (including mining, transportation, and waste disposal), but the large range of uncertainty that still attaches to nuclear safety calculations makes it difficult to provide a confident assessment of the probability of catastrophic reactor accidents. The spread of uncertainty in present estimates of the risks of both coal and nuclear power is such that the ranges of possible risk overlap somewhat. High-level nuclear waste management does not present catastrophic risk potential, but its long-term low-level threat demands more sophisticated and comprehensive study and planning than it has so far received, particularly in view of the acute public sensitivity to this issue.

The problem of nuclear weapons proliferation is real and is probably the most serious potentially catastrophic problem associated with nuclear power. However, there is no technical fix—even the stopping of nuclear power (especially by a single nation)—that averts the nuclear proliferation problem. At best, the danger can be delayed while better control institutions are put in place. There is a wide difference of opinion about which represents the greater threat to peace: the dangers of proliferation associated with the replacement of fossil resources by nuclear energy, or the exacerbation of international competition for access to fossil fuels that could occur in the absence of an adequate worldwide nuclear power program.

Because of their higher economic costs, solar energy technologies other than hydroelectric power will probably not contribute much more than 5 percent to energy supply in this century, unless there is massive government intervention in the market to penalize the use of nonrenewable fuels and subsidize the use of renewable energy sources. Such intervention could find justification in the generally lower social costs of solar energy in comparison to alternatives. The danger of such intervention lies in the possibility that it may lock us into obsolete and expensive technologies with high materials and resource requirements, where greater reliance on "natural" market penetration would be less costly and more efficent over the long term. Technical progress in solar technologies, especially photovoltaics, has accelerated dramatically during the last few years; nevertheless, there is still insufficient effort on long-term research and exploratory development of novel concepts. A much increased basic research effort should be directed at finding ways of using solar energy to produce fluid fuels, which may have the greatest promise in the long term.

Major further exploitation of hydroelectric power, or of biomass through terrestrial energy farms, presents ecological problems that make it inadvisable to count on these as significant future incremental energy sources for the

United States. (Marine biomass energy farms could have none of these problems, of course.) There is insufficient information to judge whether the large-scale exploitation of hot-dry-rock geothermal energy or the geopressured brines will ultimately be feasible or economic. Local exploitation of geothermal steam or hot water is already feasible and should be encouraged where it offers an economical substitute for petroleum.

It is too early in the investigation of controlled thermonuclear fusion to make reliable forecasts of its economic or environmental characteristics. It is not, however, an option that can be counted on to make any contribution within the time frame of this study. Nevertheless, fusion warrants sufficient technical effort to enable a realistic assessment by the early part of the next century of its long-term promise in competition with breeder reactors and solar energy technologies.

It is important to keep in mind that the energy problem does not arise from an overall physical scarcity of resources. There are several plausible options for an indefinitely sustainable energy supply, potentially accessible to all the people of the world. The problem is in effecting a socially acceptable and smooth transition from gradually depleting resources of oil and natural gas to new technologies whose potentials are not now fully developed or assessed and whose costs are generally unpredictable. This transition involves time for planning and development on the scale of half a century. The question is whether we are diligent, clever, and lucky enough to make this inevitable transition an orderly and smooth one.

Thus, energy policy involves very large social and political components that are much less well understood than the technical factors. Some of these sociopolitical considerations are amenable to better understanding through research on the social and institutional characteristics of energy systems and the factors that determine public, official, and industry perception and appraisal of them. However, there will remain an irreducible element of conflicting values and political interests that cannot be resolved except in the political arena. The acceptability of any such resolution will be a function of the processes by which it is achieved.

Sincerely,

HARVEY BROOKS
Co-Chairman

EDWARD L. GINZTON
Co-Chairman

Committee on Nuclear and Alternative Energy Systems

HARVEY BROOKS (Co-Chairman), Benjamin Peirce Professor of Technology and Public Policy, Aiken Computation Laboratory, Harvard University

EDWARD L. GINZTON (Co-Chairman), Chairman of the Board, Varian Associates

KENNETH E. BOULDING, Distinguished Professor of Economics, Institute of Behavioral Science, University of Colorado

ROBERT H. CANNON, JR., Chairman, Department of Aeronautics and Astronautics, Stanford University

EDWARD J. GORNOWSKI, Executive Vice President, Exxon Research and Engineering Company

JOHN P. HOLDREN, Professor, Energy and Resources Program, University of California at Berkeley

HENDRIK S. HOUTHAKKER, Henry Lee Professor of Economics, Department of Economics, Harvard University

HENRY I. KOHN, David Wesley Gaiser Professor Emeritus of Radiation Biology, Harvard Medical School

STANLEY J. LEWAND, Vice President, Public Utilities Division, The Chase Manhattan Bank

LUDWIG F. LISCHER, Vice President of Engineering, Commonwealth Edison Company

JOHN C. NEESS, Professor of Zoology, University of Wisconsin at Madison

DAVID ROSE, Professor, Nuclear Engineering, Massachusetts Institute of Technology

DAVID SIVE, Attorney at Law, Winer, Neuberger and Sive

BERNARD I. SPINRAD, Professor, Nuclear Engineering, Radiation Center, Oregon State University

Staff

Study Director: JACK M. HOLLANDER, Lawrence Berkeley Laboratory, University of California

Deputy Study Director: JOHN O. BERGA, National Research Council

GEORGENE MENK, Staff Asistant

KAREN LAUGHLIN, Administrative Assistant

VIVIAN SCOTT, Administrative Assistant

SANDRA JONES, Secretary

Editors

CLAUDIA B. ANDERSON

DUNCAN M. BROWN, Senior Editor

GREGG FORTE

AURORA GALLAGHER

S. EDWARD GAMAREKIAN

LYNN SCHMIDT

VIRGINIA WHEATON

Demand and Conservation Panel

JOHN H. GIBBONS (Chairman), Director, Office of Technology Assessment, U.S. Congress (Dr. Gibbons was Director of the Energy, Environment, and Resources Center, University of Tennessee, during the time of his active participation in the work of the panel)

CHARLES A. BERG, Consultant, Buckfield, Maine

NORMAN M. BRADBURN, Director, National Opinion Research Center, University of Chicago

CLARK BULLARD, Director, Office of Conservation and Advanced Energy Systems Policy, U.S. Department of Energy

ROGER S. CARLSMITH, Director, Energy Conservation Program, Oak Ridge National Laboratory

L. DUANE CHAPMAN, Associate Professor of Resource Economics, Department of Agricultural Economics, Cornell University

PAUL P. CRAIG, Department of Applied Science, University of California at Davis

JOEL DARMSTADTER, Senior Fellow, Resources for the Future, Inc.

R. EUGENE GOODSON, Director, Institute for Interdisciplinary Engineering Studies, Purdue University

LEE SCHIPPER, Energy and Resources Group, and Lawrence Berkeley Laboratory, University of California at Berkeley

ROBERT G. UHLER, Vice President, National Economic Research Associates

MACAULEY WHITING, Consultant, Dow Chemical Company

LEROY COLQUITT, JR., Staff Officer, National Research Council

STEPHEN RATTIEN, Consultant

Risk and Impact Panel

JAMES CROW (Chairman), Professor of Genetics, University of Wisconsin at Madison

CARTER DENNISTON, Professor of Genetics, University of Wisconsin at Madison

JOHN HARTE, Senior Scientist, Energy and Environment Division, Lawrence Berkeley Laboratory, University of California at Berkeley

JOSEPH M. HENDRIE, Chairman, U.S. Nuclear Regulatory Commission (Dr. Hendrie was a member of the staff of Brookhaven National Laboratory during the time of his active participation in the work of the panel)

TERRY LASH, Staff Scientist, Natural Resources Defense Council

DAVID OKRENT, Professor of Energy and Kinetics, School of Engineering and Applied Science, University of California at Los Angeles

HOWARD RAIFFA, Professor, Kennedy School of Government, Harvard University

STEPHEN SCHNEIDER, Deputy Head, Climate Project, National Center for Atmospheric Research

CARL M. SHY, Professor of Epidemiology and Director, Institute for Environmental Studies, School of Public Health, University of North Carolina at Chapel Hill

DAVID SILLS, Executive Associate, Social Science Research Council

THEODORE B. TAYLOR, Professor of Aerospace and Mechanical Sciences, Princeton University

RICHARD SILBERGLITT, Staff Officer, National Research Council

PAUL F. DONOVAN, Consultant

Supply and Delivery Panel

W. KENNETH DAVIS (Chairman), Vice President, Bechtel Power Corporation

FLOYD L. CULLER (Deputy Chairman), President, Electric Power Research Institute

DONALD G. ALLEN, President, Yankee Atomic Electric Company

PETER L. AUER, Professor of Mechanical and Aerospace Engineering, Cornell University

JAMES BOYD, Consultant, Washington, D.C.

HERMAN M. DIECKAMP, President, General Public Utilities

DEREK P. GREGORY, Assistant Vice President, Energy Systems Research, Institute of Gas Technology

ROBERT C. GUNNESS, Former Vice Chairman and President, Standard Oil Company of Indiana

JOHN W. LANDIS, Senior Vice President, Stone and Webster Engineering

MILTON LEVENSON, Director, Nuclear Power Division, Electric Power Research Institute

ERIC H. REICHL, President, Conoco Coal Development Company

MELVIN K. SIMMONS, Assistant Director, Solar Energy Research Institute

MORTON C. SMITH, University of California, Los Alamos Scientific Laboratory

A. C. STANOJEV, Vice President, Ebasco Services, Inc.

JOHN O. BERGA, Staff Officer, National Research Council

HANS L. HAMESTER, Consultant

Synthesis Panel

LESTER B. LAVE (Chairman), Senior Fellow, Brookings Institution

RICHARD E. BALZHISER, Director, Fossil Fuel and Advanced Systems Division, Electric Power Research Institute

DAVID COHEN, President, Common Cause

CHARLES O. JONES, Maurice Falk Professor of Politics, Department of Political Science, University of Pittsburgh

TJALLING C. KOOPMANS, Alfred Cowles Professor of Economics, Cowles Foundation for Research in Economics, Yale University

LAURA NADER, Professor, Department of Anthropology, University of California at Berkeley

WILLIAM D. NORDHAUS, John Musser Professor of Economics, Cowles Foundation for Research in Economics, Yale University

SAM H. SCHURR, Senior Fellow and Co-Director, Energy and Materials Division, Resources for the Future, Inc.

JON M. VEIGEL, Director of Research and Development, Energy Resources, Conservation and Development Commission, State of California

DAVID O. WOOD, Program Director, Energy Economics and Management Division, Energy Laboratory, Massachusetts Institute of Technology

BRIAN CRISSEY, Staff Officer, National Research Council

JAMES E. JUST, Consultant

Contents in Brief

Contents

6 Solar Energy 345

Preface

In June 1975 the National Research Council, at the request of the Energy Research and Development Administration, undertook a comprehensive study of the nation's energy future, with special consideration of the role of nuclear power among alternative energy systems. The Committee on Nuclear and Alternative Energy Systems (CONAES) was formed to carry out the study.

The study, in assessing the roles of nuclear and alternative energy systems in the nation's energy future, focuses on the period between 1985 and 2010. Its intent is to illuminate the kinds of options the nation may wish to keep open in the future and to describe the actions, policies, and research and development programs that may be required to do so. The timing and the context of these decisions depend not only on the technical, social, and economic features of energy supply technologies, but also on assumptions about future demand for energy and the possibilities for energy conservation through changes in consumption patterns and improved efficiency of the supply and end-use systems.

The committee developed a three-tiered functional structure for the project. The first tier was CONAES itself, whose report embodies the ultimate findings, conclusions, and judgments of the study. To provide scientific and engineering data and economic analyses for the committee, a second tier of four panels was appointed by the committee to examine (1) energy demand and conservation, (2) energy supply and delivery systems, (3) risks and impacts of energy supply and use, and (4) various models of possible future energy systems and decision making. Each panel in turn established a number of resource groups—some two dozen in all—to address in detail an array of more particular matters. (The members of each resource group are listed in Appendix C, along with contractors and consultants to the study.)

It should be emphasized that this report, although it embodies the contributions of several hundred individuals, is solely the responsibility of the committee. However, the committee was chosen to represent a wide range of viewpoints and backgrounds, and in such a group, covering so broad a topic, it is impossible to reach consensus on every issue. Committee members were encouraged, at the conclusion of the study, to submit individual statements on subjects with whose treatment in the report they were especially dissatisfied. These statements are indicated in the report by footnotes, the longer statements appearing as Appendix A.

The National Research Council customarily publishes only the final reports of its committees. However, many of the panel and resource group reports, prepared to provide information for the committee, are valuable energy documents in their own rights. They are therefore also being published. The panel reports were reviewed by designated members of CONAES under procedures approved by the Report Review Committee of the National Research Council. The resource group reports, published as supporting papers, were reviewed by less formal procedures. The findings expressed in the panel and resource group reports are those of the authors and are not endorsed by CONAES or the National Research Council; some of the conclusions are inevitably at variance with those of the CONAES report. Appendix D lists the currently available and forthcoming publications of the CONAES study.

ACKNOWLEDGMENTS

While the fourteen members of the Committee on Nuclear and Alternative Energy Systems are solely responsible for this report, many other individuals and groups contributed information and analyses. Volunteer members of the panels and resource groups were the main contributors to the body of information compiled during the study. In most cases these groups were assisted by consultants and staff assistants. The panels and resource groups in addition commissioned a number of papers and studies.

Several individuals made especially important contributions to producing the committee report. Staff officers Leroy Colquitt, Jr., Brian

Crissey, and Richard Silberglitt worked closely with the committee and individual panels (with whom their names are listed) between 1975 and 1977.

The editorial staff began its work in 1977 and carried through to the completion of the study; particular acknowledgment is due Duncan Brown and Aurora Gallagher, who were the principal editors for the committee from June 1977 to report completion. Leonard S. Cottrell III helped with background research and analysis.

All of these efforts were guided by the study director, Jack M. Hollander, who served while on leave of absence from the Lawrence Berkeley Laboratory between 1975 and December 1977, and by John O. Berga, who coordinated staff efforts in 1978 and 1979.

The staff was ably supported in processing the many manuscript drafts during most of this period by Vivian Scott, Karen Laughlin, and Sandra Jones and is particularly grateful for their efforts. Important and timely assistance was provided by the administrative units of the National Academy of Sciences, especially the Copying Service, the Manuscript Processing Unit, and the Office of Publications.

A list of individuals who made significant contributions to the work of the committee and panels is printed as Appendix C of this volume. This list is by no means complete, and the committee expresses appreciation to all the others whose efforts furthered the work of this study.

Measuring Energy

Energy is used in a wide variety of forms, with different physical and thermal qualities and different capacities for mutual substitution. It is often convenient, however, to specify the quantity of energy in terms of a common unit. For this study, and most others undertaken in the English-speaking world, that unit is the British thermal unit, or Btu (the amount of energy required to raise the temperature of 1 pound of water 1°F from 39.2°F to 40.2°F). A barrel of crude oil, for example, contains about 5.8 million Btu; petroleum as consumed averages about 5.5 million Btu per barrel. When very large amounts of energy are discussed, it is convenient to use the unit quad, defined as one quadrillion (1,000,000,000,000,000) Btu.

The following table puts these quantities into perspective.

U.S. Energy Consumption in 1978

Energy Source	Consumption: Standard Units	Consumption: Quads	Conversion Factor (values are equivalent to 1 quad)
Coal[a]	623.5 million short tons	14.09	44.3 million short tons
Natural gas	19.41 trillion cubic feet	19.82	0.979 trillion cubic feet
Petroleum[b]	6838 million barrels	37.79	181 million barrels
Hydropower[c]	301.6 billion kilowatt-hours	3.15	95.7 billion kilowatt-hours
Nuclear power[c]	276.4 billion kilowatt-hours	2.98	92.9 billion kilowatt-hours
Geothermal and other[c,d]	3.3 billion kilowatt-hours	0.07	46.3 billion kilowatt-hours
Net imports of coke	5.0 million short tons	0.13	38.5 million short tons
TOTAL[e]		78.01	

[a] Includes bituminous coal, lignite, and anthracite.

[b] Includes natural gas plant liquids and crude oil burned as fuel, as well as refined products.

[c] The conversions from kilowatt-hours to Btu's are necessarily arbitrary for these conversion technologies. The hydropower thermal conversion rates are the prevailing heat-rate factors at fossil-steam electric power plants. Those for nuclear power and geothermal energy represent the thermal conversion equivalent of the uranium and geothermal steam consumed at power plants. The heat content of 1 kilowatt-hour of electricity, regardless of the generation process, is 3413 Btu.

[d] Includes wood, refuse, and other organic matter burned to generate electricity.

[e] Details do not add to total due to rounding.

ENERGY IN TRANSITION 1985–2010

1 Overview

The energy problem now faced by the United States began to be recognized 10 years or more ago. Still, the occasional symptoms (the oil embargo of 1973, the natural gas shortage of 1976–1977, and the gasoline lines of the summer of 1979) are frequently mistaken for the problem itself. As each symptom is relieved, the public sense of crisis fades. The seeds of future crisis, however, remain.

Resolution of the problem demands a systematic examination of energy supply and demand in the context of existing policies, and articulation of a coherent set of policies for the transition to new sources of energy and new ways of using it. The essential difficulty is that these policies must be as consonant as possible with other, often conflicting, national objectives—protecting the environment and public health and ensuring national security, economic growth, and equity among different regions and classes. The nation's energy problems are exemplified by two simple facts: stagnant domestic production and rising demand. Total energy production in the United States in 1978 was about 3 percent less than in 1972, the last full year before the oil embargo and OPEC price rise of 1973–1974 (Figure 1-1). In the same period, energy consumption rose by 9 percent (Figure 1-2). The difference is made up by increasing oil imports at continually rising prices. Imports now provide about half of all the oil consumed in the United States, up from about 30 percent in 1972. The total cost has jumped from $4.77 billion in 1972 to $41.46 billion in 1978.[1]

In the meantime, total world demand for oil has risen even more rapidly[2–4] while exporting nations, with an eye to the ultimate depletion of what is in many cases the sole source of wealth, have exercised strict

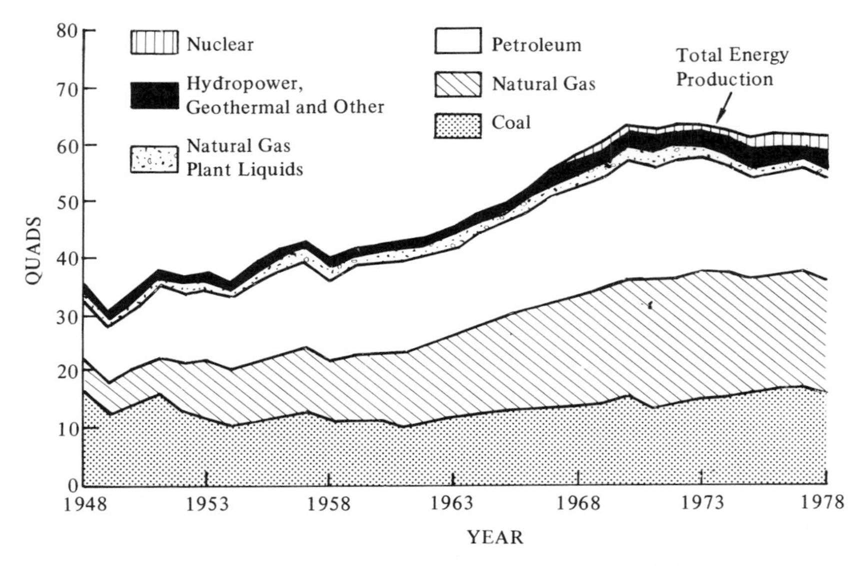

FIGURE 1-1 Energy production in the United States from 1948 to 1978, by energy source (quads). Source: U.S. Department of Energy, Energy Information Administration, *Annual Report to Congress, 1978,* vol. 2, *Data* (Washington, D.C.: U.S. Department of Energy (DOE/EIA- 0173/2), 1979).

control over production. Thus, the United States is forced to compete for supplies in an increasingly tight world market. The inevitable result is upward pressure on prices and enhanced opportunities for the control of prices by cartel.

The United States is a key factor in the world oil situation. U.S. oil consumption is huge, amounting to almost 30 percent of world consumption. At the same time, its domestic production is declining, probably irreversibly (except for some temporary help from Alaskan production, which will peak in the 1980s). Natural gas production is also on a downward trend. These production trends might be arrested by higher prices and favorable public policies, but any increase above current production levels is likely to be small and to decline after the year 2000. The only readily available large-scale domestic energy sources that could even in principle reverse the decline in domestic energy production over the next three decades—coal and nuclear fission*—face a variety of technical, political, and environmental obstacles, and will be difficult (though not impossible) to expand very rapidly.†

*See statement 1-1, by H. Brooks, Appendix A.
†See statement 1-2, by J. P. Holdren, Appendix A.

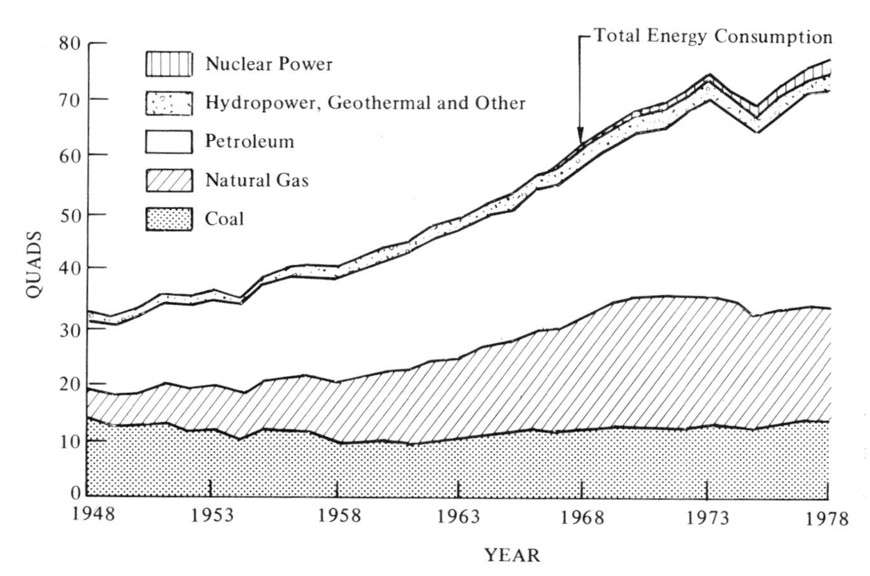

FIGURE 1-2 Energy consumption in the United States from 1948 to 1978, by energy source (quads). Source: U.S. Department of Energy, Energy Information Administration, *Annual Report to Congress, 1978,* vol. 2, *Data* (Washington, D.C.: U.S. Department of Energy (DOE/EIA-0173/2), 1979).

The implications are serious. First of all, rising dependence on increasingly costly foreign oil tends to degrade the value of the dollar and exacerbates inflation. The heavy and growing involvement of the United States in the world oil market not only worsens the domestic problem, but puts less affluent importing countries at a growing disadvantage in competing for supplies. The foreign policy consequences of this strained situation are twofold: Oil-producing countries find it increasingly feasible to exact political concessions from importers, and U.S. relations with other oil importers are weakened.

The United States has been a net importer of energy since the early 1950s. Energy was cheap, and it grew cheaper throughout the 1950s and 1960s; little concern was expressed as consumption more and more outpaced domestic production. In constant 1948 dollars, the price per barrel of crude oil at the wellhead fell from $2.50 in 1948 to $1.85 in 1972; imported oil was even cheaper. Most other forms of energy—notably electricity and coal—declined even more in price than oil. Net energy imports rose on the average more than 10 percent annually throughout the 1960s, more than doubling in that decade. Sources of supply became increasingly concentrated in the Middle East and Africa.

In 1970 domestic oil production peaked, and growth in imports accelerated. From 1970 until the fourfold OPEC price rise in 1973–1974, oil imports rose at rates exceeding 30 percent annually—almost doubling again in 3 years. The price rise brought in its wake a serious economic recession; energy consumption, and therefore imports, dipped in response. They rebounded sharply afterward, though rates of increase are now less than in the early 1970s. The nation now imports more than a fifth of its primary energy in the form of foreign oil.

The solution to this problem is not simply to produce more energy, and not simply to conserve, but rather to find a new economic equilibrium between supply and demand.* Higher prices are inevitable, and the nation must take advantage of the resulting new opportunities for both enhanced supply and greater efficiency in energy use.

Ordinary market forces will play important roles here. In some cases, however, such as the international oil market, they will be relatively ineffective and must be supplemented by government incentives to conserve and by federal aid in developing new technologies that can allow wider use of domestic resources such as coal, to allay the growth in demand for oil.

All in all, conservation deserves the highest immediate priority in energy planning. In general, throughout the economy it is now a better investment to save a Btu than to produce an additional one.† On the supply side, the most important short-term measure is to enhance domestic oil and gas production by exploiting unconventional sources and enhanced-recovery techniques. The most important intermediate-term measure is developing synthetic fuels from coal, and perhaps from oil shale, to serve where coal and nuclear power (which are most suitable now for electricity production) cannot directly replace oil and gas, as in transportation. Perhaps equally important is the use of coal and nuclear power to produce electricity for applications such as space heating, where such replacement is possible.

While these measures are being taken, the research and development necessary to bring truly sustainable energy sources—nuclear fission, solar energy, geothermal energy in places, and perhaps fusion—into place for the long term must receive continued attention. The relative merits of the principal long-term choices, and the timing of their execution, are discussed in subsequent sections of this chapter and in the body of the report.

*Statement 1-3, by R. H. Cannon, Jr.: This is too weak. Energy production increases of major proportions and vigorous conservation are both crucial to national economic viability and security. Neither alone can suffice.

†Statement 1-4, by R. H. Cannon, Jr.: Generalization unwarranted. It is often true but often not, for many energy inefficiencies have already been corrected.

MODERATING DEMAND GROWTH

Slowing the growth of energy demand will be essential, regardless of the supply options developed during the coming decades. In fact, the demand element of the nation's energy strategy should be accorded the highest priority. Some reduction in growth will inevitably result from rising energy prices, and this reduction could be accelerated by such explicit government policies as taxes and tariffs on energy and standards for the performance of energy-using equipment. In any event, studies by the CONAES Demand and Conservation Panel indicate that the growth of demand for energy in this country could be reduced substantially—particularly after about 1990—by gradual increases in the technical efficiency of energy end-use and by price-induced shifts toward less energy-intensive goods and services.[5]

In this analysis the Demand and Conservation Panel explored the dynamics and determinants of energy use by performing detailed economic and technological analyses of the major energy-consuming sectors: buildings, industry, and transportation. The projected energy intensities for each sector were based on (1) expected economic responses to price increases and income growth and (2) technical changes in energy efficiency that would be economical at the prices assumed and would minimize the life cycle costs of automobiles, appliances, houses, manufacturing equipment, and so on. No credit was taken for major technological breakthroughs; only advances based on currently available technology were considered.

A major conclusion from this analysis is that technical efficiency measures alone could reduce the ratio of energy consumption to gross national product (for convenience, the energy/GNP ratio) to as little as half* its present value over the next 30-40 years. (This conclusion is sensitive to the prices assumed in the analysis, and a result of this magnitude is attained only if prices for energy increase more rapidly than is probable in a market at equilibrium.) Similar conclusions were reached by the CONAES Modeling Resource Group,[6]† whose work suggests that such reductions are possible without appreciable impacts on the consumer market basket.

In some cases the price increases necessary to reach such reductions in demand would have to be secured by taxes that would open up a wedge between consumer prices and the costs of producing and delivering energy. Whether this would be politically tolerable or not may be open to question. It is possible, however, that if such price increases are not imposed

*Statement 1-5, by R. H. Cannon, Jr.: It would be wrong to depend on so large an improvement. Calculations using other models and assumptions predict severe economic impact for smaller energy/GNP reductions.

†See statement 1-6, by E. J. Gornowski, Appendix A.

domestically, they will be imposed by the international oil market with considerably greater abruptness.

These findings are embodied in the panel's "scenarios," or estimates of energy demand under a range of different assumed circumstances involving the price of energy and the consequent technological responses in terms of energy consumption. (A scenario is a kind of "what if" statement, giving the expected results of more or less plausible assumptions about future events, according to some self-consistent model.) The Demand and Conservation Panel's scenarios are intended to project—given certain unvaried assumptions about population growth and income growth, labor productivity, and the like—the effects on energy demand between 1975 and 2010 of various price schedules for delivered energy. The assumed prices range from an average quadrupling by 2010 to a case in which the average price of delivered energy actually decreases by one third. Table 1-1 lists the generalized assumptions and postulated prices for each of these demand scenarios. (The specific assumed prices for individual fuels in each of these demand scenarios can be found in Table 11-2 of chapter 11.) Obviously, high-priced energy evokes greater efficiency in use and thus lower consumption.

One of the key assumptions in the panel's scenarios is that the U.S. gross national product grows at an average rate of 2 percent between 1975 and 2010*; a variant of one scenario explores the implications of 3 percent growth. More rapid economic growth, as might be expected, implies higher energy consumption.†

The panel found that the economically rational responses of consumers to this range of energy prices would result in a broad range of energy consumption totals for the year 2010.‡ Figures 1-3 and 1-4 illustrate the width of this range. Chapters 2 and 11 explain more about the assumptions and methods used in making these projections.

A WORD ABOUT THE STUDY'S PROJECTIONS

The Demand and Conservation Panel's scenarios are only one of a variety of scenarios developed and used in this study to aid in visualizing the complex interplay among policies, prices, and technologies in the supply and demand of energy. Table 1-2 summarizes the main features and

*Statement 1-7, by R. H. Cannon, Jr.: Over the entire 33-yr period 1946 to present, 3.4 percent GNP growth, not 2 percent, has been consistent with a healthy economy and reasonably low unemployment.

†See statement 1-8, by H. S. Houthakker and H. Brooks, Appendix A.

‡Statement 1-9, by R. H. Cannon, Jr.: Assuming 3.4 percent GNP growth would make the 2010 quad figures (roughly) for scenario A 125, for scenario B 160, for scenario C 230, and for scenario D 270.

TABLE 1-1 Essential Assumptions of Demand and Conservation Panel Scenarios

Scenario	Energy Conservation Policy	Average Delivered Energy Price in 2010 as Multiple of Average 1975 Price (1975 dollars)	Average Annual GNP Growth Rate (percent)
A*	Very aggressive, deliberately arrived at reduced demand requiring some life-style changes	4	2
A	Aggressive; aimed at maximum efficiency plus minor life-style changes	4	2
B	Moderate; slowly incorporates more measures to increase efficiency	2	2
B′	Same as B, but 3 percent average annual GNP growth	2	3
C	Unchanged; present policies continue	1	2
D	Energy prices lowered by subsidy; little incentive to conserve	0.66	2

purposes of each set. Chapter 11 deals in some detail with all the scenario projections made in this study, but brief descriptions of the most important ones will be vital to an understanding of much of what follows.

The Supply and Delivery Panel, in its scenarios, estimated the availabilities of various energy forms between 1975 and 2010 under three progressively more favorable sets of assumed financial and regulatory conditions. These are denoted "business as usual," "enhanced supply," and "national commitment." This exercise provided the committee with an idea of the problems and potentials of the nation's major energy supply alternatives. Table 1-3 lists, as an example, the supplies of energy that might be made available if all energy sources could be accorded the incentives implied by the panel's enhanced-supply assumptions.

With the scenarios of these two panels as a basis, the staff of the study attempted to develop a self-consistent set of projections for the consumption of the various energy forms between 1975 and 2010; the method in brief was to use the demand scenarios as a framework, and to fill the demands thus established by entering the available supplies of each major energy form as given by the Supply and Delivery Panel's scenarios. Some interfuel substitutions were made, and the resulting differences in

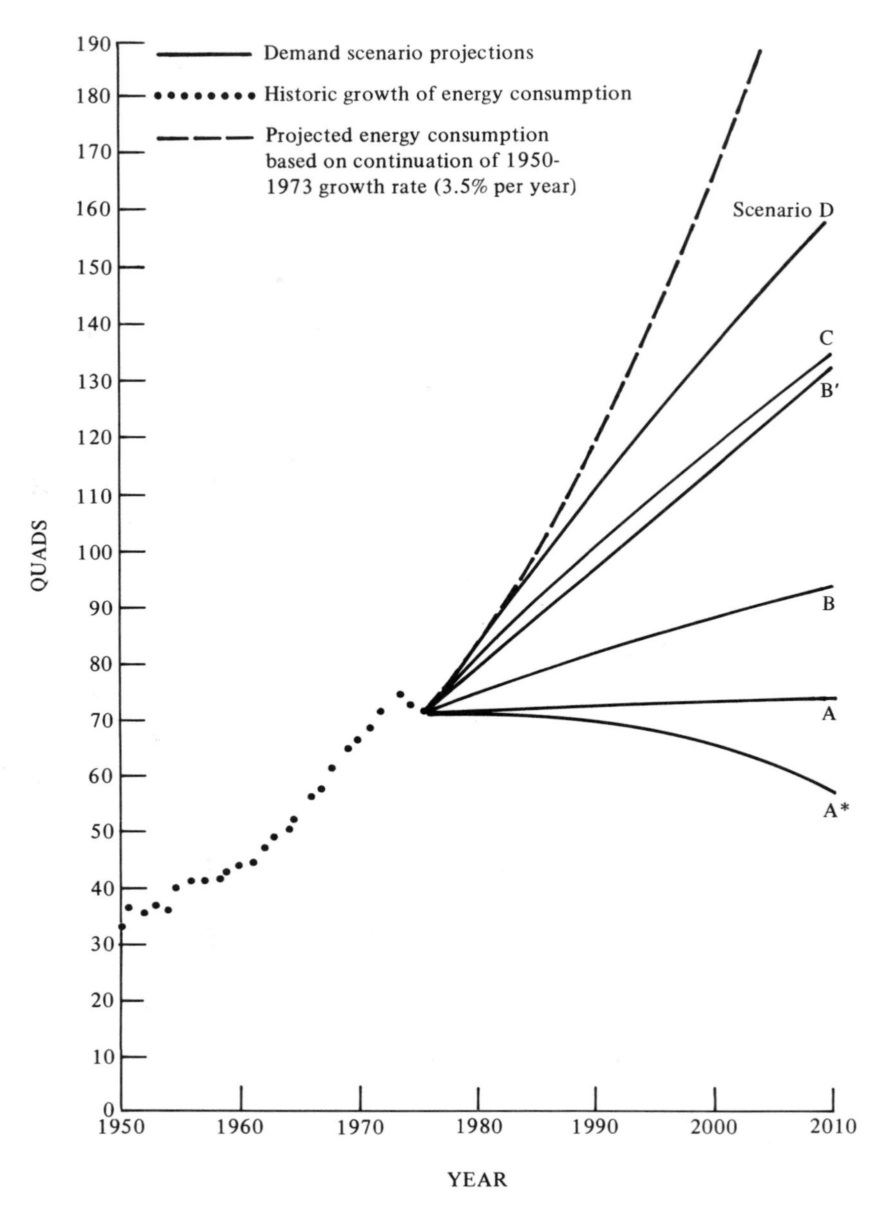

FIGURE 1-3 Demand and Conservation Panel projections of total primary energy use to 2010 (quads).

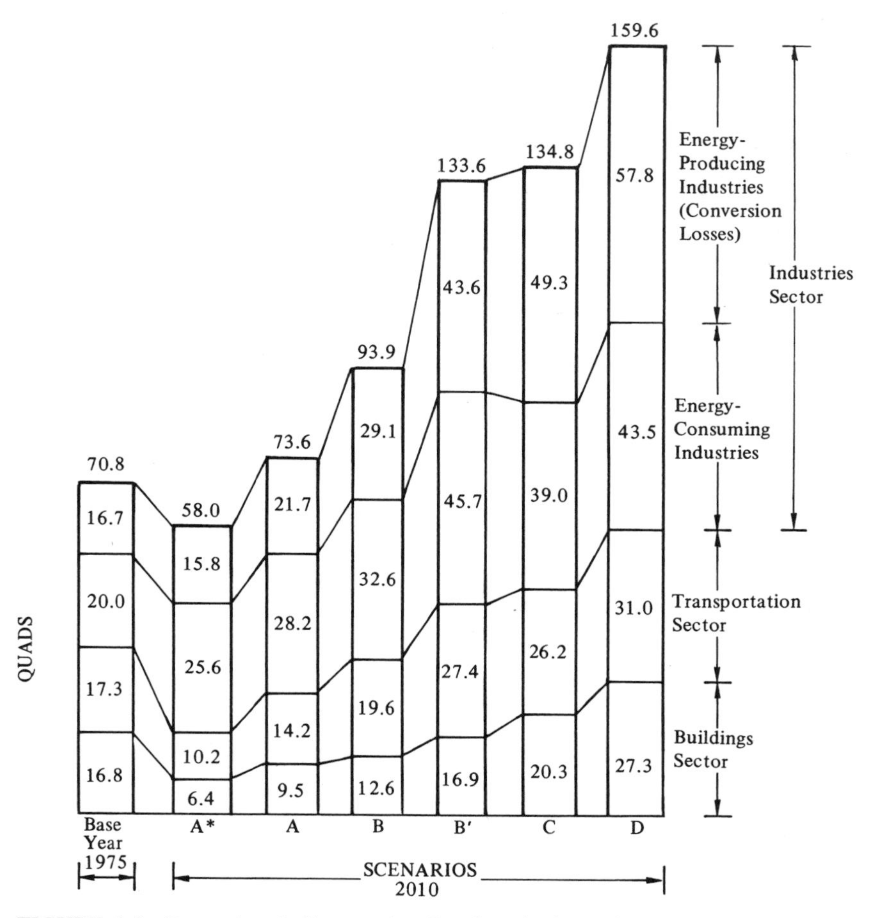

FIGURE 1-4 Demand and Conservation Panel projections of primary energy use by energy-consuming sectors to 2010 (quads). Energy demand projections for different assumptions about GNP or population growth can be roughly estimated by scaling the scenario projections. For example, for a crude idea of the effect of 3 percent average annual GNP growth (rather than the 2 percent assumed in constructing the scenarios), one would multiply the demand total by 3/2.

conversion and distribution losses and the like cause the projected totals to vary somewhat from the Demand and Conservation Panel's framework. These scenarios offer a 3 percent GNP growth variant for each of the Demand and Conservation Panel's scenarios. Figure 1-5, showing the primary energy totals for these scenarios, illustrates the difference varying GNP growth assumptions might make.

Yet another set of scenarios was developed by the CONAES Modeling

TABLE 1-2 Scenario Projections Used in the CONAES Study

Scenario	Source	Description
Demand scenarios: A*, A, B, B′, C, D	Demand and Conservation Panel	A, B, C, and D explore the effects of varied schedules of prices for energy at the point of use, from an average quadrupling between 1975 and 2010 (scenario A) to a case (scenario D) in which the average price of energy falls to two thirds of its 1975 value by 2010. Basic assumptions include 2 percent annual average growth in GNP, and population growth to 280 million in the United States in 2010. Scenario A* is a variant of A that takes additional conservation measures into account. Scenario B′ is a variant of B, projecting the effect on energy consumption of a higher annual average rate of growth in GNP (3 percent).
Supply scenarios: Business as usual, enhanced supply, and national commitment	Supply and Delivery Panel	Projections of energy resource and power production under various sets of assumed policy and regulatory conditions. Business-as-usual projections assume continuation without change of the policies and regulations prevailing in 1975; enhanced-supply and national-commitment projections assume policies and regulatory practices to encourage energy resource and power production.
Study scenarios: I_2, I_3, II_2, II_3, III_2, III_3, IV_2, IV_3 (correspondence between study scenarios and demand scenarios: I_2 = A*, II_2 = A, III_2 = B, III_3 = B′, IV_2 = C; scenario D was not used)	Staff of the CONAES study	Based on the demand scenarios; integrations of the projections of demand from the demand scenarios and projections of supply from the supply scenarios. A variant of each price-schedule scenario was projected for 3 percent annual average growth of GNP.
MRG scenarios	Modeling Resource Group	Estimates of the economic costs of limiting or proscribing energy technologies in accordance with various policies.

Resource Group in its econometric investigation of various determinants of energy supply and demand. Unlike the three sets of scenarios thus far described, those of the Modeling Resource Group do not proceed from prices (or, equivalently, policies) given at the outset. They are based instead on equilibration of supply and demand, so that prices come as outputs, rather than being given as inputs. Generally speaking, these scenarios contain much less sectoral detail than the other scenarios used in the study; in exchange for this simplification, they permit a more extensive exploration of different policies (including special constraints or moratoria on particular technologies).

It should always be borne in mind, in dealing with scenarios and other projections, that they cannot pretend to predict the future. All scenarios require great oversimplification of reality, and many judgments enter into their assumptions. The value of scenarios is in their self-consistency, which allows an approximate view of relationships between supply and demand, trade-offs among different energy sources, and the possible impacts of broadly defined policies.* The temptation to take this kind of projection too literally should be resisted, but as means of illustrating certain gross features of the nation's energy system and its possible evolution, this study's scenarios have value.

THE ECONOMIC EFFECTS OF MODERATING ENERGY CONSUMPTION

According to the analyses of the Demand and Conservation Panel, the kinds of energy conservation that offer the greatest promise of substantially moderating in the growth of energy consumption involve replacing equipment and structures with those that are more energy efficient. To avoid economic penalties, the rate of replacement must generally depend on the normal turnover of capital stock—about 10 years for automobiles, 20–50 years for industrial plants, and 50 years or more for housing—though rising energy prices will accelerate this turnover in most cases. The effects of conservation will become evident only over the long term,† but these long-term benefits require many actions that must be begun immediately, and sustained consistently over time.

As Table 1-1 and Figure 1-3 illustrate, the panel found that any of a range of primary energy consumption totals (varying by a factor of more than 2) could be compatible with the same rate of growth in GNP. Thus, energy consumption may exert less influence on the size of the economy than often has been supposed.

These findings were borne out by the work of the Modeling Resource

*See statement 1-10, by E. J. Gornowski, Appendix A.

†Statement 1-11, by J. P. Holdren: An oversimplification. Many approaches to conservation—such as retrofitting existing equipment—produce big short-term gains.

TABLE 1-3 Supply of Major Energy Forms Under Supply and Delivery Panel's Enhanced-Supply Assumptions (quads)[a]

	Annual Supply			
Energy Form	1977	1990	2000	2010
Crude oil	19.6	20.0	18.0	16.0
Natural gas	19.4	15.8	15.0	14.0
Oil shale	0	0.7	1.0	1.5
Synthetic liquids[b]	(0)	(0.4)	(2.4)	(8.0)
Synthetic gas[b]	(0)	(1.7)	(3.5)	(4.8)
Coal	16.4	26.6	37.2	49.5
Geothermal	0	0.6	1.6	4.1
Solar	0	1.7	5.9	10.7
Nuclear	2.7	13.0	29.5	41.7
Hydroelectric	2.4	4.1	5.0	5.0

[a]For specific assumptions underlying estimates, see the report by the National Research Council, *U.S. Energy Supply Prospects to 2010*, Committee on Nuclear and Alternative Energy Systems, Supply and Delivery Panel (Washington, D.C.: National Academy of Sciences, 1979) and Chapter 11, Table 11-14.
[b]Synthetic fuels are produced from coal and oil shale and are not included in totals.

Group[7]—work undertaken by different methods and for different purposes. This group sought, among other aims, a first approximation of the cost of limiting the energy available from specific technologies, the cost being measured as the size of the resulting effect on cumulative GNP. The group also assessed the feedback effect on GNP of imposing a blanket tax on all primary sources of energy to reduce energy consumption to specific levels below a base case.

The group found this feedback surprisingly small, assuming that the economy is given time to adjust by shifting capital and other resources from the processes of energy production and use to less energy-intensive processes, activities, and products. Subsequent work[8,9] has tended generally to confirm these conclusions.*

The size of the feedback depends critically, however, on the parameter that describes the quantitative effect of all these substitutions taken together: the long-term price elasticity of demand for primary energy. This value is the ratio of the percentage change in demand to the percentage change in price that evokes it. For example, if demand falls 5 percent in

*Statement 1-12, by R. H. Cannon, Jr.: Hogan confirms the trend but finds quantitatively larger GNP impact, due to less simplistic assumptions about labor productivity and capital availability.

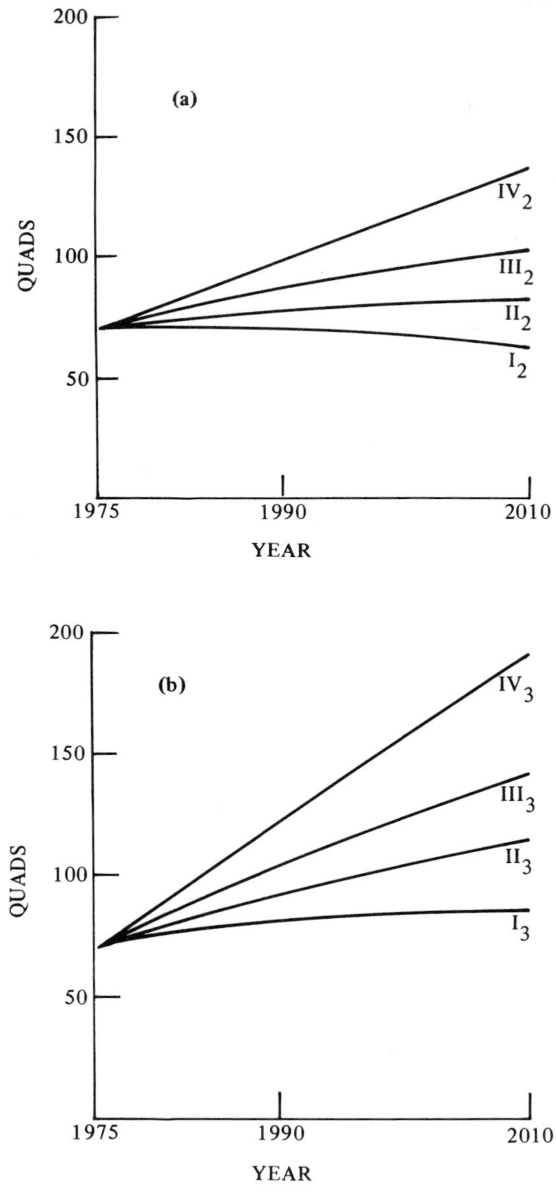

FIGURE 1-5 Projections of total primary energy consumption for CONAES study scenarios to 2010 (quads), with assumed (a) 2 percent GNP growth and (b) 3 percent GNP growth.

response to a 10 percent increase in price, the price elasticity of demand is equal to $-5 \div 10$, or -0.5.

The Modeling Resource Group reports that for the case in which primary energy consumption is reduced by 58 percent below the market-equilibrium "base case," cumulative GNP between 1975 and 2010 decreases just 2 percent if the price elasticity of demand for primary energy is -0.5, but 29 percent if the value of this parameter is -0.25. The elasticity parameter thus is a key source of uncertainty in the Modeling Resource Group's work, because its true value is not well known. More detailed discussions can be found in chapters 2 and 11.

It should be noted that even for the higher elasticity value, achieving this reduction is estimated by the Modeling Resource Group to require a tax on electricity rising by 2010 to 126 mills per kilowatt-hour (kWh) and a tax on oil and gas rising to $8.90 per million Btu (both measured in 1975 dollars). This implies a price for oil of more than 4 times the 1978 OPEC price. For electricity it implies about an eightfold increase over 1975 prices.* (See notes to Table 11-38.)

The work of the Demand and Conservation Panel and the Modeling Resource Group points up the importance of allowing the economy sufficient time to make the substitutions and institute the changes necessary to accommodate higher prices for energy or limitations on supply (or both). Sudden supply curtailments or changes in energy prices can disrupt the economy. The same changes introduced gradually over several decades may have only minor economic effects.

DOMESTIC ENERGY SUPPLIES FOR THE NEAR TERM

The supply of fluid fuels—gas and oil—which together provide about 75 percent of the nation's energy, will be critical in the 1980s and 1990s. Petroleum supplies worldwide will be severely and increasingly strained as world production approaches its probable peak near the end of the century. This probably would be true even if there were no OPEC; the possibilities of politically controlled prices and production cutbacks are greatly enhanced by such a situation. Domestic production of oil and natural gas has already peaked and begun to decline, and U.S. demand for imports already imposes rather serious strains on the world oil market. Oil production from Prudhoe Bay in Alaska will provide only temporary relief before beginning to fall off in the 1980s. Even the most optimistic projections of the CONAES Supply and Delivery Panel[10] show irreversible

*See statement 1-13, by J. P. Holdren, Appendix A.

declines in domestic oil and natural gas production in the future. Coal and nuclear power are the only large-scale alternatives† to oil and gas in the near term (before about 2000), as the use of fluid fuels begins to wane.‡ Both are best suited to the generation of electricity in this period. As such they are limited as replacements for fluid fuels, but will have uses in other applications.

A balanced combination of coal- and nuclear-generated electricity is preferable, on environmental and economic grounds, to the predominance of either. The principal points that favor nuclear electricity in its present form (light water reactors (LWR's) operated with a once-through fuel cycle without fuel reprocessing) are as follows.

- In most regions, the average cost of nuclear electricity is less than that of coal-generated electricity, and the difference is likely to continue in the future.*
- The cost of nuclear energy is less sensitive than that of coal to future increases in fuel prices and to changes in environmental standards. Because of this, the use of nuclear power could reduce future regional disparities in electric power costs.
- Nuclear fuel supplies are more readily stockpiled than coal, and nuclear electricity is thus less subject to interruption by strikes, bad weather, and transportation disruptions.
- The environmental and health effects of routine operation of nuclear reactors are substantially less than those of coal per unit of electric power produced.
- If the effect of carbon dioxide (CO_2) accumulation on climate becomes a major global environmental issue in the early years of the twenty-first century, it will be aggravated by utility commitments to the use of coal, because power plants have lives of 30–40 years.

The principal points in favor of coal are the following.

- Coal power plants and the coal fuel cycle are not subject to low-probability, high-consequence accidents or sabotage, which are inherently uncertain and unpredictable. The hazards of coal can be made relatively predictable, given sufficient research on such matters as the health effects of coal-derived air pollutants. (This research will take perhaps 15–20 years to complete, however.)

†See statement 1-14, by E. J. Gornowski, Appendix A.

‡Statement 1-15, by J. P. Holdren: My longer dissenting view, statement 1-2, Appendix A, also applies here.

*Statement 1-16, by J. P. Holdren: This point and the next one may well depend on a lower incidence of safety-related nuclear plant shutdowns than is likely.

- Coal burning in utilities has no major foreign policy implications, as does nuclear power via the problems of nuclear weapons proliferation and safeguards. The outlook for political acceptance of coal may thus be more favorable than that for nuclear energy.

- Coal is better adapted to generation of intermediate-load power, and in this sense is complementary to base-load nuclear plants. In addition, the lead time for planning coal-burning power plants is less than that for nuclear plants.

- Coal-generated electricity has a much larger resource base than light water reactors operated on a once-through fuel cycle, which will be important if fuel reprocessing and the development of more resource-efficient reactor systems and fuel cycles are further delayed.

- In the absence of a demonstrated, licensable plan for high-level waste management, the nuclear fuel cycle may be considered an incompletely proven technology, which is therefore subject to uncertainties as to whether its continued growth will be permitted. To the degree that this is so, nuclear energy runs a greater risk than coal of future capacity shortfalls due to unexpected technical developments.

After 1990, coal will likely be increasingly demanded for conversion to synthetic fuels, and nuclear generation may thus be required for continued growth in generating capacity. The amount of nuclear capacity needed is sensitively dependent on the profile of electricity growth after 1990, and especially after 2000. The several issues surrounding coal- and nuclear-generated electricity are discussed in chapters 4, 5, and 9. Chapter 11 compares various rates of installation for both coal-fired and nuclear power plants under assumed rates of growth for electricity consumption.

Electricity can be provided from almost any primary fuel and thus adds a good deal of flexibility to energy supply. However, probably even in comparison with synthetic liquids and gases, it has high capital costs.[11] There is a complex trade-off between fuel flexibility, which favors electricity, and cost, which favors fluid fuels in applications such as heating and cooling buildings and providing most industrial heat. Electricity prices are considered likely to rise less rapidly than the prices of oil, gas, and synthetic fuels, owing to technological progress in the generation of electricity and to the large fraction of electricity cost attributable to fixed capital charges, which remain constant once a plant is built but for future plants tend to increase at the same rate as the general price level. The CONAES Demand and Conservation Panel, however, assumed delivered electricity prices would rise nearly as quickly as other fuel prices. These differences may result in underestimated electricity growth in the CONAES projections.*

For the intermediate term, conservation of fluid fuels is an urgent necessity. Even in the projections embodying vigorous energy conservation, limited supplies of fluid fuels could lead to rapid price rises, especially if imports are constrained or subject to cartel pricing. If prices rise too rapidly, there will be insufficient time for development and investment to adjust, and economic dislocation will result.

The constraints on supplies of fluid fuels could probably not be fully relieved by a high-electrification policy depending on coal and nuclear fission, except at a considerably increased total cost.† However, accelerated electrification could contribute significantly to relieving future fluid fuel problems. Commitment to rapid nuclear development, for example, could be regarded as fairly expensive insurance against rapid increases in fluid fuel prices, but domestic oil and gas exploration and development of a strong synthetic fuel industry‡ should be accorded the most urgent priorities in energy supply (next in importance to conservation).

DOMESTIC OIL AND GAS

Production of both petroleum and natural gas in the United States is on the decline, and according to the analysis of this study, will continue to decline. Oil production in this country peaked in 1970 at 3.5 billion barrels, and by 1978 had fallen to 3.2 billion barrels. Domestic natural gas production shows a similar pattern; production peaked in 1973 at 21.7 trillion ft^3, and by 1978 stood at 18.9 trillion.

These trends reflect the fact that domestic oil and gas are rapidly becoming more difficult and expensive to find and produce, as development moves toward deeper wells and the exploitation of deposits in such relatively inaccessible locations as the Alaskan North Slope and the Outer Continental Shelf. Reserves of both oil and natural gas have been falling since about 1970, though exploration has expanded rapidly in that time. Reserves now equal about 10 times annual production—the lowest level since the Prudhoe Bay field was added to reserves in 1970.

*Statement 1-17, by J. P. Holdren: There is no more reason to suppose the Demand and Conservation Panel underestimated future electricity growth than to believe they overestimated it.

†Statement 1-18, by J. P. Holdren: It is completely implausible that electrification could fully relieve the fluid fuel problem in the study's time frame even at greatly increased cost.

‡Statement 1-19, by D. J. Rose and H. Brooks: An important warning has been omitted: The timing of global environmental problems from overuse of fossil fuels is uncertain, but their possible severity demands caution.

Under the policies prevailing until recently, the CONAES Supply and Delivery Panel projected that domestic production of oil would fall from 20 quadrillion Btu (quads) in 1975 to only 6 quads in 2010 (production in 1977 was 17.5 quads). Moderately enhanced conditions for oil production (including removal of price controls, accelerated offshore leasing, and somewhat advanced exploration and production technology) would bring production in 2010 to 16 quads, according to the projections, and a national commitment (relaxation of some environmental standards and permit requirements, along with federal priorities on labor and materials for oil development) might raise this to 18 quads in 2010. Under no plausible conditions does it appear possible even to maintain current domestic oil production, much less increase it.*

Gas production projections of the Supply and Delivery Panel show an even more severe decline than the oil projections. Under prevailing policies, extrapolated to 2010, gas production falls from a 1975 total of 19.7 quads to 5 quads in 2010. Moderately enhanced conditions yield a 2010 production level of 14 quads, and a national commitment results in 16 quads of gas production in 2010. Not all experts (including several participants in the CONAES study) agree with these conclusions, however. There is a considerable body of opinion that the potential for new natural gas sources, including several types of "unconventional" sources, is much higher than the study's supply projections indicate. This opinion has gained a considerable number of new adherents since 1976, when the supply projections were made.

In the light of the Demand and Conservation Panel's projections for liquid and gaseous fuels[12] (which suggest that demand is likely to continue rising until at least 2010), this outlook for production is disturbing. It suggests that the nation will become increasingly dependent on imports of oil from a world market that is already strained and that the oil situation will worsen before improving.

The situation for natural gas is not so serious, because there is a large amount of unmarketed (flared or reinjected) gas in the world. However, even sustaining current domestic natural gas consumption will probably require imports larger than the current 1 quad/yr. Most of these imports are likely to come by pipeline from Canada and possibly from Mexico, but the remainder may have to be in the form of liquefied natural gas (LNG), the landed price of which reflects the costs of liquefaction, transportation, and storage. World supplies of gas are larger compared to demand than those of oil, and their production can be expanded more readily. The international implications of importing gas are correspondingly less severe.

*Statement 1-20, by H. S. Houthakker: An increase in domestic oil production, while unlikely, cannot be ruled out if prices are high enough and new petroleum provinces are opened up.

However, the cost, and its effect on our trade balance, will not be negligible. It would be obviously unwise for the nation to become as dependent on imported gas as it now is on imported oil.*

The response of the United States to this challenge must be two-sided. Every reasonable effort must be made to conserve both oil and natural gas by using them more efficiently, by substituting alternative domestic energy forms (initially coal and conventional nuclear power for the most part, and later synthetic liquids and gases, solar energy, breeder reactors, and other long-term energy sources),† and by reducing growth in overall energy demand. An equally determined effort must be made to sustain and encourage domestic production to the extent consistent with environmental protection.

This committee does not believe that oil shale, despite the huge energy content of the domestic resource, will be a major source of energy.‡ First, the resource is concentrated in a very small and relatively primitive region, where large-scale development is likely to face resistance on environmental grounds. Second, water supplies are a serious constraint.[13] Third, the amount of solid waste that must be handled is very large relative to the energy extracted, even with in situ processing. However, these conclusions should not be interpreted as justifying the neglect of oil shale development. Every new source helps, and oil from shale will probably become economically competitive earlier than other synthetic fuels.

These efforts to deal with the problem of fluid fuels—it must be stressed—deserve high national priority in energy policy. The longer a commitment is delayed, the more likely it will be that pressures for hasty and ill-considered crash programs will build up. Such programs would involve high technological risks and possibly compromise of environmental and safety standards.

PROSPECTS FOR COAL

Coal is the nation's (and the world's) most abundant fossil fuel. Domestic recoverable reserves amount to 6,000 quads, part of a total domestic resource of about 80,000 quads and world resources crudely estimated at 300,000 quads. Of this huge supply, we consume about 14 quads each year in the United States, or less than 0.3 percent of domestic recoverable coal

*See statement 1-21, by H. Brooks, Appendix A.

†Statement 1-22, by J. P. Holdren: I reject the implication of this wording that the need to replace oil and gas justifies the use of every alternative, including breeders.

‡Statement 1-23, by R. H. Cannon and E. J. Gornowski: Despite the problems foreseen, we believe that the huge oil shale reserves in the United States will be developed to produce very large quantities of fluid fuel.

reserves. In contrast, the nation extracts almost 10 percent of its 420-quad recoverable reserves of oil and natural gas each year.

The substitution of coal for natural gas and oil on a large scale, either directly or through synthetic coal-derived substitutes, would on these grounds seem a ready-made solution to the nation's energy problems. The simple arithmetic of availability, however, does not tell the whole story. Doubling or tripling the use of coal will take time, investments amounting over the years to hundreds of billions of dollars, and coordinated efforts to solve an array of industrial, economic, and environmental problems.

Unlike oil and gas consumption, coal use is limited not by reserves or production capacity, but by the extraordinary industrial and regulatory difficulties of mining and burning it in an environmentally acceptable, and at the same time economically competitive, manner. Coal is chemically and physically extremely variable, and it is relatively difficult to handle and transport. Its use produces heavy burdens of waste matter and pollutants. Even at its substantial price advantage, Btu for Btu, it cannot compete with oil and natural gas in many applications, because of the expense of handling and storing it, disposing of ash and other solid wastes, and controlling emissions to the air. Only in very large installations, such as utility power plants and large industrial boilers, is coal today generally economic and environmentally suitable as a fuel. Domestic coal production capacity today exceeds domestic* demand, and this may well remain true until the end of the century.[14]

The health problems associated with coal affect both its production and its use. The health of underground miners presents complex and costly problems, for example, and is in need of better management; black lung is the notable instance. At the other end of the fuel cycle, the evolving state of air pollution regulations to deal with the emissions of coal combustion complicates planning for increased demand and thus in turn inhibits investment in mines, transportation facilities, and coal-fired utility and industrial boilers.

The future is obscured also by a number of more speculative problems, which may result in further regulatory restrictions on the use of coal. Chief among these is the risk that before the middle of the next century, emissions of carbon dioxide, an unavoidable (and essentially uncontrollable) product of fossil fuel combustion, may produce such concentrations in the atmosphere that large and virtually irreversible alterations may occur in the world's climate. (See chapter 9.) Also worrisome is the water-supply situation, which could limit synthetic fuel production or electricity generation unless large-scale and possibly expensive measures are taken to

*Statement 1-24, by H. S. Houthakker, D. J. Rose, and B. I. Spinrad: By the end of the century the United States may be a large exporter of coal, especially if the growth of nuclear power is impeded.

minimize water consumption and manage water supplies. The location of these industrial activities, even in the East, will require regional hydrological studies to determine where they can best be supported, with due attention to the needs of other water consumers, including ecosystems. Water shortage in the West is already a well-known difficulty. Both of these problems deserve very high research priorities.

Over the coming 10–20 years, some of these obstacles will weaken as new technologies increase the efficiency and convenience of coal use, and as the prices of oil and gas rise while their reliability of supply declines. Current expectations for some of these technologies are indicated in Table 1-4.

A number of the advanced electric power cycles for coal, now under development, would be suitable for smaller installations, and their relatively clean environmental characteristics would make it possible to locate them near users of their power. For smaller industrial users, fluidized-bed combustion and synthetic fuels could provide additional new markets for coal.

Department of Energy regulations under the Powerplant and Industrial Fuel Use Act of 1978 (Public Law 95-620), when implemented and enforced, will further improve the outlook for coal by banning oil and natural gas use in most new power plants and large industrial heating units.

This is not to imply that all the problems of coal use are solvable or that coal can become the mainstay of the domestic energy sector over the long term. Its environmental costs will remain high; mining and burning 2–3 times the present coal output, even if done efficiently and with care, will be difficult (and increasingly expensive) if the contributions of this energy source to air and water pollution and land degradation are to be kept from increasing.

With the foregoing in mind, we see the following as the prime objectives of national coal policy in the coming decades.

1. Provide the private sector with strong investment incentives to establish a synthetic fuel industry in time to compensate for declining domestic and imported oil supplies (probably some time near 1990).

2. Continue the broad federal research and development program in fossil fuel technology to widen the market for coal by increasing the efficiency and environmental cleanliness with which it can be used.

3. Improve health in the mines by strengthening industrial hygiene and by performing the necessary epidemiological research. The black lung problem especially should be clarified. (See chapter 9.)

4. Devote the necessary resources to supporting long-term epidemiological and laboratory studies of the public health consequences of coal-derived air pollutants, thus putting air quality regulation on a firmer

TABLE 1-4 Advanced Technologies for the Use of Coal

Technology	Characteristics	Status of Development	Possible Date for Introduction at Commercial Scale
Atmospheric fluidized-bed combustion	Applicable to small power plants and small-scale industrial uses	Pilot plants now operating	1980s
Pressurized fluidized-bed combustion	Applicable to larger units than atmospheric version, more efficient, better control of nitrogen and sulfur oxide emissions	13-MWe pilot plant planned	1990s
Gasification combined-cycle (gas and steam turbines) generating units	Burn medium-Btu gas produced from coal at generating site; require operation at high temperatures	Demonstration plant now being built to generate and burn low-Btu gas	1990s
Molten-carbonate fuel cells	Essentially noiseless, pollution-free, and efficient; could possibly use low- or medium-Btu gas as source of hydrogen ions for fuel	5–10 years from demonstration with synthetic gas from coal	Late 1990s; lags other fuel cell development by 5 years
Magnetohydrodynamics	Potential 50 percent conversion efficiency from coal to electricity; sulfur can be separated out in operation; high-temperature exhaust could be used directly or to generate steam	Pilot plant in U.S.S.R., fueled by natural gas; coal system still experimental	2000 or later
Synthetic gas	Low- and medium-Btu gas from coal now technically feasible, but expensive;		
	High-Btu gas (methane) also feasible, but even more expensive today; new processes now being developed	Second-generation technologies now being tested in pilot plants Third-generation technologies in design stage	1990s for second-generation processes

TABLE 1-4 (*continued*)

Technology	Characteristics	Status of Development	Possible Date for Introduction at Commercial Scale
Synthetic oil	Indirect liquefaction technology; complicated, expensive, and inefficient	Used commercially in South Africa	
	Pyrolysis: range of products, including refinable heavy high-sulfur oils and char (for which there is no ready market); not favored in current program	Small experimental unit operating since 1971	1980s
	Solvent extraction and catalytic hydrogenation: catalysts expensive; burden of hazardous wastes and control of nitrogen	Pilot plants now testing several processes	1990s

scientific basis that allows more confident and efficient setting of standards, on which industry can depend in its long-range planning. (See chapter 9.)

5. Develop a long-range plan, recognizing that coal presents some serious environmental and occupational health and safety problems, and that it does not relieve the nation of its need to develop truly sustainable energy sources for the long term.

By 1985, given reasonably coherent policy and successful research and development, domestic demand for coal should approach 1 billion tons/yr (about 20–25 quads). Some new synthetic fuel and direct combustion technologies will be on the verge of commercialization. Knowledge of the environmental and public health effects of coal production and use should be improved to the point that the current regulatory uncertainties can be reduced.

As the year 2010 is approached, coal use in the United States may reach 2 billion tons annually.* Some of the cleaner, more efficient coal-use techniques now being developed should attain full commercialization. Knowledge of the environmental and public health characteristics of coal

*Statement 1-25, by H. S. Houthakker: Exports may be of the order of 500 million tons/yr.

may be sufficient for confident standard setting. At the same time, however, water supply will be increasingly critical, and, if the hypothesis of climatic change due to carbon dioxide accumulation proves correct, the first signs of climatic effects from carbon dioxide emissions may be appearing. But it is possible that at about this time indefinitely sustainable energy sources may begin to become available.

For now, however, there is little room for maneuver. Coal must be used in increasing quantities, and mainly with current technologies, until at least the turn of the century, regardless of what happens with respect to such alternatives as nuclear fission or solar energy. However, because of the variety of environmental and social problems it presents, it cannot indefinitely provide additions to energy supply. To keep these problems under control until truly sustainable energy sources can be deployed widely, it would be wise to approach coal as conservatively as possible under the circumstances, with an eye especially to its environmental risks.

PROSPECTS FOR NUCLEAR POWER

Nuclear power could serve as both an intermediate- and long-term source of energy. Its prospects and problems are unique. For example, energy that can be extracted from the available nuclear fuel depends extremely heavily on the fuel cycle used. The light water reactors now in use in the United States, with their associated fuel cycle, make very inefficient use of uranium resources, and could exhaust the domestic supply of high-grade uranium in several decades. By contrast, if breeder reactors were to be developed and used, the domestic nuclear fuel supply could last for hundreds of thousands of years. An intermediate class of reactors and fuel cycles—advanced converters—could, under certain circumstances, extend domestic nuclear fuel supplies for perhaps a half century. These subjects are taken up in chapter 5 under the heading "Availability of Uranium."

Decisions about nuclear power have precipitated debate about the role of citizen participation in technological policy. Opposition to nuclear power in the United States has been expressed in legal and political challenges to the siting and licensing of specific power plants, and in protests over the lack of a waste disposal program and alleged deficiencies in federal regulation and management of nuclear power.[15] The resulting delays and uncertainty have contributed to rapid escalation of the capital costs of nuclear installations and to considerable difficulty in predicting their future costs and availability.

While many of these protests have centered on specific issues, social scientists suggest that the sources of public concern with the technology are broader and deeper, and thus that concern is unlikely to subside with the resolution of specific issues.[16] The technical and scientific community is itself divided, and debates among experts have heightened public awareness of the uncertainty surrounding many of the technical issues bearing on nuclear power. Very briefly, the principal issues for nuclear power as an intermediate-term energy source are as follows.

- The future role of nuclear energy, in general, and the relative roles of different nuclear options, in particular, depend on the extent of domestic and worldwide uranium resources, and on the rates at which these resources could be produced at reasonable levels of cost.
- The choice between a breeder reactor and an advanced converter reactor and the timing of development and introduction depend on a complicated integration of a number of technical factors. Most prominent among these are the rate of growth of electricity use, the supply of fuel, and the relative capital costs of advanced converters and breeders. Relatively low electricity growth rates and large supplies of low-cost uranium would generally favor the advanced converter.* It should not be forgotten, however, that the breeder and its fuel cycle are probably in a more advanced state of development worldwide than any high-conversion-ratio converter alternative, and that moderate to high electricity growth rates and/or rather limited supplies of uranium would favor the breeder alternative.
- There is a need for early action on a workable program of nuclear waste management, which has until very recently been neglected by the federal government. Adequate technical solutions can probably be found, but some particularly difficult political and institutional problems will have to be solved.
- Public appraisal of nuclear power is of vital importance. Among the most important public concerns are the potential connection of commercial nuclear power with international proliferation of nuclear weapons, the safety of the nuclear fuel cycle (a concern heightened by the recent nuclear reactor accident near Harrisburg, Pennsylvania), and the question of nuclear waste treatment and disposal.

Uranium Resources

According to the CONAES Supply and Delivery Panel's Uranium Resource Group,[17] only those uranium deposits considered, technically, "reserves" or "probable additional resources" should be taken as a basis for prudent planning. They further state that the availability of uranium ore at

*Statement 1-26, by R. H. Cannon, Jr.: Both low electricity growth rates and large supplies of low-cost uranium are highly uncertain, as noted later.

estimated forward costs (the costs of mining and milling once the ore has been found) of more than \$30/lb, is known with such little certainty that it cannot be used for planning. They estimate at about 1.8 million tons the uranium available in these categories at forward costs below \$30/lb. This committee believes that estimates of reserves and probable additional resources at forward costs of up to \$50/lb are reliable enough to plan on; according to the U.S. Department of Energy,[18] the quantity of uranium in these categories and at this forward cost is about 2.4 million tons. If, however, less reliably known uranium supplies (listed as "possible" or "speculative" additional resources) are included, the estimate would rise to about 4 million tons.

A typical 1-gigawatt (electric) (GWe) light water reactor with once-through fueling requires about 5600 tons of fuel for a 30-yr useful life. Thus, only about 400 such reactors could be built before the estimated 2.4-million-ton resource base of uranium would be completely committed. The limits on capacity could be extended somewhat (without major alterations in the fuel cycle such as recycling spent fuel) by optimizing the design of light water reactors for fuel efficiency (up to 15 percent improvement in uranium oxide (U_3O_8) consumption), and by lowering the uranium-235 (^{235}U) concentration in enrichment plant tails. The additional reactor capacity that could be available in 2000 as a result of these measures depends on how soon they could be introduced. The most optimistic estimate would probably not exceed 500 GWe (insufficient for the highest-growth projections of the CONAES study but adequate for other projections).

In brief, if the pessimistic estimates of the Uranium Resource Group are borne out by experience, more efficient reactors and fuel cycles probably will be needed in the United States by the first decade of the next century. Otherwise, the use of nuclear fission will have to be curtailed, beginning at about that time. This will occur when coal demand for synthetic fuels could be increasing rapidly to offset the decline in domestic oil and gas production, and when the first evidence of climatic change (due largely to CO_2 emissions from fossil fuel combustion) may be appearing. Unless various solar options could be introduced and spread very rapidly, this phasing out of nuclear energy would come therefore at a particulary awkward time.

Alternative Fuel Cycles and Advanced Reactors

Light water reactors with the current once-through fuel cycle use only 0.6 percent of the energy potential in uranium as mined. By contrast, breeder reactors are capable of converting the abundant "fertile" isotope ^{238}U to fissile plutonium-239 (^{239}Pu), and of regenerating more plutonium than they use. They can eventually make use of more than 70 percent of the energy potential of uranium ore. There are also conceptual reactors and fuel cycles capable of converting fertile thorium-232 (^{232}Th) to another fissile isotope of uranium, ^{233}U. These could in principle make use of nearly 70 percent of the energy in thorium, which is believed to be 4 times as abundant as uranium in the earth's crust.

Thus, the ability to unlock the energy potential of the fertile isotopes ^{238}U and ^{232}Th has a tremendous multiplying effect on available resources—much more than the approximate factor of 100 implied by the numbers just quoted. This is because the use of breeder reactors reduces the contribution of resource prices to the price of electricity by a factor of 100, thus making available ores that are too low in grade, and thus too expensive, to be used as fuel for conventional reactors. For practical purposes, the resource costs for breeders make a negligible contribution to the cost of electricity. Thus, the economics of breeders are closer to those of renewable resources than to those of nonrenewable resources.

As explained earlier, the present generation of light water reactors can be relied on as an energy source only until the early twenty-first century, even if optimized for fuel efficiency. The resource base may be extended 20–30 percent by working enrichment plants harder (to recover a larger fraction of the ^{235}U in the natural uranium). Another 35–40 percent extension could be achieved by reprocessing spent fuel in a chemical separation process to recover fissile plutonium and uranium for refabrication into new fuel elements. Either measure, however, would significantly extend the life of a nuclear industry based on light water reactors only if electricity growth leveled off after 2000.

Unfortunately, during fuel reprocessing, plutonium appears briefly in a form that can be converted into nuclear weapons much more readily than can the fissile and fertile material in the spent fuel elements themselves. This gives rise to the fear that a nation in possession of fuel reprocessing facilities might be tempted to manufacture clandestine nuclear weapons, or that a determined and well-organized terrorist group could steal enough material to manufacture a nuclear bomb. It is possible that the recycling process could be modified to make it much less vulnerable in this respect, but both the desirability and the effectiveness of such modifications are still matters of debate. (See chapter 5 under the heading "Reprocessing Alternatives.") These considerations bear heavily on decisions to deploy advanced, more efficient reactors, because all advanced reactors require reprocessing and refabrication of fuel to realize their maximum potential for more efficient resource use. (However, there are several advanced converter designs that could realize substantial, though not the greatest possible, resource savings over improved light water reactors even with a once-through fuel cycle.)

This difficulty has spurred consideration of substantial improvements in nuclear fuel use that do not require reprocessing. One option that might be available, for example, is the Canadian CANDU heavy water reactor fueled with slightly enriched uranium—perhaps 1 percent ^{235}U. (The CANDU as now operated is fueled with natural, unenriched uranium.) With a once-through fuel cycle (that is, without reprocessing), this could in principle reduce the fuel requirements per unit of power by nearly 40 percent as compared to an unmodified light water reactor of existing design. Although this might be worthwhile under some circumstances, it would still not be sufficient to preserve the option of supplying electricity by nuclear power much beyond 2000, unless the rate of growth in demand for electricity diminished greatly after that date. Uranium resources could be extended an additional 20 percent if some method such as laser isotope separation is developed for stripping the fissile material from the tailings at uranium enrichment plants (though this is unlikely before the 1990s at the soonest). The benefits of these measures would become important, however, only if the nuclear power industry were not called upon to expand significantly; growth in capacity would otherwise consume the extra supplies within a few years.

Until recently, the nuclear research and development program in this country concentrated on the liquid-metal fast breeder reactor (LMFBR) and the plutonium-uranium fuel cycle. The advantage of this approach is that the LMFBR offers the greatest degree of independence from the continuing need for natural uranium. For times of the order of hundreds of years, the LMFBR could use as fertile material the stored tails left over from the enrichment process for weapons material and reactor fuel. Such breeders could extend the life of the uranium resource indefinitely, for practical purposes, and they could be fueled initially with plutonium separated from the spent fuel of light water reactors, as well as with natural uranium. Thus, they offer electrical energy independence to the United States and other nations that have access to even small quantities of enrichment tails. (Nations that operate their light water reactors with fuel enriched in the United States are legally entitled to enrichment tails; these tails are worthless unless they can be used in breeder reactors or stripped for their remaining fissile content by laser isotope separation or another technique.)

Because the LMFBR generates almost 20 percent more fissile isotopes than it consumes, it can be used as the basis for a growing nuclear capacity without requiring the mining of new ore.* For this reason, it appears attractive for a wide range of projected growth rates in electrical capacity.

Breeders, in the course of their operation, produce more fissile isotopes

*Statement 1-27, by J. P. Holdren: Present LMFBR designs breed so slowly that capacity cannot expand rapidly without fissile material from mining-enrichment or from large numbers of LWR's.

than they consume. Converters such as light water reactors and CANDU produce a good deal less. Advanced converters produce almost as much as they consume. If their spent fuel is reprocessed and reloaded into the reactors, they can be run with much less fresh fissile material than is needed to run light water reactors or CANDU's. There are many possible advanced converters.

The principal advanced-reactor alternatives are listed in Table 1-5, along with indications of their relative developmental maturity.

Thus, as between breeders and advanced converters, the following conditions (not all of equal weight) would favor the use of fast breeder reactors over advanced converters in the United States for nuclear-generated electricity.

- The demand for electricity in the United States grows steadily after the year 2000.
- Total domestic uranium resources are found to be at the low end of recent estimates.
- Very little intermediate-grade uranium ore that can be produced at costs in the range of $100–$200/lb is found.
- The world growth of nuclear capacity in conventional light water reactors exerts pressure on the United States to export some of its uranium or enriched fuel (or both) to offset the balance-of-payments deficit from oil imports, to discourage recycling of fissile isotopes or installation of breeder reactors elsewhere, or for other reasons.

The following conditions would generally favor the use of advanced converters for nuclear-generated electricity.

- The demand for electricity in the United States grows slowly, especially after 2000.
- Sufficient uranium resources are found to fuel advanced converters at their projected rate of introduction and installation, particularly intermediate-grade ores producible at costs around $100–$200/lb.
- Capital costs of advanced converters turn out to be significantly less than those of breeders.
- The operation of advanced converters and their fuel cycles offers advantages in safeguarding against proliferation or diversion.
- New enrichment technologies that permit economic operation at low tails assays become available early.

As has been noted, economics and the type of measures adopted by the world to slow proliferation of nuclear weapons could dominate the choice. Both are highly uncertain factors; we can only estimate future costs

TABLE 1-5 Nuclear Reactors and Fuel Cycles: Development Status

Reactor Type	Fuel Cycles	Development Status	Possible Commercial Introduction in the United States[a]
Light water reactor (LWR)	Slightly enriched U (~3 percent ^{235}U)	Commercial in United States	1960
Spectral-shift-control reactor (SSCR)	Th-U[b]	Conceptual designs, small experiment run; borrows LWR technology	1990; fuel cycle, 1995 or later[c]
Light water breeder reactor (LWBR)	Th-U[b]	Experiment running; borrows LWR technology; fuel cycle not developed	1990; fuel cycle, 1995 or later[c]
Heavy water reactor (CANDU or HWR)	Natural uranium	Commercial in Canada, some U.S. experience	1990
	Slightly enriched U (~1.2 percent ^{235}U)	Modification of existing designs	1995
	Th-U[b]	Modification of designs; fuel cycle not developed	1995
High-temperature gas-cooled reactor (HTGR)	Th-U[b]	Demonstration running; related development in Germany; fuel cycle partly developed	1985; fuel cycle, 1995 or later[c]
Molten-salt (breeder) reactor (MSR or MSBR)	Th-U[b]	Small experiment run; much more development needed	2005
Liquid-metal fast breeder reactor (LMFBR)	U-Pu[b]	Many demonstrations in the United States and abroad*	1995
	Th-U[b]	Fuel cycle not developed	1995
Gas-cooled fast breeder reactor (GCFBR)	U-Pu[b] Th-U[b]	Concepts only; borrows LMFBR and HTGR technology	2000

[a] Based on the assumption of firm decisions in 1978 to proceed with commercialization. No institutional delays have been considered except those associated with adapting foreign technology. On the basis of light water reactor experience, it can be estimated that it would take about an additional 15 years after introduction to have significant capacity in place.
[b] Indicated fuel cycles demand reprocessing.
[c] Thorium-uranium fuel reprocessing is less developed than uranium-plutonium reprocessing. Indicated reactors could operate for several years before accumulating enough recyclable material for reprocessing.

***Statement 1-28, by J. P. Holdren: Fuel reprocessing with the short turnaround time, high throughput, and high plutonium recovery needed to make the LMFBR perform as advertised remains undemonstrated.**

qualitatively, and we can rely on surprises in international decision making.

This committee could not reach a consensus on whether the likelihood of the circumstances favoring advanced converters is great enough to warrant their development as insurance against difficulties and delays in LMFBR development. Nor was it able to reach agreement on how much the availability of the breeder option might be delayed by a parallel effort on advanced-converter development, and whether such a delay would be justified by a greater ultimate chance for the success of at least one advanced-reactor alternative. It did, however, reach general agreement that the LMFBR dominates the nuclear alternatives over the widest range of assumed future circumstances, provided that its cost goals and other technical objectives can be realized. Those who believe that low growth in demand for electricity is desirable and can be achieved after 1990 argue that a U.S. program to develop the LMFBR sets a poor example to other nations whose development of the LMFBR would increase the danger of proliferation. The LMFBR, they argue, would be needed only for unnecessarily high rates of growth in electricity demand, which could be avoided in this country by sensible conservation policies.* In this view, the advanced converter provides sufficiently improved resource efficiency over present reactors to fill the gap until sustainable nonnuclear long-term technologies become available. These arguments underscore the importance of energy demand considerations in planning energy supply systems for the United States.†

The Demand for Electricity

It is obvious from the foregoing that the rate of growth in electricity use will largely determine how much nuclear power is needed and will govern the strategy of nuclear development.‡ Some pertinent quantities are set out in Table 1-6, which uses the CONAES study scenarios (described in detail in chapter 11) to indicate the trade-offs between nuclear power and other sources of electricity.

Study scenario III_3, for example, shows nuclear power providing about 35 percent of the nation's electricity in 2010. Its contribution of 1670 billion kWh is about twice what the U.S. Department of Energy[19] forecasts nuclear power will contribute in 1990. Thus the scenario involves a modest rate of nuclear growth over the 20-yr period 1990–2010. Coal-generated

*See statement 1-29, by H. S. Houthakker, E. J. Gornowski, and L. F. Lischer, Appendix A.

†See statement 1-30, by L. F. Lischer and E. J. Gornowski, Appendix A.

‡Statement 1-31, by L. F. Lischer, E. J. Gornowski, and H. I. Kohn: This, in our opinion, is neither obvious nor a foregone conclusion.

electricity in this scenario is at about twice the 1978 level. Coal and nuclear power together generate some 3.8 trillion kWh.

If nuclear power were unavailable in 2010, and the entire amount of energy were generated by coal, this would represent a fourfold increase in coal-based generation over the 1978 level, approaching the threshold of serious environmental risks, and in some mining areas introducing or exacerbating problems of water supply. (See chapters 9 and 4, respectively.)

In the high-growth case represented by study scenario IV_3, 3 times the present electrical capacity would be required. Assuming that 1 GWe of nuclear capacity generates 6 billion kWh in the course of 1 year's operation, 470 GWe of nuclear capacity would be required to generate the 2810 billion kWh specified for nuclear power by this scenario. Together, nuclear power and coal generate nearly 6 trillion kWh. If coal-based generation were restricted to, say, 2 trillion (or about twice its 1978 level) and the remaining 4 trillion were supplied by nuclear power, an extraordinary national commitment to nuclear capacity additions would be necessary. With the above assumption about the productivity of 1 GWe unit of nuclear capacity, some 670 GWe of nuclear capacity would be needed, including breeders or other advanced reactors.†

These examples illustrate the limited mutual substitutability of nuclear energy and coal in the high-growth cases and suggest that if growth in demand for electricity is underestimated, shortages of energy may begin to appear during the first decade of the twenty-first century.*

Nuclear Weapons Proliferation and Breeder Development

Two interrelated issues concerning the breeder reactor are the scale and pace of development and the relationship of breeders to the problem of nuclear weapons proliferation and diversion (chapter 5). Regarding proliferation of nuclear weapons, sharply different and irreconcilable views emerged in this study. One view holds that plutonium reprocessing would be a major step toward proliferation, and advocates that the United States forgo for a considerable period the benefits of reprocessing and the breeder to demonstrate how seriously this nation regards the proliferation problem. This view acknowledges that proliferation can thus be only delayed, not prevented, but asserts that deferral of reprocessing and breeder deployment could provide time to develop international institutions and procedures to safeguard the nuclear fuel cycle. In this view, the

†See statement 1-32, by H. Brooks, Appendix A.

*Statement 1-33, by J. P. Holdren: The narrow emphasis on high-growth futures in this passage and the accompanying table is unwarranted and gives an unbalanced impression of the possibilities.

TABLE 1-6 Electricity Generated, by Source (billions of kilowatt-hours)

	Actual 1978[a]	CONAES Study Scenarios for 2010		
		II_2	III_3	IV_3
Nuclear	276	670	1670	2810
Coal	976	1460	2110	3140
Other	954	730	940	1080
TOTAL	2206	2860	4720	7030

[a] Source: 1978 data are from U.S. Department of Energy, *Annual Report to Congress 1978*, vol. 2, *Data*, Energy Information Administration (Washington, D.C.: U.S. Government Printing Office, 1979).

LMFBR should be treated primarily as a long-term technology of last resort, to be used only if research in the coming decades indicates that other long-term options are much more costly or will not be available in time to offset the phasing out of light water reactors.

The contrary view holds that the breeder has been demonstrated to be the most promising option for the long-term future, with favorable economics and minimal ecological effects, and that therefore a national commitment to large-scale development should be made now, so that LMFBR's can be available before the twenty-first century. It is argued that the commercial nuclear fuel cycle is the least likely and most expensive of several possible paths to proliferation, and that inexpensive means for producing weapons-grade material by isotope separation are likely to be widely available by the time commercial reprocessing of plutonium becomes widespread.

The response by those favoring deferral of reprocessing is that, whereas there are indeed other routes to proliferation, they require more deliberate political decisions, while a weapons capability could be "backed into" rather easily once commercial reprocessing and refabrication facilities have been installed in a given country. The critical consideration in this view is not the availability of cheaper and less elaborate routes to weapons (which certainly exist) but the reduced warning time between a decision to divert material from the commercial fuel cycle and the production of the first weapons.*

*Statement 1-34, by J. P. Holdren: Equally critical is the temptation provided by the commercial plutonium cycle, offering weapons as a "fringe benefit" of facilities justified by electricity needs.

The view that breeder development should proceed rapidly holds that deferral would increase the potential pressures of the United States on the world petroleum market and on the limited world uranium supply for light water reactors. This would in turn stimulate other countries that are much more dependent than the United States on outside energy sources to pursue the breeder reactor—the one option close to availability that promises a degree of energy independence. Moreover, this argument asserts, world conflict over limited petroleum supplies appears more likely to lead to nuclear war than weapons proliferation resulting from reasonably safeguarded commercialization of plutonium.

Management of Radioactive Wastes

The current plans for managing nuclear wastes involve underground burial. The technical aspect of the problem has two parts: first, to find the best technology for packaging and isolating the wastes and, second, to secure a geological environment that would itself be proof against the failure of containers after one or two hundred years, so that migration of the waste nuclides in groundwater would be slow enough and accompanied by so much dilution that the radioactivity of the water when it reached the biosphere would be a small fraction of the natural background.

There is no lack of potential disposal methods. There is enough knowledge about the bedded salt disposal option, for example, to warrant a full-scale engineered test of this option with an initial sample of commercial waste. The engineering of such a test would require mainly acquisition of site-specific geological and hydrological data for a few chosen sites. There is, however, no data base adequate for a final choice among the proposed solutions, nor proof that a given choice of sites and waste forms poses the lowest possible risk to the public. Waste disposal is often used as a basis for the political expression of more generalized opposition to nuclear power and to the whole decision-making mechanism for nuclear power.

Two points should be kept in mind. First, it is not necessary to look upon waste disposal as a problem to which the perfect solution must be found before any action can be taken. Caution is necessary, of course, but the risks should not be a bar to the continued use of nuclear power. The maximum hazard resulting from inadequate waste disposal is much smaller than that which could be postulated as the result of a reactor accident or sabotage. Indeed, the maximum exposures involved can almost certainly be kept below those associated with routine exposures to radioactivity in nuclear operations, which are themselves very small compared to exposure to natural background radiation. Caution is dictated not by the magnitude of the risks but by their long duration. The principal risks extend for about a thousand years, and the presence of actinides in the wastes adds a very small continuing risk for millions of years. In this respect, however, nuclear waste disposal is not entirely unique. Elevated CO_2 concentrations in the atmosphere, once established, will persist for many hundreds of years, and over this extended period could have devastating effects, if the hypothesis of climatic changes due to CO_2 accumulation proves correct.

The following specific conclusions and recommendations represent the consensus view of CONAES.

- The nature of the risks from geological disposal of nuclear waste must be clearly spelled out and publicized. The only credible mechanisms by which wastes, once emplaced, could reach the environment involve the slow return of highly dilute radioactive materials, rather than the sudden return of concentrated ones.* This could lead to small increases of environmental radiation over previous background levels, lasting for a long time and covering a large area. It could not lead to severe or acute radiation exposures.
- The federal government should immediately proceed to set criteria for geological waste disposal. These should be (1) performance criteria (i.e., leach rates, heat rates) on waste forms in categories that take account of the risks from different types of wastes and (2) site criteria (i.e., groundwater standards, seismic stability standards, resource and mining restrictions).
- The problem of disposal must be separated from the problem of spent fuel storage.
- The problem of military wastes must be settled, and the issue separated from that of commercial wastes. It may well be that long-term entombment is appropriate. If so, it should be effected. Military wastes consist mostly of fission products, and their period of high risk is therefore relatively short.
- The federal government should accept full responsibility for any radioactive wastes in existence, leaving the question of joint state-federal responsibility to be resolved for wastes generated in the future.
- Standards must be set and enforced for the treatment of abandoned mines and of tailings from mines and mills. These standards should permit disposal of low-level alpha-active wastes (i.e., alpha-active wastes which, if blended with the tailings, would not significantly increase their risk) in tailings piles.† This will require collaboration between the federal government and the uranium-mining states.

*Statement 1-35, by J. P. Holdren: To say "only credible mechanisms" bespeaks a confidence in our knowledge of the possibilities that I cannot entirely share. I would accept "most plausible mechanisms." (H. I. Kohn: I concur with the general intent of this remark.)

†See statement 1-36, by J. P. Holdren, Appendix A.

• While retrievability of waste after emplacement is a desirable feature of a test facility, and such a facility would be useful for a research and development program, retrievability ought not to be a consideration in designing a repository for actual waste disposal.

These recommendations agree substantially with those of the American Physical Society's "Report to the American Physical Society by the Study Group on Nuclear Fuel Cycles and Waste Management."[20]

Putting these recommendations into effect may involve serious political difficulty.* Most states and communities would like nuclear wastes to be disposed of elsewhere, and some have imposed virtual bans on waste treatment and other fuel cycle operations. This raises important legal and constitutional questions about the limitations of federal power to overrule state and municipal land-use laws. This committee did not consider itself competent to judge these issues.

Public Appraisal of Nuclear Power

The principal sources of public concern with nuclear power are not merely technical, but institutional and social as well. Questions about technical approaches to proliferation control, reactor safety, and waste management are largely expressions of concern about whether human beings and institutions can be relied on over the long term to manage radioactive wastes, ensure reactor safety, and secure weapons-usable material.

The accident at the Three Mile Island plant in Pennsylvania has heightened this concern. It occurred late in this committee's deliberations, and it is still too early for final judgments in detail. However, what the committee has learned about it thus far has not led it to change its assessment of the physical risks of nuclear power; chapter 9, in the section on the health impacts of energy production and use, discusses this event and its likely impact on human health (which is very small).† Public opinion of the accident and its implications, however, is vital, and it is probably too early to know how that will be expressed. Major studies of the accident and its consequences are underway throughout the world; notable in this country are an investigation by a specially appointed Presidential commission and one by the Electric Power Research Institute's newly formed Nuclear Safety Analysis Center. The Nuclear Regulatory Commission, in reaction to the accident, may impose additional safety requirements on nuclear reactors.

Other aspects of the appraisal of nuclear power reflect individual views

*Statement 1-37, by L. F. Lischer: True. But I would state the waste disposal issue thus: It is not a technical problem, it is a political problem.

†Statement 1-38, by H. I. Kohn: The adjective "small" is incorrect. Substitute "negligible."

of the social impacts of this technology. Nuclear power, for example, has become for some a symbol of large-scale, centralized technology over which citizens have surrendered control to experts who cannot be held accountable. Some feel that nuclear power, and particularly the breeder, promotes the continuation of a high-growth materialistic society that will ultimately prove disastrous to the physical and social environment. Some see nuclear power as competing for capital resources with energy systems that are more subject to local control, and thus excluding patterns of social organization that are based on such local autonomy. Many* fear that the level of social discipline necessary for adequate management and safeguarding of nuclear power will prove incompatible with democratic institutions and will erode civil liberties. They point to the growth of alienation, terrorism, and crime and to the associated vulnerability of centralized sociotechnical systems.†

Others, of course, see nuclear power as essential if people are to have enough energy to meet basic needs, live in reasonable comfort, and look forward to improving their own lives and those of their children and the underprivileged. It is clear that even in controversies over technical issues, judgments are influenced by the social and institutional values of the individuals involved. The greater the technical uncertainties, the more room there is for interpreting whatever knowledge exists to support one's subjective preferences. Not uncommonly, decisions among technological options will have to be reached—if only in the form of postponements of action—before the technical uncertainties can be fully resolved. To a great extent, therefore, technical questions as well as social and institutional ones will be decided by political processes.‡ §

INDEFINITELY SUSTAINABLE ENERGY SOURCES

Four energy sources—nuclear fission with breeding, solar energy in various forms, controlled thermonuclear fusion, and geothermal energy—offer the potential for indefinitely sustainable energy supply. That is, each could supply up to 10 times our present energy requirements for thousands of years (or much more). They differ widely in their readiness for use, in their probable side effects, and in their economics. Present knowledge is insufficient for meaningful economic comparisons and permits only limited comparisons by other criteria, such as environmental and safety risks or

*Statement 1-39, by H. I. Kohn: "Some" is a better estimate than "many."

†See statement 1-40, by B. I. Spinrad, H. Brooks, and D. J. Rose, Appendix A.

‡Statement 1-41, by H. I. Kohn: To assist these processes, the widespread dissemination of factual information must be promoted.

§See statement 1-42, by L. F. Lischer, H. Brooks, and D. J. Rose, Appendix A.

the likelihood of successful technical development. The degree of risk associated with a technology often depends on details of engineering design and on compromises between safety and economics that cannot be foreseen until the technology has been translated into full-scale designs with considerable practical operating experience to back up assessments of component reliability and the like. A technology in the conceptual stage often appears less risky than it will after the practical engineering questions have been faced.

The government's program in long-term energy supply, to allow realistic choices of long-term options, should include sustained research and development of many of these technologies. Priorities at this stage should depend more on the likelihood of significant technical progress than on economic comparisons among existing versions. New technical developments and changes in resource economics are likely to alter comparative cost assessments radically. Furthermore, a combination of long-term sources is likely to offer more flexibility and overall reliability than dependence on a single system. The ultimate total cost of deploying a new energy technology on a broad scale is so much larger than the research and development costs that maintaining an array of options in the development stage is fully justified. A cost advantage of a few percent in a deployed system would easily pay for all the research and development that produced it.

THE BREEDER REACTOR

The breeder reactor, in the form of the liquid-metal fast breeder reactor, has benefited from a sustained and relatively large federally financed research and development effort. It is also the choice of several other countries, including the United Kingdom, France, West Germany, the U.S.S.R., and Japan, all of which have large LMFBR development programs. Worldwide, about 3.8 GWe of LMFBR capacity is under construction or on order. Given the present state of breeder development worldwide, construction of a commercial breeder could begin somewhere in the world within 10 years, provided there are no unexpected technical developments or insurmountable political obstacles. Significant capacity could be in place by the year 2000. This will probably not take place first in the United States because this country has more energy options than most other countries, but it is not technically impossible. However, there are technical uncertainties related to reactor safety, capital costs, and fuel cycle safeguards that could still seriously delay the program.

Other types of breeder reactors, such as the gas-cooled fast breeder reactor (GCFBR) and the molten-salt breeder reactor (MSBR), are in much earlier stages of development but have some potentially attractive features (described in chapter 5). If the LMFBR is pursued vigorously and successfully and is required relatively soon, the other types of breeder may never be brought to the point at which they can compete. On the other hand, if breeders turn out not to be required early, these other types could prove to be realistic alternatives by the time a breeder is needed and might be superior to LMFBR's on a number of technical grounds.*

SOLAR ENERGY

In the long term, it should be possible for solar energy to provide each of the energy forms used by people: heat, electricity, and fuels.[21] In the near term, outside of hydroelectric power—included by convention with solar energy—only certain heating applications are economical.†

Assessing the long-term potential of solar energy will require an extended period of research and development. A major issue for national solar energy policy is the balance of research and development effort among the variety of solar technologies. The federal solar energy program emphasizes technologies for producing electricity, but the most important use of solar energy in the long-term future may in fact be the synthesis of fluid fuels, which could solve the problem of energy storage and make good use of the existing distribution system developed for gas and oil.

Direct Thermal Use of Solar Energy

Technologies for the direct use of solar heat are in general the most nearly economical today. Some of the methods—domestic space heating, domestic hot water heating, and production of hot water or low-pressure steam for industrial and agricultural processes—can be considered fairly well developed; they are among the most probable candidates for widespread commercialization in the intermediate term. Efficient and economical solar cooling remains a difficult problem.

The direct applications of solar thermal energy are generally more costly than conventional alternatives, Btu for Btu, and even more costly in terms of the initial investments in complete heating and cooling systems. (For a discussion of the economics of such systems see chapter 6 under the heading "Direct Use of Solar Heat.") It can be argued, however, that conventional economics do not reflect the full comparative advantage of solar applications when social costs are taken into account. Savings in imported oil may have a moderating effect on the rise of world oil prices which could generate savings elsewhere in the U.S. economy, more than

*See statement 1-43, by B. I. Spinrad, H. Brooks, and L. F. Lischer, Appendix A.

†Statement 1-44, by J. P. Holdren: Biomass (as crop, timber, and municipal wastes) is economical today for process steam and electricity generation in some U.S. localities.

offsetting the extra initial cost of solar installations. The risks of solar energy appear to be generally less than those of other energy sources, and public confidence in solar energy is strong; public controversy (which is costly in itself) can thus be avoided in deploying these technologies. These advantages strengthen the case for introducing government incentives to induce consumers to select solar systems in preference to conventional alternatives. Such measures would help solar heating for buildings and industrial processes to gain a significant market share earlier than it would otherwise. Such incentives are already widely incorporated in federal and state programs. Unfortunately, there is no agreed upon calculus by which to estimate the market penetration likely with any given level of subsidy, or with which to quantify the benefits to society of substituting solar energy for otherwise cheaper alternatives.

Solar-Generated Electricity

The amount of electricity that could in principle be generated by solar energy could more than provide for present demand. The main obstacle is cost; unless major technical breakthroughs occur, solar electricity will be expensive compared to alternatives. Four concepts under active development for generating electric power from solar radiation are: photovoltaic conversion (with so-called solar cells); solar thermal conversion, which involves concentrating sunshine to achieve high-temperature heat; wind power; and ocean thermal energy conversion, which would use floating power stations to exploit the temperature difference between the ocean's surface and subsurface waters to run heat engines.

Photovoltaic Conversion Photovoltaic conversion is a commercial technology used in space and in remote installations where performance, rather than cost, is the principal concern. Photovoltaic arrays have demonstrated adequate efficiency and reliability but at high costs—more than 20 times the prevailing cost of residential electricity. Costs have been coming down rapidly, however, and a number of unanticipated technical improvements have occurred. The economic outlook for photovoltaics is considerably more favorable than it was a few years ago. There is some debate about how the necessary additional cost reductions might best be achieved—through mass production of present technology with evolutionary improvements, or through a breakthrough in materials and device configurations resulting from exploratory research. Unlike solar thermal conversion, this is a field in which fundamental research could yield dramatic returns, and recent technical progress has been very rapid. Given the high stakes in solar energy and the long-term nature of its potential benefits, the present investment in exploratory research for photovoltaics is still inadequate, though recently much improved. CONAES is in agreement with the general assessment provided in the recent study of photovoltaics by the American Physical Society, which suggests that market penetration is unlikely to exceed 1 percent before the year 2000, and advocates the exploratory development approach in preference to the mass-production strategy.[22]

Solar Thermal Conversion The most heavily financed system for generating electricity with thermal energy from the sun is the solar tower concept, with arrays of mirrors focusing sunlight on a boiler at the top of a tower. Although this concept appears technically feasible, there is insufficient information for reliable cost estimates. Projected costs appear to lie in the range of 5–10 times the current bus-bar cost of electricity if storage costs are included. Because so much of the cost is embodied in structural materials such as concrete and steel, which represent well-developed technologies for which large cost reductions are unlikely, reducing costs will be difficult. A 10-MWe pilot plant is being constructed in Barstow, California. Photovoltaic conversion probably offers greater long-term promise and potential for improvement.*

Wind Power Wind generators constitute a form of solar energy that is already economic for a few sites and markets. However, integration of this highly variable power source into utility grids could increase total generating costs if a great deal of backup capacity were required. When used in small amounts, however, wind generators can save fuel without requiring additional capacity. Economic uses might be found in utility districts that have a high proportion of hydroelectric generating capacity, or extensive pumped hydroelectric storage, either of which could accommodate the variations in wind power output.

Sites for wind generation are limited by wind conditions and scenic considerations. The amount of land required per unit of electrical capacity is much larger than for most other forms of solar energy (although land used for wind generation is of course not completely excluded from other uses). Interference with communications can also be a problem, because television and microwave signals are reflected by the moving surfaces of wind turbines. A major environmental impact is likely to be from access roads for maintenance and construction and from electrical interconnections of numerous units.

The most immediate prospect for wind technology would be to develop a diversified design and manufacturing effort directed generally at

*Statement 1-45, by J. P. Holdren: So do solar pond collectors driving low-temperature heat engines.

machines with generating capacities of about 1 megawatt (electric) (MWe). The market potential is likely to be highly differentiated and, relative to total domestic energy demand, modest. Experience with the problems of integrating wind-generating capacity into the existing electric grid could be a valuable by-product, applicable to other solar electric technologies as they become available.

Ocean Thermal Conversion Another system of solar electricity generation is ocean thermal energy conversion (OTEC), a technology that would exploit temperature differences between surface and deep ocean water in the tropics to generate electricity at very low thermodynamic efficiency (1–3 percent). Its attractive aspect is that it would not require storage technology and thus could be directly usable for base loads. OTEC may be technically feasible, but there is not yet a basis for choice among proposed designs. Lack of knowledge and inadequate research on problems of fouling of the very large heat-transfer surfaces by marine organisms are among the uncertainties in the present program. There are also serious questions about climatic and ecological effects if OTEC stations were deployed on a scale sufficient to supply an appreciable fraction (say 10 percent) of domestic energy requirements.

Fluid Fuels

In the long term, whatever mix of sustainable energy sources is used will have to provide a large supply of fluid fuels for applications (such as transportation) that are most easily served today by oil and natural gas. The production of fluid fuels from solar energy represents a very large and promising field for basic research. Such a process would obviate the need for auxiliary energy storage, and at the same time provide fuel for the nation's existing distribution networks as natural fuels are depleted. This could provide an easier transition to the ultimate long-term energy system than a program that emphasizes electricity production alone. The federal solar energy program gives too little attention to the production of fluid fuels.

For the long term, the most attractive potential solar energy alternative for the production of fluid fuels is probably direct photochemical conversion. For example, this might involve decomposition of water to produce hydrogen, which can be used directly as a fuel or in synthesizing hydrocarbon fuels from various sources of carbon, including CO_2 from the atmosphere.

Theoretical calculations indicate the possibility of photochemical conversion efficiencies of 20–30 percent, based on incident solar energy, compared to an average photosynthetic efficiency of 0.1 percent for natural ecosystems, and up to 1.0 percent for "energy farms." A level of fluid fuels approximately equal to present consumption of oil and gas (55 quads) could be provided by efficient photochemical conversion from the solar energy falling on about 50,000 km^2, or about 1 percent of the land area of the United States. However, it must be emphasized that research on solar fuel production is at a much earlier stage than other solar energy research. There does not yet exist even a promising laboratory system worth scaling up to an engineering experiment. Thus, barring unexpected developments in fundamental research in the near future, the production of fuels from solar energy is probably much further in the future than even such sophisticated technologies as photovoltaics.

The production of fuels from biomass, a form of solar energy, also has promise in the relatively near term. CONAES has estimated that a total of 5 quads might be produced from organic municipal and agricultural wastes, from plants grown on otherwise useless land, and from seaweed. This would not be an inconsiderable contribution. Beyond this, the growth of biomass in land-based energy farms would use land that would require fertilization and irrigation for high, sustainable yields, and would compete for land and other inputs that could be devoted to uses of higher value, such as growing food. The ecological costs of such a development would be high and would rise rapidly as production requirements increased, at least in the United States. (Marine energy farming could have none of these problems. Not enough is yet known, however, to assess the potential magnitude of its contribution.)

Some Institutional Issues

A problem for many solar energy alternatives is finding ways to introduce decentralized technology into a centralized network without disrupting the economics and reliability of the network. This problem could be reduced by the development of cheap and effective energy storage systems to absorb excess energy and release it when needed.

An important institutional issue is the degree to which regulation, taxation, and subsidies should be designed to encourage market penetration of solar technologies that are uneconomic under existing circumstances. An argument in favor of this is that the social costs of solar energy are sufficiently less than those of other energy forms so that its higher economic costs should either be offset by taxes on other energy forms that are potentially more damaging to the environment, or borne in part by special government subsidies or tax benefits.*

*See statement 1-46, by B. I. Spinrad, Appendix A.

The Solar Resource Group of CONAES concluded that solar energy technologies could contribute substantially to the national energy system by 2010 if there were purposeful governmental intervention in the energy market. However, with energy prices in the range considered by the CONAES study, market penetration by solar energy (apart from biomass and hydroelectric) would be only a few quads up to 2010. One scenario was explored to see how quickly solar energy could be introduced if tax policies and economic incentives were introduced to encourage its adoption in preference to other energy forms, regardless of cost. (See chapters 6 and 11.) Under these conditions, solar technologies might provide as much as 25–30 quads of total energy needs by 2010, but the total price (at today's costs) could be enormous, running to a cumulative total of several trillion dollars—2–3 times the cost of alternatives. These costs, of course, can be expected in the future to change relative to those of alternatives. The following are the committee's main conclusions and recommendations.

1. The aim of the government's solar energy program should be to place the nation in the best possible position to make realistic choices among solar and other possible long-term options when choices become necessary. This requires continuing support of research and development of many solar technologies. Comparisons of the present costs of various solar technologies and other long-term technologies should not be regarded as critical at the present stage of development. Of more importance is the potential for significant technical advances.

2. In the intermediate-term future, the direct use of solar heat can contribute significantly to the nation's energy system. Solar heating technologies should be viewed, along with many conservation measures, as means of reducing domestic use of exhaustible resources. The role of the government program should be to support the development and assist the implementation of the most cost-effective solar techniques, used wisely in combination with energy conservation. In particular, the government should stimulate the integration of solar heating into energy-conserving architectural design in both residential and commercial construction through support and incentives for passive solar design. Since all solar energy technologies are capital intensive, uses that are distributed throughout the year, such as domestic water heating and low-temperature industrial process heating, are likely to be economically competitive earlier than uses for which there are large seasonal variations in demand.

3. Under present market conditions, solar heating systems are usually not competitive with other available technologies, and therefore market forces alone will bring about little use of solar energy by 2010—probably less than 6 quads even if average energy prices quadruple.[23] Nevertheless,

important social benefits would accrue from the early implementation of these systems: they would contribute to the nation's conservation program, they are environmentally fairly benign, and they would increase the diversity of the domestic energy supply system and its resilience against interruption. National policy should stimulate the early use of solar energy by intervening in the energy market with subsidies and other incentives.

4. Many solar energy applications require long-term development, and these technologies should properly be compared with breeder reactors or fusion. It would be unfortunate if alternatives to the breeder were rejected because too little is known about them today to count on them. It would also be wrong to assume that the choice will or should fall on a single long-term option. Diversity in the nation's long-term sources can provide valuable resilience in the face of interruptions in the supply of a single fuel or technology. Decisions that restrict the variety of our long-term options should be deferred as long as possible.

5. The cost picture for a number of solar technologies is likely to change radically in the future, with successes and failures in development. Competing technologies will display parallel trends. The costs of many factors of production are likely to change, affecting various technologies differently. In most cases, the economics of solar energy depend critically on advances in ancillary technologies, such as energy storage. It is important that the benefits of these ancillary developments be assessed for other energy technologies on the same basis as for solar, however. For example, cheap energy storage systems would benefit the economics of all systems containing capital-intensive generating technologies.

Large-scale government demonstrations of long-term solar technologies, such as the planned demonstration of a solar thermal central station power plant, could be counterproductive if undertaken prematurely. Such projects may suggest (possibly incorrectly) that the technologies could never become economically competitive, whereas waiting for additional technical developments* could result in a considerably more favorable outlook.

6. An imbalance exists in the federal solar energy program in favor of technologies to produce electricity at the expense of those to produce fuels. Much more attention should be given to the development of long-term solar technologies for fuels production, although there is at present no prime candidate besides biomass production (which is limited by ecological considerations).†

7. The diversity of solar technologies is so great that it is difficult to

*See statement 1-47, by L. F. Lischer, Appendix A.

†Statement 1-48, by R. H. Cannon, Jr.: Marine biomass, producing methane gas in situ, does not have the inherent ecological problems (or the nutrient supply problems) of land biomass referred to here.

make decisions among alternatives in a centralized way. To a great extent, the actual choice of which solar technologies to deploy should be made in as decentralized a manner as possible. In other words, the decisions should be left to private industry and individual consumers. The government's role should be development of a broad scientific and technological base in support of solar energy (much as it did for nuclear energy prior to 1960 and for aeronautics after World War I), and provision of economic incentives that favor solar alternatives.

GEOTHERMAL ENERGY

Sources of geothermal energy include crustal rocks, sediments, volcanic deposits, water, and steam and other gases at usably high temperatures that are accessible from the earth's surface. These sources of the earth's heat are not indefinitely sustainable in the same sense as solar energy. However, their total energy is sufficiently large that their potential as an energy source will depend mainly on their economic producibility, not on resource considerations.

At present, the only usable geothermal resources are deposits of hot water or natural steam. In the long-term future, it may be possible to extract heat from the natural thermal gradient in the earth's crust and from unusually hot rock formations lying close to the earth's crust. As there is no demonstrated technology for using these resources, cost and producibility can be only grossly estimated. The use of dry rock depends on developing a fracture system large enough to be economical as a source of heat. The possibilities of achieving this, and the environmental effects of doing so, are speculative.

The only widespread potential geothermal resource, the natural thermal gradient, is the most speculative in practical exploitability. As an indefinitely sustainable source, it also suffers the inherent disadvantage that the normal heat flux from the inside of the earth is only about one thousandth the solar energy flux falling on the same area.

One potentially large source of rather low-temperature geothermal energy is the geopressured brines of the Gulf Coast. These brines may also hold very large amounts of dissolved natural gas. If the heat and gas can be exploited simultaneously, this might be an attractive resource. Too little is known about it today. Considerable effort is justified in assessing its potential.

CONTROLLED THERMONUCLEAR FUSION

As a potential source of electricity, nuclear fusion makes use of deuterium—widely found in ocean water. These resources are at least equal to those upon which fission breeders depend. (However, the most likely practical fusion system will use the deuterium-tritium reaction; this requires a source of tritium, which in turn depends on lithium—which is nowhere near so abundant—as a raw material.)

Despite many hundreds of millions of dollars spent on research in its basic science and technology, fusion has yet to be demonstrated as technically feasible. There is rising optimism that a scientific demonstration will be made within the next 5 years. Until that time, little can be said about the engineering or economic feasibility of fusion as a source of power.

There are several proposed reactor configurations, and the first to demonstrate scientific feasibility may not be the most appropriate to carry forward into engineering development. For this reason, it is much too early in the development of fusion to select any single approach. The federal program should continue work on alternative approaches to plasma confinement science before attempting to move to experiments on the scale of pilot plants.

Although fusion has some of the same problems as fission, the problem of radioactive waste management is probably less severe. (The radioactive tritium fuel can pose an occupational health problem but not a waste disposal problem.) The problems associated with commercial traffic in weapons-usable fissile materials are largely absent. However, present fusion devices are prolific sources of neutrons and, if surrounded by a natural uranium blanket, could be used to manufacture plutonium and ^{233}U for weapons (or, of course, for use in fission reactors). There is general agreement that this is one of the more difficult ways of acquiring weapons-usable material and that the risk of proliferation from fusion power is not comparable to that associated with fission power. Inertial confinement approaches to fusion, though, may have an additional proliferation liability, since they may tend to spread technical insights relevant to the design of fusion weapons. The radioactivity produced in fusion devices could be from 10 to several hundred times smaller than that from fission (depending on the choice of materials), and the troublesome problem of alpha-active actinides is avoided.

Nuclear fusion is not a technology of the twentieth century and has not reached a stage of development at which it can be counted on even as a "dark horse" in meeting future energy requirements. On the other hand, the resource base is so large, and the prospects for fewer environmental, proliferation, or safety problems than with fission breeders so promising, that we must not drop it. We cannot afford to lose the momentum that has been gained through several decades of increasingly well-coordinated international research. We have not gone into a great deal of technical detail or assessment of the fusion program because it does not promise to serve as a source of energy within the period considered by this study.

The following are the committee's main conclusions and recommendations.

1. Although the development of nuclear fusion faces considerable uncertainties, it should be pursued, and reevaluated in 5 years. By that time, large scientific break-even experiments in both magnetic and inertial confinement will have been attempted. More realistic engineering designs and guidance for further research on technological obstacles should then emerge naturally.

2. Principal attention should be directed first to the problems of pure fusion reactors, before the question of fusion-fission hybrids is considered.

3. The immature state of fusion research and development offers the opportunity to give attention to environmental and safety characteristics in the earliest stages of design. Consideration of these characteristics is so important to decisions on major investments in fusion that the opportunity must not be wasted.

4. A small effort should be directed to fuel cycles other than deuterium-tritium. Pure deuterium has a much lower reaction rate, but it presents no critical tritium-regeneration problem and wreaks less structural damage from high-energy neutrons. In the so-called neutronless fuel cycles, all particles and products are electrically charged, and in theory there is no radioactivity. Smaller devices might be built, but the required plasma temperatures are much higher, and the energy balance is probably unfavorable.

5. High priority should be given to study and testing of structural materials, and assessments of their availability must be undertaken.

6. Research and development in nuclear fusion has enjoyed singularly fruitful international cooperation. This cooperation should be encouraged and extended to speed progress and reduce the cost to each individual country.

RISKS OF ENERGY SYSTEMS

All energy systems entail risks to the environment and to the health and welfare of people. It is difficult to compare such risks quantitatively, however, because our information about them is subject to great uncertainties, and because there is no widely accepted common scale of measurement for aggregating or comparing different kinds of risks and adverse effects. Furthermore, especially with centralized energy production and distribution systems, risks and benefits are not shared equally; the person who receives the benefit generally does not suffer the risk. Obviously, there are important distributional issues that complicate the weighing of risks against benefits and make social decisions about acceptable risk more difficult. There are also differences of opinion on the relative valuation of statistical and catastrophic fatalities, and of value judgments about risks to the environment—particularly to natural ecosystems, where adverse effects on human beings are less obvious and immediate than threats to health and safety.

There is danger that quantitative estimates of risks will be interpreted too literally and that their apparent definiteness will tend to outweigh qualitative and esthetic considerations. Still, it is difficult to reach and articulate meaningful conclusions without using quantitative values. It is important to realize, though, that value judgments expressed as political preferences may often predominate over quantitative technical judgments in decisions about energy systems and strategies.

Three bases for comparison of energy-related risks have been used.

1. Energy-related risks of a given kind have been compared with risks arising from background effects of the same kind; for example, the risks of cancer from the emissions of nuclear power plants can be compared to the average risk of cancer in the general population or the hypothetically estimated cancer risk associated with exposure to natural background radiation.
2. Cross comparisons have been made among alternative energy technologies, systems, or strategies with respect to similar kinds of risks; for example, comparison of the relative risks to ecosystems from coal combustion and hydropower.
3. Energy-related risks have been compared to more familiar risks; for example, fatalities from nuclear reactor accidents could be compared to fatalities from commercial airline accidents.

There are difficulties with each of these bases for comparison. In comparing energy-related risks to background effects of the same kind, the way that quantitative results are presented—in absolute or percentage terms—can influence public perception of the risk involved. If the additional risk from a particular source is very small percentagewise and the exposed population is very large, then the absolute number of deaths attributed to the source can be very large indeed, though it may constitute an infinitesimal fraction of the deaths that would have occurred anyway.

In comparing risks from different technologies, the difficulty stems from the value judgments needed in weighing the different kinds of risks. How should fatalities be compared with injury or sickness? How should immediate deaths from catastrophic events be compared to similar numbers of deaths occurring much later or in future populations? People may place quite different values on these different kinds of adverse effects, and these values may change with time.

Another problem is that the same risk is not equally acceptable under all circumstances. People accept familiar risks, such as those associated with the automobile, cigaret smoking, and industrial accidents, yet reject much smaller risks associated with new technologies. The voluntariness of risk is also important; those who voluntarily accept high risks, such as those of motorcycles or contact sports, may strongly object to the minute involuntary risk of a nearby chemical factory. Finally, the risks of an activity that provides a unique benefit—as does, for example, the automobile—are more acceptable than the risks of a technology to which there appear to be alternatives.

A general problem that arises in connection with almost all risk assessments is the significance of dose-effect relationships at very low doses, for both radiation and chemicals. The conservative assumption of a linear dose-effect relation down to zero dose leads to very large estimates of incremental threats to large populations, but such extrapolations are very uncertain. They are likely to be overestimates, but the extent of the overestimate is unknown.

One way around the problem of low-level radiation is to compare the radiation dose with that from natural background radiation. Although the effect of neither is known, one can say that a radiation dose of, say, 1 percent of the background will have an effect, if any, that is a tiny fraction of the effect of a radiation dose that the human species has experienced throughout its history. Unfortunately, no such comparison is possible with most chemical hazards.

In this study, comparison of energy-related risks to nonenergy risks was generally avoided, because it was believed to have little pertinence to energy policy decisions.*† The first two of the above-listed three approaches to risk comparison were followed, with emphasis whenever feasible on the comparison of similar types of risks from different energy technologies and strategies.

ROUTINE INDUSTRIAL ACCIDENTS AND DISEASE

Accidents are the most accurately assessed of energy-related risks. In this regard, coal is the most dangerous of major energy sources: About 10 times as many accidental deaths occur in the coal energy cycle, from mine to power plant, as in the production of an equivalent amount of power

*See statement 1-49, by L. F. Lischer, Appendix A.

†See statement 1-50, by B. I. Spinrad, H. Brooks, L. F. Lischer, and D. J. Rose, Appendix A.

from oil, gas, or nuclear energy. Most of the accident risk with coal is associated with deep mining and rail transportation. (The latter, of course, is not uniquely associated with coal.) The health of workers in the mines has been notoriously poor in the past and has led to special congressional legislation to provide benefits that now total more than $1 billion/yr. A conscientious program to improve mine safety and hygiene, especially by enforcing current regulations, and to improve railroad safety could materially improve the situation. The rising percentage of surface mining in the total of production should also tend to reduce the risk of accident and disease.

EMISSIONS

A great variety of pollutants that may affect human health as well as plant and animal life are released from the combustion of fossil fuels, especially coal. These include sulfur and nitrogen oxides, carbon monoxide, hydrocarbons, particulates, and heavy metals (in trace amounts). Local air pollution containing these substances at high levels and in varying proportions is known to have increased the incidence of discomfort and disease (especially of the respiratory system), and even death. The intent of the national ambient air quality standards is to render negligible the morbidity and certainly mortality (or so-called "premature death") from emissions.

Whether or not the standards have been set at the most efficient levels (adequately protective of health, but not needlessly restrictive or costly), and whether all toxic substances requiring regulation have been specified are topics under very active discussion and investigation. The standards themselves must be reviewed, by law, every 5 years and revised if necessary. Current interest centers on several pollutants: sulfur and nitrogen oxides, carbon monoxide, hydrocarbons, particulates, and heavy metals. Since the particulates (now regulated) comprise a spectrum of sizes, of which only those below 2 μm in size can reach the lungs, it is thought that respirable particulates may be the true measure of toxicity. A standard for sulfates had been proposed in addition to the current one for sulfur dioxide. Sulfate is a constituent of the particulates, however, so that it might be an indirect measure for them. In any event, the acidity of the atmosphere does depend on its sulfate (and nitrate) content. Hydrocarbons and heavy metals are also associated with the particulates. In setting standards, the question of whether there are thresholds (exposure levels below which there are no significant health effects from pollutants) is important. In general, standards are based on all available evidence, including that for any type of induced discomfort, promotion or induction of disease, and possible genetic effects. As a practical matter, a level at or below which measurable effects cannot be observed must be decided on, and the standard set as a matter of judgment at some level deemed to be safe. There is good reason to believe that effects, although unmeasured, do occur at levels below those set by some standards. The Clean Air Act requires that all individuals, even those unusually sensitive, be protected; other environmental statutes may have different requirements.

In discussing air pollution emissions, one should not forget that a major cause of air pollution is the automobile, which is especially responsible for carbon monoxide, nitrogen dioxide, and hydrocarbons. From a toxicological point of view, the pollutants from the automobile may interact at the biological or chemical levels with those from stationary sources such as power plants.

Standards should be regarded as reflecting the best judgment of experts at the time they are instituted, and thus subject to change (up or down) with increases in knowledge and changes in the political and social value judgments the standards reflect. In the longer term, pollution control strategies should be reassessed with a view to including greater incentives for suppliers—incentives to achieve control beyond mere compliance. The goal should be to produce the greatest environmental improvement (measured by reduction in estimated social costs) for a given overall economic cost.*

In comparing the effects of emissions from combustion and those from nuclear power plants, principal consequences are usually considered. First consider the induction of discomfort and noncancer illness (for example, that of the respiratory tract). Under routine operation there is no such risk from a nuclear plant, and there *should* be none, or practically none, from the fossil-fueled one. As noted above, however, current standards may not be sufficiently protective. The problem is under debate and is complicated by the role of automobile emissions.

Second, it is known that cancer deaths can be caused by ionizing radiation and also by emissions from certain coal-fueled industrial operations. One year's routine operation of a 1-GWe nuclear reactor (including its associated fuel supply operations) exposes a population of about one million persons and is estimated to induce eventually less than one cancer death (based on extrapolation from much higher doses on a linear dose-effect hypothesis). This compares with an annual cancer mortality rate of 1700 per million in the United States.

The cancer induced by 1 year's operation of the coal energy cycle has not been estimated. This is not to say that such a risk does not exist, nor to suggest that it might not be comparable to that of the nuclear system.

*See statement 1-51, by L. F. Lischer and H. I. Kohn, Appendix A.

Carcinogens are present in fossil fuel emissions, particularly those from coal combustion, but there is no information on their public health effects. In the past, under less stringent occupational standards, workers exposed to coal emissions suffered increases in cancer rates. In coal-based synthetic fuel processes, many carcinogens may arise, but with careful plant design it should be possible to attain a very low occupational risk. In the products themselves, most carcinogens will remain with the heavy residues, and synthetic gas and distillates should present little cancer risk to the general public. For residual liquid fuels, including those derived from shale, close control of emissions within plants and releases to the atmosphere will be necessary. Such heavy fuels would be used in large industrial boilers and power plants, where the necessary occupational safeguards could be applied.

Coal (especially certain lignites) contains varying concentrations of uranium, and its combustion releases radioactivity into the atmosphere.[24] The solid wastes from coal combustion can also be a source of radiation. These radiation effects are generally thought to be less important than those from uranium mining.

Third, too little is known about the heritable genetic effects in man of either ionizing radiation or fossil fuel emissions to permit a comparison. Both agents have demonstrable mutagenic activity in laboratory tests. By extrapolation from such results, the Risk and Impact Panel estimated that a 1-GWe nuclear plant, for each year of its operation (with the associated fuel supply) might induce 0.5 severe genetic defects, but places little confidence in the figure. No estimate is feasible for coal.*

LARGE-SCALE ACCIDENTS AND SABOTAGE

Risks of low-probability, high-consequence accidents are associated chiefly with nuclear reactors, hydroelectric dams, and transportation and storage of liquefied natural gas (LNG). The subject of nuclear reactor accidents has been extensively studied, especially by the Reactor Safety Study (WASH-1400),[25] commissioned by the Nuclear Regulatory Commission. This study concluded that over the long term, the expected health damage from nuclear accidents (treated as probability of event times consequences per event) is smaller than that from radiations emitted in routine operation. This conclusion may not be decisive in the public appraisal of nuclear power, however, because some people may have a much greater fear of very infrequent but great nuclear accidents than they have of events that cause comparable totals of illnesses and deaths spread over long periods of

*See statement 1-52, by H. Brooks and D. J. Rose, Appendix A.

time.[26] The committee is in general agreement with the appraisals of the reactor safety study conducted by the American Physical Society study group[27] and more recently by the Reactor Safety Review Group.[28] WASH-1400 contains some estimates that are excessively conservative and others that are almost certainly too optimistic. Which way this would shift the median probabilities for accidents of various severities is uncertain. The consequences of given accidents are apparently underestimated, but probably by not more than a factor of 3. However, the uncertainties in the probability estimates are almost surely several times larger than estimated in WASH-1400. If larger uncertainties are used, the mean, or expected number of fatalities from nuclear accidents, could be higher by a factor of 10 or more than the median values given by WASH-1400 (namely, 0.025 delayed deaths per reactor-year).* †

Catastrophic accidents can also occur in the case of other energy sources, especially large hydroelectric facilities. Between 1918 and 1958, an average of 40 deaths per year resulted from dam failures in the United States, though fewer in the more recent period. Some individual failures killed hundreds. Worst-case scenarios for both dams and LNG facilities lead to numbers of casualties comparable to those associated with the more severe nuclear accident possibilities. The calculated probabilities are higher, although the analyses on which they are based have been much less thorough and systematic than those for nuclear plants.

In the case of the most likely nuclear accidents, most fatalities would be delayed deaths that could not be specifically attributed to nuclear power, due to the exposure of a large population to low-level radiation (chapter 9). Casualties from dam failures and LNG accidents are immediate, with fewer delayed effects. Because such a high proportion of the reactor-related deaths are delayed, and because large populations may be at risk (even though the enhanced risk to any individual may be small), reactor accidents may create much greater apprehension than other types of catastrophic accidents that can cause the same number of fatalities.

Nuclear plants, dams, and LNG facilities are probably similarly vulnerable to sabotage, but nuclear plants are presently better guarded and may be inherently easier to guard. The consequences of sabotage of nuclear plants appear to be in about the same range as those of the severest postulated accidents discussed in WASH-1400.‡ The possible severe consequences could be much higher, though, because saboteurs could choose times and places for maximum effect. The safety analysis techniques developed for assessing nuclear reactor accidents ought to be

*See statement 1-53, by J. P. Holdren, Appendix A.

†Statement 1-54, by L. F. Lischer: Critiques of WASH-1400 have emphasized that uncertainty ranges are larger than originally stated, both higher and lower.

‡See statement 1-55, by L. F. Lischer, Appendix A.

applied to sabotage, diversion of weapons materials by terrorists, and other safeguards issues, for both nuclear power and other energy technologies.

MANAGEMENT OF WASTE

All energy systems produce wastes, and their management involves risks to health. Although coal ash and coal-mining wastes pose significant problems, nuclear waste management is considerably more difficult. The committee's view of the nuclear waste problem is discussed in detail in chapters 5 and 9. The committee's conclusions and recommendations are presented under "Prospects for Nuclear Power" in this chapter.

ECOSYSTEM EFFECTS

The adverse ecological consequences of energy production and use include loss of arable land, water resources, open space, wilderness areas, natural beauty, habitat, and wild populations or species. Among the public, there is wide divergence in judgments about the relative and absolute importance of these criteria. Some value them very highly, while others regard them as less vital than a number of other human economic and social needs. This may be partly because the long-range human consequences of the loss of ecological diversity are less well understood and much less widely appreciated than the more immediate consequences of energy development, such as direct damage to health.

By the particular criteria of damage to ecosystems, the Risk and Impact Panel judged that the energy source most destructive, per unit of energy output, is hydroelectric power* (possibly including small dams on tributaries).[29] Hydroelectric power installations destroy natural habitats in the vicinities of dams; change the health, productivity and ecological balance of downstream areas; and accelerate siltation and eutrophication in the lakes created by the dams. Nearly as destructive is the land-based production of biomass (i.e., growing crops on energy farms to be burned or converted into fuel). Among the adverse ecological effects of energy farms are land use in competition with agriculture, depletion of soil nutrients and consequent additional requirements for chemical fertilizer, and the fact that the hardy fast-growing species required for economic energy production could become widespread nuisances. So long as the use of biomass is confined to organic or agricultural wastes, or to such materials as seaweed or crops raised on wastelands, the ecological effect is minimal. It becomes a serious consideration when total use exceeds this base, and

*See statement 1-56, by H. S. Houthakker, Appendix A.

may be appreciable.[30] Among fossil fuels, shale oil and coal-derived synthetic fuels are probably the most damaging to ecosystems. The ecological implications of oil development depend on locale; offshore development in northern regions is especially risky.

For nuclear power, direct health effects are much more important than ecological impact. Nuclear power affects ecosystems less than any other source of energy, even if one considers the whole fuel cycle. Nevertheless, if the number of light water reactors built and operated begins to exhaust supplies of high-grade uranium ore, the environmental effects of mining very low grade ores could become comparable to those of coal mining. This problem would not, of course, develop with breeder reactors.

The adverse consequences of solar energy on ecosystems are poorly known, but for most applications are probably mild.[31] (Chapter 9 discusses these effects in some detail.) Significant effects, comparable to those of fossil fuels, might be encountered in extracting and processing the materials required by centralized or widespread decentralized solar installations. Large-scale use of ocean thermal conversion might pose significant hazards to marine ecosystems, owing to exchange of heat and plant nutrients beween deep and shallow water strata. These possibilities of ecosystem damage would probably arise only if the technologies were employed on a sufficient scale to provide 15–30 percent of the total national demand for energy.

WATER SUPPLY PROBLEMS

Water is potentially a limiting factor in any plan to produce and use more coal on a large scale.[32] Consumption of water in the production of electricity or synthetic fuels is many times greater than in the mining of the coal itself under current practice. Per unit output, today's conventional nuclear reactors require 50 percent more water than those burning fossil fuel; more advanced reactor designs offer the opportunity to significantly reduce water consumption, however.

We infer that a 20-quad increment in coal use for electricity production (12.5 quads) and synthetic fuels (7.5 quads) would raise water supply problems unless specific attention was devoted to solving them in advance. (The National Energy Plan of 1977 projected an 18-quad increment by 1985.) Of course, the efficiency of water use in these processes can be increased (at increased cost), now-unused sources such as brackish groundwater can be developed, and interbasin transfers might be extended. (This last may appear unlikely under general conditions of water shortage.)

On the other hand, steps can be taken to find locations where water is in fact still available, and to place increased demand at these locations, insofar as that is feasible. Study of the hydrological regions of the United States shows great disparity in the amounts of water that are potentially still available. The crucial importance of siting in relation to water supply (on both a local and regional basis) has been emphasized in the report of six national laboratories that analyzed the President's National Energy Plan of 1977.[33]

It is clear that regional and interregional, as well as local, hydrological analysis must become an integral part of national energy planning, not only to prevent water-supply failure, but especially to obtain optimal use of our hydrological resources. We recommend that all hydrological regions be studied and that a national data bank be established. Water resources are largely under the control of the states, with the result that they are controlled by different approaches in law that have long-established historical precedents; a national policy will be consequently very difficult to construct. The energy-water problem is, in fact, a part of a much broader one of water as a general limiting factor in the activities of society.

CLIMATE

Were all the world's fossil fuel resources to be burned, the CO_2 content of the atmosphere would increase by a factor of between 5 and 8. If the hypothesis of a "greenhouse effect" is correct, the climatic effects would almost certainly be catastrophic.[34] The largest uncertainties connected with the CO_2 problem pertain to the timing rather than to the existence of the problem. If the worldwide combustion of fossil fuels, particularly coal, continues to increase, the problem could begin to be perceptible as early as the first few decades of the twenty-first century, or it might not become significant until the latter part of the twenty-first century if world energy growth slows or shifts to nonfossil energy sources. Even if fossil resources were consumed at no more than the present rate, the CO_2 problem would eventually become important, though it might be postponed for a century. A serious concern is that, owing to various positive feedback mechanisms, climatic changes due to CO_2 would be irreversible by the time they were detected above natural climatic fluctuations. It needs to be emphasized that the CO_2 problem is global, not local or regional. It depends on the total world consumption of fossil fuels and not on what happens in a single nation, even one as large as the United States.

The climatic effects of increasing atmospheric CO_2 might conceivably be beneficial in some areas (for example, by lengthening the growing season in agriculturally marginal northern latitudes), but the principal effect would almost certainly be to redistribute agricultural productivity. Even with net benefits, the effects in some regions could be disastrous.* †

Solar collectors could have a global effect in the far future. If they are deployed in such a way as to alter the surface reflectivity in a sufficiently large region, they could disturb global circulation patterns and thus have climatic effects beyond the regions where they are located. Worldwide reliance on ocean thermal energy conversion could induce climatic effects by changing the average surface temperature of the tropical oceans. The possible effects of solar energy have only just begun to receive careful study.[35] They could be of no concern unless the use of solar energy becomes very large, and, in any case, there would be plenty of time to deal with the problem as it began to become important, provided it is not altogether overlooked.

Hydroelectric and geothermal sources are likely to have less serious climatic effects, although large-scale water impoundments and irrigation can affect regional hydrologic cycles and thermal balances.

Nuclear reactors, because they do not emit CO_2, will have much smaller effects on climate than fossil-fueled installations; the effects of CO_2 for the balance of heat radiation are much more important globally than are thermal releases. Should considerations of diversion and proliferation lead to the deployment of breeder reactors and reprocessing facilities in "energy parks" of more than 30-GWe total capacity, however, these might alter local or regional atmospheric circulation patterns, and even generate severe artificial convective storms in particular regions, under certain meteorological conditions.

SOCIOPOLITICAL ISSUES

The sociopolitical aspects of energy planning need to be much more thoroughly explored. For example, conventional analysis of the risks associated with energy systems and strategies gives relatively little emphasis to the distribution of risks and benefits, although from a sociopolitical standpoint, the distribution of these risks and benefits—from class to class and region to region—may be more significant than the net effects. For example, there is considerable disagreement about the distributional effects of certain energy conservation measures, such as various forms of "energy tax." Unevenness of distribution should not be used as an excuse to forgo conservation, but it must be analyzed so that it can be dealt with by compensatory measures.

*Statement 1-57, by J. P. Holdren: Even in regions where the long-term effect of CO_2-induced climate change is beneficial, the short-term effects are likely to be strongly negative.

†Statement 1-58, by H. I. Kohn: This international problem involves the automobile as well as industry. International cooperation is necessary to estimate and anticipate it.

Another sociopolitical aspect of risk is that public attitudes to risks often have symbolic and institutional dimensions that relate more to confidence in the institutions that manage the technologies than to the characteristics of the technologies themselves. This is exemplified by the wide difference in attitudes toward nuclear and solar energy. To some, nuclear power symbolizes big government, big business, and an impersonal, centralized bureaucracy unresponsive to local needs and sentiments, while solar energy represents a "natural" form of energy that can be controlled by average citizens. To others, mandated conservation measures require an intrusion of government in consumer decisions that is regarded as intolerable. Decentralized solar technologies, if deployed on a scale sufficient to provide a significant fraction of national energy needs, will require a large-scale mass production, distribution, and service industry that might not look so different from existing electric- and fuel-distribution networks. How such attitudes are likely to develop over time, or be affected by the dialog between the public and various groups of experts, is difficult to assess.

A conclusion reached in many parts of the study is that noneconomic factors will play an important, often dominant, role in influencing future energy demand and supply. Life-style, value, and welfare implications may strongly influence energy consumption patterns, and political acceptability will affect both the availability of energy resources and the conservation of energy.

Insufficient systematic attention has been given to the risks and potential consequences of energy shortages and to the vulnerability of different overall energy regimes to unexpected interruptions. Because of their importance to policy, these aspects need much more systematic study and dissemination of information to the public.

SOME GENERAL CONCLUSIONS ON RISK

Conservation

For the most part, conservation is the least risky energy strategy from the standpoint of direct effects on the environment and public health. The main reason that conservation cannot be the only strategy is that at some level of application, conservation would give rise to indirect socioeconomic and political effects, mostly through economic adversity, that would predominate over its direct benefits. We cannot be sure where that point is, but all the CONAES technical analyses suggest that it is a long way from where we are now, possibly at an energy/GNP ratio of about half its present value, given several decades for adjustment. The maximum conservation achievable without adverse socioeconomic effects will likely have health and environmental benefits and therefore should have highest priority in policies to reduce the risks of energy systems.

Fossil Fuels

Among fossil fuels, natural gas presents the smallest health and environmental risks in both production and consumption, although there is the possibility of serious accidents in the transportation and storage of liquefied natural gas. Oil is next, and coal is much higher in risk. This ranking is likely to persist, although the gap may narrow with improvements in technology. Research is most urgently needed on the health effects of coal combustion by utilities and industry, and on the possible occupational and public health hazards of producing and using synthetic fuels.

We must be prepared for the possibility that adverse health effects, global CO_2 increase and associated climatic change, freshwater supply problems, and ecological considerations will eventually severely restrict continuing expansion of coal use. These problems are likely, though not certain, to become critical at about 3 times current coal output, or less.

Nuclear Power

The routine risks of nuclear power include the induction of cancer and genetic effects by ionizing radiation released throughout the nuclear energy cycle. These risks are very small in comparison to the overall incidence of cancer and genetic effects in the general population, and they could be significantly smaller yet if the most important source of radiation in the nuclear energy cycle—uranium mill tailings—were generally better protected. There are also risks of severe accidents, whose probabilities have been estimated with a great deal of uncertainty, but whose severities could be comparable to those of large dam failures and liquefied natural gas storage system fires. There are also risks from the disposal of radioactive waste; these are less than those of the other parts of the nuclear energy cycle, but only if appropriate action is taken to find suitable long-term disposal sites and methods.

It should be clear from the earlier general discussion of risk comparisons that any ranking of the risks of technologies as disparate as coal-fired and nuclear electricity generation is subject to very broad, and in some cases irreducible, uncertainties. However, if one takes all health effects into account (including mining and transportation accidents and the estimated expectations from nuclear accidents), the health effects of coal production and use appear to be a good deal greater than those of the nuclear energy cycle. If one takes the most optimistic view of the health effects of coal-derived air pollution and the most pessimistic view of the risk of nuclear accidents, though, coal might have a small advantage in such a comparison.* †

Nuclear power is associated also with risks of nuclear weapons proliferation and terrorism, but the magnitude of these risks (and even whether nuclear power increases or decreases the risks) cannot be assessed in terms of probabilities and consequences.

Solar Energy

Several solar energy technologies appear very promising from the standpoint of health and environmental risk. Hydroelectric power (classed by convention with solar energy), however, while benign with regard to air pollution, is quite destructive of ecosystems per unit of output. Energy farms are also likely to be ecologically destructive if deployed on a scale large enough to provide more than a few percent of total energy needs. For most solar technologies, the main risks are those associated with extracting and processing the requisite large amounts of construction materials.

Public Appraisal of Energy Systems

There is an urgent need for research that will contribute to better understanding of the factors that determine public perceptions of the health and environmental risks of energy systems, and their acceptance by different subgroups within the public. No strategy for risk reduction in energy systems can be fully acceptable if it does not take into account these public perceptions and judgments, even when they are seen as irrational by experts.‡ It is unlikely that the appraisal of risk will ever be able to avoid difficult relative value judgments between different kinds of risks, as well as between risks and economic or other benefits of energy technologies. This is not to say that present methods of risk assessment cannot be improved. Nevertheless, the judgmental factor will continue to predominate in decisions among energy alternatives, and is unlikely ever to be superseded by formal analysis of risks and benefits. This underscores the importance of an informed and open public debate.

*See statement 1-59, by J. P. Holdren, Appendix A.

†See statement 1-60, by H. I. Kohn and H. Brooks, Appendix A.

‡See statement 1-61, by H. Brooks, D. J. Rose, and B. I. Spinrad, Appendix A.

INTERNATIONAL ASPECTS OF THE ENERGY PROBLEM

The energy situation of the United States is materially different from those of most other noncommunist industrial countries. The U.S. per capita energy consumption and energy/GNP ratio are, respectively, 2 and 1½ times the average for the rest of the Organization for Economic Cooperation and Development. The potential for conservation through greater efficiency is thus greater in the United States than in most other countries. Our indigenous energy resources are at the same time much greater. A world perspective obviously differs considerably from that of a purely domestic standpoint.

The committee has not undertaken the formidable task of making long-range projections of world energy markets consistent with the domestic scenarios used in chapters 2 and 11. It has drawn a few conclusions on global energy perspectives by assuming that the United States takes no new policy measures beyond those in effect in 1978, other than allowing existing price controls to expire. We shall discuss the effects of various national policies to ameliorate the impact of the United States on the world energy situation in the context of these conclusions.

In lieu of a formal presentation of alternative global projections, we confine ourselves to a few general remarks on global energy perspectives.[36]

1. The growth of world energy consumption will slow from the 5.1 percent per year recorded in 1960-1973. However, if present patterns of economic growth in the world continue, and if the aspirations of the developing countries for larger shares of economic activity are realized, the average long-term rate of energy demand growth is unlikely to fall much below 3 percent per year. Even if energy conservation in the United States accomplishes a great deal domestically, it will be more than offset by demand growth in countries at the "takeoff" stage of development. By the year 2010, world energy consumption will probably be 3 or 4 times as large as it is now. The developing countries will then have a larger share in world energy consumption than they have at present.

2. Electricity demand will probably grow more rapidly than total energy demand for two reasons. First, a large part of electricity cost is due to capital charges, and this will become more true as more capital-intensive forms of electricity generation, particularly nuclear reactors, are introduced. This means that electricity prices are less sensitive to fuel costs. If primary fuel costs rise more than capital costs, electricity would become cheaper relative to other energy forms.* Second, as societies

*Statement 1-62, by J. P. Holdren: The opposite situation—electricity becoming more expensive relative to other energy forms—seems to me at least as likely.

become more affluent they tend to prefer more convenient energy forms, such as electricity or gas, much as they convert more and more grain to animal protein in their food demand. By 2010 world electricity consumption could be 3–5 times as large as at present. If the market is the principal determinant of relative demand, and if there are no noneconomic constraints on the rate at which nuclear capacity can be expanded, then two thirds or more of electricity would probably be supplied by nuclear power, with coal a distant second, consumed mostly in the United States.† ‡ In our view, expansion of nuclear capacity at so great a rate is unlikely. Also, a breakthrough in solar electric technology, if it came soon enough, could reduce the attractiveness of nuclear power somewhat.

3. In the absence of truly spectacular discoveries elsewhere, the OPEC countries (especially those in the Middle East and Africa) will account for the bulk of the world's oil production in the early part of the twenty-first century. In addition to North America, Europe, and East Asia, even Latin America will by then probably be a large oil importer unless the Venezuelan heavy oils are fully developed. However, North American production, though smaller than at present, will still be substantial. Cumulative oil production between now and 2010 is likely to exhaust all presently proved reserves of "conventional" oil. Because of intervening discoveries, however, oil reserves should still be at least as large as they are now, but they will be high-cost reserves.

4. The Middle East and Africa will become large exporters of natural gas and uranium; U.S., Canadian, and Australian uranium will also face a considerable export demand. The degree to which these countries will be willing to satisfy this demand with political conditions acceptable to importers is difficult to foresee.

5. As oil production gradually falls more firmly under OPEC control, the opportunity for surges in oil price like those of 1973–1974 and 1979 will increase. Moreover, as OPEC's reserves of low-cost oil are depleted, the incentives to raise prices will intensify; this would be true even in the absence of a cartel. The price of uranium, increasing at an accelerating rate as the electric power industry becomes predominantly nuclear, could approach $100/lb of U_3O_8 (in 1972 dollars) by the end of this century if reprocessing is prohibited. Even with reprocessing, the uranium price may be high enough to make breeder reactors competitive with existing reactor types in some parts of the world, especially in Europe (political events and public opinion permitting). Coal and natural gas will also become considerably more expensive in real terms.

†Statement 1-63, by J. P. Holdren: Coal can be expected to play a major role in the Soviet Union, in China, and in both Germanies, as well.

‡Statement 1-64, by H. I. Kohn and H. Brooks: There is no evidence that coal would not be important to Russia, China, and Eastern Europe, nor perhaps to importing countries.

6. Because of their predominance in oil, natural gas, and uranium, the Middle East and Africa will develop an even larger surplus in their energy trades, probably running into hundreds of billions of 1972 dollars by the turn of the century. The corresponding deficits will be primarily in the industrial countries (except Canada). U.S. invisible items of trade are now quite strong and are supporting the nation's current account. A good part of this flow represents oil company earnings in the world market; this partially offsets the high costs of oil imports. In addition, new conservation efforts, new oil finds, and a high propensity to import by OPEC help keep the U.S. external position from deteriorating too much. In the United States the energy trade deficit will be somewhat reduced by the expected growth in exports of coal or uranium if such exports are permitted. If the United States were to limit uranium exports, there would be a correspondingly larger demand for U.S. coal. The main reason uranium would normally be preferred by importers is its lower transportation cost.

These projections do not take into account the trade in nuclear power plants and related facilities (and possibly other advanced energy technologies), which may offset a large part of the industrial nations' energy trade deficits but will add to the deficits of the non-oil-producing countries. In the absence of political constraints, worldwide investment in nuclear power between now and 2010 could add up to about one trillion 1972 dollars, and much of this will be supplied by North America, Europe, and Japan. Nonenergy exports of developing countries not members of OPEC would have to expand to finance their part of these investments.

CONSEQUENCES OF ACTION ON NATIONAL ENERGY POLICIES

Conservation in the United States, beyond what is induced by higher world oil prices, would reduce the growth of demand for OPEC oil and thus reduce the cartel's power to raise the price and limit production. The more the conservation effort concentrates on oil (or natural gas in uses where the two are directly substitutable), the greater will be the benefits to the rest of the world, although the magnitude of these benefits should not be exaggerated. Promotion of domestic energy production, especially of oil and gas and directly substitutable energy forms, would be equivalent to conservation in its external economic effects.

Price controls on oil and gas, or other measures shielding domestic consumers from world energy prices, would have effects opposite to those of accelerated conservation and domestic production; they would reinforce the pressure for a higher world oil price.

A tariff on imported oil would encourage conservation and domestic output by allowing the domestic price of oil to rise to match the landed price of imported oil (assuming price controls have expired). It would also enable the importing country to reduce the monopoly profit that would otherwise go to OPEC. A tariff would be particularly effective if adopted simultaneously by other major oil-importing countries. Import quotas, with competitive bidding for import licenses, would similarly reduce OPEC's power over oil prices.*

Abandoning nuclear reprocessing is likely to accelerate the rise of uranium prices. This would increase the incentives for reprocessing in uranium-importing countries. To counter this tendency, the United States (possibly in agreement with Canada and Australia), would have to keep the price of enriched uranium low enough, by subsidies if necessary, to make reprocessing uneconomic. If such a policy made a major contribution to preventing nuclear war or large-scale terrorism, the probable high cost to the United States would not be considered prohibitive. However, alternative methods of controlling proliferation (for example, international safeguards programs including international surveillance of reprocessing operations) could be cheaper and more effective, and must be explored.

Beyond all this, it must be recognized that so much attention paid to the spent-fuel end of the uranium fuel cycle tends to ignore the fact that nuclear explosives can be obtained by uranium enrichment—the so-called front end of the cycle. (See chapter 5 under the heading "Uranium Enrichment.") As years pass and new enrichment technologies appear, this front-end risk of weapons proliferation increases.

Abandonment or postponement of the breeder reactor is likely to have effects similar to the avoidance of reprocessing, raising the price of uranium, and thus strengthening the interest of other countries in the development of breeders or advanced converters. Under some plausible conditions, the United States could remain a uranium exporter through the end of this century. Hence a major delay in the domestic breeder program, rather than setting an example to others, may accelerate breeder development elsewhere, if only because it would leave less U.S. uranium available for export (or increase U.S. demand for uranium imports). In any case, European work on breeders may be too far along, and too strongly supported by energy projections, to be stopped, despite growing political opposition to nuclear power in many European countries and Japan. To the extent that public distrust of nuclear power in the industrialized countries slows its growth, the pressure on uranium supplies will decrease and the above-mentioned problems will be postponed, although the problems of the international oil market will intensify.

A slowdown in the growth of U.S. GNP would help keep down our

*Statement 1-65, by L. F. Lischer and D. J. Rose: OPEC, of course, could retaliate by stopping shipments.

energy demand and be similar in that respect to the accelerated conservation discussed earlier. However, it would also reduce U.S. demand for nonenergy imports and thus make it more difficult for other countries, especially poor ones, to finance their energy imports.

THE DEVELOPING COUNTRIES AND THE WORLD FINANCIAL SYSTEMS

As we have seen, the growing demand for energy in the developing countries will make them increasingly important in the global energy picture. Some of these countries are already considerable importers of oil, and others will become so as their transportation sectors expand. Moreover, the industrialization that is an inescapable aspect of economic development will greatly increase their reliance on electric power, of which they now have very little. Their agriculture will also shift from animal and human energy to tractors, harvesters, and trucks, and from natural to industrial fertilizers. As personal incomes rise in these countries, they will want better housing with more lighting and appliances, not to mention air conditioning. The more affluent of their citizens will demand motorcycles, automobiles, and air travel. In fact, the total demand for energy in these countries could conceivably rise faster than GNP.[37] Furthermore, we must hope that their GNP does rise at a reasonable rate, not only in their own interest but also for the sake of global political stability.

No doubt a substantial part of the required energy can be supplied from domestic sources. Oil and gas are found in many developing countries, but most of those with large resources have already joined OPEC. While there does not appear to be much coal in the developing countries, hydroelectricity could be expanded considerably, at ecologically acceptable sites, if financing were available. Sizeable quantities of uranium presumably remain to be discovered in some regions, but uranium (or thorium, of which India has large reserves) is only a small part of the cost of nuclear power.*

It is clear, therefore, that a large part of the energy needed by developing countries will have to be imported. In addition, heavy investments in electric power will be necessary even if the fuel can be obtained inside the country. Electric power, of course, is generally capital intensive, but it will be even more so if oil, gas, and coal are not available, and nuclear and hydroelectric power (or, in the more distant future, solar energy) must be used. In fact, oil is likely to be preempted by transportation uses, and in most developing countries coal would have to

*Statement 1-66, by J. P. Holdren: It is unfortunate that this passage ignores the great potential of renewables other than hydroelectricity, and the potential of geothermal energy, in many developing countries.

be imported from the United States and Australia, the countries with the greatest potentials for exports. It seems likely, therefore, that the developing countries as a whole will concentrate their investments in nuclear and hydroelectric power, at least until the end of this century, and that they will have to import increasing amounts of oil and uranium.

This prospect implies further strains in the international financial system, which is already being taxed by the aftermath of the 1973–1974 oil price increase. The developing countries generally had little leeway in their balances of payments for increased oil prices; moreover, the recession in the developed countries induced by the oil price increase had severe impacts on their export earnings. The OPEC countries on the whole did not spend much of their vast new revenue on exports from developing countries. As a result, the non-oil-producing developing countries as a group (with notable exceptions such as India) suddenly found themselves with large trade deficits whose financing continues to preoccupy the international banking community.

The difficulty is not so much that the money is not available; the OPEC surpluses remain in the world banking system and could be invested elsewhere. The problem is rather that the countries with cash surpluses (principally Saudi Arabia, Kuwait, and the United Arab Emirates) have not been willing to lend large amounts directly to the developing countries, although they have made relatively small amounts available to a few selected countries and to international organizations. These countries with surpluses have preferred to invest in short-term assets in the United States and Europe, rather than in long-term investment projects in the developing countries. Consequently, Western banks have had to assume the credit risks of loans to countries whose debt-servicing ability is heavily dependent on continued rapid economic growth. Various international arrangements are now being worked out to diversify these risks. The stakes are high, for without adequate financing the developing countries would have to curtail economic growth, to the detriment of billions of people already close to the subsistence level, and to the detriment of the international banking system's stability. The developing countries' needs for massive investments in electric power will only magnify their financial problems.

The developed countries, preferably in consultation with the OPEC countries that have cash surpluses, should give high priority to schemes for maintaining a flow of financial resources to poor countries that fosters their economic development. This means, among other things, that they should encourage imports from the poor countries even where these imports compete with domestic production. The international institutions active in this field (particularly the International Bank for Reconstruction and Development, the International Development Association, and the regional development banks) need further strengthening. Increased public awareness of the domestic aspects of the energy problem should not lead to neglect of its far-reaching international implications.*

SUMMARY

This committee has studied at length the many factors and relationships involved in our nation's energy future. It offers here some technical and economic observations that decision makers may find useful as they develop energy policy in the larger context of the future of our society.

Our observations focus on (1) the prime importance of energy conservation; (2) the critical near-term problem of fluid fuel supply; (3) the desirability of a balanced combination of coal and nuclear fission as the only large-scale intermediate-term options for electricity generation; (4) the need to keep the breeder option open; and (5) the importance of investing now in research and development to ensure the availability of a strong range of new energy options sustainable over the long term.

Policy changes both to improve energy efficiency and to enhance the supply of alternatives to imported oil will be necessary. The continuation of artificially low prices would inevitably widen the gap between domestic supply and demand, and this could only be made up by increased imports, a policy that would be increasingly hazardous and difficult to sustain.

The most vital of these observations is the importance of energy demand considerations in planning future energy supplies. There is great flexibility in the technical efficiency of energy use, and there is correspondingly great scope for reducing the growth of energy consumption without appreciable sacrifices in the growth of GNP or in nonenergy consumption patterns. Indeed, as energy prices rise, the nation will face important losses in economic growth if we do not significantly increase the economy's energy efficiency. Reducing the growth of energy demand should be accorded the highest priority in national energy policy.†

In the very near future, substantial savings can be made by relatively simple changes in the ways we manage energy use, and by making investments in retrofits of existing capital stock and consumer durables to render them more energy efficient.

The most substantial conservation opportunities, however, will be fully achievable only over the course of two or more decades, as the existing capital stock and consumer durables are replaced. There are economically attractive opportunities for such improvements in appliances, automobiles,

*See statement 1-67, by H. I. Kohn and L. F. Lischer, Appendix A.

†Statement 1-68, by L. F. Lischer and H. Brooks: To this we would add "while maintaining a healthy and growing economy."

buildings, and industrial processes at today's prices for energy, and as prices rise these opportunities will multiply.

This underscores the importance of clear signals from the economy about trends in the price of energy. New investments in energy-consuming equipment should be made with an eye to energy prices some years in the future. Without clear ideas of the replacement cost of energy and its impact on operating costs, consumers will be unlikely to choose appropriately efficient capital goods. These projected cost signals should be given prominence and clarity through a carefully enunciated governmental pricing policy. They can be amplified where desirable by regulation; performance standards, for example, are useful in cases (such as the automobile) where fuel prices are not strongly reflected in operating costs.

Although there is some uncertainty in these conclusions because of possible feedback effects of energy consumption on labor productivity, labor-force participation, and the propensity for leisure, calculations indicate that, with sufficiently high energy prices, an energy/GNP ratio one half* of today's could be reached, over several decades, without significant adverse effects on economic growth. Of course, so large a change in this ratio implies large price increases and consequent structural changes in the economy. This would entail major adjustments in some sectors, particularly those directly related to the production of energy and of some energy-intensive products and materials. However, given the slow introduction of these changes, paced by the rate of turnover in capital stock and consumer durables, we believe neither their magnitude nor their rate will exceed those experienced in the past owing to changes in technology and in the conditions of economic competition among nations. The possibility of reducing the nation's energy/GNP ratio should serve as a stimulus to strong conservation efforts. It should not, however, be taken as a dependable basis for forgoing simultaneous and vigorous efforts on the supply programs discussed in this report.

The most critical near-term problem in energy supply for this country is fluid fuels. World supplies of petroleum will be severely strained beginning in the 1980s, owing both to the expectation of peaking in world production about a decade later and to new world demands. Severe problems are likely to occur earlier because of political disruptions or cartel actions. Next to demand-growth reduction, therefore, highest priority should be given to the development of a domestic synthetic fuels industry, for both liquids and gas, and to vigorous exploration for conventional oil and gas, enhanced recovery, and development of unconventional sources (particularly of natural gas).

*Statement 1-69, by R. H. Cannon, Jr.: It would be wrong to depend on so large an improvement. Calculations using some models and assumptions predict severe economic impact for smaller energy/GNP reductions.

As fluid fuels are phased out of use for electricity generation, coal and nuclear power are the only economic alternatives for large-scale application in the remainder of this century.* A balanced mix of coal- and nuclear-generated electricity is preferable to the predominance of either. After 1990, for example, coal will be increasingly required for the production of synthetic fuels. The requirements for nuclear capacity depend on the growth rate of electricity demand; this study's projections of electricity growth between 1975 and 2010 (for up to 3 percent annual average GNP growth) are considerably below industry and government projections,† and in the highest-conservation cases actually level off or decline after 1990. Such projections are sensitive also to assumptions about end-use efficiency, technological progress in electricity generation and use, and the assumed behavior of electricity prices in relation to those of primary fuels. They are therefore subject to some uncertainty.

At relatively high growth rates in the demand for electricity, the attractiveness of a breeder or other fuel-efficient reactor is greatest, all other things being equal. At the highest growth rates considered in this study, the breeder can be considered a probable necessity. For this reason, this committee recommends continued development of the LMFBR, so that it can be deployed early in the next century if necessary. Any decision on deployment, however, should be deferred until the future courses of electricity demand growth, fluid fuel supplies, and other factors become clearer.‡ In terms of public risks from routine operation of electric power plants (including fuel production and delivery), coal-fired generation presents the highest overall level of risk, with oil-fired and nuclear generation considerably safer, and natural gas the safest.§ With respect to accidents, the generation of electricity from fossil fuels presents a very low risk of catastrophic accidents. The projected mean number of fatalities¶ associated with nuclear accidents is probably less than the risk from routine operation of the nuclear fuel cycle (including mining, transportation, and waste disposal), but the large range of uncertainty that still attaches to nuclear safety calculations makes it difficult to provide a confident assessment of the probability of catastrophic reactor accidents. The spread of uncertainty in present estimates of the risks of both coal and nuclear power is such that the ranges of possible risk overlap somewhat.

*Statement 1-70, by J. P. Holdren: My longer dissenting view, statement 1-2, Appendix A, also applies here.

†See statement 1-71, by L. F. Lischer and H. Brooks, Appendix A.

‡Statement 1-72, by R. H. Cannon, Jr., and H. Brooks: Since about 20 years will necessarily elapse between such a decision and the start of actual deployment, the decision cannot be delayed very long.

§Statement 1-73, by J. P. Holdren: My longer dissenting view, statement 1-60, Appendix A, also applies here.

¶See statement 1-74, by H. Brooks, Appendix A.

High-level nuclear waste management does not present catastrophic risk potential, but its long-term low-level threat demands more sophisticated and comprehensive study and planning than it has so far received, particularly in view of the acute public sensitivity to this issue.§

The problem of nuclear weapons proliferation is real and is probably the most serious potentially catastrophic problem associated with nuclear power. However, there is no technical fix—even the stopping of nuclear power (especially by a single nation)—that averts the nuclear proliferation problem. At best, the danger can be delayed while better control institutions are put in place. There is a wide difference of opinion about which represents the greater threat to peace: the dangers of proliferation associated with the replacement of fossil resources by nuclear energy, or the exacerbation of international competition for access to fossil fuels that could occur in the absence of an adequate worldwide nuclear power program.

Because of their higher economic costs, solar energy technologies, other than hydroelectric power, will probably not contribute much more than 5 percent to energy supply in this century, unless there is massive government intervention in the market to penalize the use of nonrenewable fuels and subsidize the use of renewable energy sources. Such intervention could find justification in the generally lower social costs of solar energy in comparison to alternatives. The danger of such intervention lies in the possibility that it may lock us into obsolete and expensive technologies with high materials and resource requirements, whereas greater reliance on "natural" market penetration would be less costly and more efficient over the long term. Technical progress in solar technologies, especially photovoltaics, has accelerated dramatically during the last few years; nevertheless, there is still insufficient effort on long-term research and exploratory development of novel concepts. A much increased basic research effort should be directed at finding ways of using solar energy to produce fluid fuels, which may have the greatest promise in the long term.*

Major further exploitation of hydroelectric power, or of biomass through terrestrial energy farms, presents ecological problems that make it inadvisable to count on these as significant future incremental energy sources for the United States. (Marine biomass energy farms could have none of this problem, of course.) There is insufficient information to judge

§Statement 1-75, by H. I. Kohn, D. J. Rose, and B. I. Spinrad: Failure of summary to mention carbon dioxide, water, and regulatory risk problems is misleading. See 'Conclusions" in chapter 9.

*Statement 1-76, by R. H. Cannon, Jr.: Two of these are marine biomass and ocean thermal energy conversion. Not enough is yet known to assess the magnitudes of their potential contributions.

whether the large-scale exploitation of hot-dry-rock geothermal energy or the geopressured brines will ultimately be feasible or economic. Local exploitation of geothermal steam or hot water is already feasible and should be encouraged where it offers an economical substitute for petroleum.

It is too early in the investigation of controlled thermonuclear fusion to make reliable forecasts of its economic or environmental characteristics. It is not, however, an option that can be counted on to make any contribution within the time frame of this study. Nevertheless, fusion warrants sufficient technical effort to enable a realistic assessment by the early part of the next century of its long-term promise in competition with breeder reactors and solar energy technologies.

It is important to keep in mind that the energy problem does not arise from an overall physical scarcity of resources. There are several plausible options for an indefinitely sustainable energy supply, potentially accessible to all the people of the world. The problem is in effecting a socially acceptable and smooth transition from gradually depleting resources of oil and natural gas to new technologies whose potentials are not now fully developed or assessed and whose costs are generally unpredictable. This transition involves time for planning and development on the scale of half a century. The question is whether we are diligent, clever, and lucky enough to make this inevitable transition an orderly and smooth one.

Thus, energy policy involves very large social and political components that are much less well understood than the technical factors. Some of these sociopolitical considerations are amenable to better understanding through research on the social and institutional characteristics of energy systems and the factors that determine public, official, and industry perception and appraisal of them. However, there will remain an irreducible element of conflicting values and political interests that cannot be resolved except in the political arena. The acceptability of any such resolution will be a function of the processes by which it is achieved.

NOTES

1. U.S. Department of Energy, Energy Information Administration, *Annual Report to Congress,* 1978, vol. 2, *Data* (Washington, D.C.: U.S. Department of Energy (DOE/EIA-0173/2), 1979).

2. American Petroleum Institute, *Basic Petroleum Data Book,* looseleaf binder, updated to April 1978 (Washington, D.C.: American Petroleum Institute, 1978).

3. J. Darmstadter and H. Landsberg, "The Economic Background," in "The Oil Crisis in Perspective," ed. R. Vernon, *Daedalus,* Fall 1975, p. 22.

4. H. Landsberg, *Low-Cost, Abundant Energy: Paradise Lost?,* annual report (Washington, D.C.: Resources for the Future, Inc., 1973), pp. 27–52.

5. National Research Council, *Alternative Energy Demand Futures to 2010,* Committee on Nuclear and Alternative Energy Systems, Demand and Conservation Panel (Washington, D.C.: National Academy of Sciences, 1979).

6. National Research Council, *Supporting Paper 2: Energy Modeling for an Uncertain Future,* Committee on Nuclear and Alternative Energy Systems, Synthesis Panel, Modeling Resource Group (Washington, D.C.: National Academy of Sciences, 1978).

7. *Ibid.*

8. Institute for Energy Studies, *Energy and the Economy,* Energy Modeling Forum Report no. 1, vols. 1 and 2 (Stanford, Calif.: Stanford University, 1977).

9. W. W. Hogan, *Dimensions of Energy Demand,* Discussion Paper Series (Cambridge, Mass.: Kennedy School of Government, Harvard University (E-79-02), July 1979).

10. National Research Council, *U.S. Energy Supply Prospects to 2010,* Committee on Nuclear and Alternative Energy Systems, Supply and Delivery Panel (Washington, D.C.: National Academy of Sciences, 1979).

11. Landsberg, *op. cit.*

12. Demand and Conservation Panel, *op. cit.*; and see chapter 11 of this report.

13. National Research Council, *Energy and the Fate of Ecosystems,* Committee on Nuclear and Alternative Energy Systems, Risk and Impact Panel, Ecosystems Impact Resource Group (Washington, D.C.: National Academy of Sciences, in preparation), chap. 6.

14. Institute for Energy Studies, *Coal in Transition: 1980–2000,* Energy Modeling Forum Report no. 1, vols. 1, 2, and 3 (Stanford, Calif.: Stanford University, September 1978).

15. See chapter 5 under "Management of Radioactive Waste"; Roger E. Kasperson *et al.*, "Public Opposition to Nuclear Energy: Retrospect and Prospect," National Research Council, *Supporting Paper 5: Sociopolitical Effects of Energy Use and Policy,* Committee on Nuclear and Alternative Energy Systems, Risk and Impact Panel, Sociopolitical Effects Resource Group (Washington, D.C.: National Academy of Sciences, in preparation); and Dorothy Nelkin and Susan Fallows, "The Evolution of the Nuclear Debate: The Role of Public Participation," *Annual Review of Energy* 3 (1978):275–312.

16. As cited by Nelkin and Fallows, *op. cit.,* pp. 275-276, these include the "powerful imagery of extinction" and "fundamental fears about the integrity of the human body" named by psychiatrist Robert Lifton, the type of surveillance and security controls that might be necessary to protect nuclear fuel cycles and installations, and mistrust of government bureaucracies. For many, they suggest, nuclear power has become a symbol of technology out of control, and of the declining influence of citizens on important matters of policy.

17. National Research Council, *Supporting Paper 1: Problems of U.S. Uranium Resources and Supply to the Year 2010,*Committee on Nuclear and Alternative Energy Systems, Supply and Delivery Panel, Uranium Resource Group (Washington, D.C.: National Academy of Sciences, 1978).

18. Assistant Secretary for Resource Applications, *Statistical Data of the Uranium Industry* (Grand Junction, Colo.: U.S. Department of Energy, 1979).

19. U.S. Department of Energy, Energy Information Administration, *Annual Report to Congress 1978,* vol. 3, *Forecasts* (Washington, D.C.: U.S. Government Printing Office, 1979).

20. American Physical Society, "Report to the American Physical Society by the Study Group on Nuclear Fuel Cycles and Waste Management," *Reviews of Modern Physics* 50 (1978): S-1–S-85.

21. See chapter 6; and National Research Council, *Supporting Paper 6: Domestic Potential of Solar and Other Renewable Energy Sources,* Committee on Nuclear and Alternative Energy Systems, Supply and Delivery Panel, Solar Resource Group (Washington, D.C.: National Academy of Sciences, 1979).

22. American Physical Society, *Principal Conclusions of the American Physical Society Study Group on Solar Photovoltaic Energy Conversion,* report prepared for the Office of

Technology Policy and the U.S. Department of Energy (New York: American Physical Society, 1979).

23. As indicated in the low-growth scenarios of chapter 11: I_2, I_3, II_2, and II_3. See also the maximum-solar supply scenario described in that chapter, and chapter 6.

24. See chapters 4 and 9.

25. U.S. Nuclear Regulatory Commission, *Reactor Safety Study: An Assessment of Accident Risks in U.S. Commercial Nuclear Power Plants* (Washington, D.C.: U.S. Nuclear Regulatory Commission (WASH-1400), 1975).

26. See chapter 9; and National Research Council, *Risks and Impacts of Alternative Energy Systems,* Committee on Nuclear and Alternative Energy Systems, Risk and Impact Panel (Washington, D.C.: National Academy of Sciences, in preparation).

27. American Physical Society, "Report to the American Physical Society by the Study Group on Light-Water Reactor Safety," *Reviews of Modern Physics,* 47 (Summer 1975): suppl. no. 1.

28. Risk Assessment Review Group, H. W. Lewis, Chairman, *The Risk Assessment Review Group Report to the U.S. Nuclear Regulatory Commission* (Washington, D.C.: U.S. Nuclear Regulatory Commission (NUREG/CR/0400), September 1978).

29. Risk and Impact Panel, Ecosystem Impacts Resource Group, *op. cit.*

30. Risk Assessment Review Group, *op. cit.*

31. *Ibid.*

32. Risk and Impact Panel, Ecosystem Impacts Resource Group, *op. cit.*; J. Harte and M. El Gasseir, "Energy and Water," *Science 1*99 (1978): 623–624; and R. F. Probstein and H. Gold, *Water in Synthetic Fuel Production* (Cambridge, Mass.: MIT Press, 1978).

33. U.S. Department of Energy, *An Assessment of National Consequences of Increased Coal Utilization, Executive Summary,* 2 vols. (Washington, D.C.: U.S. Government Printing Office (TID-29425), February 1979).

34. Risk and Impact Panel, *Risks and Impacts of Alternative Energy Systems, op. cit.,* chap. 7.

35. See, for example, J. M. Weingart, *Systems Aspects of Large-Scale Energy Conversion* (Laxenberg, Austria: International Institute for Applied Systems Analysis (RM- 77-23), May 1977).

36. More detail may be found in research inspired by the CONAES study but not conducted under the study's direction; see, for example, H. Houthakker and Michael Kennedy, "Long-Range Energy Prospects," *Energy and Development,* Autumn 1978.

37. This possibility could be offset, however, by the fact that their capital stock will be mostly new and can be designed for efficiency at present and prospective prices for energy.

2 Slowing the Growth of Energy Consumption

It is wrong to think of the nation's energy troubles as simply difficulties in energy supply. The real problem is finding a new balance between energy supply and energy demand, consistent with generally satisfactory overall economic performance.* The generally rising real price of energy lends advantage to higher investments in both supply and conservation. The economic, environmental, and political trade-offs between these two coordinate efforts are not perfectly understood, but it is clear that in many activities throughout the economy it is cheaper now to invest in saving a Btu than in producing an additional one. As prices rise in the coming decades, more such opportunities will appear.

Until only the past few years, falling real prices and the availability of new sources made energy supplies seem virtually inexhaustible. The cheapness of energy fostered buildings, consumer products, industrial processes, and personal habits that used energy in ways that are by today's standards—and by tomorrow's no doubt even more—inefficient.

During this period the prices consumers paid for energy did not fully reflect its costs to society (as they still do not, though some steps have been made). Some resource costs and diverse social costs were borne by society at large rather than directly by the producers and users of energy. Subsidies, for example, were applied to the production and use of various

*See statement 2-1, by R. H. Cannon, Jr., Appendix A.

energy resources, and many of the environmental and other social impacts of producing and using energy were simply unaccounted for.

The underpricing of energy encouraged rapid increases in the use of our most convenient and irreplaceable fossil fuels. At the same time, it provided inadequate incentives to search for additional supplies or substitutes. It is precisely because of this cheap-energy climate, and the consumption patterns it fostered, that there is now so much room for improvement in the efficiency of energy use before we shall begin to feel any serious economic penalties. Transportation (particularly via highway vehicles), comfortably warm or cool buildings, and industrial heat can be provided by much less energy than has been customary. In the future, if current trends in the cost of energy continue, the amount of energy consumed to provide a given economic product will gradually decrease. As the economy moves toward a new economic balance between the costs of energy and those of capital, labor, and other factors of production, it will surely come to produce goods and services with much improved energy efficiency.

DETERMINANTS OF ENERGY DEMAND

Energy is an intermediate good, valued not for itself but for the services it can provide. The amounts and kinds of energy needed to perform a given set of tasks are not fixed but vary from time to time and place to place, depending on a number of technological, economic, and demographic variables. The most important independent variables in this case are the level of economic activity, measured by the gross national (or gross domestic) product,[1] and the relative prices of energy in its various forms. These in turn influence the composition of GNP and the efficiency with which energy is used. Secondary influences are provided by such factors as climate and the geographical distribution of population and industry.*

A useful measure of the energy intensity of a society is the ratio of its energy consumption to its economic output, measured in dollars of constant purchasing power. This ratio is expressed for convenience as energy/GDP or energy/GNP. Among the advanced industrial societies it varies by a factor of about 2. The United States energy-to-output ratio is one of the highest.

Some, but not all, of this difference is due to geographical and demographic differences. European configurations of industry and settlement, for example, are more compact than those of this country, and

*See statement 2-2, by J. P. Holdren, Appendix A.

transport distances are correspondingly shorter. International variations in this ratio depend most importantly, however, on the price of energy, as it influences the efficiency with which energy is used. Most nations tax energy and energy-consuming equipment in ways calculated to induce efficiency, and such policies have tended to constrain energy consumption. For example, Western European countries have generally taxed gasoline to price levels double those in the United States, and have instituted purchase taxes on automobiles more or less scaled to their fuel consumption. This has resulted in cars with better fuel economy, in heavier use of public transportation, and generally in fewer miles driven per vehicle per year. Higher fuel prices have induced Western European and Japanese industries also to be relatively economical of energy, and to take advantage of technical innovations in energy efficiency that in this country have been of little or no economic benefit until recently.

From the end of World War II until the 1973–1974 OPEC price rise, the real prices of most forms of energy declined steadily, as energy producers took advantage of economies of scale and technical innovation, as well as subsidies like the oil depletion allowance. Declining real prices stimulated demand for energy in all countries, including the United States. Still, the U.S. energy/GNP ratio declined from 1920 until 1945 and then remained fairly constant until very recently, when signs of a further decline began to appear as a consequence of rising prices.

The abrupt oil price rise after the 1973 embargo and the accompanying rise in the prices of other fuels have brought about declines in the energy-to-output ratios of all the industrial nations. However, the full effects on energy consumption have not yet been seen. This is because energy, as an intermediate good, depends on durable goods such as furnaces, automobiles, refrigerators, or buildings to provide its ultimate service. The economy's adjustment to higher energy prices thus depends largely on the replacement of capital goods and consumer durables with more efficient models as the old ones reach the ends of their useful lives, and only secondarily on such immediate consumer responses as driving less or turning thermostats down.

The response of energy consumption to changes in price is usually specified by a number called the price elasticity of demand, defined as the ratio of a percentage change in consumption to the percentage change in the price of energy that evokes it. Thus, if a 10 percent price rise evokes a 5 percent decrease in consumption, we speak of a price elasticity of demand for energy of -0.5. Because it takes time for consumption to adjust fully to a new price level, economists refer to short-term and long-term elasticities, with implied lags for adjustment. The values of price elasticities are usually deduced from historical data, from international or interregional comparisons, and from microeconomic estimates and engineering analyses of the feasibility and costs of substituting new, more efficient ways of using energy. These estimates are subject to large uncertainties, and their values have been much debated.

The response of U.S. energy consumption to the OPEC price rise of 1973–1974, for example, is compatible with a wide range of different models. It is equally consistent with a small long-term elasticity and a short adjustment time or a large long-term elasticity and a long adjustment time. Yet the energy consumption for 2010 extrapolated using these two models could vary by almost a factor of 2.[2] This matter is pursued further in a later section of this chapter, which describes the CONAES Modeling Resource Group's econometric analysis of just this question. It is enough to say here that the elasticity value one chooses makes the difference between negligible and profound reduction in GNP growth as a result of large reductions in the energy intensity of the economy. The larger the price elasticity in absolute value, the more it is possible to moderate energy demand without depressing economic growth.

ENERGY PRICES AND EXTERNALITIES

Two obvious questions arise from the foregoing discussion: What is the likely future course of energy prices, and to what extent are they subject to control by policy?

In addressing these questions it is important to realize that the historical decline in the price of energy, due to technical refinements, economies of scale, and neglect of social costs, was reinforced by a variety of direct and indirect subsidies to energy producers and consumers. Examples are the oil depletion allowance, income tax treatment of foreign royalties, and special tax treatment of drilling expenses. This is a complex and controversial question, however. Some policies—notably the oil import quotas of the late 1950s and the oil production controls imposed by the Texas Railroad Commission (applied until the late 1960s)—acted as countervailing factors, tending to raise prices.

Whatever the magnitude of the effect, it is clear that the low price of energy has encouraged consumption and discouraged production and exploration for new supplies. Even after the OPEC price rise, the entitlement policies of the federal government (which spread the costs of imported oil relatively evenly over the domestic market), along with the continuation of price controls on domestic oil and gas, provided effective subsidies to imported oil.

The trend begun in the late 1960s toward incorporating more of the environmental and other social costs of energy into its price is likely to continue pushing prices up. The need to search for and produce energy resources in increasingly inaccessible areas, often in hostile environments, will also contribute to the upward trend, as will the ability of foreign producers to demand higher prices in an oil market that belongs more and more to the sellers. Energy in most forms is likely to rise in price faster than the rate of general inflation, and consumption will therefore tend to grow more slowly relative to GNP than in the past.

THE ROLE OF MANDATORY STANDARDS

Tax, tariff, and price control policies, as we have seen, are important influences on the demand for energy. But energy consumption can also be molded directly—for example, by the imposition of mandatory standards for the efficiency of energy-using equipment (miles-per-gallon standards for automobiles, thermal performance requirements for new buildings, and so on).

There are reasons to believe that such standards will be necessary in some cases to encourage economically rational demand responses to higher energy prices. This is because most energy-consuming equipment embodies trade-offs between initial costs and lifetime energy consumption; in general one must use extra insulation, larger heat exchangers, or the like to reduce the amount of energy consumed in performing the given task. This increases the initial cost, in exchange for future savings in energy costs. Consumers tend to be more influenced by first cost than by prospective operating costs. Where this is true, they have less economic incentive to purchase equipment on the basis of its energy consumption, even if that would be economically advantageous over its lifetime.* Mandatory standards applied to manufacturers and calculated to minimize life cycle costs (at some increase in initial cost) could serve the need for conservation while actually saving money for consumers in the long term.†

DEMAND AND CONSERVATION PANEL RESULTS

The Demand and Conservation Panel of this study developed a range of energy demand projections for the period 1975–2010.[3] They vary from a long-term decrease in per capita energy consumption to a large increase, and were the products of a series of scenarios embodying different assumptions about movements in energy prices. The prices assumed are those experienced by the consumer at the point of consumption and include the effects of any taxes or subsidies. These prices could also be

*See statement 2-3, by H. Brooks, Appendix A.
†See statement 2-4, by E. J. Gornowski, Appendix A.

regarded as surrogates for various nonprice conservation policies, although no correspondence between prices and policies was actually worked out.

Scenarios of this kind, it should be noted, are not predictions. They are rather the results of calculations based on simplified models of the economy and on more or less plausible and self-consistent assumptions about the future. The panel's four basic scenarios depend on a 2 percent average annual real GNP growth rate between 1975 and 2010,‡ corresponding to a real GNP in 2010 twice that in 1975. A variant explored the implications of a 3 percent growth rate, corresponding to a 2010 GNP 2.8 times as high as in 1975. (Table 2-1 traces, for reference, the growth of GNP in the United States from 1945 to 1975.)* The use of GNP as a measure of public welfare, of course, has its limitations. The appendix to this chapter discusses them briefly and offers an approach to a more accurate measure.

The basic assumptions included population growth to 279 million people in 2010, labor-force participation somewhat higher than today's, and other assumptions about key economic variables, including a substantial decline below the historical (pre-1970) rate of increase in labor productivity. Table 2-2 gives the central features of the scenarios, including price assumptions and energy consumption values resulting from them. (For a more detailed account of these assumptions see chapter 11 and the report of the Demand and Conservation Panel.)

In general the movement of primary energy prices tends to be greater than that of prices to the consumer. For example, the 100 percent rise in average primary energy prices between 1970 and 1978 corresponds to only a 30 percent average increase in final energy prices. The Demand and Conservation Panel scenario with the highest annual rate of secondary energy real-price increase uses a value of 4 percent. This is slightly less than the rate experienced between 1972 and 1978 and substantially less than that experienced between 1972 and 1979 (including the 1978–1979 price rises). However, it is hard to make a plausible case that rates of increase significantly greater than this could be sustained out to 2010.

The scenario analysis was made under the assumption that most characteristics of the national economy will behave in the coming decades much as they have in the past. The kinds of goods and services purchased by a consumer with a given income, for example, were assumed to change little, as were general attitudes and ways of life, although shifts in purchasing habits associated with increased affluence were accounted for.

‡Statement 2-5, by R. H. Cannon, Jr.: Over the 33-yr period 1946 to present, GNP follows a 3.4 percent growth line remarkably closely. Using 2 percent here predestined dangerously low energy demand projections.

*Statement 2-6, by R. H. Cannon, Jr.: Table 2-1 simply displays short-term variations. See my previous note.

TABLE 2-1 Past Economic Growth in the United States

Year	GNP (billions of 1975 dollars)[a]	Five-Year Average Growth Rate (percent)
1945	679	
		0.0
1950	679	
		4.2
1955	833	
		2.4
1960	937	
		4.7
1965	1178	
		3.0
1970	1368	
		2.1
1975	1516	

[a] Source: U.S. Department of Commerce, *The National Income and Product Accounts of the United States, 1929–1974: Statistical Tables,* Bureau of Economic Analysis (Washington, D.C.: U.S. Government Printing Office (003-010-00052-9), 1976).

Fuel price elasticities used in certain analyses of the transportation and buildings sectors are extrapolations consistent with historical data and with engineering projections for each price. No major technical breakthroughs are assumed. The panel used engineering and microeconomic methods in each of the three key sectors of energy use (buildings and appliances, industry, and transportation) to determine likely changes in energy intensities and overall sectoral energy consumption as responses to changes in price. Each of the sectors was analyzed separately, and the results were integrated into a total demand projection using input-output techniques to ensure internal consistency. Over the first decade of its projections, the panel used in addition conventional econometric analysis, which relies heavily on empirical evidence from the recent past and from international comparisons. Consumption, price, income, and other data were used in calculations that reflect the characteristics of existing capital stock, population, and economic activity.

In the panel's scenarios for 2010, the ratio of energy use to GNP ranged from half its 1975 value to about 20–30 percent higher. The panel found this entire range to be consistent with a level of GNP substantially greater than the present one.[4] In physical terms this would be accomplished by consumers' substituting capital for energy (e.g., insulation for fuel) or vice versa, depending on whether energy were cheap or costly, scarce or

TABLE 2-2 Scenarios of Energy Demand: Totals

Scenario[a]	Average Delivered Energy Price in 2010 as Multiple of Average 1975 Price (1975 dollars)[b]	Average Annual GNP Growth Rate (percent)	Energy Conservation Policy	Energy Consumption (quads)					
				Buildings	Industry	Transportation	Total	Losses[c]	Primary consumption[d]
Actual 1975	—		—	16	21	17	54	17	71
A* (2010)	4	2	Very aggressive, deliberately arrived at reduced demand requiring some life-style changes	6	26	10	42	16	58
A (2010)	4	2	Aggressive; aimed at maximum efficiency plus minor life-style changes	10	28	14	52	22	74

B (2010)	2	2	Moderate; slowly incorporates more measures to increase efficiency	13	33	20	66	28	94
B′ (2010)	2	3	Same as B, but 3 percent average annual GNP growth	17	46	27	90	44	134
C[e] (2010)	1	2	Unchanged; present policies continue	20	39	26	85	51	130

[a] Scenario D is not included in this table; its price assumption (a one-third decrease by 2010) appears implausible.
[b] Overall average; assumptions by specific fuel type were made reflecting parity and supply; price increases were assumed to occur linearly over time.
[c] Losses include those due to extraction, refining, conversion, transmission, and distribution. Electricity is converted at 10,500 Btu/kWh; coal is converted to synthetic liquids and gases at 68 percent efficiency.
[d] These totals include only marketed energy. Active solar systems provide additional energy to the buildings and industrial sectors in each scenario. Total energy consumption values are 63, 77, 96, and 137 quads in scenarios A, A*, B, and C, respectively.
[e] The Demand and Conservation Panel did not develop a scenario combining the assumptions of unchanging price and 3 percent average GNP growth. If scenario B′ is used as an approximate indicator, such an assumption would entail a primary energy demand of about 175 quads.

abundant. (Scarcity and abundance in this context can be thought of as induced by taxes, subsidies, or regulations, as well as by realizations of the actual quantities of various resources available.) Outside this range, the Demand and Conservation Panel hesitated to offer statements based on its analyses. But at some point not far below the lowest energy scenarios examined, appreciable reductions in GNP should be expected.

The panel's analysis assumed that large variations in the energy required per unit output were compatible with a given level of labor productivity. That this will be so is not certain; there is no available research to indicate the possible effect of energy intensity on labor productivity.

Capital requirements for the entire economy were shown to be relatively constant for all scenarios; investment simply shifts between energy production and energy conservation. (In practice, if there are large shifts in the required allocation of capital, there may be temporary bottlenecks of capital availability to particular sectors for institutional reasons.) Actually, the panel found that at present it takes considerably less capital to save a Btu than to produce one. As the more productive opportunities for saving energy are exploited, this will become less generally true.

The following sections, based on the work of the Demand and Conservation Panel and others, describe this study's assessments of the opportunities for conservation over the next several decades within the framework of the demand scenarios.

TRANSPORTATION

Energy savings in transportation can be achieved through modest investments in existing technology and improved management. The greatest such savings can be realized in automobiles, aircraft, and freight trucks. Because of the typically dispersed U.S. settlement patterns, the energy-saving potential of mass transportation, particularly fixed-rail transit, is far less, except over periods longer than the 35 years considered in this study.

In total, it may be possible to halve the energy requirement per passenger- or ton-mile in the United States over the next 25–35 years. This could offset the large expected increase in the amount of transportation, and in turn lead to total energy consumption of, for example, 27 quadrillion Btu (quads) for transportation in the Demand and Conservation Panel's scenario B′, in which real energy prices double and GNP rises by an annual average of 3 percent (corresponding to a total GNP 2.8 times higher in 2010 than in 1975). Under the scenario A assumptions that real energy prices quadruple and GNP growth averages 2 percent, total energy consumption for transportation in 2010 could be as low as 14 quads, below the present total of 17 quads.

Automobiles

The automobile offers the greatest single opportunity to improve the energy efficiency of the U.S. transportation system. The fuel economy of the automobile fleet could be raised to 30–37 miles per gallon by 2010 for less than a 10 percent increase in manufacturing costs (in constant dollars).[5] The fuel-efficient automobile fleet for 2010 would consist of light, energy-conserving, 2-, 5-, and 6-passenger cars with performance, comfort, and safety similar to those of 1975 cars except for lower acceleration and somewhat smaller interior size. Energy savings much beyond this, however, would involve major advances in technology, compromises in performance, or higher costs. The fuel savings themselves should be considered in context; steady improvements in fleet fuel economy over the next few years must be achieved, but at the same time efforts must be made to meet increasingly stringent standards for engine emissions. Gains toward one make gains toward the other more difficult.

The potential for energy conservation in replacing today's automobiles with more efficient ones after their 10–15 years on the road is so great that total annual fuel consumption by automobiles could remain relatively constant over the next 30 years, over which time the total national mileage driven can be expected to increase by 50–100 percent.

The total cost of owning and operating an automobile, however, is not very sensitive to fuel economy, and even a well-informed buyer may find little to prefer in a fuel-efficient car. Thus, fuel economy standards must augment the incentives of the marketplace if the potential energy savings are to be realized.* Other policies might include one or more of the following: allowing the price of gasoline to rise toward its free-market value; advertising accurate miles-per-gallon figures for each model; setting yearly registration fees by vehicle efficiency or weight; taxing new cars according to fuel efficiency; and enacting a gradual rise in gasoline taxes.

Electric vehicles offer some opportunity to moderate the demand for petroleum in the transportation sector, if the electricity is generated from sources other than oil. They may offer other advantages too—for example, shifting pollution from the vehicle to the power plant and raising the off-peak demand for electricity. Available electric vehicles have important limitations (such as range), but may with improvement find an appropriate market, such as driving within metropolitan areas. The energy-conserving potential of electric vehicles depends on the availability and costs of liquid fuels, on institutional and environmental issues, and on the development of high-energy-density batteries. Their attractiveness for a growing number of

*Statement 2-7, by E. J. Gornowski: My statement 2-4, Appendix A, also applies here.

applications may be considerably enhanced in the future if (as is likely) the cost of liquid fuels rises more rapidly than the cost of electricity (on a percentage basis). This advantage is not fully reflected in the scenario projections presented here, which assume that the price of electricity rises almost as rapidly as the price of liquid fuel.

Air Travel

The efficiency of the passenger aircraft fleet could be improved 40 percent by 2010.[6] This would require aircraft industry investments in development; airline purchases of more efficient airplanes; scheduling and pricing policies that increase load factors (passengers per plane trip); and additional improvements in traffic management. Some of these developments are already occurring as a result of recent airline deregulation and the consequent fleet expansion and rise in load factors.

An interagency task force of the federal government estimates that the research and development required to introduce fuel-conserving jet and turboprop engines, lighter structures with associated control systems, and improved design for laminar-flow control and aerodynamics, all of which would cost $670 million, could make possible savings of 2 billion barrels of oil between 1980 and 2005—$25 billion at 1976 oil prices.[7] The airlines' incentive to invest in fuel-efficient aircraft and management practices depends on favorable long-term prospects for air traffic and profit, as well as on forecasts of the price of fuel and attendant policies.

If the design and load factor improvements described were introduced, new aircraft in 2010 might consume 3180 Btu per passenger-mile, as against 7630 Btu in 1975. If a passenger load factor 44 percent above the 1975 values is included,[8] then even at quadrupled energy prices (scenario A), the per capita air travel demand would increase 58 percent. The total energy consumed in scenario A would be 1.0 quad, compared to a 1975 consumption of 1.2 quads.[9] For a doubling of energy prices (scenario B), demand would increase from 745 to 1800 passenger-miles per capita, and total energy consumption would be 1.6 quads.

Truck and Rail Freight

Freight-hauling trucks built in the future can be made 30 percent more fuel-efficient by using turbocharged diesel engines, improved axles, radial tires, and declutching fans.[10] Some of these options are now being installed in freight trucks. Lighter tractors designed for aerodynamic efficiency would represent an additional 10 percent gain. The fuel economy of the truck fleet can be improved by installing diesel engines in medium- and light-duty trucks, but this opportunity to conserve fuel depends on resolution of the air quality issues associated with diesel emissions.

The 7–14 percent savings in fuel that can be realized in trucks by conservative driving may not be achieved even at much higher fuel prices, since the costs of labor and capital substantially outweigh that of fuel in the truck freight business, where both wages and depreciation of equipment are time dependent and revenue is distance dependent. Any fuel-conserving measure that increases the time per trip incurs costs in wages and capital that fuel savings would be hard pressed to match.[11]

Removal of Interstate Commerce Commission restrictions on commodities, routes, and backhauls would allow the trucking industry to plan for fully loaded round trips, improving the industry's load factor.[12]

Railroads, which account for about 1 percent of national energy consumption, are highly energy efficient for long-haul freight. A shift of long-distance freight transportation from truck to rail could be accomplished by major changes in regulatory policies, to allow, for example, the formation of integrated transportation companies free to seek optimum combinations of truck and rail freight through competition.

Energy Demand Projections for Transportation in 2010

Table 2-3 sets out demand for energy by the transportation sector in 2010 under the Demand and Conservation Panel's scenario assumptions. Improvements in energy intensity assumed under the conditions of each scenario are included for ready comparison.

In scenario C, the real price of energy remains constant, but even so the energy intensities of all forms of transportation drop, consistent with historical patterns under falling real prices for energy. Prevailing standards for improvement in automobile fuel economy are assumed to be met, and as automobile travel begins to reach saturation (in cars per owner, minutes spent in automobiles daily, etc.), air travel claims a larger percentage of expenditures for passenger transportation.

In scenario B, prices for energy have climbed steadily to twice the 1975 levels by 2010, and in response the fuel efficiency of automobiles has doubled. Rail freight has expanded by 30 percent, truck freight has grown by 40 percent, and air freight has tripled.

In scenario A energy prices quadruple by 2010, and public policies accelerate the response to this energy conservation incentive. As a result, present federally mandated standards for automobile fuel efficiency are met and superseded by new standards. By the year 2000, advanced fuel-conserving technology (perhaps Brayton and Stirling engines) would begin to be used in new automobiles. Air travel would increase by about 60 percent, and under this intensified demand as well as public policies

TABLE 2-3 Scenarios of Energy Demand for Transportation

Scenario[a]	Average Delivered Energy Price in 2010 as Multiple of Average 1975 Price (1975 dollars)	Energy Conservation Policy	Energy Intensity by Application		Total Demand (quads)
			Application	Intensity[b]	
A*	4	Very aggressive, deliberately arrived at reduced demand requiring some life-style changes			10
A	4	Aggressive; aimed at maximum efficiency plus minor life-style changes	Automobile	37 mpg	14
			Light trucks and vans	30 mpg	
			Air passenger	0.42	
			Truck freight	0.60	
			Air freight	0.60	
			Rail freight	0.91	
B	2	Moderate; slowly incorporates more measures to increase efficiency	Automobile	27 mpg	20
			Light trucks and vans	21 mpg	
			Air passenger	0.45	
			Truck freight	0.80	
			Air freight	0.60	
			Rail freight	0.97	
B′	2	Same as B, but 3 percent average annual GNP growth			27
C	1	Unchanged; present policies continue	Automobile	20 mpg	26
			Light trucks and vans	16 mpg	
			Air passenger	0.50	
			Truck freight	0.90	
			Air freight	0.60	
			Rail freight	1.00	

[a] Scenario D is not included in this table; its price assumption (a one-third decrease by 2010) appears implausible.
[b] Figures not followed by "mpg" refer to the ratios of new-equipment energy intensities in 2010 to those of 1975.

encouraging conservation, perhaps two new generations of aircraft would be introduced by 2010, the last consuming only about half the fuel per seat-mile as the 1975 average. Airplane load factors would reach 70 percent (considered the maximum achievable). Truck freight would improve substantially in fuel efficiency and load factors. Despite significantly higher per capita use, the transportation sector would consume 18 percent less energy in 2010 than in 1975.

Scenario A* was used to test the possibility of reducing energy consumption to levels below those of scenario A, using the same price assumptions. This scenario includes assumed changes in tastes and preferences that produce reductions in energy consumption beyond those available from technological efficiencies. One change of this sort would be a shift to a more service-oriented economy, emphasizing lasting, repairable goods over disposable ones. Another would be a strong trend toward living and working in the same area, reducing by 10 min the time the average person spends in cars each day and yielding energy savings of 1 quad. Under the assumptions of this scenario, freight transportation would also be significantly reduced.* This scenario also includes improvements in the fuel efficiencies of military vehicles, which are not considered in other scenarios.

Although the energy consumption of our principal means of transportation can be moderated by technology and management for improved efficiency, another opportunity to achieve long-term conservation in transportation presents itself in new patterns of living, working, and recreation. These patterns will change regardless of energy prices and policy over the next 35 years, but comprehensive land-use policies combined with incentives and penalties could promote energy-conserving patterns. However, the concentration of metropolitan populations in the United States has been thinning since World War II, along with the concentration of jobs. These patterns have contributed to a general decline in the use of public transit as well as to increased automobile ownership and longer average trips. This trend would have to be reversed if significant energy savings were to be realized from new living patterns.[13]

It is difficult for a dispersed population to achieve significant energy savings through public transit. Direct shift of travel demand from the private auto to public transit has relatively little benefit in terms of energy conservation. Only when public transit, by altering settlement and industrial location patterns, reduces total travel demand (e.g., as measured by passenger-miles of travel per capita) can it have a significant impact on energy consumption.[14]

*See statement 2-8, by H. Brooks, Appendix A.

Today, for example, 98 percent of the urban passenger-miles covered in this country are by private automobile, and urban travel accounts for half the fuel consumed by automobiles. Even if the use of public transit were 15 times greater and consumed no fuel at all, the overall savings in fuel would amount to only 15 percent if total travel demand remained the same.[15] However, with the more fuel-efficient automobiles assumed in all scenarios except scenario IV, the difference in efficiency between private automobiles and rail or even bus transit per passenger-mile is small. The savings in energy consumption that might be achieved by fixed-rail mass transit depend directly on patterns of settlement and land use that run counter to the recent locational trends described above. Such changes could probably be realized only over a period of time well beyond that addressed in this study. Today buses, van pools, and car pools, because they can make flexible use of an already existing network of roads and highways, are the most effective means of reducing energy consumption in commuter travel.[16]

BUILDINGS AND APPLIANCES

The demand for energy by residential and commercial buildings is expected to grow more slowly from 1975 to 2010 than in the past few decades, as population growth slows and some demands become saturated. Rising energy prices, aided by mandatory building and appliance standards, could foster wider use of such well-known measures as heat pumps, better insulation for buildings, larger heat-exchange surfaces for air conditioners and refrigerators, and passive solar building design. Through these and similar conservation measures, the energy demand for buildings in 2010 could be below today's level of 16.8 quads, despite a projected 30 percent increase in population and 63 percent increase in residential buildings.[17]

Existing technology could be incorporated in new buildings and appliances to reduce energy consumption substantially. For example, gas and oil heating systems have been built to use 30 percent less gas or oil than conventional designs, through improved combustion and heat transfer. Electric heat pumps for space heating deliver about 3 times more heat than electric resistance heaters per unit of electricity consumption, and can also be reversed for cooling. The energy consumption of refrigerators can be economically reduced (even at present electricity prices) by 50 percent through better design and construction. Well-insulated new single-family houses need 40 percent less heating than the average house built before 1970.[18] An experimental program of retrofitting existing housing has demonstrated that in a New Jersey community nearly one third of the energy required annually for space conditioning can be saved.[19]

More comprehensive energy-saving measures, such as community-based utility systems,[20] are economically feasible but impractical today for a variety of reasons, including environmental restrictions, the difficulty of making connections with utilities, and developers' reluctance to assume the burden of resolving technical complexities. Concentrated efforts to remove or reduce these institutional barriers could result in substantial additional energy savings in the buildings sector.

Economic considerations are likely to weigh more heavily than others in the decision to improve thermal integrity[21] in buildings. But because many homeowners move frequently, and because the consumer who pays the bills for operating and maintaining a building does not usually make the design and construction decisions, any decision to improve efficiency must be encouraged by building code standards, financial incentives, and information campaigns. Builders have inadequate incentives to minimize life cycle costs; in fact, they tend to favor low initial costs instead. (Of the 1,455,000 single-family homes built in 1977, 904,000 were erected by developers for sale on the open market.)[22]

Retrofitting an existing structure for greater efficiency in energy use will not appear wise to many consumers, even as energy prices rise. Existing buildings offer less scope for energy conservation than do new ones, but some retrofit measures are generally economical. Caulking, increasing attic insulation to 6 in. and wall insulation to 3 1/2 in., adding foil insulation in the floor, and installing storm doors and windows are the best energy conservation investments. Together, they can reduce heating requirements by as much as 50 percent; savings on air conditioning would be somewhat less. Figure 2-1 illustrates the energy savings possible in new and retrofitted single-family residences as a function of incremental capital costs.

The American Society of Heating, Refrigeration, and Air Conditioning Engineers (ASHRAE) developed a new building code in 1975, setting out construction guidelines for energy-efficient buildings (ASHRAE-90-75). The Energy Policy and Conservation Act of 1975 (Public Law 94-163) requires that states adopt ASHRAE-90-75 or similar standards to be eligible for federal funding of state energy conservation plans. More stringent standards, based on performance, are being prepared and discussed.

High on the list of priorities in any program to accelerate improvements in existing buildings is accurate information for lending institutions, homeowners, and homebuyers on the advantages of retrofitting. Readily available loans, subsidies, and tax incentives can also stimulate retrofitting. A justification for such measures is that the whole society benefits from reduced dependence on oil imports. An effective means of reducing energy consumption by existing buildings would be to require that they meet thermal efficiency standards at the time of resale. Such a measure has been introduced in Congress but not adopted.*

Energy Demand Projections for Buildings and Appliances in 2010

Table 2-4 sets out the improvements in energy intensity per unit of service that might be achieved under each of the Demand and Conservation Panel's scenarios, along with the total energy demand by buildings and appliances.

In scenario C, real energy prices remain constant, with the exception of natural gas prices, which double to compensate for past underpricing. The lagged response to public policies now in force brings about improved efficiency in new buildings and appliances through 1980, and to some extent thereafter as older stocks are replaced. Small improvements are introduced in gas appliances, but electric resistance units (rather than heat pumps) continue to dominate electric space heating.

Under the conditions of scenario B, prices for energy double by 2010, with substantial effect in the buildings sector. Energy consumption in this sector decreases at an average annual rate of 0.6 percent (compared to an annual growth rate of 3 percent from 1950 to 1975), mainly because of the reduction in space heating requirements. Because of differences in assumed energy prices, the relative market share of electricity increases from 21 to 51 percent, while the natural gas share declines from 53 to 21 percent. Electric heat pumps become cheaper and more efficient because of large-scale production and find widespread use. Solar energy finds a market toward the end of the period for water heating, space heating, and air conditioning. The energy efficiency of new air conditioners in 2010 is close to 10 Btu per watt-hour (compared to 6 Btu in 1975). Higher energy prices translate into increased expenditures for energy in buildings. However, the percentage of personal income spent for household fuel increases only moderately—from 3.1 in 1975 to 3.9 in 2010. The technologies to produce the higher efficiencies in this scenario are either on the market or achievable by well-known means.

Under scenario A, as energy prices quadruple by 2010, energy consumption in the buildings sector declines to 11 quads, from 16 quads in 1975. New energy-efficient appliances find ready markets, and solar energy begins to make a significant contribution near the end of the period: 25 percent of new air conditioners, 50 percent of new space heaters, and 70

*Statement 2-9, by E. J. Gornowski: My statement 2-4, Appendix A, also applies here.

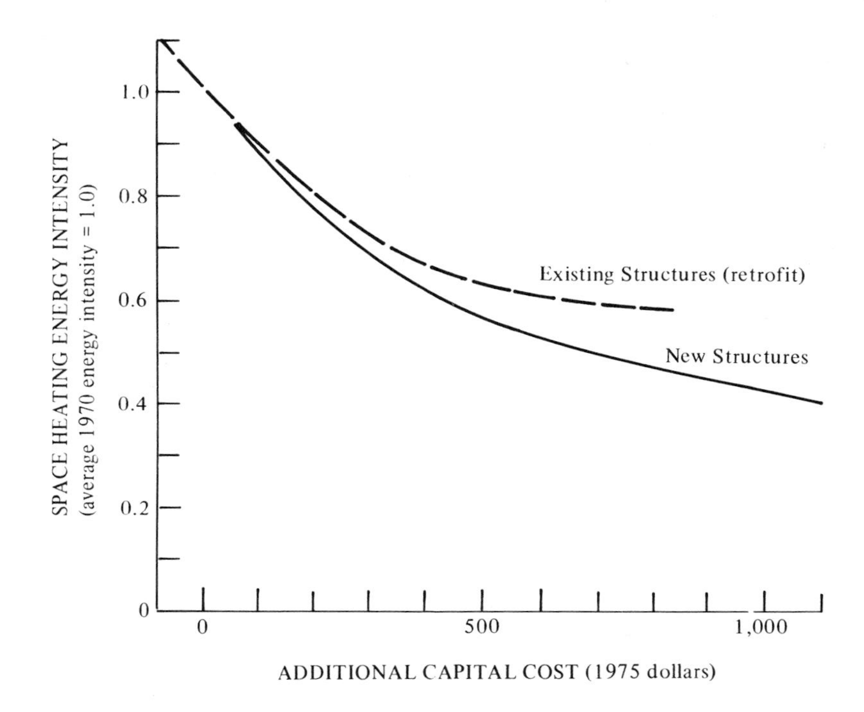

FIGURE 2-1 Space heating energy intensity of single-family residences (average 1970 energy intensity equals 1.0) as a function of increases in capital costs. Source: E. Hirst, J. Cope, S. Cohn, W. Lin, and R. Hoskins, *An Improved Economic-Engineering Model of Residential Energy Use* (Oak Ridge, Tenn.: Oak Ridge National Laboratory (ORNL/CON-8), 1977).

percent of new water heaters in 2010. Improved retrofit measures and construction practices contribute to the energy savings use in this scenario. Increasing income over the period helps ease the pinch of higher expenditures for fuel, but residential fuel expenditures rise from 3.1 percent of personal income to 4.9 percent.

Along with the improvements noted in scenario A, scenario A* assumes continued migration to sun belt states and acceleration of trends to multifamily residential units. These changes reduce heating and cooling requirements by about a third. Additional savings are achieved by the use of integrated utility systems in residential complexes to cogenerate electricity and heat.[23]

TABLE 2-4 Scenarios of Energy Demand for Buildings Sector

Scenario[a]	Average Delivered Energy Price in 2010 as Multiple of Average 1975 Price (1975 dollars)	Energy Conservation Policy	Energy Intensity by Application		Total Demand (quads)
			Application	Intensity[b]	
A*	4	Very aggressive, deliberately arrived at reduced demand requiring some life-style changes			6
A	4	Aggressive; aimed at maximum efficiency plus minor life-style changes	Thermal integrity (heating)		10
			Residential	0.63	
			Commercial	0.42	
			Government and education	0.35	
			Space conditioning		
			Air conditioning	0.66	
			Electric heating	0.52	
			Gas and oil heating	0.72	
			Refrigeration and freezing	0.58	
			Lighting	0.60	

B	2	Moderate; slowly incorporates more measures to increase efficiency	Thermal integrity (heating)		13
			Residential	0.63	
			Commercial	0.60	
			Government and education	0.45	
			Space conditioning		
			Air conditioning	0.75	
			Electric heating	0.63	
			Gas and oil heating	0.75	
			Refrigeration and freezing	0.68	
			Lighting	0.70	
B′		Same as B, but 3 percent average annual GNP growth			17
C	1	Unchanged; present policies continue	Thermal integrity (heating)		20
			Residential	0.76	
			Commercial	0.70	
			Government and education	0.50	
			Space conditioning		
			Air conditioning	0.94	
			Electric heating	0.90	
			Gas and oil heating	0.80	
			Refrigeration and freezing	0.92	
			Lighting	0.70	

[a] Scenario D is not included in this table; its price assumption (a one-third decrease by 2010) appears implausible.
[b] Figures give the ratios of energy intensities in 2010 to those in 1975.

INDUSTRY

Industrial energy consumption per unit of output in the United States fell at an annual rate of about 1.6 percent from 1950 to 1970.[24] Rising prices for energy will very likely accelerate this trend, as it becomes economical to apply known energy-saving technology in new industrial plants. (Overall, the energy required per unit output in U.S. industry fell by nearly 20 percent between 1973 and 1979.) Where no change in a basic industrial process is possible, somewhat smaller savings can be realized by building more efficient facilities. For example, a new cement manufacturing plant can be made 25 percent less energy intensive than today's 20-yr-old average plant by capturing and using the high-temperature waste heat now released to the atmosphere.[25]

The study's assessment indicates that new plants in 2010 could consume an average of 40 percent less energy than existing plants with the same outputs.[26] However, a third of today's plants will probably still be in operation by 2010, and it is much more costly to retrofit existing plants than to build the same features into new plants. Furthermore, some of the industries that consume the most energy per unit output—such as petrochemicals—also show the greatest increases in demand (a trend that is likely to continue).[27]

In part this increased demand will be the result of substituting materials for one another to improve energy efficiency in other products. For example, increasing the use of aluminum and plastics in automobiles will result in lifetime fuel savings considerably greater than the additional energy required to produce the materials, but this will, of course, be reflected in increased output of such materials. If the costs of improving old plants and the increasing demand for energy-intensive industries are taken into account, the overall energy savings that industry can achieve by 2010 will probably be no more than 30 percent, even for scenario A. Greater savings would require public policy measures to compel increased durability of equipment, so that a given amount of consumer service would be provided by a lower output of manufactured goods than at present.

Opportunities for industry to conserve energy can be grouped into four categories.[28]

- Improved housekeeping—operating and maintaining equipment at peak performance, turning out unneeded lights, and reducing heat losses—can quickly reduce energy use per unit of industrial output by 5–15 percent in a non-energy-intensive industry.
- Waste-heat recovery and insulation—modest plant improvements to reduce heat loss and recover the higher-temperature waste-heat streams for use—typically can improve thermal efficiency by 5–10 percent in an

energy-intensive industry. Most of the nation's industry could fully institute these improvements in less than 10 years.

- Introducing new processes would improve the efficiency of energy use anywhere from 10 to 90 percent in aluminum smelting, uranium enrichment, refining operations, steel making, and other industries; because they generally require new production facilities and equipment, these measures will be introduced only gradually over several decades.
- Recycling saves energy in factories (for example, by burning wood and scraps in paper mills). Materials such as aluminum, paper, glass, and steel can be recovered for reprocessing, and this becomes economically more attractive as energy costs rise. A number of tax and regulatory policies, however, still favor the use of materials from virgin sources rather than recycled materials; a systematic effort to identify and eliminate such economic distortions would result in additional energy savings that are difficult to estimate at present.

Many industries can reduce their requirements for purchased electricity by the use of cogeneration. This involves the use of the high temperatures available from fuel combustion to generate electricity, with the lower temperature exhaust heat from the generator used for industrial process heat. This can be done, for example, by producing high-pressure steam to operate a steam turbine generator and then using the lower pressure exhaust steam as process steam, or by using combustion gases to operate a gas turbine or diesel generator and then using the exhaust gases to produce low-pressure steam in a waste-heat boiler. If the high-temperature heat can be generated from coal rather than oil or natural gas—for example by on-site medium-Btu coal gasifiers (chapter 4)—then not only is fuel used more efficiently, but scarce natural fluid fuels are conserved.

Cogenerating electricity and steam presents an attractive opportunity to conserve energy, because about 40 percent of industrial energy is used to produce low-pressure process steam. The additional fuel required for cogeneration is about half that required by the most efficient single-purpose utility plant. Cogeneration also produces fewer air pollutants and lower thermal discharges than conventional single-purpose systems, which consume and reject more energy for equivalent service. A number of existing cogeneration systems can be applied in a variety of plants and modified to accept any of a number of fuels.[29] As fuel prices rise, the fuel savings resulting from cogeneration will offset the higher investment and operating costs in an increasing number of situations.

Economic and regulatory factors, together with industry reluctance, have discouraged cogeneration over the last 50 years. If all large-scale industrial demands for low-pressure steam were provided through cogeneration in 2010, around 10 quads of by-product electricity could be generated.[30] Nevertheless, we estimate that only 2–5 quads of electric power would actually be supplied by cogeneration in 2010.[31] Indeed, strong new incentives and other public policies would be necessary to achieve even this level. The technological, economic, and institutional barriers to the replacement of existing systems with cogeneration, or even to the construction of new cogeneration systems, are formidable.

In spite of its advantages, industrial cogeneration has declined in the United States from supplying 30 percent of industrial electricity in the 1920s to about 8 percent today. Other industrialized countries practice more extensive cogeneration: West German industries, for example, supply 12 percent of their own electricity through cogeneration.[32]* Cogeneration declined in the United States as the real price of utility-generated electricity fell. Consequently, a huge stock of industrial equipment that cannot be adapted for cogeneration was installed. Today's typical primary industrial boiler is fabricated inexpensively and fired by natural gas. It requires little or no operating attention or maintenance to supply reliable low-pressure steam; the design and operation of cogeneration systems, by contrast, require highly skilled personnel. Thus, the personnel costs associated with cogeneration, especially in smaller installations, tend to offset the fuel savings and make them economically less attractive.

Furthermore, the close coordination with an electric utility that would be needed by most cogeneration systems, for the sale of surplus cogenerated electricity and the purchase of backup power, might seem disadvantageous to the utility, since demands for heat and electricity often do not balance. Cogeneration may appear disadvantageous to industries that pay up to 3 times as much for the electricity they buy as the utilities pay them for the surplus electricity they sell. Some industries fear that selling cogenerated power may bring them under regulation by the federal and state agencies that regulate power generation, limiting their flexibility to make quick business decisions.

The other cogeneration possibility, so-called district heat—selling heat from utility stations—faces an uncertain market. Many customers would have to be found quickly to justify the commitment in additional capital for by-product steam marketing. The distribution system, particularly if it supplies steam, is difficult to operate, maintain, and meter. And just as with electricity, backup capacity must be provided.†

The economic penalty for scrapping existing energy systems to make

*Statement 2-10, by B. I. Spinrad: Most West German "cogeneration" is done by the coal industry. Electricity is the primary product, steam is the by-product, and the example is inappropriate.

†See statement 2-11, by B. I. Spinrad, Appendix A.

way for cogeneration can be minimized by careful planning and timing, taking into consideration the expected lifetime of existing equipment, tax advantages, and decisions to build new plants. Other barriers to cogeneration could be overcome by incentives like the following.

- Priority consideration of cogeneration projects for scarce gas and oil when these fuels are required, and exemption from mandatory conversion to coal from natural gas for cogeneration facilities.†
- Favorable financial incentives to install cogeneration facilities.
- Revision of the Federal Power Act and Public Utility Holding Company Act to exclude or limit regulation of industries that sell surplus electrical power.
- Commercialization of on-site medium-Btu coal gasification in conjunction with gas turbine generators to allow environmentally clean use of coal in cogeneration, thus enlarging the market for cogeneration facilities.

State regulatory commissions can identify opportunities for cogeneration and bring utilities and industries together to explore them. Educational programs could interest students and practicing engineers in cogeneration and train them to design and operate these systems.*

Energy Demand Projections for Industry in 2010

The Demand and Conservation Panel's projections of total demand for energy in the industrial sector in 2010, exclusive of the energy-producing sector itself, are summarized in Table 2-5. The last column shows the total energy consumed by the industrial sector, measured at the entrance to the factory, as it were, and ignoring losses in the energy sector associated with providing the energy. These losses depend on the particular ways the energy is provided from the primary sources, and are shown in Tables 11-6 through 11-10 in chapter 11 for particular assumed supply mixes, as described in detail in that chapter. Table 2-5 shows only the mix of coal, oil, gas, and electricity consumed in the factory. The electricity component is computed under the assumption that no cogeneration over that used at present will be used in 2010.

†Statement 2-12, by B. I. Spinrad: So we solve our oil problem by not using coal (at admittedly lower efficiency). I consider this recommendation counterproductive for the problem that exists now.

*See statement 2-13, by H. Brooks, Appendix A.

TABLE 2-5 Scenarios of Energy Demand for Industry

Scenario[a]	Average Delivered Energy Price in 2010 as Multiple of Average 1975 Price (1975 dollars)	Energy Conservation Policy	Energy Intensity by Industry		Total Demand (quads)
			Industry	Intensity[b]	
A*	4	Very aggressive, deliberately arrived at reduced demand assuming some life-style changes			26
A	4	Aggressive; aimed at maximum efficiency plus minor life-style changes	Agriculture	0.85	28
			Aluminum	0.55	
			Cement	0.60	
			Chemicals	0.74	
			Construction	0.58	
			Food	0.66	
			Glass	0.69	
			Iron and Steel	0.72	
			Paper	0.64	
			Other industry	0.57	

B	2	Moderate; slowly incorporates more measures to increase efficiency	Agriculture	0.85	33
			Aluminum	0.63	
			Cement	0.63	
			Chemicals	0.78	
			Construction	0.65	
			Food	0.76	
			Glass	0.76	
			Iron and Steel	0.76	
			Paper	0.71	
			Other industry	0.75	
B′	2	Same as B, but 3 percent average annual GNP growth			46
C	1	Unchanged; present policies continue	Agriculture	0.95	39
			Aluminum	0.79	
			Cement	0.75	
			Chemicals	0.84	
			Construction	0.73	
			Food	0.86	
			Glass	0.82	
			Iron and steel	0.83	
			Paper	0.76	
			Other industry	0.85	

[a] Scenario D is not included in this table; its price assumption (a one-third decrease by 2010) appears implausible.
[b] Figures give the ratios of energy intensities in 2010 to those in 1975.

TO WHAT DEGREE WILL MODERATING THE CONSUMPTION OF ENERGY AFFECT THE NATIONAL ECONOMY?

One frequently encounters references to the fixed relationship between energy use and GDP or GNP, with the implication that any reduction in energy use entails a loss of output and hence a lower standard of living. Such an impression has been fostered by the fact that during the 20 years before 1973, energy consumption and economic output followed each other very closely in both Europe and the United States.

In the short run, energy demand and the output of the economy are indeed tightly coupled. Sudden shortages of energy are translated immediately into reduced output, as illustrated by the 1973 oil embargo, the 1977 natural gas shortage, and the 1978 coal strike. At a given time the structure of the economy is characterized by a production function, or input-output matrix, that relates outputs to inputs in a deterministic way. However, GNP includes only the final goods and services demanded by consumers and the capital equipment demanded by producers. Intermediate goods—the materials, commodities, and services used by industry in producing its final outputs—are not included in GNP, and in the long run the mix of these intermediate goods can change substantially in response to changes in prices and technology without changing the composition of final demand nearly as much. Thus, over the long term the same mix of outputs or GNP may be provided through a considerable variety in the mix of inputs, including energy.

The historical record of the energy/GNP ratio in the United States reveals that the relation has not been constant over time.[33] Comparing the prevailing ratios of different industrialized nations also presents some intriguing similarities and differences.[34] The reverse causal link—from energy use to GNP—has been directly examined through the work of the Modeling Resource Group of the Synthesis Panel.[35] The work of this group has been extended by the Energy Modeling Forum of the Electric Power Research Institute[36] and others.[37] This work supports qualitatively the conclusions of the Modeling Resource Group, with certain qualifications that will be mentioned later.

THE INFLUENCE OF GNP ON ENERGY USE

The direct effect of GNP on the consumption of energy is obvious. If the output of goods and services increases in all industries during a period of little change in relative energy prices and technology, the needed input of energy increases as well. The increase is roughly proportional to the increase in GNP if total expenditures remain distributed in the same proportions among the various goods and services produced. In mature economies such as that of the United States, energy consumption may in fact grow slightly less than proportionately to GNP because of the gradual shift toward services, which are less energy intensive than material goods. Conversely, in developing economies, energy consumption tends to grow faster than GNP because of the increasing proportion of energy-intensive heavy industries. Thus the composition of consumption tends to change with changes in GNP, and energy intensity is not uniform. Even if energy prices remain constant, therefore, energy consumption and GNP do not necessarily change proportionately as GNP grows.

The economy can adjust over time to reduce energy consumption per unit of GNP significantly, if the reduction is accomplished by improving the energy efficiency of a unit of economic output. Changes in the price of energy are important influences in inducing higher efficiencies. Thus, the causal link from GNP to energy use may be modified by cost-saving technological change, often driven by changes in the real (constant-dollar) price of energy. If the incremental cost of the capital and labor required to save energy is less than the cost of the energy saved (suitably discounted to present value), the corresponding technological change will be made; the result will be less energy consumption per unit of GNP. This effect, of course, is subject to a considerable time lag, corresponding to the time it takes for capital equipment to reach the end of its useful life and be replaced. (Of course, the "useful life" of a piece of energy-consuming equipment is in part determined by the price of energy.) Such cost-saving innovation may also come about simply because a new technology is more cost effective for other reasons, with the energy savings an incidental by-product. It is this latter effect that probably accounts for the decline of the U.S. energy/GNP ratio between the 1920s and the 1950s, which occurred despite declining real energy prices. A substantial increase in the real price of energy evokes an additional conservation effect by increasing the prices of energy-intensive goods and services seen by the ultimate consumers, who in response tend to curtail their direct or indirect use of energy. Consumers will shift away from goods and services that are energy intensive in their production or supply, toward others that serve similar purposes with less energy input. The extent of these substitutions will in general be larger the higher the component of energy cost in the price of the good or service. Some substitutions may be made very quickly, but again most depend on some turnover of capital or consumer durables.

THE HISTORICAL TRENDS

The historical pattern in consumption of purchased energy per dollar of U.S. GNP is shown in index number form (with 1900 as 100) in Figure 2-2 for the period 1885–1975. Throughout this period, the value of energy resources at the point of production has generally accounted for only a few percent of total GNP. Thus, the input-output relationship involved is between a small input (3.1–4.5 percent of GNP) and a very large—and heterogeneous—output. Under these circumstances, it would not be surprising if the relation of energy to national output were to exhibit major changes, especially over long periods. From the late nineteenth century to the second decade of the twentieth century, energy consumption persistently grew more rapidly than GNP. From the end of World War I until the mid-1940s, it grew persistently more slowly than GNP. From 1945 to 1973, numerous short-term fluctuations occurred, with no persistent upward or downward trend.

Four groups of factors should be considered for their parts in explaining these long-term trends: the composition of national output, the thermal efficiency of energy use, the composition of energy consumption, and the behavior of energy prices in relation to those of consumer durables, producer durables, and other energy-using goods.

Composition of National Output

Energy requirements per dollar value of product vary for different types of production. It is therefore reasonable to expect that if the goods and services constituting national output undergo significant change, energy consumption per unit of national product will change with it. To measure the historical effects of this factor in detail requires a quantitative record well beyond anything that is now in hand. Nevertheless, some general points can be made. First, in the period prior to World War I, development of such basic industries as the railroads and iron and steel production dominated national economic growth. Heavier energy consumption relative to national output was thus associated (as might be expected) with the stronger influence of heavy industry. The 1920–1945 period was one in which lighter manufacturing and the broad services component of national output grew rapidly, and the corresponding decline in energy consumption relative to national product was a result.

Thermal Efficiency of Energy Use

The term "thermal efficiency" is used here to mean the ratio between primary energy input and useful energy output. Two cases of increased thermal efficiency are of considerable importance. One is the sharp

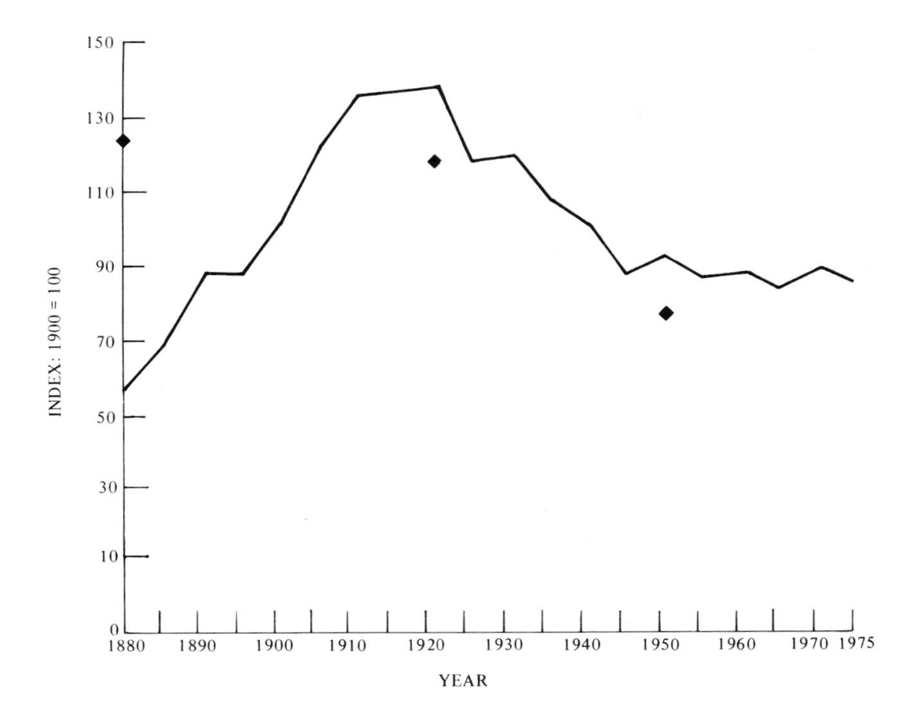

FIGURE 2-2 An index (1900 = 100) of energy consumed per dollar of real gross national product for the United States from 1880 to 1980 shows successive trends of rise, decline, and stability. This plot excludes fuel wood, whose consumption exceeded that of coal into the 1880s. Single-year points that do include fuel wood are indicated for 1880, 1920, and 1950. Source: Adapted from Sam H. Schurr, Joel Darmstadter, Harry Perry, William Ramsay, and Milton Russell, *Energy in America's Future: The Choices Before Us,* Resources for the Future (Baltimore, Md.: Johns Hopkins University Press, 1979). Copyright 1979 by Resources for the Future, Inc.; all rights reserved.

increase over time in the efficiency of converting fuels into electricity, in which there was a decline from almost 7 lb of coal per kilowatt-hour generated in 1900 to 3 lb in 1920, and less than 1 lb by the mid-1950s (but little change since then). The other striking case is the replacement of steam power by diesel power in railroad locomotives in the 1940s and 1950s, a change characterized by an almost sixfold improvement in the efficiency of fuel use.

Measurements of changes in thermal efficiency on an economy-wide basis are not available; however, these two cases cover major areas of energy consumption. Similar changes, not so readily documented, affected space heating as coal furnaces were replaced by oil- and gas-burning furnaces in residential and commercial use. Railroads, electric power plants, and space heating together accounted for large amounts of energy consumption, and improvements in their thermal efficiencies must have been very important in molding the trends shown in Figure 2-2.

Composition of Energy Consumption

Changes in energy consumption characterized by the expanding use of new energy forms appear to be of major importance. Growth in the use of electricity and internal combustion fuels appears to have been particularly significant.

During the period of declining energy consumption relative to GNP, the consumption of electricity grew far more rapidly than that of other energy forms. From 1920 to 1955, for example, electricity consumption increased more than tenfold while that of all other energy only doubled.

Two subtle points about electrification are essential to a proper understanding of the relationship between energy and GNP. One pertains to the question of thermal efficiency as compared to the economic efficiency of energy use. Compared to other ways of using fuel, generating electricity is thermally inefficient; it takes an average of 3 Btu of fuel to produce 1 Btu of electricity. However, the peculiar characteristics of electricity have made possible the performance of tasks in entirely new ways.

Electrification during the 1910s, 1920s, and 1930s increased the overall productive efficiency of the economy, particularly in manufacturing, where the use of electric motors freed the manufacturing industries from the rigid conditions imposed by the steam engine. (For example, electric motors can be switched on only when actually needed.) It allowed factories in addition to be reorganized into more logical sequences and layouts. This greatly enhanced manufacturing productivity—yielding greater returns per unit of labor and capital employed—and thus the productivity of the total economy.

Through its positive effects on productivity, electrification also enhanced the productivity of energy use, resulting in a decline in the amount of raw energy required per unit of national output. Taken together, improvement in the thermal efficiency of electricity generation and the enhancement of productivity due to the use of electricity were major factors in the decline of the energy/GNP ratio following World War I.

The internal combustion engine, powered by liquid fuels, played a somewhat analogous role by allowing mechanization of agriculture, a development that contributed significantly to the rising productivity of the American economy. The growth of truck transportation at the same time made possible the movement of industry away from sites dictated by the location of railroad facilities or waterways.

Changes in production techniques made possible by changes in the composition of energy output have thus enhanced the efficiency with which energy itself has been employed as a factor of production. The broad downward trend in the energy/GNP ratio characterizing much of the historical record would, of course, have been even sharper if the thermal efficiency with which energy had been used in automotive transportation had also shown sharp increases.

The Behavior of Prices

It is of interest to examine the behavior of energy prices during the period of persistent long-term decline in the amount of energy consumed in relation to GNP. In general, the prices of the basic fuels in constant dollar terms fell during the two decades following 1920. Thus, energy consumption relative to GNP was declining even while energy prices were falling. This runs counter to the kind of behavior that would be predicted if the price of energy were the primary factor in explaining how much energy is used in relation to other factors of production in yielding the national output. Obviously, other forces have been of overriding importance, particularly changes in technology and in the composition of economic output.

Some General Conclusions About the Historical Record

Since the end of World War II, the numerous short-term fluctuations have more or less canceled to yield a record of comparative constancy in the ratio over the entire period. In the 35-yr period from 1920 to 1955 (35 years is chosen because this study tried to look ahead 35 years), the decline in energy consumption relative to GNP was about 35 percent. This decline was accompanied by fundamental transformations in the energy forms used in such major applications as railroad transportation, electric power generation, and space heating. A case can be made that the 35 percent decline would have been even sharper if energy prices had been rising at the same time instead of generally falling.

INTERNATIONAL COMPARISONS

Figure 2-3 displays correlations between energy use and gross domestic product[38] in different nations. In general, countries with high GNP's use large amounts of energy, and those with low GDP's use small amounts.

GDP is composed of inherently different elements in different countries,

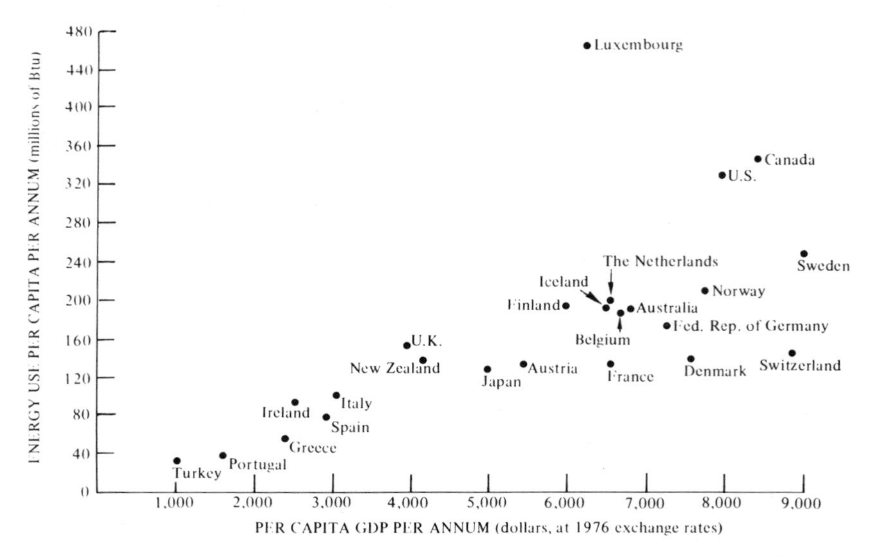

FIGURE 2-3 Correlation between per capita energy use and gross domestic product per annum for various industrialized countries. Source: Data from *OECD Observer*, OECD Member Countries 1978 Ed., 14th Year, no. 91 (March 1978): 20–21.

and these differences as well as others are apparent in the fine structure of the figure, where some of the variation in energy/GDP ratios from country to country can be found within the gross overall correlation already noted between energy and GDP. Sweden, for example, has a per capita GDP comparable to that of the United States, but consumes only 75 percent as much energy per capita, despite the fact that it specializes in inherently energy-intensive industry (steel, paper, metals). Canada and the United States, where energy prices have in the past been markedly lower than elsewhere, are well above the line for most industrialized countries, while Sweden, West Germany, and Japan are well below.[39] This variation is important, because reducing the energy/GDP ratio of the United States by, say, 30 percent could make an important contribution to reducing imports of oil.

Comparative international energy analysis points to a variety of reasons for differences between the United States and countries such as West Germany and Sweden. Some, such as patterns of automotive use, appear to be significantly influenced by differences in land areas and the locational patterns of population and industry—particularly the generally greater distances between homes and workplaces or markets in the United States. Other differences stem from years of response to low energy prices in the

United States and higher prices elsewhere. Overall, about 40 percent of the difference between the United States and other industrialized countries can be explained by structural and geographical differences, and the remainder by lower end-use efficiency in the United States.[40]

Transportation

The greatest difference in energy use among nations is in the transportation sector, especially in the use of automobiles. For example, not only do American passenger cars use about 50 percent more fuel per passenger-mile than European cars, but Americans also drive more than Europeans. A greater percentage of total miles driven in the United States can be attributed to urban, rather than intercity, driving; the foreign energy mix includes more public transit. These differences result partly from the much higher cost (because of taxes) of buying and operating cars abroad, and partly from the compact structure and organization of European cities. European countries have followed policies of high gasoline taxes as well as excise taxes on automobiles assessed by weight or fuel consumption. Their response to the oil embargo was to raise these taxes.

Freight transportation also contributes to the higher energy/GDP ratio of the United States (though the U.S. freight system is the world's most efficient in terms of energy used per ton-mile) because of the large size of the country and the need for long-distance moving of bulk commodities, such as ores, coal, and grains. In West Germany, for example, the ratio of ton-miles of freight to dollars of GNP is only 40 percent of that in the United States.[41]

Industry

The U.S. industrial sector forms a smaller fraction of total GNP than those of most other industrial countries, but the overall energy efficiency of its processes is relatively low. Sweden uses about 85 percent as much energy per unit of production, reflecting newer technology and design attention to historically higher fuel prices.[42]

Households

American households consume more energy than foreign ones relative to income, even if one adjusts for climate. The difference is in space heating and cooling. The single-family homes typical of the United States, and practices that developed in periods of low fuel prices—for example, heating unoccupied rooms and maintaining higher winter temperatures—account for some of the difference. Per square foot of living space, Sweden appears to consume half as much energy in space heating to achieve similar internal comfort under comparable climatic conditions.[43]

Implications

Although the influences of a great many factors have been left unexamined, these international comparisons suggest that the energy policies and practices in other countries deserve a closer look, to determine which differences are inherent in geographical and structural characteristics and which might be subject to policy influence, at least in the long run. Because of grossly different geographical circumstances and patterns of industrial and residential location in North America, it would be misleading to conclude that we could simply imitate European per capita energy consumption, but close study of energy consumption in other countries suggests that somewhat more than half the difference is amenable to influence by policy.

ECONOMETRIC STUDIES BY THE MODELING RESOURCE GROUP

The Modeling Resource Group of the CONAES Synthesis Panel, in examining the relation between energy consumption and economic growth, experimented with six computer models that attempt to characterize the response of the economy to various kinds of energy policy.[44] The policy variables considered included moratoria or production limits on various energy sources, such as might be fostered by environmental concerns; relaxation of environmental standards now applied to coal and nuclear technologies; and "non-price-induced" energy conservation (due to mandatory standards, public education, and so on). The results suggest substantial flexibility in the ratio of energy consumed to economic output, and thus little relative reduction in economic growth over the next 35 years if energy conservation measures take effect smoothly and gradually enough for the economy to respond rationally.

As inputs to the models, the Modeling Resource Group used "base-case" values (with higher and lower variants) for GNP growth, the costs of major energy supplies (and the availability of the resources and technologies on which they depend), the price and income elasticities of demand for energy, discount rates, and the like.

Of crucial importance in estimating the feedback between energy consumption and economic output are the price and income elasticities of demand, which specify the assumed responses of energy consumers to changes in their incomes and in the price of energy. Energy, like most other commodities, is in greater demand as incomes rise, other things being equal; when its price rises, demand generally falls off. Economists quantify these responses by means of ratios with the percentage change in income or price as the denominator and the resulting percentage change in consumption as the numerator. Thus, if a 10 percent rise in income produces a 10 percent rise in energy consumption, we speak of an income elasticity of demand for energy equal to 1. Similarly, if a 10 percent rise in the price yields a 10 percent decline in the demand for energy, the price elasticity of demand is equal to −1. Because the full response of consumers to changes in incomes or prices takes time to develop, economists speak of short- and long-run elasticities, with specified lags in response.

It is also important to distinguish between the elasticity of demand measured in terms of end-use energy prices and that measured with respect to the prices of primary energy inputs. Because of conversion and distribution costs, as well as taxes, subsidies, and direct price regulation, consumer prices generally change by smaller proportions than the corresponding primary fuel prices. For example, between 1972 and 1978 the consumption-weighted average primary fuel price increased in real terms by about 100 percent, but the price of net delivered energy averaged over all end-uses increased by only 30 percent. Roughly speaking, price elasticities measured relative to end-use would be expected to be 2–3 times those measured relative to primary energy (but this would tend to fall as primary energy prices rise to constitute a larger fraction of delivered prices).

A comprehensive review of the statistical evidence indicates that the overall income elasticity of demand for energy is slightly less than 1,[45] but the possible range is very large. This is important, because small variations in elasticity can have rather large effects on energy consumption. Simple arithmetic, for example, shows that projected energy demand in 2010 will vary within a range of about 20 percent if estimated long-term income elasticities are varied between 0.9 and 1.1, a range that is compatible with historical analysis.

Similarly, small differences in price elasticities can have large effects on the energy-GNP feedback, and the price elasticity for energy is known with even less certainty than the income elasticity. The overall long-term price elasticities assumed in the models used by the Modeling Resource Group ranged between −0.25 and −0.5, measured in terms of primary energy prices. (Corresponding elasticities measured with respect to final energy are between −0.7 and −1.2.)[46]

Other important uncertainties are whether the capital necessary for the technological responses to energy curtailment will be available, how much more the second and third increments in fuel efficiency would cost than the first, and how the price elasticity of demand may change over long periods of time and large changes in price. Exploring these questions will require more experience and data gathered over a long time.

To obtain some first estimates of what might happen over a range of assumptions, the Modeling Resource Group made a number of simulation runs under the simple assumption that the price elasticity of primary energy is constant over the entire range of GNP, energy consumption, and price levels realized in the model. It was also assumed that the proportion of GNP going into capital investment would be independent of the level of energy consumption. While the results obtained are certainly speculative, particularly when they address more than small excursions from experience to date, they indicate a range of future states of affairs that might accompany one or another set of choices.

From these studies emerges the important conclusion that the rate at which energy use grows may be substantially reduced over several decades with only minimal effect on GNP. Thus, in the Modeling Resource Group's base case (no major policy changes, approximately doubled energy prices by 2010, GNP assumed to have tripled over the same period—a 3.2 percent average annual GNP growth) the ratio of energy to GNP is calculated to decline to about two thirds its present value by 2010.

For the extreme policies (represented by a strong energy tax) that are assumed necessary to lead to zero energy growth, two of the models used calculated that a 50–60 percent reduction in energy use below the base-case level would produce cumulative GNP reductions (between 1975 and 2010) ranging from about 30 percent to only a few percent for assumed price elasticities of demand of -0.25 and -0.5, respectively.

These results thus show that the relations between energy use, price elasticity, and curtailment of GNP are nonlinear. For the same level of energy consumption, a price elasticity of demand equal to -0.5 has a small effect on GNP while a price elasticity half as big has an effect many times as great.

Implicit in these calculations is the important difference between short-term effects and effects from a long-term adjustment process. Even in a slowly growing economy, the 35-yr period from 1975 to 2010 would be long enough for major changes in the physical composition of the nation's capital stock, as existing plants and equipment for the production, conversion, and use of energy are replaced. In a more rapidly growing economy, the composition of energy-consuming stocks would change more rapidly. The most important reason for the small feedback effect is that capital and labor can be shifted from the energy supply sector to production of other goods and services without altering the quantity or mix of final goods or services which comprise GNP. In other words it is only the intermediate goods, including energy inputs, that are changed.

One question arises naturally: How are curtailments of energy use by the amounts implied in some scenarios to be brought about if environmental or other noneconomic considerations compel one or another? In the Modeling Resource Group's analysis, the question was not addressed as a policy problem; rather, the convenient device of a "conservation tax" on primary energy was employed strictly for the purpose of introducing hypothetical reductions in energy consumption into models of consumers' and producers' quantitative response to prices. This tax could be defined as the additional amount that must be added to the price of energy to reduce demand the required amount below the supply that would be elicited at the price. The term conservation tax is used in that sense here, although any non-price-induced reduction in energy use that can be mandated or achieved by other means will correspondingly reduce the tax actually needed to balance supply and demand.

Limitations of the Modeling Resource Group Results

The results of the Modeling Resource Group's work are evidently very sensitive to the value of price elasticity of demand that is assumed. This parameter represents the local slope of a log-log curve (in this case, one giving the relation between two rates); when large changes are being contemplated, it is inherently a rough approximation. Particularly when the changes will be far beyond experience, it has value mainly as an exploratory tool rather than as a basis for prediction.

However, support for using a constant elasticity, over even as wide a range as that in the Modeling Resource Group's models, comes from econometric evidence (data for which is relatively inexact and covers a fairly small range), and to a larger extent from technological analysis of economic opportunities for improving efficiency.[47] Unless the log-log slope is highly nonlinear, the conclusions are not very sensitive to this assumption; the arbitrary conservation tax can simply be adjusted to give the desired energy consumption in the models.

Four other qualifications must be entered regarding these first-approximation results. First, the calculation constraining energy use by a blanket tax minimizes the feedback effect. Calculations experimenting with alternative constraints on one or two specific energy supply technologies show a significantly greater effect on GNP than that induced by a price increase or a conservation tax. In one calculation, a 10 percent reduction of energy use achieved by restricting particular energy supplies had about the same feedback effect—a 1 percent decrease of GNP—as a roughly 50 percent reduction from imposing a conservation tax. This difference reflects greater flexibility to adjust by substitution when no constraints are placed on particular energy sources.

Second, the availability of capital to be substituted for energy is assumed to be constant. The assumption of a GNP growth rate averaging 3.2 percent over the period 1975–2010 is based on projections of growth in population, changes in labor-force participation, increases in productivity, and private decisions and public policies that influence the allocation of each year's GNP between consumption and investment. Should it be found that reduced growth in energy consumption will in some way influence this allocation to decrease the amount of savings that go into capital formation, then the GNP in 2010, and the portion of the GNP trajectory nearer to the year 2010, would be somewhat lower. The estimates assume this effect can be disregarded or counteracted by government policy. This effect has been roughly estimated by the Energy Modeling Forum,[48] and the results are discussed in chapter 11 under the heading "Work of the Modeling Resource Group: Feedback from Energy Use to GNP."

Third, the assumption that choice among fuels is determined exclusively by relative price, in every case where substitution of one fuel for another is technologically possible, may understate differences in the income elasticities of demand for different forms of energy. Specifically, such an assumption may underestimate the demand for electricity and gas, which are more convenient and flexible in use than other forms of energy.*

Fourth is the assumption that channeling innovative effort and investment into materials and energy conservation will not decrease the contributions of technical progress to improving labor productivity. In other words it is assumed that labor productivity growth will be independent of energy growth. If in fact labor productivity growth declined with increasing energy conservation, this would be equivalent to a negative feedback of energy consumption on GNP growth. The same would be true, of course, if energy conservation affected other relevant variables such as labor-force participation or population growth. There is at present no basis for even speculation on the direction or magnitude of these effects. They are clearly an important topic for future research because they could have great impacts on future economic welfare.

Thus, while the calculations made by the Modeling Resource Group involve major simplifications and assumptions, they suggest that a strong economy could well exist three or four decades hence with a ratio of energy use to GNP as low as half the present value. This conclusion is substantially reinforced by similar findings from the Demand and Conservation Panel's work, conducted by quite different methods. It is also important to realize, however, that the large reductions in energy consumption, while probably not affecting economic growth, imply substantial shifts in the structure of employment and capital investment,

*See statement 2-14, by H. Brooks, Appendix A.

principally from the energy-producing sector to the nonenergy sectors of the economy. They also imply large price increases induced by taxes, as well as other public policy measures whose political acceptability may be questionable. CONAES has made no judgment on this; that is for debate in the political arena.* Energy taxes, of course, produce revenues that can be invested elsewhere in the economy. To the extent that these revenues are used to rebate high energy costs to certain regions or groups for equity reasons, the impact of prices on energy consumption will be reduced.

RESEARCH AND DEVELOPMENT FOR ENERGY CONSERVATION

Energy conservation technology characteristically must be closely adapted to the particular circumstances of individual end-uses of energy. One would hope that the fact and prospect of higher energy prices would lead the private sector to develop and commercialize most of the technology required to achieve the energy efficiencies projected in the preceding sections of this chapter.

Thus there is a question as to the appropriate roles of the federal government in stimulating research and development to promote energy efficiency and encouraging adoption of energy-efficient technologies. Just as we saw that mandatory performance standards might be necessary to ensure and accelerate the achievement of energy efficiencies that would be economically rational given projected future prices, so there may be a case for government showing the way in energy conservation technology. In the first place, because of institutional inertia, ingrained patterns of consumption, and uncertainties about how prices will rise in the future, prices alone may not provide sufficient incentive, especially for individual consumers.† There may be a strong case for government-supported research to demonstrate new energy conservation technologies in order to reduce the uncertainties about their potential benefits and hence their market potential. In the second place, the political risks and economic impacts of large petroleum imports may justify looking upon energy conservation as partially a public good, eligible for public subsidy.‡

Despite these justifications, however, there is a question whether government-supported research and development will result in technologies sufficiently well adapted to the true requirements of end-uses. Experience with other government programs to develop civilian technologies, in which the technological choices were made by government itself,

*See statement 2-15, by H. Brooks, Appendix A.
†Statement 2-16, by E. J. Gornowski: My statement 2-4, Appendix A, also applies here.
‡See statement 2-17, by E. J. Gornowski, Appendix A.

has not been a very happy one. A more efficient government strategy may be to provide conservation incentives that leave the choice of technologies as much as possible to the pluralistic workings of the private sector.

However, government-supported research that contributes to general understanding of the workings of the energy sector of the economy can have several benefits. In the first place, since government-mandated energy performance standards may be a necessary component of national energy policy,‡ research that improves our ability to predict the effectiveness of standards and other government policies designed to influence patterns of energy consumption will be of obvious value. Any knowledge that improves forecasts of the economy's responses to prices and policies with respect to aggregate demand for various fuels can also help to guide policy. Even engineering or physical research that helps to evaluate the feasibility of various equipment performance standards that are under consideration could be justified, although the actual development of commercial equipment should probably be left to private initiative and funding. Close monitoring of the progress of energy-conserving technology in the private sector may demand some publicly supported research simply to properly inform government policy making. Examples of opportunities for research and development in support of energy conservation policy are listed in Tables 2-6 and 2-7, taken from the work of the Demand and Conservation Panel. Table 2-6 deals with the technological research that could be supported by industry, with government monitoring and disseminating progress data to help accelerate adoption. Table 2-7 lists examples of economic, social, and behavioral research that may provide guidance for federal policy and help inform the private decisions of entrepreneurs and consumers.

CONCLUSIONS AND RECOMMENDATIONS

The economic and engineering analyses of this study in the area of energy demand disclose several principal results. The methods for projecting energy demand as a function of price and public policy are described in chapter 11; here we only summarize the findings.

- Taking a long view of the historical record, it is evident that the ratio of energy consumed to GNP has varied considerably in this country, depending on the composition of national production, the thermal

‡Statement 2-18, by E. J. Gornowski: My statement 2-4, Appendix A, also applies here.

TABLE 2-6 Opportunities for Technological Research and Development in Support of Energy Conservation

	Buildings and Appliances	Transportation	Industry
Basic studies	Properties of materials Automatic control technology	Materials properties, e.g., strength-to-weight Thermodynamics of internal/external combustion engines Chemical energy storage Automatic control technology	Materials properties at high temperatures Characteristics of industrial combustion Heat transfer and recovery methods Automatic control technology
Near-term energy-use patterns	Automatic set-back thermostats Pilot/burner retrofit	Specific data on factors that influence fuel economy of existing cars	Improved methods for energy monitoring and housekeeping Improved methods for scrubbing combustion gases
Intermediate-term retrofit	Reinsulation methodologies Solar water heating and passive design Metering for time-dependent utility pricing Automatic ventilation control for building and appliances (e.g., clothes dryers)	Improved power-to-weight ratios, as well as interior volume-to-weight ratio Instrumentation to provide driver with real-time data on fuel efficiency Improved intermodal freight and passenger terminals Improved traffic control	Process retrofit technologies Improved combustion of marginal fuels Cogeneration of heat and electricity Automated monitoring of energy performance Low-temperature heat utilization
Long-term technologies	High-performance electric and heat-driven heat pumps Solar space cooling Sophisticated appliance controls and integrated appliance design More sophisticated design of buildings to provide desired amenities at low energy demand	New motors Improved aerodynamic design for cars, trucks New primary energy sources (liquid, electric) Improved intermodal transfer technology Technology for improved energy efficiency in air transport	Basic new processes that reduce overall requirements for energy and other resources (e.g., recycling, durability) per unit output Modification of material properties to enable replacement of energy-intensive materials with less energy-intensive material in specific applications

TABLE 2-7 Opportunities for Economic, Social, and Behavioral Research in Support of Energy Conservation

	Buildings and Appliances	Transportation	Industry
Basic studies	Consumer motivation and decision processes Demand as a function of price, income, and demographic factors Influence of taxes, social factors on first cost or life cycle cost decisions Standardized data (e.g., performance data on new stocks) Analysis of effects of initial cost and energy use on purchase patterns for consumer durables	Basic consumer purchase motivations Standardized data (e.g., fleet average miles per gallon for cars and trucks, by function) Effect of relative prices on choice of transportation mode Identification of energy subsidies and effects	Relationship between industrial demand, national economics, and demographic features of our society Consequences for energy use of alternative long-term raw material programs Standardized data (e.g., energy per unit output statistics)
Near-term energy-use patterns	Identification of information useful in modifying consumer behavior to reduce energy waste (e.g., effects of thermostat setbacks) Ways to induce near-term commitment to solve long-term energy problems (investment choices and behavior patterns)	Ways to influence drivers' habits that affect energy consumption in existing cars Ways to induce actions in the near term in response to a long-term problem	Development and demonstration of improved techniques for corporate energy management (e.g., energy audits, energy accounting, energy information systems, etc.) Evaluation and demonstration of mechanisms, motivation factors, and procedures for minimum life cycle cost

TABLE 2-7 (*continued*)

	Buildings and Appliances	Transportation	Industry
			decisions in procurement (e.g., lightweight steel versus plastic for lighter automobiles) Development and demonstration of systems to increase recycling of basic materials (e.g., steel, glass)
Intermediate-term energy-use patterns	Improved information on benefit/cost of energy performance in new and retrofitted housing, appliances	Intermodal shifts of freight Reform of driver-training programs Identification and removal of regulatory constraints to more efficient passenger and freight transport	Ways to encourage energy-related corporate investment decisions on the basis of applicable marginal or future energy costs
Long-term technologies	Improved quality and craftsmanship in manufacturing and construction Matching desired level and comfort to minimal resource demands	Increased use of mass transportation Improved design for energy consumption Zoning and land-use strategies that sustain and enhance mass transportation, modal shifts	Experimentation and analysis of incentives for innovation in industrial energy use Identification and evaluation of incentives to increase product durability

efficiency of energy conversion and end-use, the composition of energy consumption, and movements in energy prices.

- The energy/GDP ratios of industrialized countries with similarly high levels of GDP per capita vary over a significant range. Although many differences help account for this variance, past energy prices in the different countries are major factors.
- The analyses of the Demand and Conservation Panel indicate that the energy necessary to produce a given unit of economic output can be substantially reduced by (1) lowering the energy intensity of production, (2) changing the composition of energy consumption, and (3) shifting to an inherently less energy-intensive mix of goods and services.
- The critical parameter for estimating the extent to which these energy-conserving measures will actually be used in any assumed scenario of future energy prices, and the feedback of this reduction in energy use on GNP, is the long- term price elasticity of demand for energy. For small reductions in energy consumption—10–20 percent below the Modeling Resource Group's market-equilibrium base case value for 2010—assumed price elasticities between -0.25 and -0.50 yield approximately the same feedback effect on cumulative GNP between 1975 and 2010 (a 1 percent decline). For larger reductions, the high and low elasticity values yield widely different results.
- This uncertainty in estimating the price elasticity of demand for energy produces a considerable range in estimates of energy demand over the next three decades. Nevertheless, the projections by the Modeling Resource Group all imply a gradual long-term decline in the energy/GNP ratio, without substantial impacts on economic growth.
- The most powerful influence acting to moderate the growth of energy consumption in the analyses of this study is smoothly rising prices for energy, although realizing the effect of prices may require supplementation by regulations and minimum standards of performance for energy-using equipment.*
- The economy will need time to change. Opportunities to reduce the consumption of energy are realized primarily through the replacement of present capital stock with less energy-intensive stock; this will generally take 10–40 years even under the pressure of rising energy prices and other incentives.

In sum, it appears that the energy-to-economic-output ratio in the U.S. economy can be lessened, over the long term, and that prudent, sustained

*Statement 2-19, by E. J. Gornowski: My statement 2-4, Appendix A, also applies here.

policies can help the economy continue growing with constrained growth of energy consumption.

POLICY RECOMMENDATIONS

The analyses described in this chapter depend generally on the plausible assumption that energy prices will continue rising to reflect scarcity and intensified world competition for supplies; increasingly expensive discovery, extraction, and conversion of energy resources; the costs of environmental protection and repair; and so on. The willingness to invest in capital substitutions for energy and to practice energy conservation clearly rises or falls with changes in the anticipated price of energy. Conservation of energy represents a middle- to long-range investment; if the investment is to be made, the signals the economy reads from prices for energy must be unambiguous, and the trends reasonably predictable over the lifetimes of normal investments.

However, because even accurate, widely noted market signals are sometimes insufficient to guide market decisions in the direction of energy conservation—as, for example, when the total cost of owning and operating a particular facility, appliance, or process is relatively insensitive to energy efficiency—price alone cannot carry the burden of effective conservation policy.*

At a minimum, energy prices should rise smoothly to levels that reflect the incremental cost to society of producing and using additional secure sources of energy. Environmental costs—coal mine reclamation, emission controls, and the like—must be incorporated in the price of energy. Subsidies to energy users, such as price controls and crude oil entitlements, should be eliminated over time.

For such a pricing policy to have its greatest effect, consumers must be provided with the most accurate possible information on its implications. That is, the energy costs of appliances, building features, and industrial equipment must be as clearly as possible referrable to the corresponding initial costs, so that consumers can make the necessary cost trade-offs with ease. Labeling appliances to indicate their energy consumptions is a good first step for the benefit of individual consumers; publicizing the energy trade-offs in various forms of insulation and other building improvements would also yield substantial benefits.

For many industrial processes, adoption of energy-conservative technology is balked by uncertainty about its costs and benefits, and most industrial establishments tend to be conservative in making such decisions

*Statement 2-20, by E. J. Gornowski: My statement 2-4, Appendix A, also applies here.

in the face of uncertainty (though when the benefits are obvious, industrial energy consumers are probably quicker to seize them than household consumers). It might therefore be beneficial to carry on a few government-supported demonstrations of promising technologies in actual industrial situations. Investment tax credits to encourage conservation investments would also be useful, especially for inducing more efficient use of oil and natural gas, as in cogeneration or integrated utility systems.

Where energy prices are insufficient to induce the appropriate, economically rational responses from consumers—as they are, for example, in the case of the automobile—they could be supplemented by nonprice measures.* Mandatory fuel economy standards, installation of peak load charges or cutoff devices on certain energy-intensive equipment, thermal integrity standards for buildings, and other similar measures may all be useful policy instruments.

The scope of this study did not allow us to explore deeply the potential conflicts between energy policy and other national goals. We are particularly concerned that higher prices for energy may affect inequitably those least able to pay—low-income households in rental housing, small businesses—or particular regions of the country. Public policy must attend to the untoward consequences of higher prices. However, compensation to disadvantaged consumers should take the form of discretionary funds rather than being tied directly to energy (as it would be, for example, with "energy stamps"). Otherwise such consumers would have no incentive to make energy-conserving investments.

Several economic analyses suggest that policies tending to reduce energy consumption per unit of output also tend to increase labor inputs. Thus, investing capital to increase energy efficiency is likely to generate more employment than investing the same capital to increase energy supply. It has been suggested that taxing energy to increase energy prices and devoting the revenue thus derived to reducing employment taxes is likely to reinforce the substitution of labor for energy.[49] The committee has not discussed these arguments in detail, nor has it addressed the political feasibility of such adjustments. The several implications for policy justify further investigation and evaluation. Federal and state programs to provide financial resources for conservation investments also encourage job creation.†

While conservation measures can be selected with an eye to other national objectives—the most important of which may be social equity—they should not be unduly encumbered by competing demands. The

*Statement 2-21, by E. J. Gornowski: My statement 2-4, Appendix A, also applies here.

†Statement 2-22, by E. J. Gornowski: The paper hypothesizes—but does not assert—that conservation investments create more employment than production investments. This has not been substantiated.

resolution of social inequity, for example, deserves its own instruments. The amount of energy consumed in the future will be determined by economic evolution in the market place, and by events and forces many of which cannot now be foreseen. General directions can be selected for the growth of consumption, but attempts to plot the future in detail are likely to founder on the unexpected cumulative effects of many small decisions. We have attempted here only to sketch possible patterns of growth, and to indicate how these might be affected by various near-term choices.

APPENDIX: A WORD ABOUT GNP

GNP is a composite of many items that mean different things to different people—the summation of apples, can openers, bus rides, homes, and other things. Its calculation is made possible by the use of their market prices, a rough reflection of their economic costs of production in quantities determined by the preferences and purchasing power of consumers. Statistical observation of these quantities and market-balancing prices yields GNP estimates for many nations. It also results in a bias toward items whose prices are determined by the market at the expenses of items, perhaps at least as important, for which no market prices exist.

Several corrections have been made or proposed for a more ample concept that expresses aggregate economic welfare. One such attempt, for example, is the recent proposal by Nordhaus and Tobin,[50] for a measure of economic welfare (MEW), illustrated by their estimates for 1 year as shown in Table 2-A1.

The corrections and extensions made in this table are of two kinds. Some eliminate the cost of necessary activities that do not directly add to welfare but are aimed at preventing its impairment; these are more correctly regarded as inputs than as outputs. Other corrections add nonmarket activities and uses of time that contribute very substantially to welfare. The economic values of these additions are estimated by the market prices on alternative uses of the time or other resources devoted to these activities.

There is one specifically "environmental" effect in the list of corrections, that for disamenities of urbanization, estimated from the wage differential for comparable work that holds people in the cities. The energy problems that are the focus of the CONAES study involve a great many more environmental effects that are difficult to assess economically. The costs of abating air and water pollution by standards already enforced are included in current GNP estimates as costs of production and consumption. These

TABLE 2-A1 Gross National Product (GNP) and Measure of Economic Welfare (MEW), United States, 1965 (billions of dollars at 1958 prices)

Gross national product		618
Allowance for depreciation of capital	−55	
Net national production		563
Regrettable necessities (defense, police, sanitation, etc.)	−94	
Value imputed to desired leisure (at wage rate)	627	
Value imputed to housework and other nonmarket work	295	
Disamenities of urbanization	−35	
Services of public and private capital, not included in GNP	79	
Additional depreciation of capital	−93	
Measure of economic welfare for 1 year		1343
Investment needed to sustain per capita MEW over subsequent years as population grows	−102	
Sustainable MEW		1241

costs have therefore been reclassified from environmental to economic costs. Instances are the cost of stack gas scrubbers in coal combustion, or the higher price of low-sulfur coal and oil.

Direct economic assessment of the health, safety, and other risks not removed by a given level of protection is much more difficult than estimating the cost of that level of protection. As more experience in measurement and policy is gained, greater consistency between policies addressed to different environmental effects can be expected to help in the measurement of the remaining effects. It is, of course, hard to imagine a market in alleviation of environmental damage. However, one may expect that the inevitable policy debates and struggles will, as time goes on, settle into a more balanced pattern of policies, reflecting with some internal consistency the people's willingness to pay for the various forms and degrees of protection. If and when this comes about, the economic cost of a given decrease in the ambient standard for sulfur dioxide, or of the further decrease by 1000 tons of the inventory of radioactive waste awaiting safe ultimate disposal, may allow an extrapolation that estimates the remaining unabated detriment as perceived by its representatives. These estimated detriments would be among the items still to be subtracted from GNP or from MEW to form an index of economic welfare. Any particular standards would from there on be further improved if and when the cost of doing so were to fall below the resulting estimated gain in welfare. The balance of policies might then take the place of the market mechanisms in providing weights by which environmental benefits are combined to measure quality of life.

Meanwhile, the state of the art of energy scenario projection uses the GNP as a stand-in for MEW or quality of life. CONAES has made use of this stand-in, recognizing the limitations it imposes on the findings reported.

NOTES

1. The gross national product (GNP) of a nation is the total market value of the goods and services produced in the national economy, during a given year, for final consumption, capital investment, and government use. (Note that GNP does not include the value of intermediate goods and services sold to producers and used in the production process itself.) Gross domestic product (GDP) is defined as GNP minus net factor payments abroad (such as income from foreign investments and wages paid to foreign workers). GDP is generally the preferred measure for international comparisons. For the United States the difference between the two measures is for most purposes insignificant—of the order of 1 percent.

2. W. W. Hogan, *Dimensions of Energy Demand,* Discussion Paper Series (Cambridge, Mass.: Kennedy School of Government, Harvard University (E-79-02), July 1979).

3. National Research Council, *Alternative Energy Demand Futures,* Committee on Nuclear and Alternative Energy Systems, Demand and Conservation Panel (Washington, D.C.: National Academy of Sciences, 1979).

4. The GNP level assumed for 2010 was twice the 1975 level. If faster economic growth were assumed, the spread could be wider, because a smaller percentage of 2010 energy-consuming capital stock would be of pre-1975 vintage. In addition, one could expect that with higher incomes people would tend to consume more energy.

5. Demand and Conservation Panel, *op. cit.*, chap. 5.

6. U.S. Department of Transportation, *Report to the Congress by the Federal Aviation Administration: Proposed Programs for Aviation Energy Saving* (Washington, D.C.: U.S. Government Printing Office, 1976).

7. Federic P. Povinelli, John M. Lineberg, and James J. Kramer, "Toward a National Objective: Improving Aircraft Efficiency," *Astronautics and Aeronautics,* February 1976, p. 30.

8. The 1975 load factor was 52.2 percent; the projected load factor is 75 percent.

9. Demand and Conservation Panel, *op. cit.*, chap. 5.

10. *Ibid.*

11. *Ibid.*

12. "Cain's Trucks Ended 40-Year-Old I.C.C. Rule," *New York Times,* November 22, 1978, p. D-2.

13. Brian L. Berry, *Demographic and Settlement Trends: Their Transportation Implications—A Background Paper Prepared for the U.S. Department of Transportation* (Washington, D.C.: U.S. Department of Transportation, June 1, 1976).

14. M. Pikarsky, "Land Use and Transportation in an Energy Efficient Society," in National Research Council, *Transportation and Land Development,* Conference Proceedings, Special Report 183, Commission on Sociotechnical Systems, Transportation Research Board (Washington, D.C.: National Academy of Sciences, 1978).

15. E. Hirst, *Energy Intensiveness of Passenger and Freight Transport Modes, 1950-1970* (Oak Ridge, Tenn.: Oak Ridge National Laboratory (ORNL-NSF-EP-44), 1973).

16. Demand and Conservation Panel, *op. cit.*, chap. 5.

17. *Ibid.*, chap. 3.

18. *Ibid.*; and Marquis R. Seidel *et al., Energy Conservation Strategies* (Washington, D.C.: U.S. Environmental Protection Agency, July 1973).

19. R. Socolow, "Twin Rivers Project on Energy Conservation and Housing: Highlights and Conclusions," *Energy and Building* 1, no. 3 (April 1978):207-243.

20. In community-based utility systems, fuel is converted locally into electricity, space heating and cooling, and water heating for a large number of commercial or residential units, or both.

21. Thermal integrity refers to a building's ability to retain its heated or cooled interior temperature.

22. National Association of Home Builders, personal communication, November 28, 1978.

23. For example, district heating or cogeneration. See, for example, Governor's Commission on Cogeneration, 152 *Cogeneration: Its Benefits to New England* (Commonwealth of Massachusetts, October 1978).

24. The Conference Board, *Energy Consumption in Manufacturing* (Cambridge, Mass.: Ballinger Publishing Co., 1974).

25. G. A. Schroth, "Suspension Preheater System Consumes Less Fuel," *Rock Products,* May 1977.

26. Composite average improvement of 40 percent based on calculated potential improvements in energy intensity for new plants built in each of 10 industry classifications. For details, see Demand and Conservation Panel, *op. cit.*, chap. 4.

27. *The American Economy—Prospects for Growth to 1988* (New York: McGraw-Hill Publications Economics Department, 1974).

28. For a detailed discussion of each of these estimates, see Demand and Conservation Panel, *op. cit.*, chap. 4.

29. R. S. Spencer *et al., Energy Industrial Center Study,* prepared for the National Science Foundation (Midland, Mich.: Dow Chemical Company, June 1975). See also S. E. Nydick *et al., A Study of Inplant Electric Power Generation in the Chemical, Petroleum Refining and Paper and Pulp Industries,* monograph prepared for the Federal Energy Administration (Waltham, Mass.: Thermo-Electron Corporation, July 1976).

30. Spencer *et al., op. cit.* See also F. Von Hippel and R. Williams, *Energy Work and Nuclear Power Growth,* draft report of the Center for Environmental Studies (Princeton, New Jersey: Princeton University, August 1976).

31. Demand and Conservation Panel, *op. cit.*, chap. 4.

32. Governor's Commission on Cogeneration, *Cogeneration: Its Benefits to New England* (Commonwealth of Massachusetts, October 1978).

33. For a detailed analysis of historical trends in the relationship of energy consumption to GNP (or GDP), see Jack Alterman, *The Energy/Real Gross Domestic Product Ratio—An Analysis of Changes During the 1966–1970 Period in Relation to Long-Run Trends,* Bureau of Economic Analysis Staff Paper 30 (Washington, D.C.: U.S. Department of Commerce, October 1977).

34. This subject is explored in Joel Darmstadter, Joy Dunkerley, and Jack Alterman, *How Industrial Societies Use Energy: A Comparative Analysis* (Baltimore, Md.: The Johns Hopkins University Press, 1977).

35. National Research Council, *Supporting Paper 2: Energy Modeling for an Uncertain Future,* Committee on Nuclear and Alternative Energy Systems, Synthesis Panel, Modeling Resource Group (Washington, D.C.: National Academy of Sciences, 1978).

36. Institute for Energy Studies, *Energy and the Economy,* Energy Modeling Forum Report no. 1, vols. 1 and 2 (Stanford, Calif.: Stanford University, 1977).

37. Hogan, *op. cit.*

38. See note 1 above.

39. Demand and Conservation Panel, *op. cit.*; and see note 34 above.

40. S. Schurr, J. Darmstadter, H. Perry, W. Ramsay, and M. Russell, *An Overview and Interpretation of Energy in America's Future: The Choices Before Us* (Washington, D.C.: Resources for the Future, June 1979).

41. See note 33 above.

42. See note 34 above.

43. Lee Schipper, "Raising the Productivity of Energy Utilization," in *Annual Review of Energy,* ed. Jack M. Hollander, vol. 1 (Palo Alto, Calif.: Annual Reviews, Inc., 1976), pp. 455–517.

44. Synthesis Panel, Modeling Resource Group, *op. cit.*

45. L. A. Taylor, *The Demand for Energy: A Survey of Price and Income Elasticities,* working paper prepared for CONAES. (Available in CONAES public file, April 1976.)

46. Hogan, *op. cit.*

47. Demand and Conservation Panel, *op. cit.*

48. Institute for Energy Studies, *op. cit.*

49. D. Chapman, "Taxation, Energy Use, and Employment," statement presented at hearings, Subcommittee on Energy, Joint Economic Committee, U.S. Congress, 95th Cong., 2d Sess., March 15–16, 1978.

50. W. Nordhaus and J. Tobin, "Is Growth Obsolete?" in *Economic Growth,* National Bureau of Economic Research (New York: Columbia University Press, 1972).

3 Oil and Gas Supply

As we have seen in chapters 1 and 2, the primary source of the energy problem lies in the peaking of U.S. oil and natural gas production around 1970 while domestic demand was still growing rapidly. This rapidly accelerated the growth of oil imports, threatening the nation's economic and political security and also placing stresses on world oil markets that had political, financial, and economic implications in all the countries of the world. As we shall show further in this chapter, the likelihood of reversing the slow decline in domestic oil and natural gas production is quite small, and the prospect of compensating for this decline by continued growth of oil imports is equally small, at least beyond a few years in the future.

In 1978 the United States consumed 37.8 quads of oil and 19.8 quads of natural gas, about three fourths of the 78-quad total domestic primary energy consumption. Of this, 17.5 quads of oil (13.25 quads as crude oil and 4.23 quads as refined products) were imported, while all of the natural gas, except about 0.5 quad, was produced domestically. U.S. energy imports were greater than those of any other single country, although constituting a much smaller fraction of total energy consumption than for most of the other industrial countries in the noncommunist world. Future actions of the United States that affect its needs for oil imports have a tremendous impact and are viewed with intense interest and concern outside the United States—much more so, apparently, than by our own citizens except in occasional crisis situations. In the next several years all the problems, domestic and international, stemming from U.S. oil imports are likely to intensify unless we succeed in moderating our demand for both oil and gas, which translates into increased oil imports, at least in the short term. In the longer term the development of other domestic energy forms (at first mainly electrical generation by coal and nuclear power, later synthetic liquids and gases derived from coal and oil shale, and finally solar and other long-term energy sources) will also contribute increasingly to the moderation of oil imports.

In general it is to be expected that the demand for fluid fuels in the rest of the world will increase faster in percentage terms than in the United States, especially if the developing countries are to realize their aspirations for rapid economic development. Given the probability that world oil production will peak in the 1990s and decline gradually thereafter, it is thus extremely unlikely that the United States will be able to offset its declining domestic production of fluid fuels by increasing its share of world imports. Instead, political and economic pressures on the United States to decrease its share of imports will steadily mount.

DOMESTIC OIL AND GAS PRODUCIBILITY

RESOURCES AND RESERVES

The availability of minerals is stated in terms of "resources" and "reserves." Resources include all deposits known or believed to exist in such forms that economic extraction is currently or potentially feasible. Reserves are that part of the identified resources that can be economically extracted with current technology and at prevailing prices.

The price dependence of reserve estimates is important. As returns to producers rise, reserves in previously discovered fields increase, because more of the minerals underground become economically recoverable. An improvement in recovery technology may similarly add to reserves by making more of the resource available at the prevailing price. Policy changes can also affect reserves; a change in environmental regulations, for example, may make more or less of the basic resource economically recoverable.

Table 3-1 lists the estimates of the Supply and Delivery Panel's Oil and Gas Subpanel of U.S. and world oil and gas resources and reserves. It can be seen that the United States, which consumes more than a fourth of the world's oil production and about half the world's production of natural gas, has only about a twentieth and a tenth, respectively, of the world's proved oil and gas reserves. It should be noted, however, that the reserve figures in Table 3-1 do not reflect oil and gas in existing fields that have become economically recoverable as a result of the large price increases starting in 1973. The figures are therefore understated. Calculations to

TABLE 3-1 Estimates of U.S. and World Resources and Reserves of Crude Oil and Natural Gas as of 1975 (quads)

Location	Crude Oil[a]	Natural Gas	Total
Resources			
United States[b]	667	716	1,383
Other market economies	6,412	3,975	10,387
Centrally planned economies	2,760	2,881	5,641
Total world	9,839	7,572	17,411
Reserves			
United States[b]	201	221	422
Other market economies	3,122	1,225	4,347
Centrally planned economies	619	841	1,460
Total world	3,942	2,287	6,229

[a]Includes natural gas liquids.
[b]Source: U.S. estimates from B. M. Miller, H. L. Thomsen, G. L. Dolton, A. B. Coury, T. A. Hendricks, R. E. Lennartz, R. B. Powers, E. G. Sable, and K. L. Varnes, *Geological Estimates of Undiscovered Recoverable Oil and Gas Resources in the United States*, U.S. Geological Survey Circular 725 (Washington, D.C.: U.S. Geological Survey, 1975).

adjust estimates of existing reserves to the current price situation started only recently and so far have covered only a small part of the United States.

PROJECTIONS OF FUTURE DOMESTIC PRODUCTION

Reserve and resource estimates do not in themselves reveal long-term production potential. They must be seen in the light of the rate at which additions to reserves are made. In the United States, oil production has outpaced reported additions to proved reserves since the late 1960s if one excludes the Prudhoe Bay discovery because of its exceptional size. (It is estimated to contain nearly twice as much recoverable oil as the next largest field, East Texas, did in 1930 when it was discovered.)[1] If one includes Prudhoe Bay, proved reserves are now about the same as they were 10 years ago. Domestic consumption shows an even greater disparity with reserve additions, by the amount of net imports, which have risen fairly consistently, from virtually zero in 1947 to about 15 quads in 1976.

A reserves-to-production ratio of 8:1 or 10:1 is generally considered necessary to sustained production of oil; present U.S. reserves of both oil and gas are roughly 10 times annual domestic production. Thus, to sustain present production rates the United States must add about 20 quads each of oil and gas (the equivalent of current production) to reserves each year.

This may be difficult. New discoveries are now often made in inaccessible locations, where exploration is more expensive and time consuming than it was when the large oil and gas fields in the interior of the United States were opened in the first half of this century. Major new finds are increasingly made in places as remote as the Alaskan North Slope or in offshore areas. Moreover a large proportion of recent drilling in the United States has been directed at relatively small onshore prospects whose aggregate contribution to total reserves is quite limited. Higher oil and gas prices, of course, provide a strong incentive for exploration of these prospect areas, even though the cost of drilling has also risen rapidly. (Between 1972 and 1976 the cost per foot drilled, for all wells including dry holes, almost doubled, from $20.76 to $40.46, in current dollars; the prices of new oil and intrastate gas, however, approximately quadrupled.)[2]

One way to understand the trend of exploration is to calculate the number of barrels of oil or cubic feet of gas added to reserves per foot drilled in exploratory and development wells. Data available for 1960–1974 (excluding Prudhoe Bay) indicate that approximately 15 barrels of crude oil and 92 thousand ft^3 of natural gas were added to reserves for each foot drilled in that period. If one assumes, with most petroleum geologists, that the average finding rate will be smaller in the future (say 10 barrels of oil and 60 million ft^3 of gas per foot drilled), then it would be necessary to drill some 360 million ft each year to maintain present oil and gas production. (Twenty quads is equivalent to about 3.6 billion barrels of oil or 20 trillion ft^3 of gas.) This is not impossible; drilling has been rising rapidly, and the 1978 figures approached the mid-1950s peak of 230 million ft. However, it is quite possible that the finding rate will drop further as exploration moves toward yet more difficult areas and still smaller fields. In fact, preliminary figures for the most recent years suggest a sharp fall in the finding rate. The minimum annual need for drilling may therefore be larger than calculated above.

The rate at which oil and gas will actually be produced depends on both cost-price considerations and the general financial and regulatory climate in which resource development is carried out. The Supply and Delivery Panel, concluding that oil and gas producibility depend more on institutional variables than on price, did not attempt to derive price-versus-supply curves for oil and gas, but instead focused on financial, institutional, environmental, and regulatory constraints on production.[3] It described three scenarios, based on three postulated sets of assumptions about the climate for development. These scenarios reflect the judgments of a body of experts on the potential of oil and gas production under various conditions. They do not, of course, represent the only possible future conditions.

Business-as-Usual Scenario

This scenario projects future oil and gas production as if the uncertain policy and financial conditions of 1976 remained in effect through 2010. It depends on the following assumptions.

- Price controls on domestic oil and gas remain in effect and keep domestic oil and gas prices below world market prices.
- Current environmental considerations, including requirements for environmental impact statements and strict environmental quality standards, remain in effect.
- Public lands continue to be withdrawn from exploration.
- Development of the outer continental shelf continues to be a two-stage process, involving both exploration permits and federal and state production permits.
- Exploration and production technology continue to evolve at current rates.

Moderately Enhanced Conditions Scenario

This scenario projects oil and gas production under a set of assumed government incentives. The assumptions are the following.

- Federal offshore areas are leased at an accelerated rate.
- The offshore permit process is streamlined.
- Wellhead prices for "new" gas are decontrolled.
- Evolution of exploration and production technology accelerates modestly, mainly because decontrolled prices make enhanced recovery economical.
- Transportation of natural gas from the Alaskan North Slope is available by 1985.
- There is no change in land availability.
- Environmental regulations, including environmental impact statement requirements and environmental quality standards, remain in effect.

National-Commitment Scenario

This scenario projects oil and gas production if the government encourages production to the maximum extent, at the expense of other energy supplies and other social and political goals. It has some of the features of a national emergency program. Most constraints on production are lifted, and the federal government fosters capital formation and renders much other assistance to industry, essentially guaranteeing a return on investment. The assumptions are the following.

- Implementation of the Clean Air Act is relaxed, and the requirement for environmental impact statements is eliminated. Most other environmental standards are retained.

- Significant incentives for technology development, including loan guarantees, are established.
- More public land is made available for oil and gas exploration, including some wilderness areas and single-purpose tracts that had previously been withdrawn or had been otherwise unavailable.
- Needed goods, services, and personnel are given priority by the federal government. For example, steel intended for use in oil development might be given first-priority status.
- Tertiary oil recovery and production of natural gas from tight formations become economical, mainly due to price increases.

Table 3-2 illustrates the estimated effects on domestic oil and gas production of the conditions of the three supply scenarios. By 2010, production is on the decline in all cases. Legislation and regulations in force as of 1979 appear to imply future conditions somewhere between the business-as-usual and the moderately enhanced scenarios. Subsequent legislation may change the outlook.

TABLE 3-2 Estimated Production of Domestic Oil and Natural Gas Under the Conditions Assumed in the Supply Scenarios (quads)

Year	Present Conditions	Moderately Enhanced Conditions	National Commitment
Oil			
1975	20	20	20
1985	18	21	21
1990	16	20	21
2000	12	18	20
2010	6	16	18
Gas			
1975	19.7	19.7	19.7
1985	13.5	16.1	18.5
1990	10.3	15.8	18.0
2000	7.0	15.0	17.0
2010	5.0	14.0	16.0

Source: National Research Council, *U.S. Energy Supply Prospects to 2010*, Committee on Nuclear and Alternative Energy Systems, Supply and Delivery Panel (Washington, D.C.: National Academy of Sciences, 1979).

OIL

As noted in Table 3-2, domestic oil production will decline over the coming decades, under the assumed conditions of the three supply scenarios. In the business-as-usual scenario, production drops to less than one third of its 1975 level by 2010. Even with the incentives of the moderately enhanced conditions and national-commitment scenarios, oil production peaks before 1990 at little more than its present level and then declines. Production under the conditions of the national-commitment scenario is only 10 percent higher than under those of the moderately enhanced conditions scenario. At the same time, liquid fuel demand, as projected by the Demand and Conservation Panel,[4] is likely to increase from the present 35 quads to as much as 50 quads even in the middle-range energy demand scenarios.

These projections imply that some combination of imports and domestic substitutes will be needed for the indefinite future. The following discussions of future imported and synthetic oil supplies focus on the difficulties in supplying the large amounts called for in the middle-range demand scenarios.

FUTURE AVAILABILITY OF IMPORTS

The oil import requirements of the United States already severely strain world market oil supplies, to the detriment of other importing countries. If U.S. imports continue to grow, they may be available to the United States only at very high prices. The political costs may be even less acceptable.

Future U.S. demand for imported oil cannot be estimated with any precision. However, the trend toward increased imports will be very hard to reverse within the next several years, because of the time required for new conservation and production policies to produce their full effects. Beyond 1985, the deficit between domestic oil production and consumption will depend heavily on government policy with respect to supply, demand, and price.

For example, by 2010 the need for oil imports would be very small if liquid fuel demand were held to 20–30 quads by the higher energy prices and the conservation policies assumed in the low-growth demand scenarios (chapters 2 and 11),[5] if the policies implicit in the moderately enhanced conditions or national-commitment oil production scenarios were adopted and the upper estimates of coal-based synthetic fuel production[6] were realized. Obviously, these many "if's" make such a prospect unlikely. No less unlikely, since it calls for a probably unattainable and almost surely unsustainable level of imports (over 50 quads a year), is a case in which the oil demands of the highest-growth study scenario might be combined with the business-as-usual oil supply scenario and with slow development of a synfuels industry. A more likely middle case would still entail 15–30 quads of imports.

While the exact values from these illustrative scenarios should not be taken literally, the trends are clear. Even in the middle case, the demand for imports in 2010 would be greater than it is today. This is clearly undesirable; U.S. demand on this scale, particularly if combined with the widely expected decline in world oil production near the turn of the century, would severely strain the world oil market. Supply and Delivery Panel estimates of future world production suggest that even this middle case would entail a U.S. demand for a larger percentage share of the world oil supply than at present; this would intensify the already great pressures on consuming countries less able than the United States to pay the resulting higher prices.

Assuming for the moment that the United States were willing and able to pay the economic and political prices of larger imports, which countries would be able to supply them? Although growth in world oil output may be widely dispersed, the growth in export potential over the next two decades is likely to be concentrated in a few countries. About a third of the output growth will take place in the nonmarket economies (chiefly the Soviet Union and China); but growing demand in those countries will probably leave little or none available for net exports. Another third of the growth in production is likely to occur in Latin America (particularly Mexico), the North Sea, and possibly the United States if a large synthetic fuel industry is established. The rest will take place mostly in the Arab members of the OPEC cartel, with Saudi Arabia the most important source of additional supply. At present it appears that the only producers, outside the nonmarket economies, able to increase production enough to meet world demand will be Saudi Arabia, Iraq, and Mexico. Venezuela could also resume its former importance as an exporter by undertaking vigorous development of the Orinoco tar belt.

Saudi oil policy is likely to be critical by that time, since for various reasons other producers are unlikely to produce at maximum projected capacities. Some countries whose petroleum resources will be approaching exhaustion will probably restrict production to extend the lives of the resources, as Canada, Venezuela, and Kuwait are doing even now. Members of the cartel, which will probably still be in existence, may attempt to restrict production to achieve higher prices rather than higher output. Few countries, in fact, have adopted government policies conducive to maximum oil production. Under these circumstances the demand for Saudi oil could exceed 50 quads by 1995. (The Saudis now, at temporarily enhanced production levels, produce about 19 quads annually.)

It is easy to think of reasons why Saudi Arabia may be unwilling to increase production and may actually freeze or roll back production. Among these are the following.

- An inability to make productive use of the massive amounts of foreign exchange that higher levels of exports, at a possibly much higher price per barrel, would generate.
- Saudi concern about excessively rapid reserve depletion and exploitation of oil that could be more valuable to future generations.
- A shift to a less pro-Western political alignment.
- Dissatisfaction with the U.S. role in Middle East politics.
- The possibility of future conflict in the Persian Gulf area, which would disrupt oil supplies for long periods.
- Technical problems that could limit production.

None of these can be considered implausible. Even if fortune favors the oil importers, Saudi production is very unlikely to exceed 28 quads in 1985 and 40 in 1995. Thus, the margin of excess capacity, under even the most optimistic assumptions about future world oil demand, will be thin indeed—and perhaps nonexistent.[7]

Could the World Oil Situation Improve?

The preceding discussion has emphasized the unfavorable side of the world oil outlook. This is appropriate since the preponderant risks appear to be on that side. Nevertheless the probability of an improvement in the world oil situation, while small, is not negligible. Until one or two years ago, for instance, it was not realized that Mexico's oil resources could be as large as Saudi Arabia's. (For that matter, the substantial oil potential of the North Sea was discovered less than 10 years ago.) With exploration in many parts of the world as active as it is, the possibility of additional major discoveries cannot be excluded, though it would be imprudent to count on it. We should not forget that the past literature on world oil supply is replete with authoritative predictions of imminent exhaustion. The failure of these predictions in the past does not imply that they will fail in the future, but neither can it be taken for granted that oil production must decline in the near future.

It should also be recognized that the natural desire of oil exporters to raise the price by restricting output often conflicts with other goals. All oil exporters have increased their imports rapidly, and some (e.g., Indonesia, Nigeria, and Venezuela) have actually run into balance-of-payments problems. Between 1974 and 1977 the dollar value of imports rose by over 500 percent in Saudi Arabia, 400 percent in Nigeria, 150 percent in Iran, and 134 percent in Venezuela. For all oil-exporting countries together this value rose by 163 percent. In 1974 only 27 percent of these countries' exports were covered by imports; by 1977 the corresponding figure was 59 percent. The total trade surplus of these countries fell from $86 billion in 1974 to $60 billion in 1977, and was no doubt reduced further in 1978.[8] An obvious way out of such problems is to export more oil, which is what Indonesia and Nigeria have done, though Venezuela has preferred to borrow abroad. In any case, decisions to leave oil in the ground are essentially speculative and therefore liable to disappointment.[9] Few (if any) attempts to keep mineral prices at an artificial level have been successful for more than a few years in the past, but OPEC may well be the exception that proves the rule.

In this connection it is also relevant that the demand for oil is sensitive to its price, although the necessary adjustments in demand (particularly in the stock of energy-using equipment) may take considerable time. In chapter 10 it is shown that oil demand outside of the United States has already been lowered by higher prices; in the United States this effect has been thwarted to some extent by price controls but is nevertheless detectable.

None of these observations, we repeat, weakens the serious concern expressed earlier. On balance the outlook is unfavorable, and vigorous measures are necessary. A rational energy policy, however, should take account of all contingencies rather than concentrate on those that appear to be of immediate concern. Until the early 1970s U.S. oil policy was aimed at protecting the domestic industry from cheap imports, which were seen as the principal energy-related threat facing this country; the possibility of much higher oil prices was simply ignored. In retrospect that policy appears foolish indeed (being in effect a "drain America first" policy), but at the time it was defended as a reasonable response to the most likely contingency. The least to be learned from this experience is that we do not know enough about the future to direct policies at the most visible threat only. (In the unlikely event of a decline in the world oil price, the question of protection would no doubt come up again; the more high-cost energy sources are developed in the meantime, the more insistent would the calls for protection be.)

Conclusions on Oil Imports

The gap between domestic supply and demand for oil will be equilibrated, of course. However, as the preceding discussion has shown, it may be difficult to equilibrate it entirely by increased imports. If the United States does not significantly increase its energy supplies and rather sharply reduce its demand for oil, the world price of petroleum is likely to rise rapidly in the last 10 or 15 years of this century. The probable result would be much higher prices for consumers in this country and curtailed consumption for those in less affluent consuming countries. Some importers might be forced to default on their debt obligations. In addition to these financial strains, a tight supply situation would increase the threat and the potential effectiveness of politically motivated supply interruptions. While oil imports will be important to this country for some time to come, national energy policy must take account of these international stresses. U.S. energy imports must be considered not merely in terms of their price in dollars, but also in the context of global economics and politics.

OIL SHALE

U.S. oil shale resources are estimated to contain 3660 quads of recoverable oil. Technologies for exploiting them exist and, if the price were right, the necessary production facilities could begin construction immediately. The Supply and Delivery Panel reports, however, that oil from shale cannot compete with natural petroleum until oil prices reach $21.50–$27.50 (in 1978 dollars) per barrel, if the venture is financed entirely with equity capital. This is somewhat less than the estimated price at which synthetic crude from coal would become competitive, but it is higher than the current price of "new" oil in the United States.

Furthermore, a number of important technical and environmental problems must be considered in planning and developing a shale oil industry. The severities of these problems vary considerably depending on the process used for extracting the oil from the shale rock. The important considerations include the mass of material that needs to be handled, the toxicity and leachability of spent shale, the water requirements, the air and water pollutants released in retorting, and the socioeconomic implications for the localities involved. A number of experimental projects have helped identify these problems and possible solutions.

To ensure that oil from shale can be extracted successfully and in an environmentally acceptable way, the private sector should be encouraged to begin pioneer projects immediately. Incentives such as price subsidies or market guarantees are probably sufficient to attract the necessary investment. All the facilities should be of commercial scale, and everything needed for technical, economic, and environmental success should be included. These projects would help determine whether solutions to all the problems that would be faced by a commercial industry are available. One or two years of successful pioneer operation would lay the foundation for rapid but orderly growth of a commercial oil shale industry.

Pioneer projects could be based on underground or surface mining with surface retorting, or possibly on a modified in situ operation. Surface-retorting techniques have been tested more completely than the modified in situ method and therefore are more assured of success. They also handle better grades of shale and offer higher production rates. In the modified in situ process, part of the shale is mined to open a suitable underground retorting cavity, and then retorted at the surface; the bulk of the shale is retorted in place. This process appears promising in its potential for reducing costs and the difficulties of materials handling and waste disposal.

The nation should not lightly forgo the opportunity to obtain this large additional source of badly needed hydrocarbon liquids. Environmental considerations are critical, and pioneer projects would permit impacts to be fully assessed at relatively low total risk and cost, particularly relative to the potential long-term benefits. Pioneer projects offer a step-by-step way to show whether solutions to these and the remaining technical problems are possible and acceptable.

At present it is difficult to predict the maximum production potential of oil from shale. According to the Supply and Delivery Panel[10] it is quite modest—a maximum of 3 quads annually by 2010—even under national-commitment conditions.

COAL-DERIVED SYNTHETIC LIQUID FUELS

Coal can be converted into liquid and gaseous fuels comparable to those in common use today. Even at present (1979) oil prices, synthetic oil from coal would not be competitive; the Supply and Delivery Panel estimates the cost of synthetic liquids at $30–$36 (1978) per barrel, or almost twice the price of imported oil. However, if depletion of worldwide reserves drives up the price of petroleum and improved technology lowers the relative price of synthetics, the prices of imported and synthetic oil will tend to converge.

The government has financed a number of experimental synthetic fuel development projects at the pilot plant level. However, the government's role with respect to technologies that are near to commercialization can take either of two directions. On one hand, it can create a healthy climate for private investment by assessing the nation at large. This can be done directly or contingently through market guarantees, completion assurances, subsidies, grants, loans, and significant tax and capital recovery policies. On the other hand, private investors could be encouraged through a regulatory process that assesses a narrower community—the consumers of the products. This can be done with tariffs that cover the risks of project completion, price, and investment return. Other possibilities include all-events tariffs, surcharges, fuel adjustment clauses, and escalation provisions. In any case the potential investors would have to be assured that whatever regulation or policy is used has the virtue of constancy.

This study's estimates of the potential availability of coal-derived synthetic liquid fuels under three sets of assumptions about government incentives indicate that a significant contribution from such synthetics is at least 15 years away. At best a few quads of synthetic liquids will be produced from coal by 1990, although up to 12 quads could be available by 2010 if environmental problems can be solved. Coal conversion, like shale oil development, presents serious environmental and health hazards that will probably limit the ultimate attainable production level and slow the buildup of production. The most severe hazard is probably the strain a large conversion industry would exert on already scarce water resources, in western states especially but also in many parts of the East. (See chapter 4 for more detail on water-supply concerns.)

Coal conversion also produces a number of degradation products (organic and inorganic) that could contaminate the air and, especially, water. It is too early to state how easy or expensive their control will be, but there is no case for overriding pessimism. For a more complete treatment of the technology and risks of coal conversion, see chapters 4 and 9.

PROJECTIONS OF TOTAL DOMESTIC LIQUID FUEL PRODUCTION

Table 3-3 summarizes the Supply and Delivery Panel's estimates of domestic production of all liquid fuels, both natural and synthetic, under the conditions of the three supply scenarios.

GAS

In 1975, about 20 quads—or 27 percent of U.S. primary energy consumption—were supplied by natural gas. Over half the country's households and a great majority of commercial and industrial premises are connected to the gas network.

In 1976–1977 (perhaps a somewhat exceptional period), natural gas demand exceeded supply by an estimated 3.8 quads; the difference was met by curtailments of supply. At present only about 0.3 quad of gas is imported, relatively little synthetic natural gas is being produced from petroleum liquids, and essentially no synthetic gas is produced commercially from coal. Because these products are still more expensive than domestic natural gas, their wide introduction in the near future depends on federal pricing policy and other regulations, applied consistently and predictably enough to assure potential investors of reasonable returns.

Because of the relative attractiveness of gaseous fuels, as gas prices rise the declining supply of domestic natural gas will probably be supple-

TABLE 3-3 Projected Domestic Liquid Fuel Supply (quads per year)

Scenario and Energy Source	1975	1985	2000	2010
Business as usual				
Petroleum	20	18	12	6
Coal-based liquid fuels	—	0.1	2.3	6.1
Shale oil	—	—	—	—
TOTAL	20	18.1	14.3	12.1
Enhanced supply				
Petroleum	20	21	18	16
Coal-based liquid fuels	—	0.1	2.4	8.0
Shale oil	—	0.2	1.0	1.5
TOTAL	20	21.3	21.4	25.5
National commitment				
Petroleum	20	21	20	18
Coal-based liquid fuels	—	0.1	4.7	12.9
Shale oil	—	0.7	2.5	3.0
TOTAL	20	21.8	27.2	33.9

Source: Compiled from National Research Council, *U.S. Energy Supply Prospects to 2010,* Committee on Nuclear and Alternative Energy Systems, Supply and Delivery Panel (Washington, D.C.: National Academy of Sciences, 1979).

mented by increased imports and conversion of coal into synthetic gas. Table 3-4 lists projections of the availability of domestic and imported natural gas and of synthetics from coal, under the varying assumptions of the Supply and Delivery Panel's three supply scenarios. In the business-as-usual scenario, supply would be curtailed in the intermediate term, largely because of the time it would take to build up a synthetic gas industry.*

IMPORTS

Transportation costs are more important for natural gas than they are for oil. Large quantities of gas in other countries are flared, reinjected, or simply left in the ground for lack of local markets. The most economical means of moving large amounts of gas over very long distances, particularly across oceans, is to liquefy it by cryogenic techniques and transport it by special tankers. The technology for doing so has been available for years, but the high cost and concerns about the safety of

*See statement 3-1, by H. Brooks, Appendix A.

TABLE 3-4 Projected Gas Availability (quads)

Scenario and Energy Source	1975	1985	2000	2010
Business as usual				
Domestic natural gas	19.7	13.5	7.0	5.0
Imported natural gas[a]	0.2	2.3	3.1	3.1
Synthetic gas from coal[b]	0	0.3	3.5	4.1
TOTAL	19.9	16.1	13.6	12.2
Moderately enhanced conditions				
Domestic natural gas	19.7	16.1	15.0	14.0
Imported natural gas[a]	0.2	2.5	5.2	5.2
Synthetic gas from coal[b]	0	0.5	3.5	4.8
TOTAL	19.9	19.1	23.7	24.0
National commitment				
Domestic natural gas	19.7	18.5	17.0	16.0
Imported natural gas[a]	0.2	2.7	7.4	7.4
Synthetic gas from coal[b]	0	0.7	4.5	7.9
TOTAL	19.9	21.9	28.9	31.3

[a] Includes gas from Canada plus liquefied petroleum gas and liquefied natural gas from other nations.

[b] Does not include low-Btu synthetic gas.

Source: Data are from National Research Council, *U.S. Energy Supply Prospects to 2010*, Committee on Nuclear and Alternative Energy Systems, Supply and Delivery Panel (Washington, D.C.: National Academy of Sciences, 1979).

handling and storing the highly volatile and flammable liquefied natural gas (LNG) have limited the growth of this trade. A small fraction of a quad of LNG is now being imported to this country to serve peak loads, and the scarcity of domestic natural gas could make LNG more attractive if there were no alternatives.

However, for shorter distances, especially on land, pipelines are less costly than LNG transportation. There are prospects of importing gas by pipeline from Canada and Mexico. Negotiations are under way, and it is reasonable to assume that some quantities of natural gas will be imported from these countries. The timing of these imports will depend largely on the pace of the negotiations.

NATURAL GAS IN UNCONVENTIONAL FORMATIONS

Very large amounts of natural gas are dispersed in unconventional geological formations such as the tight brown shales of several Appalachian and midwestern states, coal seams, and the geopressured thermal brines of the Gulf Coast. Very small amounts are now being produced from the shales, and some gas is drawn from coal seams before mining, mainly as a safety measure. The geopressured deposits are not exploited, and it is likely that the gas cannot be produced economically unless the heat and mechanical energy in the brines can be exploited at the same time, which is by no means certain. (See chapter 8.)

None of these is an economic source of gas at present prices, but the expected rise in energy prices between now and 2010 will make such low-grade sources more attractive. Most such deposits are unlikely ever to be produced at rates comparable to those attained in current gas fields, but for local use they could become important.

SYNTHETIC GAS

Coal can be converted to a variety of gaseous fuels, by a number of fairly well-developed technologies (chapter 4). In principle, a coal conversion industry could be developed to the point of producing as much as 8 quads of gas by 2010 (Table 3-4). However, as discussed in the earlier treatment of synthetic liquid fuels from coal, this depends on a number of considerations. First, the supply of water to meet the requirements of the conversion process is in question, and supplies must be considered and managed carefully. In addition, there are the problems of assessing and controlling the risks of air and water pollution. The technology is unfamiliar and untried on the requisite scale, and the rate at which these problems can be understood and controlled will largely determine the rate of development of a coal conversion industry.

THE OUTLOOK FOR OIL AND GAS IN THE INTERMEDIATE TERM

The analyses underlying this chapter suggest on balance that domestic supplies of liquid and gaseous fuels will decline further while demand continues to increase. The potential for greatly increased imports, especially of oil, appears to be limited. In fact, the study's scenarios (chapter 11) suggest the possibility, under even moderate energy growth assumptions, of brief but recurrent shortages of oil and perhaps of gas. Such shortages could have the same transient character as the episodes following the 1973 OPEC embargo, those experienced during the cold winter of 1976–1977, and the gasoline shortages of 1979. Whether this problem materializes will depend to a large extent on the maintenance of price controls, which discourage production and encourage consumption. If it does materialize, the government must see to it that available supplies of specific fossil fuels go to those applications for which they are best suited and most needed. Usually, though not invariably, this will mean reserving natural gas for space heating and special applications in industry, oil for transportation and petrochemicals, and coal for generating steam and electricity. While the allocation of increasingly scarce fuels could probably not be left entirely to the market in an emergency situation, care must be taken to prevent distortions of the price mechanisms that result in false signals to consumers and producers and thus aggravate the long-term problem. Any allocation system should preserve a reasonable balance between household use and industrial or commercial use; giving households absolute priority in natural gas, for instance, may lead to avoidable unemployment and loss of nonenergy output.

It has already been argued that a vigorous program of conservation and enhanced domestic production is required, with the overall aim of reducing U.S. dependence on imported oil and gas as much as possible over the coming decades. However, even those members of CONAES who are convinced that the oil and natural gas era is coming to an end feel that improved domestic production is essential to permit a transition to sustainable long-term energy sources. For reasons of national security, minimizing dependence on imported oil should be accorded prominence in national energy policy comparable with other foreign policy goals related to energy, such as avoiding the proliferation of nuclear weapons. The social, economic, and political prices this country pays for oil imports are likely to become less and less acceptable over the next 10–20 years. How critical this problem may yet become will depend in large part on how seriously it is taken by the government and the people of this country.

NOTES

1. Richard Nehring, *Giant Oil Fields and World Oil Resources* (Santa Monica, Calif: Rand Corporation (R-2284-CIA), June 1978).
2. American Petroleum Institute, *Basic Petroleum Data Book,* Section III, Table 10, Supplement (Washington, D.C.: American Petroleum Institute, 1978).
3. National Research Council, *Energy Supply Prospects to 2010,* Committee on Nuclear and Alternative Energy Systems, Supply and Delivery Panel (Washington, D.C.: National Academy of Sciences, 1979).
4. National Research Council, *Alterative Energy Demand Futures to 2010,* Committee on Nuclear and Alternative Energy Systems, Demand and Conservation Panel (Washington, D.C.: National Academy of Sciences, 1979).
5. *Ibid.*
6. Supply and Delivery Panel, *op. cit.*
7. William A. Johnson and Richard E. Messick, "The Supply and Availability of Imported Oil Through 2010," preliminary report for the Supply and Delivery Panel (Available in CONAES public file, April 1977).
8. International Monetary Fund, *International Financial Statistics* (Washington, D.C.: International Monetary Fund, January 1979), pp. 36–37.

9. Keeping a mineral off the market is profitable, generally speaking, if its price rises more rapidly than the rate of interest. Studies of nonfuel minerals suggest that over long periods of time this condition has rarely been satisfied. For a recent analysis along these lines see G. Anders, W. P. Gramm, and S. C. Maurice, *Does Resource Conservation Pay?,* Original Paper 14 (International Institute of Economic Research, July 1978). Available from Green Hill Publishers, Inc., P.O. Box 738, Ottawa, Ill. 61350.

10. Supply and Delivery Panel, *op. cit.*

4 Coal

INTRODUCTION

Coal is the nation's (and the world's) most abundant fossil fuel. Domestic recoverable reserves amount to 6000 quadrillion Btu (quads), which is part of a total domestic resource of about 80,000 quads and a world resource crudely estimated at about 300,000 quads. Of this huge supply, we currently consume about 14 quads each year, or less than 0.3 percent of domestic recoverable coal reserves. In contrast, as noted in chapter 3, the nation extracts each year almost 10 percent of its 420-quad recoverable reserves of oil and natural gas.

The substitution of coal for natural gas and oil on a large scale, either directly or through synthetic coal-derived substitutes, would on these grounds seem a ready-made solution to the nation's energy problems. The simple arithmetic of availability, however, does not tell the whole story. Doubling or tripling the use of coal will take time, investments amounting over the years to hundreds of billions of dollars, and determined efforts to solve an array of industrial, economic, and environmental problems.

Unlike oil and gas consumption, coal use is limited not by reserves or production capacity generally, but rather by the extraordinary industrial and regulatory difficulties of burning it in an environmentally acceptable and, at the same time, economically competitive manner. Coal is chemically and physically extremely variable, and it is relatively difficult to handle and transport. Its use produces heavy burdens of waste matter and pollutants. Even at its substantial price advantage, Btu for Btu, it cannot compete with oil and natural gas in many applications because of the expense of handling and storing it, disposing of ash and other solid wastes, and controlling emissions to the air. Only in very large installations, such as utility power plants and large industrial boilers, is coal today generally economic and environmentally suitable to burn. Domestic coal production capacity today exceeds economic demand, and this may well remain true until the end of the century.[1]

The health problems associated with coal affect both its production and its use. The health of underground miners presents complex and costly problems, for example, and is in need of better management; black lung is the notable instance. At the other end of the fuel cycle, the evolving state of air pollution regulations to deal with the emissions of coal combustion complicates planning for increased demand, and thus in turn inhibits investment in mines, transportation facilities, and coal-fired utility and industrial boilers.

The future is obscured also by a number of more speculative, less easily surmountable problems, which may result in further regulatory restrictions on the use of coal. Chief among these is the risk that before the middle of the next century, emissions of carbon dioxide, an unavoidable (and essentially uncontrollable) product of fossil fuel combustion, may produce such concentrations in the atmosphere as to produce large and virtually irreversible alterations in the world's climate. (See chapter 9.) Also worrisome is the water-supply situation, which could limit synthetic fuel production or electricity generation unless large-scale and possibly expensive steps are taken to minimize water consumption and manage water supplies. This is already a limiting factor in some western locations, but the eastern coal regions may be approaching trouble too.

Coal-fired power plants burned nearly 70 percent of U.S. coal production in 1977, producing more than 45 percent of the nation's electricity. Most of this was in large, centralized facilities with generating capacities of 300 megawatts (electric) (MWe) or more, designed to produce base-load power. (Smaller, less efficient oil- and gas-fired units and small, older coal units serve intermediate and peak loads.) Industry used one fifth of national production, slightly more than half of that in coke plants and the rest to produce steam and dry process heat. Almost all the rest (about 8 percent) was exported, mainly to make coke. Imports were less than 1 percent of U.S. production.

In the future, the market for coal can be widened. Development of efficient, relatively clean coal power cycles for use in smaller electricity-generating units decentralized to serve local loads, for example, will be attractive to industry and to utilities with power plant siting problems. Coal use for industrial process heat and chemical feedstocks will be harder to stimulate, especially in smaller installations, because of the expense and difficulty of handling the coal and the various wastes and emissions from its combustion. In small units, the capital and operating costs of dealing with this inconvenient fuel are proportionally larger than in larger ones. The domestic and foreign market for metallurgical coal (used in blast furnaces, smelters, and chemical plants) amounts to 15–20 percent of the nation's coal production; it is tied primarily to the world steel industry and is not expected to grow rapidly in the near future.

Over the next 10–20 years some of the obstacles to increased demand will weaken as new technologies increase the efficiency and convenience of using coal, and as the prices of oil and gas rise while their reliability of supply declines. A number of the advanced electric power cycles for coal, now under development, would be suitable for smaller installations, and their relatively clean environmental characteristics would make it possible to locate them near users of their power. For smaller industrial users the fluidized-bed combustion and synthetic fuel processes now undergoing development would be especially valuable.

Department of Energy regulations under the Powerplant and Industrial Fuel Use Act of 1978 (Public Law 95-620) will, when implemented and enforced, further improve the outlook for coal by banning oil and gas as fuels in most new power plants and large industrial heating units.

This is not to imply that all the problems of coal use are solvable, or that coal can become the mainstay of the U.S. energy sector over the long term. Its environmental costs will remain high; mining and burning 2 or 3 times the present output of coal, even if it is done efficiently and with care, will be difficult (and increasingly expensive) if coal's contributions to air and water pollution and land degradation are to be kept from increasing.

With the foregoing in mind, we see the following as the prime objectives of national coal policy in the coming decades.

1. Provide the private sector with strong investment incentives to establish a synthetic fuel industry in time to compensate as domestic and imported oil supplies begin to decline (probably some time near 1990).

2. Continue the broad research and development program in fossil fuel technology to widen the market for coal by increasing the efficiency and environmental cleanliness with which it can be used.

3. Improve health in the mines by strengthening industrial hygiene and by performing the necessary epidemiological research. The black lung problem especially should be clarified. (See chapter 9.)

4. Devote the necessary resources to supporting long-term epidemiological and laboratory studies of the public health consequences of coal-derived air pollutants to put air quality regulation on a firmer scientific basis that allows more confident and efficient setting of standards, on which industry can depend in its long-range planning. (See chapter 9.)

5. Develop a long-range plan, recognizing that coal presents some

serious environmental and occupational health and safety problems, and that it does not relieve the nation of its need to develop truly sustainable energy sources for the long term.

By 1985, given reasonably coherent policy and successful research and development, annual domestic demand for coal should approach 1 billion tons (about 20–25 quads). Commercial techniques for using coal will have changed little, but some synthetic fuel technologies will be on the verge of commercialization, and improved techniques for direct coal combustion will begin to enter the market. Knowledge of the environmental impacts of coal production and use (especially the public health consequences of coal-derived air pollutants) should be improved to the point that the current regulatory uncertainties can be reduced.

As the year 2010 is approached, coal use in the United States may reach 2 billion tons annually. By then, some of the clean, efficient techniques of coal use now being developed should attain full commercialization, and knowledge of the environmental and public health characteristics of coal may be sufficient for confident standard setting. At the same time, however, water supply will be increasingly critical, and the first signs of climatic effects from carbon dioxide emissions may be appearing. It is at about this time that truly sustainable energy sources must begin to become available, to provide new flexibility for energy policy and to relieve some of the pressure on coal.

For now, however, there is little room for maneuver. Coal must be used in increasing quantities, and mainly with the current technologies, until at least the turn of the century, regardless of what happens with respect to such alternatives as nuclear fission or solar energy. However, because of the variety of environmental and social problems it presents, it cannot indefinitely provide additions to energy supply. To keep these problems under control in the meantime, it would be wise to approach coal conservatively, with an eye especially to its environmental risks.

RESOURCES

RANK AND QUALITY

Coal is an extremely complex and variable material whose structure is inadequately understood despite centuries of use. The increased demand for coal is accelerating the effort to better understand its behavior and the nature of its effluents and by-products, under both direct use and conversion to synthetic fuels.

Coals are classified by rank according to the state of "coalification" they have reached.[2] The categories of rank, which roughly indicate the stage of coalification, are, from lowest quality to highest, lignite (two varieties), subbituminous (three varieties), bituminous (five varieties), and anthracite (three varieties). (See Figure 4-1 and Table 4-1.)

Low volatility and high carbon content together make anthracite the slowest and cleanest burning coal, qualities which made it the most desirable residential and commercial heating fuel before the advent of heating with oil and natural gas. However, for making coke the best coal is low-volatile bituminous. Per pound, the percentage of ash is lower and the Btu content is higher in low-volatile bituminous than in anthracite. The agglomerating or caking nature of bituminous coal, however, makes it ill suited for certain types of coal gasification processes, which become clogged by the fused carbon material. These processes require noncaking coal, which is usually subbituminous—a rank found almost exclusively in the West. Some gasification processes have been developed to use either type of coal.

The quality of coal is determined in general by two classes of material: (1) the organic remains of plants and (2) inorganic substances contributed by the plants, by water seepage, and by the surrounding geological matter, generally referred to as mineral matter. The organic portion of coal consists of carbon rings linked by chains that contain nitrogen, hydrogen, sulfur, and oxygen. Many products therefore arise when coal is heated. These products may be commercially useful, but may also be potentially hazardous; they include small amounts of carcinogens, mutagens, and respiratory irritants. The inorganic or mineral portion of coal usually constitutes from 9 percent to about 30 percent of the coal by weight. It includes up to half the sulfur and small, potentially toxic amounts of antimony, arsenic, beryllium, cadmium, mercury, lead, selenium, zinc, heavy radionuclides, and asbestos.[3]

Sulfur products are currently the most important of all pollutants released by coal combustion (chapter 9); their health effects are the most widely discussed, and they are by far the most costly to control. Sulfur in coal occurs mostly in a form that is either combined with the coal material ("organic") or attached to, but physically distinct from, the coal ("inorganic" or "mineral"). (An additional small amount occurs as a sulfate, a product of weathering.) The distinction between organic and inorganic sulfur is important because inorganic sulfur can be removed in large part by washing prior to combustion; the organic form, under most washing processes in use today, cannot. In coals with very low total sulfur content by weight (less than 0.6 percent), most of the sulfur is organic; but when total sulfur makes up a greater percentage of the coal, the amount of inorganic sulfur is greater, commonly about 50 percent.

Emissions of sulfur dioxide (SO_2) are measured in terms of pounds per

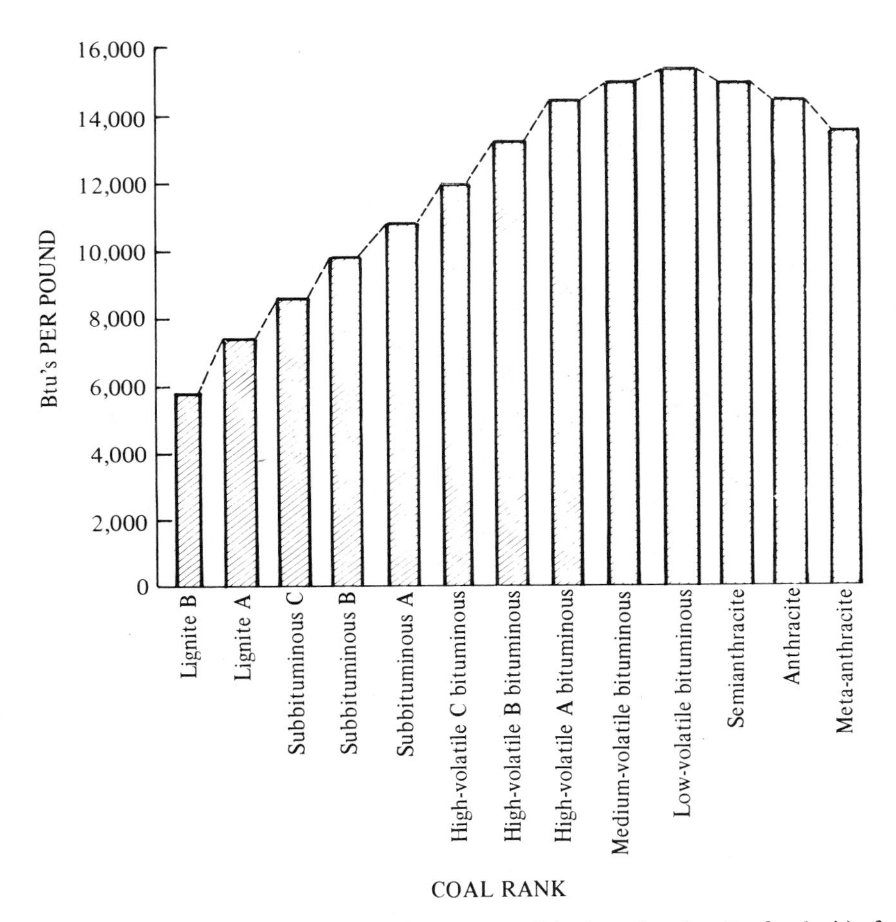

FIGURE 4-1 Comparison of heat values per pound (moist, mineral-matter-free basis) of coal of different ranks (Btu's per pound). Source: Adapted from Paul Averitt, *Coal Resources of the United States, January 1, 1974,* U.S. Department of the Interior, Geological Survey Bulletin 1412 (Washington, D.C.: U.S. Government Printing Office (Stock No. 024-001-02703), 1975), p. 17.

million Btu of fuel burned. For enforcement purposes, the Environmental Protection Agency (EPA) assumes that all the sulfur in the coal is released as sulfur dioxide, the weight of which is twice that of sulfur. The ceiling on emissions is 1.2 lb of sulfur dioxide per million Btu, which under earlier EPA regulations could be met by burning coal with 0.6 lb or less of sulfur per million Btu.[4] In May 1979, however, the EPA imposed additional control requirements on new plants (construction or alteration begun after

TABLE 4-1 Composition of Some American Coals

	Proximate Analysis[b] (percent)				Elementary Analysis[c] (percent)					Heating Value[d] (Btu per pound)
Type[a]	Moisture	Volatile Matter	Fixed Carbon	Ash	S	C	H_2	O_2	N_2	
Anthracite	2.5	6.2	79.4	11.9	0.60	93.5	2.6	2.3	0.9	14,500
Bituminous										
Low volatile	1.0	16.6	77.3	5.1	0.74	90.4	4.8	2.7	1.3	15,600
Medium volatile	1.5	23.4	64.9	10.2	2.20	87.6	5.2	3.3	1.4	15,600
High volatile A	2.5	36.7	57.5	3.3	0.70	85.5	5.5	6.7	1.6	15,000
High volatile C	12.2	38.8	40.0	9.0	3.20	79.2	5.7	9.5	1.5	12,400
Subbituminous										
A	14.1	32.2	46.7	7.0	0.43	80.9	5.1	12.2	1.3	12,100
C	31.0	31.4	32.8	4.8	0.55	74.0	5.6	18.6	0.9	8,800
Lignite	37.0	26.6	32.2	4.2	0.40	72.7	4.9	20.8	0.9	7,600

[a] The analyses are meant to give an idea of the range of variation among coal types. There is considerable variation within each type, as well.

[b] Proximate analysis totals 100 percent by weight.

[c] Elementary analysis totals 100 percent by weight on a dry, ash-free basis.

[d] Heating value is on a moist, mineral-matter-free basis.

Source: Adapted from U.S. Environmental Protection Agency, *Electric Utility Steam Generating Units: Background Information for Proposed Particulate Matter Emission Standards*, Office of Air Quality Planning and Standards (Springfield, Va.: National Technical Information Service (PB-286-224), 1978), Table 3-6.

September 18, 1978) that limit the possibility of using low-sulfur coal as the sole means of meeting the standard. (The sulfur content of coals in the major U.S. producing areas ranges from 0.3 to 4.7 lb per million Btu.)[5] The regulations and their impact are discussed in more detail in the section of this chapter entitled "Air Pollution Regulation and Control."

DISTRIBUTION OF RESOURCES

While recoverable quantities of coal are found in at least 32 states (see Figure 4-2), 90 percent is found in only 10 states, distributed among four regions. These regions and states are the Appalachian basin (chiefly, in order of deposit tonnage, West Virginia, Pennsylvania, Ohio, and eastern Kentucky), the Illinois basin (also called the Eastern Interior region—Illinois, western Kentucky, and Indiana), and the Northern Great Plains and Rocky Mountain regions (Montana, Wyoming, Colorado, and North Dakota).

A broad distinction is drawn between types of U.S. coal based on whether they are located east of the Mississippi (Appalachian and Illinois basins) or west (the Western Interior, Northern Great Plains, Rocky Mountain, Gulf, and Pacific regions). There are major differences between eastern and western coals, based on their different conditions of formation. The major deposits east of the Mississippi are generally the oldest and deepest, and are therefore of highest rank. Most western coals were formed about 150 million years later than eastern coal, at shallower depths; their coalification is less advanced, and their energy and carbon content are therefore generally lower, than those of eastern coal. Created from different plants and under different geological conditions, western coal has a carbon structure and mineral content generally different from that of eastern coal; these characteristics must be allowed for and sometimes counteracted in designing boilers, controlling effluents, and choosing coal for coking or for conversion to synthetic fuels.[6] (Boilers, for instance, suffer an efficiency loss when a nondesign coal type is substituted for the type originally intended.) Lastly, much eastern coal was formed under salt water, giving it a high sulfur content compared to most western coal, which formed under fresh water.

U.S. coal is generally in huge beds, ranging in thickness from a few inches to more than 100 ft, that extend with some interruptions for up to 30,000 mi^2. The major Appalachian deposit is the Pittsburgh bed, which, because of size, the high heat value of the coal it contains, and its use in the creation of the Pennsylvania iron and steel industry, has been called the most important single mineral deposit in the United States. Fairly uniform and continuous over 6000 mi^2 of western Pennsylvania, West Virginia, Ohio, and Kentucky, the Pittsburgh bed yielded through 1973 an

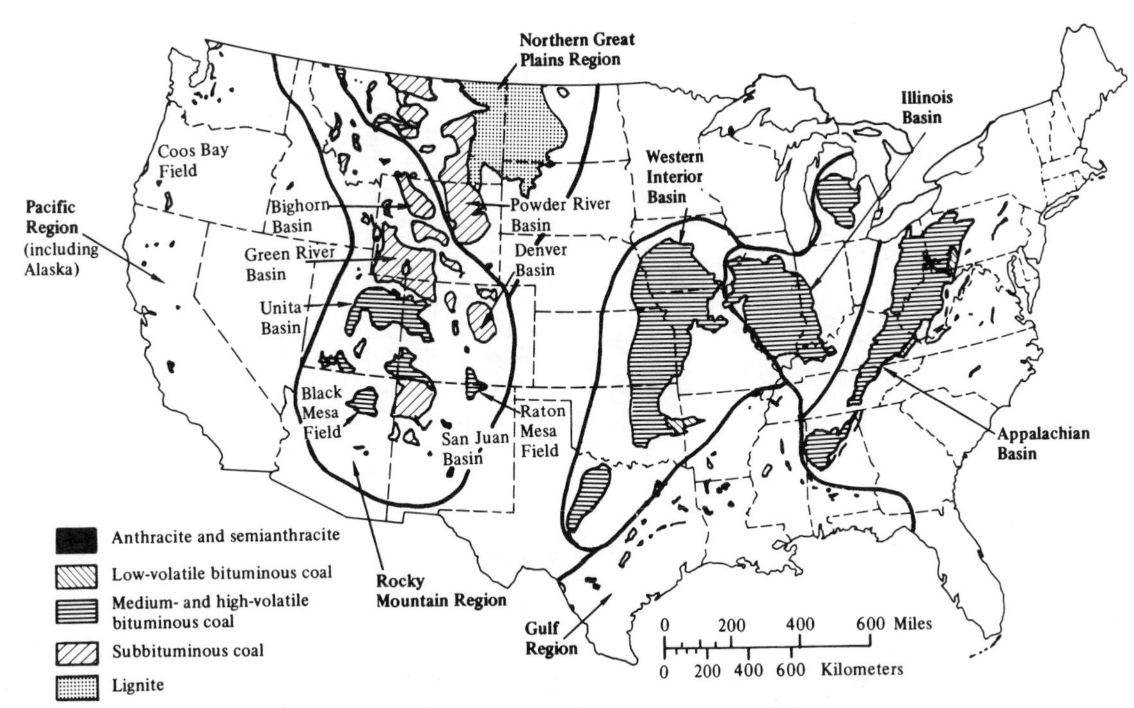

FIGURE 4-2 Coal fields of the coterminous United States. Source: Adapted from Paul Averitt, *Coal Resources of the United States, January 1, 1974,* U.S. Department of the Interior, Geological Survey Bulletin 1412 (Washington, D.C.: U.S. Government Printing Office (Stock No. 024-001-02703), 1975), p. 5; and U.S. Department of Energy, Energy Information Administration, *Coal Data* (Washington, D.C.: U.S. Government Printing Office (DOE/EIA-0064), 1978), p. 1.

estimated 9 billion tons or about one fifth of total U.S. production to that date, with much still remaining.[7]

The largest unbroken concentration of coal in the United States is the remarkable Wyodak bed, centered in Cambell County, Wyoming. This bed penetrates the surface continuously along a line extending for 120 miles at thicknesses ranging from 25 ft to 150 ft. To a depth of 200 ft, it contains an estimated 15 billion tons of strip-minable coal (or about 20 times total U.S. output in 1976). To a depth of 2000 ft, it contains perhaps 100 billion tons of coal.[8] (Given the low heating value of this coal and current prices, however, these deeper deposits are not now economically recoverable.)

Coal resources are generally reported in short tons. The latest U.S. Bureau of Mines estimate of the tonnage in U.S. coal beds that are legally and economically accessible (called the reserve base) is 438 billion tons, about 32 percent of it accessible by surface mining. The reserve base is part of a total U.S. coal resource estimated, much less reliably, at 4 trillion tons.

In terms of the reserve base, practically all anthracite is found in Pennsylvania underground-minable deposits; all but 16 percent of bituminous deposits are found east of the Mississippi; all subbituminous deposits are found in the West; all lignite is found in surface deposits, practically all of it in Montana, North Dakota, Texas, and Colorado. The distribution of the U.S. reserve base by coal rank, type of mining, and region is shown in Table 4-2.

The distribution of coal energy resources is also of interest. The Bureau of Mines reserve base tonnages were converted to recoverable reserves and to corresponding average energy content figures by the Congressional Office of Technology Assessment using a recovery factor of 57 percent in underground mines and 80 percent in surface mines.[9] The resulting regional distribution of recoverable coal reserves by energy content and method of mining is summarized in Table 4-3.

Table 4-3 shows a distribution of recoverable reserve energy that is not far different from the distribution of reserve base tonnages. This is because, in comparison to the eastern coal, the lower energy of western coal is compensated for by its greater recoverability.[10]

DISTRIBUTION OF PRODUCTION

In 1976 (the most recent year of output unaffected by the 1977–1978 coal strike), surface mines yielded 57 percent of U.S. coal production tonnage. In that year, Appalachia supplied 60 percent of U.S. coal, with 45 percent of the region's output produced by surface mines; the Interior supplied 24 percent, about two thirds of it surface mined; and the West supplied 16 percent, practically all of it surface mined.

The most important of the 26 producing states in 1976 were all east of

TABLE 4-2 Distribution of U.S. Coal Reserve-Base[a] Tonnage by Coal Rank and Type of Mining (percent)

Region	Coal Rank: Anthracite		Bituminous		Subbituminous		Lignite		Total Surface	Total Deep	Grand Total[b]
	Surface	Deep	Surface	Deep	Surface	Deep	Surface	Deep			
East[c]	<0.5	2	9	35	0	0	<0.5	0	9	37	46
West[d]	<0.5	<0.5	2	6	14	25	7	0	23	31	54
TOTAL[b]	<0.5	2	11	42	14	25	8	0	32	68	100

[a]Legally and economically accessible deposits reported by the U.S. Bureau of Mines as of January 1, 1976, to be 438.3 billion short tons. Figures shown here are independently rounded.

[b]Totals may not add due to rounding.

[c]Appalachian and Eastern Interior basins (see Figure 4-2).

[d]Western Interior, Gulf, Northern Great Plains, Rocky Mountain, and Pacific (including Alaska) coal regions (see Figure 4-2).

Source: U.S. Department of Energy, *Coal Data—A Reference*, Office of Energy Data and Interpretation, Energy Information Administration (Washington, D.C.: Energy Information Administration Clearinghouse (DOE/EIA-0064, Order no. 704), 1978), p. 2.

TABLE 4-3 Distribution of U.S. Recoverable Reserve Coal Energy[a] by Method of Mining (percent)

Region	Surface	Deep	Total
East[b]	11	36	47
West[c]	25	28	53
TOTAL	36	64	100

[a]Reserve base is made up of economically and legally accessible deposits. Recoverable reserves calculated at 80 percent of surface, and 57 percent of deep, reserve base. Recoverable reserve coal energy is an estimated 6334 quads.
[b]Appalachian and Eastern Interior basins (see Figure 4-2).
[c]Western Interior, Gulf, Northern Great Plains, Rocky Mountain, and Pacific (including Alaska) coal regions (see Figure 4-2).

Sources: Calculated from Office of Technology Assessment, *The Direct Use of Coal: Prospects and Problems of Production and Combustion* (Washington, D.C.: U.S. Government Printing Office (052-003-00664-2), 1979), p. 63. For the combined deep and surface reserve base energy in Alaska and Washington, multiplied by 60 percent for an estimate of total recoverable reserve energy in those states, Francis X. Murray, ed., *Where We Agree: Report of the National Coal Policy Project*, 2 vols., sponsored by the Center for Strategic and International Studies, Georgetown University (Boulder, Colo.: Westview Press, 1978), vol. 2, p. 395.

the Mississippi: Kentucky (21 percent of the U.S. total), West Virginia (16 percent), Pennsylvania (13 percent), Illinois (8 percent), and Ohio (7 percent), all of which are states with large reserves. Two states with small reserves but fairly large outputs were Virginia (6 percent of the 1976 output) and Alabama (3 percent). Two states with very large reserves but comparatively small outputs were Montana and Wyoming (each with about 4 percent of the U.S. total). Montana and Wyoming were the largest western producers and have the fastest growing productions in the nation. On a Btu basis, the East's share of total output would be larger, and the West's smaller, than the tonnage shares shown here.[11] Table 4-4 shows the share of recoverable reserves, production, and consumption for 1976, by census region.

The extent to which U.S. coal production will shift to the West's easily mined, surface, low-sulfur deposits depends on the willingness of the federal government, which owns perhaps 55–60 percent of western deposits, to lease new, large tracts of land assembled into "logical mining units."[12] It will also depend on water-use policy in the West, on transportation rates, and in the longer run, on the progress of coal liquefaction and gasification technology. The mining industry estimates that the current short-term U.S. coal production capacity is roughly 15–20 percent greater than output, which reflects the fact that the most critical short-term constraints on increased coal use are in the demand sector.[13]

In 1977, coal provided about 14 quads (or 18.4 percent) of the 76.6 quads of energy consumed in the United States. Electric utility boilers consumed 10.7 quads, or about 77 percent of total coal consumption; about 13 percent, or 1.7 quads, was consumed in the manufacture of coke for steel; and other industries consumed 1.4 quads, or 10 percent, primarily as boiler fuel. Less than one fifth of a quad was distributed at retail, primarily as a commercial and residential fuel.[14] In the utility market, coal competes with oil, natural gas, and nuclear power; in the industrial boiler market, it competes with oil and natural gas.

CONSUMPTION

INTRODUCTION

The expansion of coal consumption is and will continue to be limited by constraints on demand. Under the National Energy Plan (NEP) of 1977, which called for a near doubling of coal use by 1985, industrial coal consumption was projected to increase at nearly 12 percent annually, and electric utility coal consumption at about 6 percent.[15] The nation's utilities may be able to meet the forecast rate of increase, because under the Powerplant and Industrial Fuel Use Act of 1978 (Public Law 95-620) they are limited almost entirely to coal and nuclear power for new base-load generating capacity. The industrial projections of the NEP appear optimistic, however, given the small scale of many industrial installations now using oil and gas, and the consequently large investments required in new coal-handling facilities, expensive exhaust-gas scrubbers, and the like. (It should be noted that Congress failed to enact some of the plan's key coal-use incentives.)

In addition, rapid changes in the regulation of coal mining and combustion render the future economics of coal use rather uncertain. New surface-mining legislation, for example, will no doubt decrease somewhat the current cost advantage enjoyed by surface-mined coal over deep-mined coal. Newly proposed air pollution regulations, which would require all new large coal-fired units to install desulfurization equipment, will raise the cost of burning coal and reduce the economic advantage of low-sulfur western coal as a means of meeting air quality requirements. Until the economic implications of these and other regulatory actions are fully understood, coal will not find ready acceptance in many applications, especially in smaller industrial applications, where the costs per unit of heat tend to be greater than in larger operations such as utilities and large industrial establishments.

TABLE 4-4 1976 Recoverable Reserves, Production, and Consumption of U.S. Coal, by Census Region (percent of total tons)[a]

Census Region[b]	Recoverable Reserves[c]	Annual Production[d]	Annual Consumption[e]
New England	0	0	0.1
Middle Atlantic	6.3	13.4	12.8
East North Central	21.9	19.0	30.4
West North Central	5.1	2.6	8.9
South Atlantic	9.4	22.1	15.5
East South Central	6.9	25.5	11.8
West South Central	1.3	2.7	2.6
Mountain	47.5	13.8	7.1
Pacific	1.7	0.7	1.2

[a] Columns may not add due to rounding.
[b] Census regions divide the states as follows: New England includes Maine, Massachusetts, Rhode Island, New Hampshire, Vermont, and Connecticut; Middle Atlantic includes New York, New Jersey, and Pennsylvania; East North Central includes Ohio, Indiana, Illinois, Michigan, and Wisconsin; West North Central includes Minnesota, Iowa, Missouri, North Dakota, South Dakota, Nebraska, and Kansas; South Atlantic includes Delaware, Maryland, Virginia, West Virginia, North Carolina, South Carolina, Georgia, and Florida; East South Central includes Kentucky, Tennessee, Alabama, and Mississippi; West South Central includes Arkansas, Louisiana, Oklahoma, and Texas; Mountain includes Montana, Idaho, Wyoming, Colorado, New Mexico, Arizona, Utah, and Nevada; and Pacific includes Washington, Oregon, California, and Alaska.
[c] Source: Calculated from Office of Technology Assessment, *The Direct Use of Coal: Prospects and Problems of Production and Combustion,* (Washington, D.C.: U.S. Government Printing Office (052-003-00664-2), 1979), p. 63.
[d] Sources: Calculated from U.S. Department of Energy, *Coal—Bituminous and Lignite in 1976,* Office of Energy Data and Interpretation, Energy Information Administration (Washington, D.C.: Energy Information Administration Clearinghouse (DOE/EIA-0118/1 [76], Order no. 703), 1978), p. 11; and U.S. Department of Energy, *Coal—Pennsylvania Anthracite in 1976,* Office of Energy Data and Interpretation, Energy Information Administration (Washington, D.C.: Energy Information Administration Clearinghouse (DOE/EIA-0119, Order no. 705), 1978), p. 3.
[e] Sources: *Coal—Bituminous and Lignite in 1976* (see footnote *d*), p. 58; and *Coal—Pennsylvania Anthracite in 1976* (see footnote *d*), p. 25. Calculated as percentage of total shipments by region of destination. Remainder of total shipments, 9.6 percent, allocated to export, dock storage, and rail and lake vessel fuel.

It seems almost certain that the 1985 target of the 1977 National Energy Plan—1.2 billion tons of coal per year—will not be met. Electric utilities will probably continue to be the major consumers, with industrial consumption of secondary importance as a source of demand. How much the demand for input energy in these two sectors will grow depends generally on the level of economic growth, the impact of conservation efforts, the pace of substitution of electricity for other forms of energy, and the amount of nuclear generating capacity that is installed.

Given recent trends, a plausible assumption about electricity production is that it will grow from 2.2 trillion kilowatt-hours (kWh) in 1978 at a modest 4 percent annually to 2.9 trillion kWh in 1985. If recent utility plans for a roughly 50-50 split in the distribution of new capacity between coal-fired and nuclear units holds true, the additional 170 million tons of demand for utility coal in 1985 will be about 35 percent higher than the 1978 level, requiring a 25 percent boost in total output. Few power plants now burning oil or gas will convert to coal, except for some that were originally designed to burn coal.

Large increases in industrial coal demand for steam and process heat are not likely through 1985. The coal demand of this sector has shrunk more than 10 percent since 1972, and most existing plants face many difficulties in attempting to convert to coal and in burning it.

Beyond 1985, the possibilities of course become more varied. To begin to grasp their range, we will again assume that through 1985 electricity output rises at a 4 percent annual rate, with the total increase shared equally by coal and nuclear fission. Thereafter, we assume a 3 percent annual growth rate for electricity output. If we assume that 1 trillion kWh will be supplied by oil, gas, hydroelectric, wind, solar, geothermal, and tidal generating plants by 2010, then 5 trillion of the 6 trillion kWh being generated in that year will come from nuclear and coal-fired plants (with the distribution of new capacity again split equally between the two plant types).

Nuclear capacity contributed 0.275 trillion kWh in 1978, which would rise to about 0.6 trillion kWh for 1985, and 1.7 in 2010, and require that 3.3 trillion kWh be produced by coal in 2010. Based on utility coal conversion efficiency measured in 1978, this would require 1640 million tons of coal, 3.4 times as much as utilities consumed in 1978 and 2.5 times total 1978 domestic coal production. Any shortfall in nuclear capacity would of course increase this demand.

By 2010, however, utilities will not be the only significant source of demand for coal. Advanced coal-based generation technologies are now undergoing development. Atmospheric fluidized-bed combustion may make coal use more economic for industry, and fuel cells using gas from coal may be economic in decentralized load-following applications. The successful development of technologies to extract gases and liquids from coal could prove to be the biggest source of demand for coal as an indirect industrial substitute for oil and natural gas. It would also yield fuels attractive to utilities as alternatives to the direct use of coal.

ELECTRICITY GENERATION FROM COAL

The following discussion uses utility generating stations as a model for large-scale coal combustion in general. Utility boilers fired by coal are quite similar to industrial ones, though generally larger. Air quality regulations apply in essentially the same ways to both. The Powerplant and Industrial Fuel Use Act of 1978, when its regulatory provisions are formalized, will impose similar, though probably somewhat less restrictive, constraints on industrial oil and gas burning as on consumption of these fuels by electric utilities. The industrial nonenergy uses of coal—as a chemical feedstock and in metallurgy—will be discussed later.

Most large utility power plants use pulverized coal, burned in a suspension of fine particles to generate steam to drive the electric generator's turbine. A 1000-MWe coal-fired power plant consumes about 2.5 million tons of coal each year and produces approximately 6.5 billion kWh of electricity. With flue gas desulfurization equipment it occupies an area of about 600–700 acres, which houses the plant and allows for disposal of the various solid wastes produced in the course of 35–40 years of useful life. From site selection to power production, it takes about 7–8 years to obtain regulatory clearances, build the plant, and make final operating adjustments.

Before the advent of sulfur dioxide controls, coal-fired base-load power plants could convert about 40 percent of the coal's chemical energy to electricity; the rest was rejected to the environment as heat. The power demands of sulfur dioxide scrubbers and other control equipment that will be required on new plants will reduce coal-to-electricity efficiencies to about 38 percent. In addition, the EPA-proposed increases in regulatory requirements for emission control and waste disposal are estimated to add over $150 per kilowatt (electric) (kWe) of generating capacity to the capital costs of large plants, which without such measures would cost $550/kWe (1976 dollars). The pollution control costs are dominated by the costs of sulfur dioxide scrubbers and waste disposal; a great deal of development work, with the aim of increasing efficiencies and decreasing costs of power plants that are acceptably clean and efficient, is being carried on. Advanced methods of coal combustion will be discussed later.

Air Pollution Regulation and Control

The policy problem of greatest significance to the near future of coal consumption is undoubtedly the question of air pollution and its control. Under the authority of the Clean Air Act of 1970 and its 1977 amendments, the federal government directly and indirectly regulates the siting of fossil-fueled utility and industrial boilers, as well as their permissible emissions of particulates, sulfur oxides, and nitrogen oxides. Other considerations may arise under the Clean Water Act, the National Environmental Policy Act, the National Historic Preservation Act, the Endangered Species Act, and other regulations, but it is the provisions of the Clean Air Act that apply most consistently and with most force.

The Clean Air Act sets national ambient air quality standards (NAAQS) for atmospheric concentrations of sulfur dioxide, nitrogen dioxide, and particulate matter. These are enforced in a variety of ways to control emissions from different stationary and mobile pollution sources. The requirements are different not only for different types of sources, but also for different regions. The act divides the nation into 247 air quality control regions, and charges the states with developing strategies for meeting the applicable ambient pollutant standards. In regions that are already in violation of the ambient standards, any new pollution source must meet the lowest attainable emission rate and at the same time secure from other sources in the region reductions in emissions that more than offset its own. In areas where the air is cleaner than required by the NAAQS (so-called "prevention of significant deterioration" areas), emissions are regulated to prevent significant increases in ambient concentrations of particulate matter and sulfur dioxide. Regulations covering certain other pollutants (including nitrogen dioxide and hydrocarbons) will be promulgated in late 1979.

In addition to the standards regulating ambient air quality, and thus the siting of plants, the Environmental Protection Agency promulgates and enforces regulations limiting the emissions of individual plants. Under the authority of the 1977 amendments to the Clean Air Act, EPA proposed new, tighter standards governing the sulfur dioxide, nitrogen dioxide, and particulate emissions of all new or substantially modified fossil-fueled steam power generating units that consume more than 250 million Btu of fuel energy per hour.[16] This is equivalent to about 10 tons of coal per hour—enough to fuel a 10-MWe unit. The standards apply, regardless of the ambient air quality, to plants whose construction or alteration begins after September 18, 1978.

The standards, made final in May 1979, set ceilings on the emissions of the three above-mentioned pollutants, and in the case of sulfur dioxide impose the further requirement that "potential" emissions (those that would prevail if completely uncontrolled) be reduced by certain percentages, depending on the sulfur content of the coal. The percentage reduction requirement in effect means that all regulated coal-fired power plants must install flue-gas scrubbers to remove sulfur. The use of low-sulfur coal thus is no longer acceptable as the sole means of reducing sulfur emissions at a power plant; this is expected to reduce the market for low-sulfur western coal in the Midwest and East, but the magnitude of the reduction and its precise effect on western coal production are very uncertain.

The new standards, and for comparison the previous ones, are briefly as follows.

- For sulfur dioxide, the previous standard simply limited emissions to 1.2 lb per million Btu of fuel consumed. The new standard retains this upper limit, but imposes a percentage reduction requirement of 70–90 percent, depending on the sulfur content of the coal.
- For particulates, the old emission limit was 0.1 lb per million Btu. The new standard limits particulate emissions to 0.03 lb per million Btu, and in addition limits the opacity of exiting flue gas.
- For nitrogen dioxide, the previous standard limited emissions to 0.7 lb per million Btu. The new standard allows emissions of between 0.5 and 0.8 lb per million Btu, depending on the type of coal burned.

Discussions to follow describe these standards and their implications in more detail.

A recent econometric study[17] suggests that, while the tightened emission standards will raise the cost of coal-generated electricity by about 15–20 percent at power plants that install sulfur oxide scrubbers, the overall price of electricity would not rise by more than about 4 percent even by the year 2000, given a number of countervailing assumptions. For example, utilities with flue-gas desulfurization equipment could switch to cheaper, higher-sulfur coal; other fuels, including nuclear, would be available and increasingly competitive; and some plants built before the standards took effect would still be in operation.

Although nuclear generation capacity will be somewhat cheaper than coal-fired capacity in many locations, it will be installed at a rate limited by regulatory and associated investment risk considerations, with coal making up the balance of the installed generating capacity, regardless of emission standards. Up to at least the mid-1990s no electrical energy source other than these two will make appreciable contributions to the nation's new generating capacity. Therefore, according to this econometric study, the emission standards will have little effect on total coal demand. Of course, these results are sensitive to variations in electricity growth rates and the future role of nuclear power. Shortfalls in nuclear projections would tend to raise the average electricity price by requiring more new coal-fired plants subject to the new emission standards.

Given the assumptions of the model—that coal and nuclear are the only options for new generating capacity and that the availability of nuclear power will be limited by noneconomic factors—information on the environmental and health effects of coal emissions will have enormous economic value, determining whether there will be investments of many billions of dollars for emission control equipment. The range of scientific uncertainty and controversy over the impacts of emissions is much wider than for the much more widely publicized case of low-level radiation in connection with nuclear power (chapter 9). Yet, while the potential costs to the economy, and even the costs in additional energy and water resources required, are hundreds of times the cost of acquiring the information that is needed to make better-informed decisions, past and present expenditures in assessing the impacts of coal combustion are minuscule compared with the corresponding efforts in the nuclear field. Unfortunately, it will be nonetheless necessary to expand the use of coal greatly in the immediate future; in fact the largest expansion may well be required in the next 10–15 years, during which we will not have an adequate base for definitive decisions. In the absence of such a base we are forced to pay the price for standards set at levels that may prove overly conservative in the light of improved knowledge.

Sulfur Oxides Under the authority of the 1977 Clean Air Act Amendments, the Environmental Protection Agency has proposed new, more stringent standards for reducing sulfur and other emissions. The standards apply to new and substantially altered (after September 18, 1978) fossil-fueled boilers burning more than 250 million Btu/hr (about 10 tons of coal, 40 barrels of residual fuel oil, or 250 thousand ft^3 of gas). For coal or other solid fuel, the standards limit sulfur dioxide emissions to a maximum of 1.2 lb per million Btu consumed. In addition, the proposed regulations would require sulfur emissions to be reduced by 70–90 percent from the uncontrolled levels, with the precise reduction determined by the effort required to reduce emissions to the threshold value of 0.6 lb per million Btu. These percentage reduction requirements could be met by any combination of flue-gas desulfurization, coal cleaning, conversion to synthetic fuels, or other means of removing sulfur before, during, or after combustion. Sulfur removed by a coal pulverizer or in bottom ash or fly ash would also be included in the calculation. The percentage reduction requirement in effect mandates the use of sulfur dioxide scrubbers on all plants to which the standards apply.

At present, there are two commercial means of removing sulfur from coal. First, a good deal of the inorganic sulfur, included as pyrites, is removed by mechanical cleaning and washing. This removes about 70 percent of the coal's inorganic sulfur (which may range from 60 percent of total sulfur in Appalachian coals to very small amounts in some western coals) and is almost entirely ineffective in removing organic sulfur. Thus, ordinary treatment of this type can remove at best about 40 percent of the total sulfur, too little to meet the proposed requirements.

The second means, upon which most operators of large boilers will have to rely in the future, is to "scrub" the combustion gases with a technique known as flue-gas desulfurization (FGD). In essence, FGD involves bringing the sulfur-oxide-laden exhaust gases into contact with some substance with which the sulfur oxides can react so that they can be removed from the gas stream. The various FGD schemes are distinguished from one another by whether a wet or dry scrubbing agent is used and whether the used scrubbing agent is thrown away or regenerated for reuse, with the captured sulfur as a separate by-product.

About 11.5 gigawatts (electric) (GWe) of coal-fired generating capacity have been fitted with FGD systems. Another 17–18 GWe of capacity now under construction will include FGD systems,and additional capacity amounting to about 27 GWe is scheduled to be in operation with scrubbers by 1985.[18] Thus, by 1985, about 56 GWe of coal-fired generating capacity (out of a total of 300 GWe) will include FGD systems.

At present, the predominant commercially available FGD process is wet scrubbing with lime or limestone. Figure 4-3 is a schematic diagram of such a system. The lime or limestone absorbent is introduced into the reaction tank along with scrubber effluent and water recycled from the dewatering section. The mixture is pumped to the top of the scrubber and sprayed down into the gas. A stream of effluent from the scrubber is drawn off and dewatered, producing a sludge that must be disposed of as solid waste. A typical unit using such a system and burning 2 percent sulfur coal will produce about 200 lb of sludge (dry weight) for each ton of coal burned. A 500-MWe plant equipped with a lime or limestone FGD system requires over its lifetime a 560-acre sludge disposal pond 40 ft deep.

These systems, in their 15 years of service in the electric utility industry, have been plagued by design and operating problems that have limited their reliability. Scaling and corrosion of pipes and valves have been especially important problems, necessitating frequent maintenance and careful chemical control. These problems are gradually being brought under control, but the devices are by no means simple to operate, and they raise the operating costs of the power plants significantly. They consume large amounts of electrical and thermal energy (thus decreasing the overall coal-to-electricity efficiency of the power plants in which they are installed

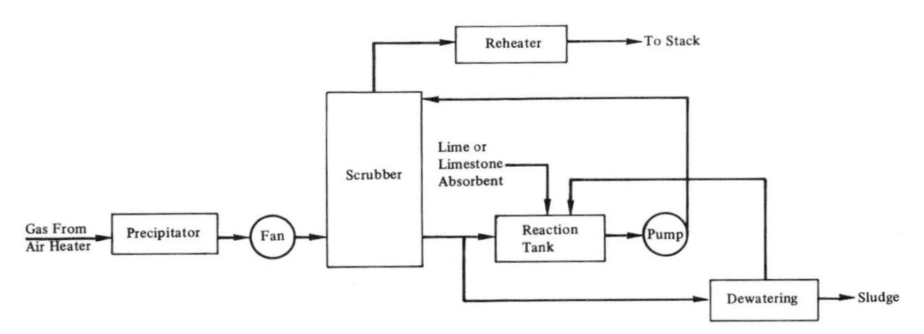

FIGURE 4-3 Generalized lime or limestone flue-gas desulfurization system for coal-fired utility plant.

by about 5 percent); and they consume large amounts of water (0.5–1.5 tons per ton of coal),[19] a feature which could raise serious siting problems, especially in the West. The costs of specific installations vary from place to place, but the capital costs typically may be $80–$120/kWe of generating capacity (1975 dollars), with annualized operating costs (capital charges plus operating and maintenance costs) of about 5.5 mills/kWh.[20] This would add perhaps 15–20 percent to the typical power plant's capital cost, and a slightly smaller percentage to annualized operating costs.

Although lime or limestone scrubbing is the most fully developed FGD technology, better processes may be available soon. The nearest to commercial adoption appears to be the regenerable double-alkali process, which uses a recyclable alkali metal hydroxide to neutralize the sulfur oxides; its main advantage is that it lessens the problem of scaling. The Wellman-Lord process, which uses a combination of sodium scrubbing and thermal stripping, is also fairly near to commercial maturity. Several other techniques show somewhat longer term promise.

Fluidized-bed combustion, now under development for use in industrial and utility boilers, holds the potential for somewhat more economical sulfur oxide control, and in addition should reduce nitrogen oxide emissions. It is described below in the section entitled "Advanced Technologies for Coal Combustion."

Nitrogen Oxides The 1977 Clean Air Act Amendments, as interpreted by the Environmental Protection Agency in the proposed New Source Performance Standards, limit nitrogen oxide (NO_x) emissions in new steam boilers consuming more than 250 million Btu of fuel per hour to 0.50–0.80 lb per million Btu, depending on the rank of coal used. These standards are based, according to the EPA,[21] on the emission levels achievable with combustion modification techniques (that is, without the use of scrubbers or chemical systems).

Nitrogen oxides formed during coal combustion draw nitrogen from both the coal and the air in which it burns. The amount formed depends on flame temperature, the amount of excess air in the flame, the length of time the combustion gases are maintained at the elevated temperature, and the rate of quenching. Higher flame temperatures and excess air foster nitrogen oxide production, as does rapid cooling.

At present, the only available means of lowering NO_x emissions in coal-fired boilers is to modify the design and operating characteristics of combustion equipment. One successful approach is to design boilers for two-stage combustion, with the first stage deficient in air and the second causing complete combustion by the injection of additional air. Burner arrangements that yield lower combustion temperatures are also effective, as are restricting air intake, recirculating flue gases, and injecting water into the fire box. Combinations of these methods are reported to lower NO_x emissions from larger utility boilers by 40–50 percent.[22]

Atmospheric fluidized-bed combustion, because of its lower combustion temperatures, will produce less than half the NO_x per ton of coal produced by conventional combustion. Pressurized fluidized-bed combustors should do even better.

Particulates The new EPA proposed standards for utility boilers would limit overall emissions of particulate matter to 0.03 lb per million Btu and at the same time set standards for the opacity of existing flue gases. They would require that particulate emissions be reduced by over 99 percent of the uncontrolled emissions. These standards are based on the use of "well designed and operated" baghouses or electrostatic precipitators.[23]

There are four types of particulate control systems—mechanical collectors, electrostatic precipitators, wet scrubbers, and fabric filter baghouses. They vary greatly in cost and effectiveness; sometimes two types are used in series, with a cheap but relatively inefficient system serving as a first stage to lower the load on the more efficient and costly second stage. All methods of control are more effective in capturing the larger particles; it is probably the smaller, respirable particles that pose the greater danger to health, however, as explained in chapter 9.

Mechanical collectors use gravity, inertia, or centrifugal forces to separate the particles from the gas. Some, for example, consist simply of enlarged chambers in the gas stream, which slow the gas flow and allow the heavier particles to settle. Others depend on induced centrifugal swirling motions in the gases; the particles in these devices are moved by centrifugal force to the outer walls and drop to the bottom of the chamber, where they can be removed. The units are simple and relatively inexpensive to install and maintain. Because they depend on the masses of the particles, they work best on the larger particles. Because of their low cost and relative ineffectiveness against small particles, they are often used as first-stage cleaners in connection with more efficient (and more expensive) devices.

Electrostatic precipitators consist of high-voltage discharge electrodes and grounded collection plates, between which the flue gases pass. Particles are charged by ions emitted by the discharge electrode and move toward the collection plate, where they can be collected and removed. The devices can remove as much as 99.9 percent (by weight) of the particles in the flue gas, but they too are most effective on larger particles. Capital costs of such systems for large boilers burning medium- or high-sulfur coal are about $25/kWe of generating capacity. With low-sulfur fuel, they are more expensive, because the particles produced in burning such coal tend to be electrically more resistive. To compensate, the precipitators can be enlarged, they can be placed in a hotter part of the flue gas stream (where the high temperatures reduce the particles' resistivity), or the flue gas can be treated with some substance that reduces resistivity. The third of these options may, however, add some pollutants to the gas, and it is generally used only as a last resort.

Wet scrubbers use water to wash solid particles out of the gas stream. These devices are rather expensive to install and operate and are not widely used for particulate control, though some installations exist.

Fabric filter baghouses are the most effective of all particulate control methods against small particles, but the pressure drop involved in forcing the gas through the necessary fine filters increases operating costs. The devices have been used in industrial particulate control for many years, but the high temperatures and corrosive chemicals in coal combustion gases, among other problems, have limited their use with utility boilers. Improved, heat- and chemical-resistant filters have been developed recently, making this option more attractive to utilities. In view of the importance of small, respirable particles in the health effects of coal combustion, it is likely that filter devices will see increased use in the future, and they warrant vigorous development.

Advanced Technologies for Coal Combustion

A number of new approaches to electricity generation, aimed at improved efficiency, lower costs, or reduced environmental impacts, are under development. None of those about to be described is yet available for commercial use, but all show sufficient promise to support important research and development efforts.

Fluidized-Bed Combustion The technique known as fluidized-bed combustion is a promising alternative to scrubbers for meeting sulfur oxide emission standards. It involves burning the coal in a mixture with limestone (to absorb sulfur), suspended turbulently in the stream of combustion air rising from beneath the bed. This method of combustion allows the coal to be burned at lower temperatures than are normal in today's pulverized coal burners at 1500°F–1600°F rather than about 3000°F. The lower combustion temperatures substantially reduce emissions of nitrogen oxides. The process is under development now, particularly for small units; the current development goals are high efficiency, durability, low emissions of nitrogen and sulfur oxides, and reliability. These seem attainable in the next few years, and it is conceivable that fluidized-bed combustion systems could be in use for industrial boilers in little more than a decade. Their use in large utility boilers will come later.

There are two approaches to fluidized-bed combustion, atmospheric and pressurized. The first, combustion at atmospheric pressure, is the simpler in design and operation. It is considered a first step toward more economical use of high-sulfur and slagging coals, and because of its relative simplicity it would be especially suitable for small-scale use in commercial and industrial boilers and cogeneration systems. It is on the verge of wide commercialization now; the first few vendors are appearing.

The pressurized (perhaps 6–16 atm) version of this technology is much more complicated and much less well developed. It promises, however, higher efficiencies and lower sulfur and nitrogen oxide emissions than atmospheric fluidized-bed combustion. It holds high promise for larger scale use, as in utility power generation.

Combined-Cycle Generation The combined-cycle approach to power generation holds a good deal of promise in the relatively near term as a means of using coal efficiently and cleanly. The combined-cycle configuration—using the exhaust heat of a gas turbine first stage to raise steam for a steam turbine second stage—is inherently the most efficient power cycle available. This makes it potentially feasible to extract power efficiently and economically from coal even through the intermediary of synthetic fuels, with their inherent conversion losses.

At present there are about 15,000 MWe of combined-cycle capacity in the Western world; the average unit capacity is about 150 MWe.[24] Most of this capacity has been installed in the last 10 years. Almost all the plants are fueled by petroleum distillate or natural gas, though the use of heavier fuels is becoming more common as the ability of gas-turbine components to withstand fuel impurities improves.

The combined-cycle configuration provides a likely application for pressurized fluidized-bed combustion systems. The hot combustion gas from the combustor could be cleaned of most particulates and expanded through a gas turbine, the exhaust from which could raise steam for a steam turbine. The high efficiency of the pressurized combustor would allow overall efficiencies somewhat higher than now achievable in even the best conventional plants. A 13-MWe demonstration of this approach, funded by the Department of Energy, is to begin operation in the early 1980s.[25] The fluidized-bed approach could be modified if the problem of particulate carry-over to the turbine proves serious. Combustion gases, instead of being sent directly to the turbine, could be sent instead to a high-temperature heat exchanger using an inert working fluid such as helium, which in turn could heat a closed-cycle gas turbine first stage. This would protect the gas turbine blades but entail a loss of efficiency because of the temperature drop in the heat exchanger.

Another strong possibility is the so-called gasification-combined-cycle approach. In this configuration, a medium-Btu gasifier would be integrated with a combined-cycle power plant, essentially as a means of precombustion emission control. Eventually, coal-to-electricity efficiencies of up to 45 percent (far above the 35 percent typical of standard scrubber-equipped steam turbines fed by pulverized coal) should be attainable, though this will require a good deal of work, especially in developing materials and cooling methods for turbines capable of operation at inlet temperatures well over the 2000°F now possible. Refinement of the gasification system will also be necessary. Such highly efficient systems are unlikely to come into wide use before the late 1990s, but they could be very important for industrial cogeneration, as well as utility power generation, in the decade after 2000.

The nearer term demonstration of gasification-combined-cycle plants is also of some interest, however. The first integrated gasification-combined-cycle plant in this country is being designed as a demonstration funded by Texaco, Southern California Edison, and the Electric Power Research Institute. Construction is intended to begin in 1981, and operation in 1984. The unit as planned will have a capacity of 110 MWe, with a 75-MWe gas turbine and a 35-MWe steam turbine, and will consume 1000 tons of coal per day.[26] The turbine components will be essentially standard commercial units, and the gasifier will be a scaled-up version of a second-generation, medium-Btu Texaco gasifier that has been operated in Germany for about a year at 150 tons/day. The main problems at this early stage in the project are integrating the gasifier fully with the generating components so that gas can be produced as needed during load fluctuations; improving the operating characteristics of the gasifier's waste-heat boiler; and modifying the combustion process to control nitrogen oxide formation.

Sulfur oxide and particulate control in gasification-combined-cycle plants should be relatively simple matters of scrubbing the gas before combustion (far cheaper than scrubbing the emissions of conventional power plants). This is one of the potential attractions of the concept: Even tightening the current sulfur emission standards by an order of magnitude is estimated to increase generating costs in such a system by only 7–8 percent.[27]

Fuel Cells Fuel cells convert fuel energy to electricity electrochemically rather than thermally. They are essentially nonpolluting, practically noiseless, and potentially very efficient. Furthermore, they are modular in construction and can be installed in units ranging in capacity from a kilowatt or so to several megawatts. They could be used in small installations distributed near load centers, since siting problems should be minimal even in densely populated urban areas. A 4.5-MWe demonstration will take place in New York City in the next few years.

In concept and operation, a fuel cell is very simple. It is essentially an electrolyte between two electrodes. A hydrogen-rich fuel is exposed at one electrode and an oxygen-rich gas (such as air) at the other. The hydrogen at one electrode splits into hydrogen ions and electrons, which move separately toward the oxygen, with which they combine as water. The hydrogen ions move through the electrolyte, and the electrons through an external circuit as electrical current.

Current fuel cell technology uses phosphoric acid as the electrolyte; however, carbon monoxide in the fuel would interfere with cell performance, so that relatively cheap low-Btu gas from coal is not a suitable fuel. Second-generation fuel cells will use a molten carbonate electrolyte, which can tolerate carbon monoxide; they will be more efficient than the first-generation fuel cells (about 45 percent, as compared to 40 percent), and are expected to produce power at lower cost.

Fuel cells can maintain this efficiency over a wide range of loads, unlike conventional power generation units, which are markedly less efficient at part power than at their rated capacities. Fuel cells have nearly constant efficiency from 25–100 percent of rated power output, and can respond within seconds to large changes in load. Pollutant emissions (mainly from the fuel processing section) would be an order of magnitude below the levels set by current emission regulations.

The ease of siting, the modular construction, the environmental characteristics, and the efficiency and operating flexibility of this concept mean that fuel cells could be a major source of coal-derived energy by the early twenty-first century, especially for intermediate load following. It may be easy to apply fuel cells to those processes which can make effective use of the waste heat, such as industrial cogeneration and integrated utility systems to provide electricity and heat for on-site residential and commercial use.

Magnetohydrodynamic Generation Like fuel cells, magnetohydrodynamic (MHD) technology is a means of producing electricity without the need for a turbine—in this case by expanding hot, electricity-conductive gas (such as fuel combustion gases treated to conduct electricity) through a magnetic field. The interaction of the accelerated, conducting gas with an intense transverse magnetic field induces an electric field in the gas. If electrodes are present to collect the current, then electric power can be supplied to an external load. The emissions from an MHD plant would be almost free of sulfur, even if raw coal combustion gases were used, because the potassium salt used to render the gas conductive would combine with the sulfur, which would be separated out as the potassium was recycled. Efficiencies of as much as 50 percent, from coal to electricity, could be achieved using raw coal combustion gases.[28] Exhaust gases are at high temperatures, and their heat could be captured and used in a steam turbine bottoming cycle. MHD as a commercial technology for using coal to generate electricity is more than 20 years away, and will make no major contribution to energy supply in the period with which this study is concerned.

Energy Storage Electric utilities for years have used energy storage to derive benefit from off-peak power generated by base-load plants. It is obvious that wide adoption of any suitably efficient energy storage procedure would allow large coal-fired and nuclear base-load stations to serve increasing amounts of the load; and that this would provide further scope for the use of coal in utilities, while minimizing the need for additional generating capacity. The savings in oil and gas, which now fire most of the intermediate and peaking generators, could be very great.

Except in utilities well endowed with hydroelectric dams, the only method of bulk power storage now used—and the most economical source of peaking power in many situations—is pumped hydroelectric storage. This technique entails pumping water uphill with off-peak power, then running it downhill through turbines to help meet intermediate and peak loads. This system is economical and efficient, but takes up large land areas (mainly for the reservoirs); in many places sites are unavailable, though systems with their lower reservoirs underground have been suggested. The storage alternative most nearly available as a supplement is to compress air in underground caverns, then extract energy by running the air through turbines. If fuel is fed into the air stream before it enters the turbine, a combustion turbine can be used and the result is very efficient and economical peaking power. The use of medium-Btu gas from coal would obviate the need for additional oil and gas. Sites for such plants are less limited by topography than those for pumped hydroelectric installations. A 290-MWe compressed-air storage plant burning natural gas, the first of its kind, is being used in Huntdorf, Federal Republic of Germany, for peaking. It has been in operation only since November 1978, but so far has proved very economical as a source of peak power.[29]

A number of other approaches, including batteries, thermal storage, flywheels, superconducting magnets, and hydrogen electrolyzed from water with off-peak power, are being developed.[30] None is sufficiently advanced to assume commercial service over the coming 10 years.

Cogeneration The cogeneration of electricity and heat is very attractive as a fuel conservation measure, as explained in chapter 2. The development of small- and medium-sized coal-fired units, with suitable environmental characteristics, would allow more flexibility in matching the demands for steam and electricity; it would also make coal increasingly suitable as a replacement for oil and gas in cogeneration projects or decentralized integrated utility systems for apartment or office complexes, hospitals, and the like. There would be some difficulties in fuel delivery and ash removal, but this option should be studied carefully.

SYNTHETIC GAS FROM COAL

Coal gasification has been a commercial operation for more than 150 years. It was carried on in many cities in the United States until the years after World War II, when cheap natural gas began to be transported through the long-distance pipelines built to carry petroleum for the war effort. By the mid-1950s the domestic industry had virtually disappeared. Europeans continued to gasify coal and substantially improved the gasification technology until the 1950s, when Middle Eastern oil and Dutch natural gas began to displace the European coal-derived gas industry; very few new coal-to-gas plants have been constructed.

None of the gasification technologies currently available for commercial use produces gas with a heating value equivalent to that of natural gas, which is essentially all methane. The gases from the conversion processes are mixtures of carbon monoxide and hydrogen, with small amounts of methane distilled from the coal; these "low-Btu" gases have heating values of 150–300 Btu/ft^3, as compared with natural gas at 1000 Btu/ft^3. Some changes in process conditions (for example, substituting oxygen for air during the partial combustion that produces the gas) can increase the heating value to about 500 Btu/ft^3; gas of 300–500 Btu/ft^3 is known by convention as "medium-Btu gas." Medium-Btu gas, however, can be converted to high-Btu gas, fully equivalent to natural gas, though no commercial installations are in use.

Virtually all coal gasification processes entail similar steps. First, the coal is crushed. (If coking coal is to be used in certain gasification processes, it is necessary to pretreat it after crushing to destroy its fusing properties, generally by heating it in a mixture of steam and air or oxygen.) The coal is fed to a gasifier where it is heated in the presence of air or oxygen. Depending on the type of gas to be made, the resulting low- or medium-Btu gas may be treated with steam to adjust the ratio of carbon monoxide and hydrogen (the so-called shift reaction), after being cleaned of impurities. This is followed by a second cleaning to remove carbon dioxide and hydrogen sulfide. If high-Btu gas is desired, the clean medium-Btu gas is passed through a methanation step, in which the raw gas is reacted over a catalyst to produce essentially pure methane.

Low- and Medium-Btu Gas

The low-Btu gas produced by an air gasification system has a heating value too low to be economically transported or used in equipment designed for natural gas. It could, however, be used to power on-site industrial boilers and electric power generating equipment. As explained earlier, combined-cycle generating plants fueled by coal gasifiers appear to offer the potential of efficient and relatively low-cost electric power generation as well as environmental benefits over direct coal firing.

When oxygen is used in the gasifier instead of air, the gas produced can have a heating value 30–50 percent that of natural gas, essentially because it is not diluted with nitrogen from the air. Costs are increased by the need for an oxygen-generating plant, but the higher quality medium-Btu gas requires less cleaning than low-Btu gas and can in addition be transported economically for distances of 150 miles or more. Industrial gas users who must convert from natural gas may find medium-Btu gas preferable, especially if their access to petroleum is restricted. Medium-Btu gas may also serve as a fuel for combined-cycle generating plants, as explained earlier.

The commercially available processes for low- and medium-Btu gas production were all developed in Europe. The predominant processes are the Lurgi, the Winkler, and the Koppers-Totzek, all of which use oxygen and steam but differ in the method of bringing these reactants into contact with the coal. A number of improved processes are undergoing development here and abroad.

High-Btu Gas

High-Btu gas, equivalent in heating value to natural gas at 1000 Btu/ft^3, could serve as a direct replacement for natural gas in existing pipelines and distribution systems. The projected price (roughly $4–$5 per million Btu (1977 dollars)) is about half that of electricity (though much more than that of natural gas). The choice between high-Btu gas and electricity rests on the final cost to the consumer per useful Btu, that is, the cost of the delivered gas or electricity times a factor to allow for the efficiency of end-use. In residential services the advantage of gas for cooking is clear. For space heating, particularly with heat pumps, and for air conditioning, the case is not clear. The alternative of using high-Btu gas is a very important one, because there are some 40 million residential gas customers, 85 percent of whom use gas for space heating. Furthermore, there is a large and efficient gas distribution system already in place, representing an investment of several billion dollars. The allocation of emphasis between gas and electricity as the ultimate products of coal use must be decided carefully for each region, considering the value of using gas distribution systems where they already exist.

The most nearly available high-Btu coal gasification technique is the methanation of medium-Btu gas. The Lurgi process has been tested as a source of the raw gas, and some improvements in this process have been made in the past few years with the aim of better suiting it as a component of a high-Btu gas system. There is need, however, for more economical gasification processes than the "first-generation" Lurgi technique. Thus, a number of second- and third-generation processes are under development, with some of the more advanced projects under combined industry-government funding.

Of the second-generation processes now under investigation, a number have been subjected to pilot plant tests over the past few years. Among the prime objectives is to produce raw gases of high methane content and thus to reduce the need for oxygen in the gasifier. Among the federal development projects are the HYGAS process, which has been operated in a 3-ton/hr pilot plant by the Institute of Gas Technology, and the two-stage Bi-Gas process, which can use all types of coal and is being tested in a 120-ton/day pilot plant at Homer City, Pennsylvania. Private industry is investigating many other approaches in smaller scale facilities.

Third-generation processes promising much improved thermal efficiencies and lower costs are also being investigated through feasibility studies and very small experimental reactor tests. The two such processes emphasized in the federal research and development program are hydrogasification, which produces high-Btu gas by direct reaction of coal with hydrogen, and catalytic gasification, which eliminates the shift reaction and methanation steps entirely.[31]

Underground Coal Gasification

Production of low- or medium-Btu gas from coal in situ, by injecting air or oxygen into coal seams, has a number of attractive features. First, more than 90 percent of the nation's coal resource is inaccessible to mining; some of this could be gasified in place. Of particular interest in the federal research and development program are "steeply dipping beds," or those that slope more than 35 degrees from the horizontal; such beds in the United States are estimated to contain more than 100 billion tons of coal,[32] and coal in these formations may be easier to gasify in situ than that in conventional beds.

Underground coal gasification has several potential health, safety, and environmental advantages over conventional mining followed by gasification. It requires less land disruption, uses less water, brings less solid waste to the surface, and probably creates less air pollution. The most important environmental uncertainties are the potential for contamination of groundwater, and in some cases the hazard of subsidence of the surface over gasified seams. This kind of development must be accompanied by intensive environmental research and monitoring.

Several different methods are being investigated. The most advanced is the linked vertical well method, used to produce low-Btu gas from western coal by air injection. Work on this process began in 1972 and several pilot tests have been quite successful. The Department of Energy estimates that commercialization of the process could begin by 1987 if the product is competitive in price.[33] Experiments have also been performed in the production of medium-Btu gas by oxygen injection.

The gas could be used for electricity generation at the gasification site, for the local generation of process heat (for which medium-Btu gas would be especially well suited because it can be economically transported moderate distances), as a raw material for chemical feedstocks, or for conversion to high-Btu gas. One estimate suggests that in situ gasification could eventually provide as much as 1.5 quads of energy in the region surrounding the Rocky Mountains.[34]

Product Costs

After a comparison of many synthetic gas cost estimates, one authority calculated that a Lurgi high-Btu gas plant yielding 10.4 million standard cubic feet per hour could operate at a cost (1977 dollars) of roughly $4.50 per million Btu.[35] However, all such estimates are highly speculative and varied. There is as yet no domestic industry and the experience in Europe is not much help—the facilities there were built long ago and the processes are not necessarily those that will be used in this country. Capital costs will account for half to two thirds of the product price, and estimates are therefore quite sensitive to assumptions about construction costs and methods of financing. Inflation in heavy construction, for example, has been rapid over the past 10 years, and some estimators do not state explicitly whether they use constant or current dollars. Construction costs also vary markedly from region to region. The debt-equity ratio in financing is another source of confusion, along with the assumed rate of return to investors. The costs of coal and water, as well as emission and waste control, are also important variables.

All in all, firm estimates of production costs must await the construction and operation of the first few plants. As the supplier industries develop an infrastructure, experience is gained, and technical refinements are made, costs should fall, as they have in other similar developing industries.

Research and Development Priorities

Economic Comparison of Gasification Options Pilot plants already constructed should be operated to supply data complete enough to permit capital, operating, and product costs for all types of commercial plants to be estimated and compared. The operational data so developed should then be examined to assess the best combinations of gasification and purification processes and other promising economic possibilities for the following applications.

1. Use in combined cycle electric power operations, both utility and industrial.
2. Expanded use in metallurgical industries, particularly for direct reduction of iron ore.
3. Use as feed for petrochemical industries, particularly for alcohol and fertilizers.
4. Expanded use in industry for process heat and process steam raising.
5. Use in integrated utility systems to provide cogenerated electric power and heat for apartment complexes, hospitals, and the like.
6. Domestic use of high-Btu gas as a replacement for natural gas.

Technological Assessment and Development In parallel with the economic assessment of various processes, a major program of research and development is needed to accomplish the following tasks.

1. Using existing basic gasification technology, build and operate

commercial pioneer projects to obtain data about costs and operational problems in U.S. commercial markets.

2. Continue development of alternative in situ, high pressure, and catalytic processes.

3. Develop an improved process for gasification as a basic step required for indirect liquefaction for both fuel and chemicals synthesis.

4. Explore the use of by-product pyrolysis liquids since the value of this material is a major factor in determining the cost of the primary product gas.

5. Develop long-range water-use plans to help assess the value of water-conserving technologies.

6. Assess more completely the health and environmental impacts of gasification plants including ash disposal, handling and use of pyrolysis products, and pipeline shipment of gas containing carbon monoxide.

7. Increase knowledge of solids handling. Particular attention should be given to systems that feed coal into and remove coal from high-pressure reactors.

LIQUID FUELS FROM COAL

Many different coal liquefaction processes are under development, but there is no fully developed new technology suitable for large-scale commercial use. Several processes have advanced, under federal sponsorship, to the pilot plant stage, however, and at some time in the future can be expected to become commercially competitive. None will see wide commercial use before the late 1990s without large government subsidies.

The approaches to coal liquefaction fall into four categories: indirect liquefaction, pyrolysis, solvent extraction, and catalytic liquefaction.

Indirect liquefaction is the only liquefaction process in commercial use; the SASOL plants in the Union of South Africa will soon be producing about 40,000 barrels of gasoline and fuel oil per day. In this process the coal is first gasified; the resulting synthesis gas is cleaned of impurities and converted over a catalyst to any of a range of products, from gasoline to heavy oils and waxes or methanol, depending on the catalyst and process conditions. Indirect liquefaction processes have low thermal efficiencies (40–45 percent), and for this reason may not be economically suitable for use in the United States.[36] In producing substitutes for highly refined petroleum products such as gasoline, however, indirect liquefaction, which would obviate the heavy demand for hydrogen in refining the heavier liquids from direct approaches, may have significant cost advantages.

In pyrolysis, coal is heated in the absence of oxygen to obtain various liquid and gaseous products and char. Hydrogenation is accomplished either by relying on hydrogen from the coal itself, or by carrying out the process in a stream of hydrogen gas to improve the yield of liquids. Thermal efficiencies can exceed 80 percent, though the liquids are in poor quality and of insufficient value to justify the cost of further processing. Pyrolysis is not a serious candidate for further development as a source of liquid fuels.

In solvent-extraction processes the pulverized coal is mixed with hydrogenated solvents that help to dissolve it and at the same time contribute hydrogen to the hydrogenation reaction. The reaction is catalyzed to some extent by inorganic substances in the coal itself. Thermal efficiencies in this group of processes are 60–65 percent in experimental reactors. One of the more difficult technical problems is how to separate the undigested coal and ash from the liquid products.

Catalytic liquefaction processes have similarly high efficiencies and similar problems in separating solids from the liquid products. Catalyst deactivation is difficult in processes in which the catalyst actually touches the coal. This results in very large requirements for replacement catalysts and consequently rather high potential costs. Some catalytic processes therefore suspend the coal in heavy oil and pass it over a catalyst bed, thus avoiding intimate contact between coal and catalyst; the oil serves as a hydrogen donor after being hydrogenated at the catalyst.

Of the many approaches that have been evaluated, three are at present favored; they enjoy roughly equivalent stages of development. These are two solvent-extraction techniques—Gulf's Solvent Refined Coal-II (SRC-II) and the Exxon Donor Solvent (EDS) process—and one catalytic liquefaction process (the H-Coal processes of Hydrocarbon Research, Inc.). Each yields about 2.5–3.0 barrels of liquid per ton of coal. The EDS process (which can be considered fairly typical), yields 0.1 barrel of liquid petroleum gas (LPG), 1.0 barrel of naphtha, and 1.5 barrels of fuel oil (as well as 78 lb of sulfur, 12 lb of ammonia, and 223 lb of ash).[37] All of these processes are in relatively good positions to become commercial in the next decade or so. The production of methanol from coal by the indirect method is also well advanced technically.

Gulf's SRC-II process is based on work done in Germany about 50 years ago. It has benefited from a good deal of effort in this country during the past 15 years, first as a means of providing a high-grade, clean-burning solid fuel (SRC-I) for use in utility and industrial boilers, and more recently as a source of liquid fuels for a broader market. The SRC-II process is closest to commercial scale. In Tacoma, Washington, a Department of Energy facility is consuming about 30 tons of coal per day,[38] and a full-scale demonstration building on this experience might be achievable somewhat earlier than demonstrations of the other two processes. A 6000-ton/day plant is proposed for operation by 1985. The solid-product process has been more extensively examined; its further development to provide clean boiler fuel is also being supported by the utility industry. There is doubt, however, that the sulfur content of the solid fuel can be lowered economically to the levels required by the new sulfur emission regulations.

The EDS process uses an indirectly catalyzed reaction, in which a hydrogen-rich solvent stream is produced over a catalyst and then mixed with coal and additional hydrogen in a second reactor, where the solvent releases its hydrogen in reaction with coal. A 250-ton/day pilot plant is under construction in Texas[39] and could be scaled to commercial size.

The H-Coal process is an extension of a process used to desulfurize heavy petroleum oils. Its development is being fostered by the Department of Energy, the Electric Power Research Institute, and a consortium of oil companies led by Ashland Oil. A 600-ton/day plant is now being constructed in Kentucky,[40] and Ashland Oil has proposed construction of a 20,000-ton/day plant for operation in 1985.

Besides these processes, a number of others are undergoing experimental work; a few are being tested in laboratory-scale facilities, but none on the scale of the three described above.

Product Costs

The costs of products from these processes have been estimated by many different organizations at different times, using different assumptions about capital costs and financing, coal prices, production levels, specific liquefaction techniques, and products. The National Research Council's Committee on Processing and Utilization of Fossil Fuels[41] was able to make estimates "merely indicative of the general level of cost." Its report states that product costs (in January 1977 dollars) might be as little as $24 per barrel of oil equivalent (at 6 million Btu per barrel) with coal costing $0.40 per million Btu (about $8.00/ton), or as much as $41 per barrel with coal costing $1.40 per million Btu (about $28/ton). This spans most of the range of current coal costs. Obviously, though, it will be impossible to make firm cost estimates until the technologies are much closer to commercial status than they are today.

These estimates do indicate, however, that the price of oil will have to rise a great deal before synthetic products will be able to compete on the market. Arguments can, however, be made for subsidizing synthetic fuels as a means of decreasing demand for imported oil, and a number of means of applying such subsidies have been suggested. In one scheme, the federal government would require that a certain proportion of the oil to refineries be of synthetic origin, and that the higher price of synthetic crude be distributed through the price of all oil. Thus, if 10 percent of the nation's oil were of synthetic origin, and this synthetic oil were twice as costly as natural petroleum, the net increase in oil prices due to synthetics would be only 10 percent. As the synthetic industry grew, and the availability of oil diminished, the mandated proportion of synthetic crude could be gradually increased. At some point the prices of synthetic and natural oil would presumably approach parity, and further such subsidies would be unneeded. The main advantage of such a scheme is that it leaves to industry the choice of technology. There are other approaches to commercializing synthetic liquid fuel technology by guaranteeing a market or otherwise subsidizing the industry; this committee has not examined any in detail.

Refining

The solvent extraction and catalytic liquefaction processes that are now emphasized in development are well adapted to produce a variety of fuel oils as their principal products. These are generally fairly low in sulfur, but rather high in some other troublesome elements. Their generally high nitrogen contents, for example, render them rather unstable in storage and also produce nitrogen oxide emission problems. Some special refining will likely be necessary to denitrogenate the liquids; this will require improved denitrogenation catalysts. Those available now are unselective in removing nitrogen, since they merely hydrogenate the liquids to saturation before removing nitrogen. This increases the expense of hydrogen, which is a key part of the cost of producing and refining coal liquids.

Pollution Problems

The environmental and health benefits of using coal liquids as substitutes for coal in power plants and large industrial boilers are not clear. While these products can be "cleaner" in combustion than coal itself, their use may merely shift the sources of pollution from power plants to liquefaction plants. The data base for evaluating the environmental impacts of coal liquefaction plants is, however, very limited.

Animal testing and occupational health records from related processes such as coking indicate that the heavier fractions of coal liquids—particularly those boiling at temperatures over 560°F—would pose a substantial threat of cancer to workers chronically exposed. This is due to the concentration of polycyclic organic matter in these fractions. (See chapter 9.) Modern industrial hygiene can probably eliminate this hazard in the liquefaction plants and the large utility and industrial boilers where these fractions would be most appropriately burned.

Research and Development Priorities

Coal-derived liquid fuels will become increasingly important over the coming decade or two, as the world oil-supply situation tightens. Because the current processes for coal liquefaction promise prices more than double the present world price of crude, lowering costs must be a prime objective of the synthetic fuel research and development program. Even a small-scale demonstration of a more economical process could have a salutary effect on the world petroleum market. Additional research and development should be conducted, including the following tasks.

1. For the more promising processes, build commercial-scale pioneer plants.
2. Continue to look for better processes (including indirect liquefaction).
3. Assess the technologies for using less (or poorer-quality) water.
4. Examine carefully the health and environmental impacts of the processes. It will be necessary to have solutions to the major impacts before large amounts of the liquids are produced.
5. Demonstrate the processes required to separate undigested solids from the liquids in direct liquefaction units.
6. Evaluate refining techniques for the raw liquids to determine product quality and costs.
7. Develop catalysts to synthesize various hydrocarbons from carbon monoxide and hydrogen for use as fuels or chemicals.
8. Continue the current broad program of fundamental research. Examples of the fundamental questions vital to coal liquefaction are the chemical structures and compositions of different coals, the effects of process variables on reaction paths and rates of hydrogen transfer with and without catalysts, the roles of coal's mineral and nitrogen content in the catalytic hydrogenation processes, and catalytic mechanisms in general.

COAL FOR INDUSTRIAL PROCESS HEAT

U.S. industry burned nearly 60 million tons of coal (about 1.5 quads) to generate steam and process heat in 1976. This amounted to nearly 10 percent of industrial primary energy use. Coal use in industry has been dropping for the past 30 years in this country, and there is no very great incentive for industry to reverse the decline. Even at oil and gas prices higher than those that prevail today, burning these fuels is almost always cheaper and more convenient to the small user than burning coal, because of the extra capital and labor costs of handling the coal and treating wastes and emissions.

Technically, however, there are significant opportunities for conversion to coal as existing oil- and gas-fired boilers, kilns, and furnaces are replaced. One recent study,[42] for example, examined the most likely prospects for industrial use of coal as a source of direct heating, and found opportunities for saving almost 1 quad of oil and gas in six energy-intensive industries—cement, lime, iron and steel, copper, glass, and ceramics—by using coal where it is environmentally and economically most acceptable. The prospects for coal use in boilers are potentially much greater than this, although the cost penalties—due largely to the capital and operating costs of pollution control and coal-handling equipment—would be significant. According to a recent report,[43] a coal-fired boiler and its ancillary equipment cost from 2 to almost 5 times as much as an equivalent oil- or gas-fired boiler. Fuel savings must be sufficient to return these costs in a timely fashion if coal use in industrial boilers is to be a good investment. At the current relative prices of coal, oil, and gas, this is not usually the case except in units consuming very large amounts of fuel.

However, the Powerplant and Industrial Fuel Use Act of 1978 will, when its regulatory provisions are formalized by the Department of Energy, impose rather strict limits on the use of oil and natural gas in new "major fuel burning installations," defined as those consuming more than 100 million Btu/hr (about 4 tons of coal) per individual unit or 250 million Btu/hr for each multiunit facility. Such installations will in general be forbidden to burn oil or gas; the rules for exemptions and certain further prohibitions of oil and gas use are still under consideration, so that the ultimate effect of the act is impossible to predict. The Department of Energy itself projects industrial noncoke coal use without the act to rise by 1990 to 5.6 quads/yr (about 235 million tons). According to the Department of Energy's projections, the most stringent of the restrictions under consideration would raise this to 7.4 quads, or about 315 million tons. Because the act depends for its effect mostly on the replacement of existing units, replacement of oil and gas by coal would presumably accelerate in the 1990s. In 1974, 90 percent of the oil and 70 pecent of the gas burned in boilers was burned in major fuel-burning installations and therefore would be potentially subject to the act.[44]

The many smaller burners scattered throughout industry will present far smaller opportunities for coal consumption, because at their rates of fuel use the capital and operating cost penalities of coal use will far outweigh the potential fuel savings. The research and development efforts aimed at small-scale coal burning (see "Advanced Technologies for Coal Combustion" above) will probably make coal use a better investment for small users by the 1990s, but it is impossible to predict their impact very precisely.

COAL AS A CHEMICAL FEEDSTOCK

The pressures on oil and natural gas prices, together with the development of coal gasification and liquefaction processes, should make coal an increasingly attractive source of petrochemical feedstocks, especially after 1985. There are often a number of fuels besides oil or natural gas that could serve as petrochemical feedstocks. The factors that determine the choice of fuel are generally its delivered price, its efficiency in the production process, and its associated capital and operating costs. Given comparatively high efficiency and low capital and operating costs, one fuel could be cheaper to use at twice the cost of another fuel per million Btu, as is roughly the case now with natural gas compared to coal in the production of synthetic ammonia.

The great upward pressure on the costs of today's two principal petrochemical feedstocks, natural gas and petroleum, should begin to force a switch to coal products within a decade. The domestic reserves and production of natural gas and oil have already sharply declined from their respective peaks, and their prices will continue to rise, even if pressure on world prices is eased by a reduction of imports. By 1985, moreover, natural gas prices will have become deregulated under the Natural Gas Policy Act of 1978 and will probably be at least equal to world oil prices. Resource depletion will not, however, be a large factor in the U.S. use of coal for the foreseeable future. And, while coal costs will be raised by regulatory restrictions on the mining and use of coal, as well as, perhaps, by poor productivity gains and increasing concentration in the industry, these increases are not expected to match or approach the real-cost increases of oil and gas.

As a result, there will likely come a time—for some applications the middle to late 1980s—when the premium paid for the use of oil and gas as simple, efficient petrochemical feedstocks will become too high.

The amount of coal potentially involved in the substitution for oil and gas feedstocks is difficult to estimate given the different rates of conversion of the various feedstocks and the varying rates of conversion of one feedstock when used in different conversion processes. In the case of ammonia, for instance, it takes 34.5 million Btu of natural gas as a source of fuel and feedstock to produce 1 ton of ammonia, while it takes 41 million Btu of coal to produce the same result. (At current rates of ammonia production—about 18 million tons/yr—it would take 0.740 quads of coal to meet the needs of the ammonia industry, or about 32 million tons—5 percent of 1978 production.)[45] While coal is in fact likely to replace oil and gas in many feedstock operations, it will not do so uniformly throughout the country. Rather, since there are large regional differences in the delivered price of coal, it will become competitive as a feedstock in some regions earlier than it will in others.

COKE PRODUCTION

In 1976, 84 million tons of bituminous coal and 400,000 tons of anthracite were carbonized into 58 million tons of coke. Almost all of the production took place in coke ovens operated by steel and smelter companies to feed their blast furnaces and foundries. About 25 percent of the cost of the coal was recovered from chemicals yielded as a by-product of the carbonization process: about 10 percent from coke oven gas not used to fuel subsequent carbonizing but used elsewhere in the plant or sold; and about 10 percent from tar and its derivatives, including naphthalene. Other chemicals produced at coke plants include ammonia, light oil, and light oil derivatives.[46]

Most of the 60 million tons of coal exported in 1976 was of coking quality, bringing total U.S. production of coking coal that year to about 140 million tons, or about one fifth of total U.S. coal production. As of 1974, there were an estimated 20 billion tons of coal in the U.S. reserve base that were of potential coking quality (i.e., of rank higher than high-volatile C bituminous, with an ash content not greater than 8 percent and sulfur content not greater than 1 percent).[47] Discounting mining losses puts recoverable coking coal reserves somewhere in the neighborhood of 12 billion to 16 billion tons. More than half of these resources are in West Virginia and another fourth are in eastern Kentucky; West Virginia has practically all of the "strongly" coking low-volatile bituminous coals, of which there are roughly 2 billion recoverable tons and of which approximately 18 million tons were consumed in 1976. Coking coal supplies could begin to tighten toward the middle of the next century, depending on the future of exports, of the steel industry (the direct reduction of iron ore to steel is now seeing some commercial use), and the ability to use blends of lower-quality coals to make coke.

PRODUCTION

INTRODUCTION

If coal demand rises significantly, a concomitant rise in coal production will not be automatic, despite coal's abundance and physical accessibility. A variety of institutional issues surrounding coal mining could, unless resolved, take precedence in the 1980s over today's questions of demand constraint.

Coal output is likely to rise in the coming decades, by varying amounts, in all coal-producing regions. The result will be to intensify all currently outstanding production problems and controversies, both East and West. The Environmental Protection Agency has estimated new power plant construction and regional coal production by 1995 to total 220 new coal-fired power plants in the East and Midwest, and 130 in the West and in the West South Central region, for a total of 258 GWe.[48] Coal production would increase over 1975 production by 1.1 billion tons—65 percent from the West, 27 percent from the Midwest, and only 8 percent from Appalachia. CONAES has made no independent studies of the regional distribution of coal consumption. We use the EPA estimates for illustrative purposes only, noting that the total coal consumption used by the EPA for 1995 corresponds quite closely with CONAES study scenario III_3 (Table 11-22), but the EPA estimate of coal-fired electricity generation is about 20 percent less than this study scenario's estimate (Table 11-28).

As discussed earlier, the new EPA policy requires scrubbers on all new coal-fired power plants, regardless of the sulfur content of the coal. While this requirement will put demand for western coal significantly below the amount that would have been generated by its use as "compliance" coal, the EPA still expects some extra demand for western coal as a means of permitting less-costly light scrubbing (with experimental "dry scrubbers"). This will contribute to the rise in the amount of western coal shipped eastward, the EPA estimates, from 21 million tons in 1975 to 70 million tons in 1995. (EPA expected 100–120 million tons shipped eastward in 1995 if no scrubbing were required for low-sulfur coal.)[49]

Under the EPA estimates, western production will rise sharply to meet needs west of the Mississippi, regardless of the fate of eastern demand for low-sulfur coal. If western mining does continue to expand, western production controversies concerning federal leasing, land reclamation requirements, water rights, and the local and regional impacts of boomtown development could become at least as important as the traditional eastern ones concerning labor-management relations (although unionization of western coal mines could also become significant if production continues its massive growth there). Conversely, if there is a stalemate on these western issues, or if eastern labor and management achieve a gradual rapprochement under pressure from western competition, there could be a larger role than is now projected for mining east of the Mississippi.

LABOR FORCE

There were 237,000 coal workers in 1977, including about 213,000 miners, at least 7000 coal mine construction workers, and others who worked in coal preparation plants and in repair shops. Roughly three fourths of mining employment is in underground mines, which yield about 40 percent of the nation's coal. Of the 160,000 members of the United Mine Workers of America (UMW), 100,000 are underground miners, 63 percent of all such workers. Since the UMW-organized underground mines tend to be the largest ones, the union probably mined more than 63 percent of all deep coal. About 11,500 UMW members were employed at surface mines, representing 21 percent of all surface miners. Since the UMW-organized surface mines tend to be the smaller ones, their workers probably produced less than 21 percent of all surface coal. Overall, UMW workers probably mined about 50 percent of total output.[50]

The problems of using coal, discussed in the previous section, make it appear unlikely that rapid increases in coal demand will occur through 1985. However, to convey a sense of the impact that a rapid escalation of output would have on labor supply and labor-management relations, we will assume a near doubling of current output to 1.2 billion tons in 1985. If output reached 1.2 billion tons at the 1976 rate of labor productivity, the labor requirement would be 368,000 workers. In addition to the 130,000 employment increase over current levels, approximately 60,000 additional workers would be needed to compensate for attrition at a 2.5 percent turnover rate.[51]

An engineering labor requirement of 3.5 percent of total mining employment (it was 3.2 percent in 1976) would call for 6,200 additional mining engineers in coal production by 1985 and 900 mining engineering graduates per year to supply all mining industries. The number of bachelor and postgraduate mining engineering degrees rose nearly fourfold in the 1973–1976 period, from 160 to 625; the rate was nonetheless inadequate to meet the needs of the coal companies, which attracted engineers from other industries. These figures indicate that a rapid buildup of production would require intensive efforts by industry and government to promote engineering careers and support engineering education.[52]

As for the miners themselves, there is generally expected to be adequate supplies of labor to meet even steep increases in output, both east and west of the Mississippi. The question mark in the general area of mine labor concerns the future of labor-management relations in the East and the extent to which western mines will see increasing unionization.

Although it affected all regions, the surge of coal production and employment in the 1970s centered in western, nonunion, surface mines, thus reducing the UMW's role in the industry. The period also brought into the UMW ranks a heavy influx of young, inexperienced miners concurrent with the departure—due to retirement or disability—of 50,000 experienced miners, a demographic change of far-reaching effect. This influx of inexperienced miners is thought to have contributed directly to the rise in mine accidents per worker-hour and the decline in output per worker-hour that began in the late 1960s.

A second contributor to the productivity decline was the flare-up of traditional labor-management conflict that began in the late 1960s.[53] Per worker-day, deep-mine productivity dropped from about 16 tons in 1969 to about 8 tons in 1978. Surface mine productivity dropped, in the latter part of this period, from about 37 tons to about 25 tons in 1978 (see Figure 4-4).[54]

A third source of the productivity decline was the opening of more marginal mines in the 1970s to meet growing demand. Just as productivity was aided in the previous decade by stagnant demand and the closing of marginal mines, it was reduced by the expansion of mining, especially in the East, to more difficult seams.

The passage of the 1969 Coal Mine Health and Safety Act (amended in 1977), which set more stringent health and safety standards for mining, was a fourth source of the productivity decline. A measure of the effectiveness of the act was the 60 percent cut in the number of deaths per million hours of exposure in underground mining (and in absolute numbers from 220 in 1970 to 100 in 1977). However, an improvement in the historical rate of disabling injuries (i.e., those causing the loss of at least 1 day's work following the day of the injury) appeared only in 1974 and 1975. The 1977 rate was 7 percent higher than the 1969 rate, and total annual injuries in underground mining remain in the 9000–11,000 range. Fatalities and disabling injuries per hour in surface mines occur about half as frequently as in underground mines; moreover, the fatality rate in surface mines has fallen to about one third of the 1970 level, and the disabling injury rate has declined about 13 percent.[55]

The Office of Technology Assessment (OTA) has estimated that if output rises from about 0.7 billion tons in 1977 to 1.5 billion tons in 2000 at constant labor productivity and injury rates, total annual fatalities would rise from 139 to 259 and disabling injuries would grow from 15,000 to 29,000. The estimate shows almost the entire increase in each category would occur in underground mines. To reduce injury rates further, the OTA urged that improved engineering and safety standards be extended to more aspects of mining machinery, that mine inspections be made more frequently, and that labor, industry, and government make stronger efforts at safety training, especially at mines with high accident rates.[56]

The 1969 act also limited the amount of respirable dust to be allowed in the mines, established enforcement procedures, and granted compensation to miners totally disabled by any of the ailments with respiratory symptoms that are collectively called "black lung." From 1970 to 1977, 421,000 miners have received $5.6 billion in black lung compensation.[57]

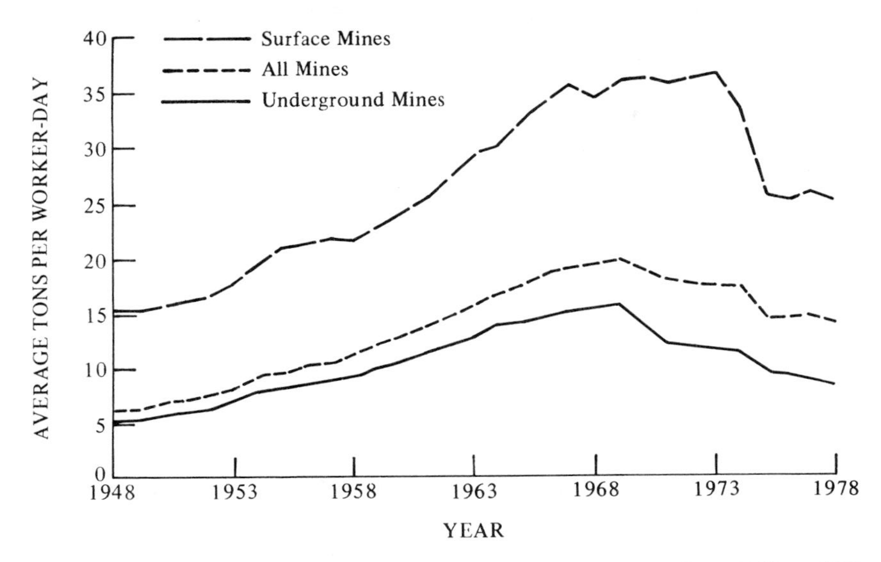

FIGURE 4-4 Productivity in bituminous coal and lignite mining from 1948 to 1978 (average tons per worker day). Source: U.S. Department of Energy, Energy Information Administration, *Annual Report to Congress, 1978,* vol. 2, *Data* (Washington, D.C.: U.S. Government Printing Office (061-000-00288-0), 1979), p. 96.

The extent to which coal mining per se is responsible for all of this disability, however, is a question that has not been precisely evaluated.

The occupational health problems relating to coal mining have not been satisfactorily analyzed, in part owing to the many factors involved, sociological as well as occupational. Although the respirable dust standard—if it can be effectively enforced—should provide very significant protection against certain conditions (the pneumoconiosis-fibrosis categories), it is not known to what extent it will protect against other conditions in the very broad black lung category, or against other conditions that could be specifically associated with working in the mines. The problem is discussed in chapter 9. The Office of Technology Assessment[58] has emphasized the potential threats to health from other dusts, noise, mine gases, and engine emissions. The whole problem requires intensive study and control that will entail the active cooperation of the workers and management.

It would appear that the broadest relevant generalization concerning the question of labor and working conditions is that the recent period of rapid growth has brought both upheaval in labor-management relations and progress in working conditions. The coming sharp increase in national reliance on coal—whenever it comes—will further force the attention of labor, government, and management on resolving the outstanding issues of health and safety, benefits, work rules, grievance procedures, and community development.

MINING TECHNOLOGY

Underground Mining

In underground mines, greater use of longwall mining and the removal of technical constraints on the use of the continuous-mining machine in room-and-pillar mining could produce a sharp improvement in both productivity and miner safety.[59] Longwall mining requires an extensive, level bed of coal, with suitable roof conditions. One edge of the coal seam is exposed underground and sheared away by a machine making successive passes across the coal face. As the wall of coal retreats, roof supports are advanced, allowing the roof to collapse in the mined-out section. Ninety percent or more of the coal can be extracted using the longwall method, a rate of recovery that can equal or surpass that of surface mining. Longwall mining is used widely in Europe and is gradually seeing greater use in the United States.

Room-and-pillar mining uses pillars of coal to help support the roof; this method, used for 95 percent of U.S. deep-mine production, leaves 45 percent or more of the coal in place and provides less certain roof control than the longwall method. In the conventional room-and-pillar method, one machine cuts away coal from the bottom edge of the face, another machine drills holes in the coal above the cutaway portion, and then either chemical explosives or compressed air is used to blast the overhanging coal. Several sections of the mine are worked simultaneously at one or another stage of this process. The continuous miner is a machine that eliminates several stages in the conventional room-and-pillar extraction method, cutting the coal and loading it on a conveyor in one operation. Its use is constrained by the need to halt it while workers extend ventilation ducts, install roof bolts, and clear away mined coal that builds up at the face. There appears to be considerable scope for further improvement of this technology, increasing both productivity and safety. Moreover, given prospective demands for growth, the financial resources for rapid development of and investment in new mining technology are likely to be much more available than in the recent past. High petroleum and natural gas prices could provide increasing leeway for investment in improved coal mining technology without eroding the competitive position of coal.

Surface Mining

Four basic surface-mining techniques are used to extract coal from beds lying as deep as 150 ft under level terrain and from outcropping coal seams along hillsides. These four methods are area (the major technique in the West and parts of the East), open pit (also used in the West for very thick beds), contour, and auger (the latter two used mainly in Appalachia). The average ratio of overburden thickness to coal thickness in 1970 surface mines was 11:1, and in some locations exceeded 30:1. Rising coal prices since then have probably pushed the ratio higher.[60]

The equipment used in most surface mining includes conventional earth moving equipment and adaptations such as the "dragline," an enormous shovel with a boom 180–375 ft long and a bucket up to 200 yd^3 in volume. One reason that a labor force is more easily recruited for surface mining than for deep mining is that the surface-mining skills are transferable from the established machine operators' trade.

Area mining involves removing the topsoil and other overburden in strips usually 1 mile long and about 100 ft wide. Topsoil and overburden are kept separate, and blasting is sometimes used to loosen both the overburden and the underlying coal, which is then loaded on trucks. Reclamation is accomplished by replacing the overburden and topsoil and revegetating.

A larger strip is excavated by open pit mining than by area mining, perhaps 1000 ft wide, with the overburden shifted within the pit to uncover the coal. This method is used in the thickest western beds.

On the hillsides of mountainous coal lands, the edge of coal beds extending horizontally through the mountain are often exposed on the slopes of hills. In contour mining, a ledge is cut along the side of the hill, with the overburden cut away to reveal the coal. The process continues inward until the volume of overburden exceeds the amount of coal to an uneconomic degree.

The use of huge augers, or drills, can extend access to contour-mined seams by boring up to 200 ft into the hillside, with the coal removed by the turning of the drill, without removing overburden.

In the absence of careful and well-enforced regulations, surface mining often has caused erosion, water pollution, and the loss of the land for any other use after mining was completed. The Surface Mining Control and Reclamation Act (SMCRA) was passed in 1977 to promote reclamation and prevent mining in areas that could not be reclaimed. The act requires states to develop licensing procedures as outlined in the act; until they do so, the federal government retains enforcement authority. To open a new surface mine in an area not designated unsuitable, a company must present enough data to allow regulators to assess the probable impact of the operation on the surrounding area's water and land resources and the likelihood of reclamation success. Underground mines are also regulated under the act to control subsidence and prevent toxic drainage. During mining, inspectors are to ensure that there is minimal disruption as defined by the permit.

Prior to the act, operators estimated their reclamation costs at $0.20–$1.00/ton mined, depending on seam thickness. Estimates of costs under the new act vary widely.[61] Small companies, say many operators, will be unable to meet filing requirements; for firms that can, lead times for opening mines can be extended for years (especially in combination with the requirements of the Mine Safety and Health Act, the Resource Recovery and Conservation Act, and the Clean Water Act), and costs will be increased sharply. The failure of states to issue and enforce regulations under the SMCRA could severely limit enforcement of the act unless federal regulatory resources were greatly increased.[62] While the new federal regulations governing mining have increased capital and operating costs, most observers do not anticipate supply restrictions to ensue. The regulations could, however, help concentrate output in those firms best able to meet the greater initial investment required by the law.

FEDERAL LEASING POLICY

In 1971 the federal government imposed a moratorium on leasing of federal lands for coal production, to gain time for an examination of comprehensive changes in its procedures. At that time, outstanding leases included 17 billion tons of recoverable reserves (about 5 percent of federally owned coal reserves). Leases on Indian lands encompassed another 3.5 billion tons. The moratorium had been imposed because the leases, for the most part, were being held but not used, leading to charges of speculation. Under the moratorium, rising coal prices have tripled western output, and federal leases contributed about 30 percent of western coal in 1977. Output on 67 active leases rose 240 percent between 1973 and 1977 to 52 million tons and coal output of Indian lands doubled in the same period to about 23 million tons.[63]

As a result of a 1977 suit by the Natural Resources Defense Council, a federal court ordered an environmental impact statement on leasing, due this year from the Department of the Interior.

While there is probably enough coal under federal lease in the West to support production goals there through 1985, it is possible that too much of the leased portion is in uneconomical parcels or on restricted land to fully support production beyond that date. Restricted parcels include valley floors and prime agricultural land, locales that were disqualified for mining under the Surface Mining Control and Reclamation Act of 1977.

With as many as 5 years required to open a new surface mine, some leasing should probably be instituted soon to meet post-1985 production goals, if only to fill out present leases to sufficient size to support 20–30 years of production. Railroads are the second largest holders of western coal resources. If further leasing is undertaken, the widespread, checkerboard pattern of railroad land holdings may interfere with the assembling of "logical mining units" and require that railroads be included in the planning of a leasing program.

INDUSTRY STRUCTURE

As oil and gas assumed the major share of U.S. energy consumption, the coal market shifted from a large number of small- or medium-sized independent producers selling to many buyers, to a small number of large companies or their subsidiaries selling to a few large buyers.[64]

In 1977, the largest 15 producers yielded 40 percent of national output; two were independent, five were owned by oil companies, five by steel companies or utilities, and three by conglomerates.

In 1977, utilities mined about 14.5 percent of their own coal consumption, up from about 9 percent in 1973. Moreover, 86 percent of all coal shipped to utilities in 1976 was under long-term contracts. Long-term contracts are usually written for 2 years' duration or more, with "market reopener" clauses allowing for price adjustments as the market price for comparable coal rises or falls, and with *force majeur* clauses allowing for termination in the face of uncontrollable or unforeseen events such as strikes, new regulations, or coal seam depletion. From December 1972 to September 1978 the average price per ton of contract steam coal (for use in steam-raising boilers) rose from $8 to $22.66.[65] The average price per ton delivered in 1977 to utilities of at least 25-MWe capacity varied from $4.97 for 6600-Btu/lb, low-sulfur Texas surface coal delivered on contract in Texas to $49.56 for 13,000-Btu/lb, high-sulfur West Virginia coal sold in the spot market in Minnesota. The average coal price for these selected utilities was $20.37 or about 89 cents per million Btu, with 70 percent of it surface coal and 80 percent of it on contract.[66]

Utilities are acquiring coal mines to gain more control of the supply and price of the coal they burn. In 1977, the Federal Power Commission predicted that by 1985 utility coal consumption would rise to 770 million tons and that the amount produced by the utilities themselves ("captive coal") would triple from the 1975 level, reaching 145 million tons, or almost 19 percent of utility coal consumption. Some maintain that this vertical integration allows utilities to manipulate supplies and production costs to justify rate increases and expand profits. The utilities counter that mine ownership makes them better able to hold down the cost of electricity to their customers.[67]

Horizontal integration, the acquisition of coal mines by gas and oil companies, has proceeded faster than vertical integration and has received more attention. Fifteen oil and gas companies mined 161 million tons of coal in 1976, including 125 million tons of steam coal. This amounted to 24 percent of national production and 35 percent of all noncaptive steam coal. The principal investors in coal reserves and leases in recent years have been oil and gas companies, which now make up 6 of the top 10 reserve holders. Oil and gas companies account for about 40 percent of planned new capacity, and while they are not expected to undertake all of the planned expansion, they are expected to produce 260–360 million tons of new steam coal in 1986, bringing their share of nonmetallurgical, nonexport production to 48 percent. Seven companies account for two thirds of the capacity additions planned by oil and gas firms.

The phenomenon of horizontal integration raises charges and counterclaims similar to those in the case of utilities. Critics maintain that control of competing energy sources by a small group of firms allows self-interested manipulation of supplies and prices. Oil and gas companies maintain that their presence in the coal mining industry brings needed management expertise and capital.

By and large, coal is sold not nationally, but rather in regional markets according to type of coal. When broken down by region and type, coal becomes a concentrated industry. Although more than 3000 companies operated more than 6000 mines in 1976, the top four companies in the Midwest produced 65 percent of the coal there; in the Northern Plains, 38 percent; in the Southwest, 64 percent; and in Appalachia, 22 percent (a figure that is probably understated since northern and southern Appalachia are actually two separate markets). By 1985, estimates the Department of Energy, the top four noncaptive producers will supply 47 percent of new, noncaptive utility supply now under contract, and the top eight will supply 71 percent.[68]

The level of concentration in Montana and Wyoming in 1975 was estimated to reach 70 percent of output in those two states for the top four producers and 92 percent for the top eight. In addition, the boards of the top coal companies exhibit widespread interlocking with the boards of other coal companies and with the boards of major banks and coal consumers.

The dominance of long-term contracts, concentrated markets, and board interlocks carry with them possible implications of noncompetitive power and behavior that could be damaging to the economy. However, the detailed study required to explore these implications has not been done.

The rapid transformation of the structure of the coal industry was accompanied in the early 1970s by an even more rapid change in the industry's gross income and profitability. While output rose 14 percent from 1972 to 1976, the mine-mouth value of coal produced rose 187 percent, from $4.6 billion in 1972 to $13.2 billion in 1976.[69] From an average of less than 11 percent in the 1950–1970 period, coal companies' return on net worth rose to an estimated 30 percent in 1974 and was 21 percent in 1976.[70] The labor costs of coal dropped from 58 percent of total value received in 1950 to 20 percent in 1974; the labor cost share is estimated to lie in the 25–30 percent range today. Conglomerates do not show separate earnings for their coal operations; however, the financial statements of the two major independent coal companies (the Pittston Corporation and North American Coal) show that after 1973, despite their erratic output and declining labor productivity, the companies' total net income and net income per employee increased sharply. The net coal income of 24 large private producers rose 400 percent from $128 million in 1971 to $639 million in 1974.

It is difficult to foresee the implications of the changes in industrial structure or of the rising profitability of the coal industry. Undoubtedly concentration and horizontal integration, operating under the shelter of world oil and gas prices governed by the petroleum cartel, will lead to higher prices for coal. Moreover, producers will be much less willing to resist increased labor demands than in the past. Thus, coal prices will tend to follow closely world petroleum prices, suitably discounted for the greater cost of burning coal as compared with fluid fuels. On the positive side, continued high profitability of the industry, and capital availability from the oil and gas companies, will make possible a higher rate of introduction of new technology, and thus rapid improvements in productivity, safety, and responsible environmental behavior could ensue. It will also be easier to enforce environmental, health, and safety regulations against a concentrated industry, both because large units are easier to monitor, and because the companies will be able to pass along internalized social costs to consumers. It may be that these beneficial effects will outweigh the greater ability of large companies to resist regulation through the courts and through lobbying in the political process. Obviously, however, the coal industry is undergoing dramatic structural changes, and these are likely to accelerate in the future as the coal market inevitably expands. The situation will bear close monitoring.

TRANSPORTATION

Coal transport is dominated by railroads. Rail-only shipments account for about half (by weight) of coal haulage. Another 20–25 percent travels at least part of the way by water, and trucks carry 10–15 percent. Almost all of the rest is used at the mine. Forty percent of 1976 rail tonnage was hauled by unit trains—100-car trains that carry only coal and usually make successive round trips between one mine and one power plant to service long-term contracts. Most unit-train traffic originates in Appalachia and Illinois, although it is increasing in the West. Transportation represents about 20 percent of the delivered cost of coal hauled by rail.[71]

About 5 million tons of coal per year is carried 270 miles from Arizona to southwestern Nevada by coal slurry pipeline, perhaps the most economical coal transportation method for 300- to 400-mile distances. The average ton of coal moved by rail traveled about 300 miles. The obstacles facing greater use of pipelines to haul coal are water-use restrictions and the refusal of railroads to grant rights-of-way to a competing mode. Efforts to grant slurry pipelines right of eminent domain have failed in Congress, and the future of this transportation mode remains in doubt.

Barge transport is cheaper than rail, but its use is limited by the geography of waterways and the size of river locks. Although truck transport can be as flexible as rail, it is roughly estimated to be 5 times more costly and is also limited by the deteriorating condition of mine roads, especially in Appalachia.[72] In the absence of slurry pipeline development, railroads will continue to offer the best combination of cost and flexibility.

Railroad capacity should be able to keep pace with long-term growth of coal output, since the lead time necessary for acquiring new rolling stock is no longer than that required for starting a large mine or power plant. Short-term surges of output, however, could exceed the railroads' short-run ability to add to their rolling stock.

The models used in one transportation study[73] showed that Appalachian coal production is more sensitive to coal mining costs than to transportation costs, while western coal production exhibited the opposite tendency—it would be stimulated more by holding transport costs down than by holding western mining costs down.

The physical conditions of many Appalachian roads and the financial condition of some major eastern coal-hauling railroads pose problems for greatly expanded coal transport in the East. The problems, however, would be greatly reduced if the expansion were gradual.*

Given the right combination of siting, cost, and federal regulatory conditions, a utility could find a financial advantage in locating a new power plant near a mine site and transmitting power to a distant load center over high-voltage lines. It has also been argued, however, that precisely because of the problems associated with their use, coal-fired

*See statement 4-1, by H. Brooks, Appendix A.

power plants should be located in the vicinities of those who use the power in order to allocate costs and benefits consistently.[74]

WATER SUPPLY AS A LIMITING FACTOR

Water is potentially a limiting factor in any plan to produce and use more coal on a large scale.[75] Plans to triple or quadruple coal production and use during the transitional period of the next 20 years must include precise projections of how water requirements will be met. The problem is recognized at present, although a consensus has not been achieved. Much will hinge on what can be done in arid regions, but water concerns elsewhere are also of specific importance.

Potentially most troublesome will be providing water needed in using coal (primarily to produce electricity and synthetic fuels), rather than in mining it. To service a 35- to 45-quad increment will be difficult, even with optimal planning at both regional and local levels. The ultimate capacity of the coterminous 48 states to supply water, and the fraction of water supply that could be devoted to power supply systems, have not been estimated, but should be for national energy planning. Realistic analysis requires the development of two tools: (1) a national water-data bank and (2) practical indices to gauge water-supply potential regionally and locally, taking into account possible environmental damage resulting from excessive withdrawal.

PRECIPITATION, RUNOFF, AND CONSUMPTION

The total annual precipitation for the coterminous 48 states averages about 4.5 billion acre-ft, of which 30 percent is runoff.[76] Nationally in 1975, about 9 percent of the total runoff was consumed but, within the 18 hydrological regions into which the 48 states are divided, consumption varied from less than 1 percent in New England to more than 200 percent (owing to imported water) in the Lower Colorado region (Table 4-5). The Missouri, Rio Grande, Lower Colorado, California, and Great Basin hydrological regions consumed more than 30 percent of their runoff.

As shown in Table 4-6, the consumption of water by use varies tremendously. Irrigation is by far the heaviest consumer, taking 76 percent of the total consumption; this figure rises to 88 percent when combined with evaporation from man-made reservoirs. Coal mining at present takes less than 0.2 percent, and when this is combined with power plant cooling needs (both fossil and nuclear), the total is still less than 2 percent.

The regional and end-use differences in water consumption suggest that tradeoffs could be made, so that one region or industry might receive less water (in exchange for something else) to permit another one to receive more. However, undoing the present patterns of distribution (including allotment plans in water-short states) would be very difficult, if not impossible, today. Planning for future developments with the whole country's interest in mind will have to involve such difficult considerations.

Figure 4-5 is a map of the nation's hydrological regions superimposed on a map of the nation's major coal deposits. The eastern group of mines is composed largely of the Upper Mississippi, Ohio, and Tennessee basins; the western group occurs mainly in the Upper Colorado and Missouri basins (including the Powder River).

In the western group, water consumption in 1975 was 20 percent of runoff, and water there is at a premium. In the eastern group, consumption amounts to somewhat more than 1 percent of runoff, and there should still be room for expansion. The nation has generally depended heavily on the eastern group of hydrological regions, which supply about 80 percent of the nation's coal and about 30 percent of its electric power.

Looking toward a future that probably will involve large increases in the production of coal, electricity, and coal-derived synthetic fuels, what projections might be made about water supply as a limiting factor?

PROJECTED WATER CONSUMPTION

In the West, the stringency of water supply is well known. The projection of what additional consumption should be allowed there will have to depend on a case-by-case analysis. The pressure for such additional consumption derives from the large stores of low-sulfur coal that can be extracted relatively cheaply by surface mining.

For the eastern group of mines, the relative abundance of water would seem adequate to support increased electricity production and coal-derived synthetic fuel production, as well as coal mining. We first note that the production of coal and the reclamation of mined land requires less than 5 percent of the water used (per quad) in the production of electricity or synthetic fuel from that coal. In general, then, the water requirements for any large-scale expansion of coal production will stem primarily from the use of the coal and not from its mining. Furthermore, electricity production from oil and natural gas has the same water consumption as from coal; with nuclear fuel and current nuclear reactor technology, the consumption is about 50 percent greater, given similar cooling practices.

Consider the water requirements associated with the incremental consumption of 40 quads of coal (about 2.5 times present production). This amount could support tripling the present steam-generated production of electricity, or the synthetic production of 75 percent of today's domestic liquid fuel consumption, or the production of synthetic gas equal

TABLE 4-5 Regional Runoff and Consumption Statistics for 1975[a]

Region	Mean Annual Runoff (millions of acre-feet per year)	Data for 1975: Consumption (millions of acre-feet per year)	Per Capita Runoff (acre-feet per person per year)	Consumption as a Fraction of Mean Annual Runoff
New England	75.3	0.49	6.40	0.0066
Mid-Atlantic	97.2	1.78	2.43	0.018
South Atlantic Gulf	218.7	4.13	8.26	0.019
Great Lakes	81.0	1.22	3.65	0.015
Ohio	137.7	1.38	6.48	0.01
Tennessee	46.2	0.32	13.77	0.0068
Upper Mississippi	72.9	1.05	3.73	0.014
Lower Mississippi	81.0	6.16	13.77	0.069
Souris-Red Rainy	7.0	0.14	9.72	0.016
Missouri	60.8	19.44	6.80	0.32
Arkansas	81.0	12.96	12.96	0.16
Texas Gulf	35.6	10.53	3.40	0.30
Rio Grande	5.6	4.86	2.84	0.87
Upper Colorado	14.6	2.75	32.40	0.19
Lower Colorado	3.6	8.10	1.38	2.3
Great Basin	8.1	4.46	5.67	0.55
Pacific Northwest	234.9	14.58	35.64	0.062
California	69.7	27.54	3.32	0.40
Alaska	648.0	0.0062	1620.00	9.6×10^{-6}
Hawaii	14.6	0.62	17.82	0.043
United States	2001.5	122.31	8.91	0.060
United States excluding Alaska and Hawaii	1338.9	121.50	6.32	0.091

[a] Totals may not add due to rounding.

Source: Adapted from John Harte and Mohamed El-Gasseir, "Energy and Water," *Science* 199 (1978):624.

to the nation's natural gas consumption. The quantities may seem large today, but considered in the light of prospects for other sources of energy to replace dwindling oil and gas supplies in 2010, they are not implausible. Suppose that 30 quads of this increment are directed to the production of electricity (yield, about 11.5 quads) and 10 quads to synthetic liquids (yield, about 6.5 quads). The total water consumption would be about 4.2 million acre-ft, on the basis of the factors in Table 4-7.

How should such a burden be distributed? It might be argued that the nearer consumption is located to production, the more efficient it will be.

TABLE 4-6 1975 U.S. Consumption of Freshwater by Different Categories of Use (millions of acre-feet)

Use	Consumption
Domestic and commercial use	7.45
Industrial mining and manufacturing	4.54
Coal mining	0.16
Power plant cooling (fossil and nuclear)	2.11
Irrigation	93.15
Evaporation from artificial reservoirs	14.58
TOTAL	121.99

Source: Adapted from John Harte and Mohamed El-Gasseir, "Energy and Water," *Science* 199 (1978):624.

In the case of the eastern group of three river basins, which in 1975 had a water consumption of 2.8 million acre-ft, use of the 40 quads would increase water consumption to 7 million acre-ft, or about 2.7 percent of runoff.

Although such gross analysis appears favorable, other considerations are important. The gross runoff may include water that is available only at great cost, or water to which there is no longer access at desirable locations. The runoff may be subject to seasonal variations that affect the constancy of supply. The withdrawal of large amounts of water (added to such variations in flow) may so diminish flow that it will no longer support the ecological integrity of the river and its banks, in turn leading to other biological and environmental effects.

ESTIMATING PERMISSIBLE CONSUMPTION LEVELS

Clearly, some measurement of permissible flow is needed. A stringent one based on ecological considerations, proposed by Samuels,[77] takes as a baseline the minimum weekly flow that can be expected each 10 years (in hydrologists' terminology, ${}_7Q_{10}$) and proposes that, annually, no more than 10 percent of it be used ($0.1 \times 52 \times {}_7Q_{10}$).

The Risk and Impact Panel analyzed the eastern and western groups of basins referred to above and for simplicity pooled the entire water supply within each group: this is equivalent to assuming that the distributions of water supply and demand are optimally matched. Under the 40-quad scenario, consumption in the western group is 20 times greater than the Samuels-derived criterion of 1.1 million acre-ft; for the eastern group, consumption is four times as great as the criterion of 1.8 million acre-ft.

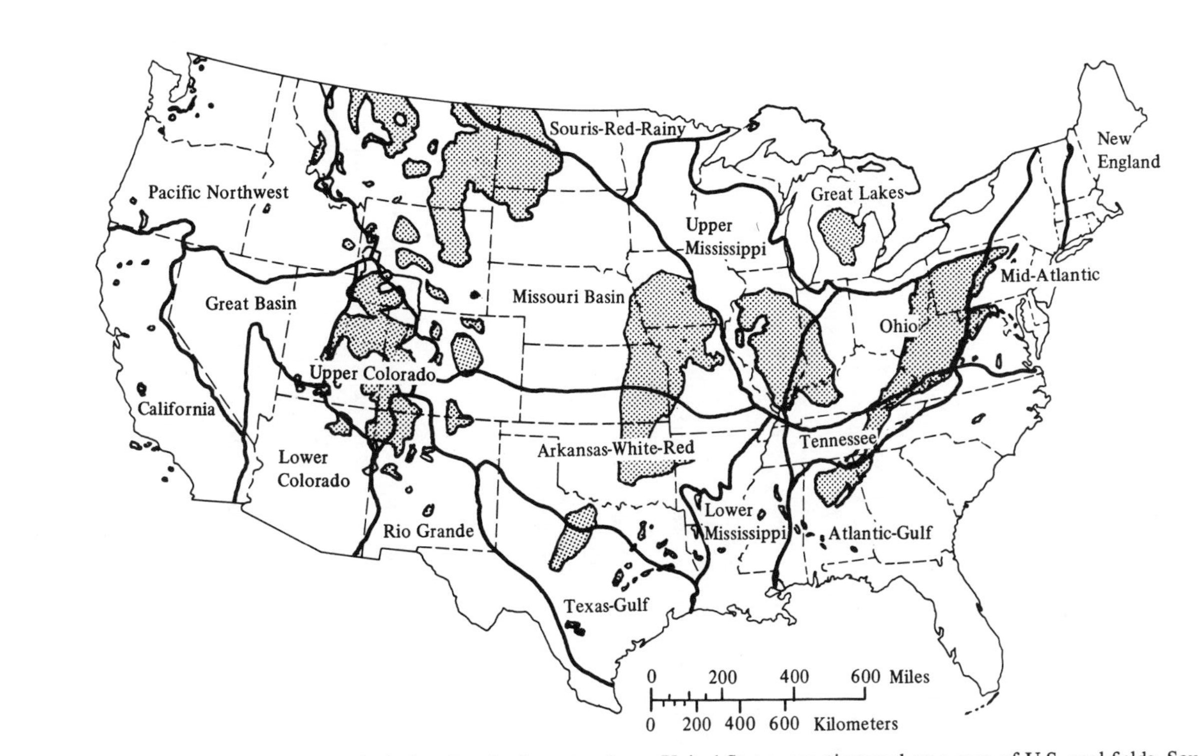

FIGURE 4-5 Water Resources Council hydrological regions in the coterminous United States, superimposed on a map of U.S. coal fields. Source: Adapted from Paul Averitt, *Coal Resources of the United States, January 1, 1976,* U.S. Department of the Interior, Geological Survey Bulletin 1412 (Washington, D.C.: U.S. Government Printing Office (Stock No. 024-001-02703), 1975), p. 5, for coal fields; and U.S. Water Resources Council, *The Nation's Water Resources* (Washington, D.C.: U.S. Government Printing Office, 1968), for hydrological regions.

TABLE 4-7 Estimated Freshwater Consumption Factors for the Coal Fuel Cycle (millions of acre-feet)

	Water Consumption	
Product	Per Quad of Coal	Per Quad of Product
Electricity[a]	0.12	0.31
Synthetic gas[b]	0.05	0.07
Synthetic liquids[b]	0.06	0.09

[a] Assumes that production of electricity from fossil fuels is to be 38 percent efficient and that 17 percent of the waste heat is dissipated directly to the atmosphere along with stack gases. Water consumption for the major cooling modes are (in millions of acre-feet per quad of electricity): once-through (no storage), 0.17–0.34; once-through (with storage), 0.43–1.28; wet-tower cooling, 0.34–0.51. Estimates apply to the following distribution of cooling modes: one-third once-through (no storage), one-third once-through (with storage), and one-third wet-tower cooling. Other combinations, of course, are possible, but for the kind of gross estimate with which this discussion is concerned, this distribution provides a useful example.

[b] Assumes an average efficiency of conversion of 68 percent.

Source: Adapted from John Harte and Mohamed El-Gasseir, "Energy and Water," *Science* 199 (1978):627–628. Harte and El-Gasseir give a large range of values for each item based on known practice or estimates given in environmental impact statements. Their minimal estimates have been used in each case since engineering practice will tend to improve.

In the examples given above, only two groups of hydrological regions were considered. It would appear that the water problems associated with the assumed 40-quad increment could be mitigated by employing other hydrological regions, those either with large freshwater supplies (Table 4-5) or with access to the ocean (Figure 4-5). Detailed studies at both the regional and local levels, however, will be needed to evaluate the resources.

An equivalent problem has been studied in great detail by the six national laboratories, which analyzed the water requirements of the President's National Energy Plan of 1977.[78] That plan called for an additional 18 quads of coal—13.5 for electricity and 4.5 for industrial use—and the findings were considered to apply by and large to the plans under the subsequent National Energy Act of 1978. Using a less demanding water shortage criterion (critical surface supply) than that employed by the Risk and Impact Panel, the report concludes that such an increase is feasible, provided that great attention is paid to the many siting problems that will occur. The problem will be not in mining the coal, but in its use.

In addition to the control of siting, engineering practice may contribute to easing the problem. Dry cooling (or a mixture of wet and dry cooling) could be instituted at all new power plants, for example, and synthetic fuel plants could recycle water more thoroughly. Maintenance of plants could be scheduled for the dry seasons. In places where water is fully allocated, or nearly so, supplies might be bought from current holders of water rights—not all of whom now use water very efficiently. Water could in principle be moved by pipeline from basins in which it is relatively abundant to those where it is scarce, though political opposition by those in the donor basins would likely be serious. Brackish and otherwise unusable groundwater supplies could be tapped in some parts of the country.

All of these would involve increases in the costs of the electricity or fuels produced. As energy development expands, the cost of water will become increasingly important as a factor in optimizing the design of facilities, and therefore as a component in the price of the products.

Economics aside, in some places the arrival of new, large water consumers may be effectively barred by state water allocation systems, which can be very strict in their standards for use and strongly favor established users over newcomers. This means that even where water allocations are unused, it may be impossible for new facilities to obtain them.

In conclusion, the analysis shows the increasing importance of water as a potential limiting factor for the increased production of electricity from fossil or nuclear fuel, and of synthetic fuels from coal. We judge that tripling present coal production for these ends will be contingent on facing the water problem squarely. The technical means of reducing water consumption should be stressed. Siting must be carefully planned, not only to prevent water-supply failure, but especially to obtain optimal use of our water resources. We recommend that all hydrological regions be studied, and that a national data bank be established. We note that water resources are largely under the control of the states, that two different approaches in law have been used to control them (the riparian doctrine and the appropriation doctrine), and that their use in national planning will not be a simple matter. The energy-water problem is, in fact, a part of a much broader one of water as a general limiting factor in the activities of society.

GENERAL CONCLUSIONS

Coal will remain a key element of U.S. energy policy well beyond the end of this century, regardless of the development of other energy sources. Its prime virtue is availability; it will be used increasingly over the next several decades to make up for delays in the development of other energy sources. In this role, it will help tide the nation over until truly sustainable energy sources can be used to replace dwindling supplies of oil and natural gas. It is impossible to predict very precisely how much coal will be needed annually as we approach the end of the century, but it is likely that demand, rather than supply, will be the limiting factor on coal use for as long as this study looks into the future.

In making projections, it is important to distinguish between the near term—up to the late 1980s—and the longer term. First, since the lead time for construction of both coal-fired and nuclear power plants is of the order of 10 years, the electricity-generating capacity of these sources in the near term is already fairly well determined.

Second, it is likely that the use of coal will be determined by the availability of facilities that can burn it in an environmentally acceptable manner rather than by the ability to produce or transport coal in sufficient quantities; any shortfalls in electrical output will have to be made up largely from oil- and gas-fired utilities, usually by delaying their phase-out or their transfer from base to intermediate and peak load use.

Third, if the expansion of nuclear power is constrained by safety and related considerations, then the demand for coal in the near term will be little influenced by economic considerations or by its competitiveness with nuclear power.

For the period beyond 1990 the situation becomes more complex. By then, oil and natural gas will be making smaller contributions to electricity generation, and coal and nuclear fission will tend to be more directly substitutable. If fears about safety, waste disposal, or related issues continued to constrain nuclear power, then coal demand would depend largely on the total demand for electricity and would be relatively insensitive to the competitiveness of its price or the severity of environmental regulations. If concerns about nuclear fission subside, then the choice between coal and nuclear power will likely be made increasingly on an economic basis; the demand for coal would be more sensitive to environmental standards and to the cost and reliability of pollution controls.

In both the short and long term, emission standards will have an impact on the regional distribution of coal production and hence on transportation requirements. Expansion of production is not likely to be a limitation unless there is substantial vacillation and uncertainty about environmental requirements. Because of the regional character of coal markets, uncertainty about environmental standards could affect the regional distribution of coal output while total supply remained the same.

Beyond the turn of the century the situation will change again for several different reasons. First, coal will be increasingly required to produce synthetic fuels to substitute for declining oil and gas production. To the extent that gases and liquids could be produced in situ from coal seams not otherwise accessible to mining, this particular competition with direct combustion could be decreased.

Second (particularly if electricity growth is high), expansion of nuclear power will be increasingly limited by the availability and price of uranium unless advanced reactors and fuel recycling are permitted and well established by that time. If nuclear capacity is restricted to light water reactors on a once-through fuel cycle, the demand for coal in the early decades of the twenty-first century could accelerate.

Third, if the carbon dioxide problem (chapter 9) proves serious, as seems quite probable, it would begin to become apparent shortly after the turn of the century—the same time at which total coal use would have reached the level where it could strain water resources.

Thus, the first few decades of the twenty-first century could be a very critical time in balancing coal use with the exploitation of other alternatives, the principal one of which is likely to be nuclear fission. Among other things, this points to the importance of having as thorough a knowledge as possible of all aspects of the environmental, health, and climatic effects of coal use by the time the choices have to be made.

On balance, it seems unwise to depend on coal use to increase much more than threefold in this country by the end of the century, though it would be technically possible to produce a good deal more than this. At about this level, the problem of water supply could become pressing, and the difficulty of dealing with air pollutant emissions is likely to be great. The climatic effects of carbon dioxide emissions may well become the overriding consideration at about this time. All of these factors, with their complex political and economic interactions, will combine to slow the growth of coal demand.

Even so, at 3 times today's production rate, or about 45 quads annually, coal is likely to supply perhaps one third to one half of the nation's energy by the year 2000. This level of use, with all its costs and potential dangers, will have served its purpose if the intervening years are used to develop alternative, safe, and sustainable energy sources.

NOTES

1. Energy Modeling Forum, *Coal in Transition: 1980–2000,* 3 vols. (Stanford, Calif.: Institute for Energy Studies, Stanford University, 1978).

2. Plant matter that sank to the bottom of swamps 1 million to 600 million years ago formed peat which, when subsequently covered by sedimentation and rock, underwent chemical change. Heat in the absence of oxygen and the passage of time caused a progressive reduction in the amount of moisture and volatile matter in the coal and a progressive increase in the carbon and energy content. The amount of heat (which was a function mostly of depth of burial) and the length of time were the primary determinants of how far the process advanced.

3. H. M. Braunstein, E. D. Copenhaver, and H. A. Pfuderer, eds., *Environmental, Health, and Control Aspects of Coal Conversion: An Information Overview,* 2 vols., prepared for the U.S. Energy Research and Development Administration (Oak Ridge, Tenn.: Oak Ridge National Laboratory (ORNL/EIS-94), 1977), pp. 2-25, 30, 31; 4-138, 139; 5-5, 6, 7.

4. The 0.6 lb or less of sulfur per million Btu is equivalent to coal with 10,000 Btu/lb and 0.6 percent or less sulfur by weight. Each 1000-Btu increase in the coal's energy allowed an additional 0.06 percent sulfur in the coal to meet the ceiling requirement.

5. New emissions standards in U.S. Environmental Protection Agency, news release on "New Standards for Coal-Fired Power Plants" (Washington, D.C.: U.S. Environmental Protection Agency (R-R-90), May 25, 1979). Distribution of coal by sulfur content in Francis X. Murray, ed., *Where We Agree: Report of the National Coal Policy Project,* 2 vols., sponsored by the Center for Strategic and International Studies, Georgetown University (Boulder, Colo.: Westview Press, 1978), vol. 2, pp. 291, 333, 393.

6. Braunstein, Copenhaver, and Pfuderer, eds., *op. cit.*, pp. 2–32.

7. Paul Averitt, *Coal Resources of the United States, January 1, 1974,* U.S. Department of the Interior, Geological Survey Bulletin 1412 (Washington, D.C.: U.S. Government Printing Office (024-001-02703-8), 1975), pp. 63–75.

8. *Ibid.*, pp. 71–72.

9. When the amount of coal in the *reserve base* is reduced by the amount expected to be lost in mining, it is called the *reserve* or the *recoverable reserve*. While the rate of coal recovery varies widely (from 25–30 percent in some deep mines to more than 90 percent in some deep and surface mines), 50 percent recovery has been estimated as the country's average historical rate. This average is now rising as more surface deposits are mined and advanced underground technology is used to a greater extent.

10. The National Coal Policy Project (NCP) argues that the Montana and Wyoming underground subbituminous coal included by the U.S. Bureau of Mines in the U.S. reserve base is currently uneconomical to mine and should not be counted. If those deposits are subtracted, the West's share of recoverable coal energy would drop from 53 percent to about 45 percent (Murray, ed., *op cit.*, p. 286; calculated by taking 57 percent of NCP's estimate of energy lost and subtracting it from the estimate of the Office of Technology Assessment (see Table 4-3) of western and total recoverable reserve energy).

11. U.S. Department of Energy, *Coal—Bituminous and Lignite in 1976*, Office of Energy Data and Interpretation, Energy Information Administration (DOE/EIA-0118/1[76]) (Washington, D.C.: Energy Information Administration Clearinghouse (703), 1978), p. 11; and National Coal Association [The first edition of *Coal Facts* published since the 1974–1975 issue], *Coal Facts 1978–1979* (Washington, D.C.: National Coal Association, n.d.), p. 80, for rates of change.

12. Averitt, *op. cit.*, p. 88, for federal government data.

13. The industry estimates that mines could produce 800–850 million tons in 1980 if there were demand for it, although there would be a short-term shortage of rail cars at that level of

output. Constance Holmes, Vice President for Economics and Director of Foreign Trade, National Coal Association, personal communication, May 1, 1979.

14. U.S. Department of Energy, in National Coal Association, *op. cit.*, p. 60; and U.S. Department of Energy, *Monthly Energy Review, March 1979,* Office of Energy Data, Energy Information Administration (DOE/EIA/0035/3 [79]) (Springfield, Va.: National Technical Information Service (NTISUB/E/127), 1979), p. 10.

15. Executive Office of the President, *The National Energy Plan,* Office of Energy Policy and Planning (Washington, D.C.: U.S. Government Printing Office (040-000-00380-1), 1977).

16. Environmental Protection Agency, "Electric Utility Steam Generating Units: Proposed Standards of Performance and Announcement of Public Hearing on Proposed Standards," Part V, *Federal Register* 43 (September 19, 1978): 42154–42184.

17. Energy Modeling Forum, *op. cit.*, vol. 1.

18. Office of Technology Assessment, *The Direct Use of Coal: Prospects and Problems of Production and Combustion* (Washington, D.C.: U.S. Government Printing Office (052-003-00664-2), 1979), p. 96.

19. *Ibid.*

20. Energy Modeling Forum, *op. cit.*, vol. 1, p. 35.

21. Environmental Protection Agency, "Electric Utility Steam Generating Units," *op. cit.*

22. Office of Technology Assessment, *op. cit.*, p. 99.

23. Environmental Protection Agency, "Electric Utility Steam Generating Units," *op. cit.*

24. A. Wunsch, "Combined Gas/Steam Turbine Power Plants: The Present State of Progress and Future Developments," *Brown Boveri Review* 65 (October 1978): 646.

25. U.S. Department of Energy, *Fossil Energy Program Summary Document* (Washington, D.C.: U.S. Department of Energy (DOE/ET-0087), 1979), p. 267.

26. Ralph M. Parsons Co., *Preliminary Design Study for an Integrated Coal Gasification Combined Cycle Power Plant,* prepared for the Southern California Edison Company (Palo Alto, Calif.: Electric Power Research Institute, 1978).

27. Fluor Engineering and Constructors, *Effects of Sulfur Emission Controls on the Cost of Gasification Combined Cycle Power Systems* (Palo Alto, Calif.: Electric Power Research Institute (AF-916), 1978).

28. William D. Jackson, "MHD Electrical Power Generation: Program Status Report," in *Scientific Problems of Coal Utilization,* DOE Symposium Series no. 46 (Washington, D.C.: U.S. Department of Energy Technical Information Center, 1978).

29. Robert Farmer, "NWK 290-MWe Air Storage Plant at 5,300-Btu Heat Rate," *Gas Turbine World,* March 1979, pp. 32–38.

30. Electric Power Research Institute, *An Assessment of Energy Storage Systems Suitable for Use by Electric Utilities,* prepared by Public Service Electric and Gas Co., Newark, N.J. (Palo Alto, Calif.: Electric Power Research Institute (EPRI-EM 264), 1976).

31. U.S. Department of Energy, *Fossil Energy Research and Development Program* (Washington, D.C.: U.S. Department of Energy (DOE/ET-0013[78]), 1978).

32. U.S. Department of Energy, *Fossil Energy Program Summary Document, op. cit.*, p. 154.

33. *Ibid.*, p. 152.

34. A. J. Molland and D. L. Olsen, "Preliminary Evaluation of Western Market for UCG-Derived Fuel Gas," SRI International, Proceedings of the Fourth Annual Underground Coal Conversion Symposium, Steamboat Springs, Colo., 1978.

35. Harry Perry, "Clean Fuels from Coal," in *Advances in Energy Systems and Technology,* vol. 1 (New York: Academic Press, 1978), p. 307.

36. National Research Council, *Assessment of Technology for the Liquefaction of Coal* (Washington, D.C.: National Academy of Sciences, 1977).

37. L. E. Swabb, Jr., "Coal—Conversion to Fluid Fuels" (Paper presented at Exxon Energy Research and Development Symposium, New York, N.Y., September 27, 1978).

38. U.S. Department of Energy, *Fossil Energy Research and Development Program, op. cit.*, p. 78.

39. Perry, *op. cit.*, p. 282.

40. *Ibid.*, p. 283.

41. National Research Council, *Assessment of Technology for the Liquefaction of Coal*, Commission on Sociotechnical Systems, Committee on Processing and Utilization of Fossil Fuels, Ad Hoc Panel on Liquefaction of Coal (Washington, D.C.: National Academy of Sciences, 1977), pp. 131–152.

42. Institute of Gas Technology, *Assessment of Applications for Direct Coal Combustion*, prepared for the National Science Foundation (Springfield, Va.: National Technical Information Service (PB-263651/AS), 1976).

43. Congressional Budget Office, *Replacing Oil and Natural Gas with Coal: Prospects in the Manufacturing Industries* (Washington, D.C.: U.S. Government Printing Office, 1978).

44. Executive Office of the President, *Replacing Oil and Gas with Coal and Other Fuels in the Industrial and Utility Sectors* (Washington, D.C.: Office of Energy Policy and Planning, 1977).

45. D. E. Nichols, P. C. Williamson, and D. R. Waggoner, "Assessment of Alternatives to Present Day Ammonia Technology with Emphasis on Coal Gasification" (Paper presented at the Symposium on Nitrogen Fixation, Madison, Wisc., June 12–16, 1978).

46. U.S. Department of Energy, *Coke and Coal Chemicals in 1976*, Office of Energy Data and Interpretation, Energy Information Administration (DOE/EIA-0120/76) (Washington, D.C.: Energy Information Administration Clearinghouse (706), 1978); and U.S. Department of Energy, *Coke Producers in the United States in 1977*, Office of Energy Data and Interpretation, Energy Information Administration (DOD/EIA-0122/1) (Washington, D.C.: Energy Information Administration Clearinghouse (708), 1977).

47. National Coal Association, *op. cit.*, p. 76.

48. See note 5.

49. See note 5.

50. Office of Technology Assessment, *op. cit.*, pp. 121–122.

51. National Research Council, *Coal Mining*, Commission on Sociotechnical Systems, Committee on Processing and Utilization of Fossil Fuels, Ad Hoc Panel on Coal Mining Technology (Washington, D.C.: National Academy of Sciences, 1978), pp. 55–60.

52. *Ibid.*

53. Office of Technology Assessment, *op. cit.*, pp. 121–146.

54. U.S. Department of Energy, *Coal Data—A Reference*, Office of Energy Data and Interpretation, Energy Information Administration (DOE/EIA-0064) (Washington, D.C.: Energy Information Administration Clearinghouse (704), 1978), p. 14.

55. Office of Technology Assessment, *op. cit.*, pp. 276, 278–289.

56. *Ibid.*, pp. 282–286, 289.

57. *Ibid.*, pp. 259–263.

58. *Ibid.*, pp. 265–275.

59. In 1976, 63 percent of underground production came from continuous-mining machines and 4 percent from longwall mines. U.S. Department of Energy, *Coal—Bituminous and Lignite in 1976, op. cit.*, p. 30.

60. Averitt, *op. cit.*, p. 55.

61. Office of Technology Assessment, *op. cit.*, p. 147.

62. *Ibid.*, p. 382.

63. *Ibid.*, p. 110.

64. The following material draws on Office of Technology Assessment, *op. cit.*, pp. 111–120.

65. *Ibid.*, p. 112; and U.S. Department of Energy, *Monthly Energy Review*, March 1979, *op. cit.*, p. 95, for 1978 data.
66. Federal Energy Regulatory Commission, Washington, D.C., unpublished data, 1978.
67. Thomas Petzinger, Jr., "Captive Customers? Utility-Owned Mines, Meant to Assure Fuel, Often Lift Power Cost," *The Wall Street Journal*, May 10, 1979, p. 1.
68. Alexander Gakner, Chief, Branch of Fuel and Environmental Analysis, Federal Power Commission, in testimony, *The Energy Competition Act*, as cited in Office of Technology Assessment, *op. cit.*
69. U.S. Department of Energy, *Coal—Bituminous and Lignite in 1976, op. cit.*, p. 4.
70. Charles River Associates, *Coal Price Formation*, prepared for Electric Power Research Institute (Palo Alto, Calif.: Electric Power Research Institute (EA-497, Project 666-1), 1977), pp. 4–40; see data for companies issuing annual reports in Office of Technology Assessment, *op. cit.*, p. 120; and Fred Dunbar, Charles River Associates, Cambridge, Mass., personal communication, May 1, 1979.
71. U.S. Department of Energy, *Coal—Bituminous and Lignite in 1976, op. cit.*, pp. 49–51; and U.S. General Accounting Office, *U.S. Coal Development—Promises, Uncertainties* (Washington, D.C.: U.S. General Accounting Office, (EMD-77-43), September 22, 1977), p. 5-5.
72. Office of Technology Assessment, *op. cit.*, p. 160.
73. Teknekron, Inc., *Projections of Utility Coal Movement Patterns: 1980–2000* (Washington, D.C.: Office of Technology Assessment, 1977).
74. Office of Technology Assessment, *op. cit.*, p. 160; and National Coal Policy Project, "Summary and Synthesis," in *Where We Agree: Report of the National Coal Policy Project*, sponsored by the Center for Strategic and International Studies, Georgetown University (Washington, D.C.: Automated Graphic Systems, 1978), p. 35.
75. See John Harte and Mohamed El-Gasseir, "Energy and Water," *Science* 199: 623–634; and Ecosystems Impact Resource Group, "Energy and the Fate of Ecosystems," in National Research Council, *Risks and Impacts of Alternative Energy Systems*, Committee on Nuclear and Alternative Energy Systems, Risk and Impact Panel (Washington, D.C.: National Academy of Sciences, in preparation), chap. 6.
76. An acre-ft is the amount of water that would cover an acre to a depth of 1 ft. It is equivalent to 325,829 gal. Runoff is water from precipitation that is not lost through evaporation or transpiration by plants.
77. G. Samuels, *Assessment of Water Resources for Nuclear Energy Centers* (Oak Ridge, Tenn.: Oak Ridge National Laboratory (ORNL-5097 UC-80), 1976).
78. U.S. Department of Energy, *An Assessment of National Consequences of Increased Coal Utilization, Executive Summary*, vols. 1 and 2, report prepared by the staff of the six national laboratories: Argonne, Brookhaven, Lawrence Berkeley, Los Alamos, Oak Ridge, and Pacific Northwest (Washington, D.C.: U.S. Department of Energy (TID-29425), February 1979).

5 Nuclear Power

Nuclear power could make a substantial contribution to the base-load electrical system of the United States in the intermediate term. Advanced converters or fission breeders could enlarge this contribution, and extend it many decades or thousands of years. Nevertheless, the expansion and further development of nuclear power face uncertainties and controversies.

- The demand for electricity is difficult to predict.
- The amount of uranium that will be available to fuel the present generation of reactors at economical prices is uncertain.
- The safety of nuclear reactors is a controversial topic.
- Policies for disposal of radioactive waste have not been developed, and delay in their development has heightened concern about the efficacy of proposed methods.
- The possibility that terrorists or other groups might divert nuclear materials is a matter of concern. The degree of protection that can be achieved against diversion has been discussed and argued without resolution.
- The contribution nuclear power might make to increasing or decreasing the risks of nuclear weapons proliferation and nuclear war is controversial, and the obvious importance of this issue makes it a matter of urgent concern.*

*See statement 5-1, by E. J. Gornowski, Appendix A.

These and related issues are addressed in this chapter. We first present a summary statement and principal conclusions. The balance of the chapter takes up these items in detail.†

SUMMARY

Nuclear power contributes to diversity in the sources of energy on which the United States can draw. In 1978, 66 light water reactors (LWR's) supplied close to 13 percent of the electricity generated in the United States. In some regions of the country, the share of electricity generated by nuclear power exceeded 40 percent. The distribution of nuclear plants is illustrated in Figure 5-1. The generating capacity of nuclear power plants totals 52 gigawatts (electric) (GWe).[1]

Of the energy sources that can be used to generate large amounts of electricity, only coal and nuclear power offer reasonably assured ability to support significant expansion in electrical generating capacity over the next few decades. The costs of electricity produced from coal and nuclear power are roughly comparable and depend on plant location and financing conditions. Nevertheless, new orders for nuclear power plants were offset by cancellations of previous orders the past 3 years, and this will create a pause in the expansion of nuclear capacity after 1985 unless the licensing of nuclear plants is accelerated and their construction time reduced. The nuclear industry in the United States can produce at least 500 GWe of nuclear generating capacity for installation by the year 2000, and more than 750 GWe by 2010. The actual rate at which this capability will be called upon depends on several factors.

First, there is the question what the demand for electricity will be. The Supply and Delivery Panel evaluated a number of projections and concluded that an annual average growth rate of 4 percent to the year 2010 represents a reasonable figure for planning the growth of electrical capacity.[2] This would lead to a total demand for just under 2000 GWe of capacity in 2010, if the total system's capacity factor is unchanged. In the scenario of highest energy consumption considered by CONAES (assuming constant real prices and 3 percent annual average rate of growth in gross national product (GNP)), the required electrical capacity falls below 1500 GWe in 2010. Assuming a higher rate of electrification, the required capacity might be about 1750 GWe in 2010. (See chapter 11.) Although these projected rates fall below the historical rate of growth, they may still

†See statement 5-2, by E. J. Gornowski, Appendix A.

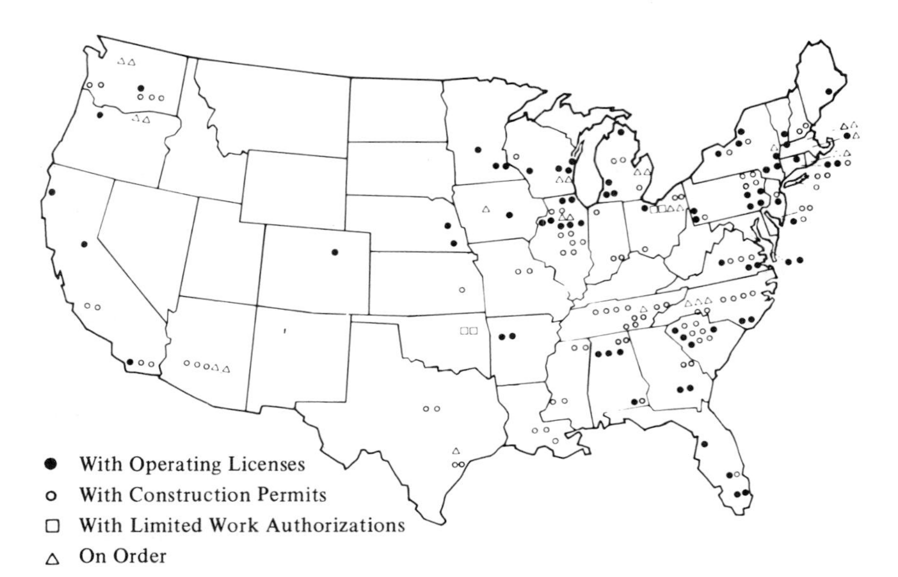

72	Reactors with operating licenses	52,396 MWe
92	Reactors with construction permits	101,148 MWe
4	Reactors with limited work authorizations	4,112 MWe
30	Reactors on order (including 4 units not sited on map)	35,082 MWe
198	Total	192,738 MWe

January 1, 1979

FIGURE 5-1 Nuclear power plants in the United States as of January 1, 1979. Source: Atomic Industrial Forum, *Electricity from Nuclear Power* (Washington, D.C.: Atomic Industrial Forum, 1979).

be unrealistically high. Models constructed for the CONAES study project lower and declining rates of growth in GNP than the rates experienced in the past.[3] The CONAES models have also explored the effects of higher and increasing prices for energy, or equivalent policies. These assumptions lead to scenarios in which the demand for electricity ranges from below present values to just under 3 times present values (2.8 times) by 2010.[4]

Correspondingly, utility capacity would be between about 400 and 1450 GWe of installed central station power (1978 capacity was 560 GWe). These estimates assume that the fraction of total energy demand satisfied by electrical end-use remains constant.* In the past, electricity has tended to displace the direct use of fuels at the point of consumption, and the fraction of total energy demand met by electricity has increased.

How much of this demand for electricity will be met by nuclear power is also uncertain. Nuclear power has a slight economic advantage over coal. This advantage has good prospects for enhancement, but also has some chance of reversal. Prudent utility planners are likely to plan mixed systems of nuclear power and coal, given these contingencies, but the proportion of each can only be guessed. In addition to cost, planners must also consider the reliability of supply, the stability of regulatory requirements, and prospective public policy. Some considerations will favor nuclear power, others, coal.

A major reservation against too great a reliance on nuclear power may arise from uncertain availability of natural uranium, the primary resource for nuclear fuel. The Uranium Resource Group of this study[5] concluded in 1977 that not more than 1.8 million tons of minable domestic uranium oxide (U_3O_8) reserves and probable resources should be considered as a basis for prudent planning. CONAES has revised its own figure to 2.4 million tons, reflecting higher estimates recently published by the U.S. Department of Energy. (Table 5-1, under the section "Availability of Uranium," sets out the pertinent estimates.) Translating these figures into nuclear power capacity, 2.4 million tons of U_3O_8 would meet the lifetime fueling requirements of about 400 GWe of installed capacity, assuming the continued use of light water reactors on once-through fuel cycles. The total nuclear capacity in operation, under construction, or planned in the United States in 1979 amounts to 193 GWe.[6] According to the Supply and Delivery Panel, the uranium production rates required to reach installed nuclear capacities much above 200 GWe by 2010 would demand a national commitment to uranium resource exploration and extraction.[7]

Further expansion and continuation of nuclear power could be accommodated if fuel reprocessing were permitted. The industrial position is that expansion much beyond current commitments would not be undertaken unless the durability of nuclear power were confirmed by commitment to a breeder reactor (or to equivalent fuel production systems, such as accelerator breeders, or fusion-fission devices).† Without firm plans for reactor designs to follow light water reactors, or for fuel reprocessing and recycle, nuclear capacity would have to be gradually phased out as reactors were retired, beginning early in the twenty-first century. However, if (as some resource economists believe) considerably more uranium is found as the price rises, then nuclear capacity could be

*See statement 5-3, by L. F. Lischer, Appendix A.

†Statement 5-4, by E. J. Gornowski: It is unlikely that there is unanimous opinion that no new LWR's would be built if the breeder were forever excluded.

expanded even if the introduction of new reactors and fuel cycles were to be postponed.

Some expansion of light water reactor capacity (with a once-through fuel cycle) could also be achieved by reconfiguring the light water reactor to minimize U_3O_8 consumption, and also by lowering enrichment tails to 0.1 percent or less (see "Uranium Enrichment"). This might raise the allowable capacity in the year 2000 for the same resource base by nearly 25 percent, to 500 GWe.

Another possibility for a more durable industry is to switch from the present generation of light water reactors on the once-through cycle (no reprocessing or other reuse of spent fuel) to reactors and nuclear fuel cycles that make more efficient use of uranium. Under present conditions, only about 0.6 percent of the fission energy potentially available is used. The fission of uranium-235 (^{235}U) contributes 0.4 percent, and the fission of plutonium-239 (^{239}Pu) created in the reactor contributes 0.2 percent. If the spent fuel removed from the reactor were reprocessed, and the ^{235}U and ^{239}Pu recycled in fuel, the use of uranium could be raised to 0.9 or 1 percent. Such reactor types as the Canadian CANDU or advanced high-temperature gas-cooled reactors (HTGR's) could be designed and operated to use up to 2.0 percent of the energy embodied in uranium on a once-through cycle. Combining lower enrichment tails and the possible stripping of existing accumulated tails with the use of the enriched CANDU once-through cycle might further increase the capacity that could be safely committed by 2000, perhaps to more than 525 GWe.

By loading uranium and plutonium into breeder reactors, and recycling the load many times through similar reactors after reprocessing, it is possible to recover perhaps 70 percent of the energy in the original uranium ore—an improvement in energy recovery by about a factor of 100 over light water reactors. This possibility not only multiplies the energy from existing resources (including existing enrichment plant tails), but permits economic recovery of energy from much less concentrated and more widely distributed uranium ores, essentially making uranium a potential source of energy for hundreds of thousands of years.

In addition to recovering a large fraction of the energy in ^{238}U, it is possible to recover the energy in another element, thorium, that is probably 4 times more abundant in the earth's crust than uranium. The single isotope of thorium, thorium-232 (^{232}Th), can be converted to another fissile isotope of uranium, ^{233}U, in nuclear reactors. Various combinations of thorium-uranium and uranium-plutonium fuel cycles can greatly multiply energy resources.

Making more efficient use of nuclear fuel resources depends on using new designs for reactors and operating these reactors in combination with fuel reprocessing.[8] These reactor designs may be divided into two classes:

advanced converters designed for the use of thermal neutrons and generally operating on the thorium cycle, and fast breeders designed for the use of fast neutrons that can generate more plutonium from ^{238}U than they consume in generating power. Breeders can also generate ^{233}U from thorium. Advanced converters using thorium and ^{233}U can be designed to function as thermal breeders. With sufficiently careful design and frequent fuel reprocessing, they can operate without additional fissile isotopes from nature. However, these conditions are not likely to yield economical power generation.[9]

The breeder design closest to commercial status in the United States and elsewhere is the liquid-metal fast breeder reactor (LMFBR). In the most resource-efficient version, this reactor would be fueled with plutonium separated from the spent fuel of light water reactors and with depleted uranium left behind in the enrichment process for today's light water reactor fuel. The energy available from uranium already mined and stored as depleted tails from domestic enrichment plants, if used in LMFBR's, could provide one third to one half of the energy recoverable from domestic coal reserves and resources.

Advanced converters can also extend resources, but unless they are fueled with plutonium from the spent fuel of light water reactors, their operation will require some additional uranium feed. The amount of this required feed can be minimized by frequent reprocessing and by features in the converter designed to hold down the loss of neutrons to fission products, control rods, and structural materials. The advanced converter most widely used in the world is the natural-uranium, heavy water CANDU, developed in Canada. The advanced converters closest to commercial status in the United States are the high-temperature gas-cooled reactor and the light water breeder reactor (LWBR). They both use the thorium-uranium cycle with enriched ^{235}U feed. Both require more uranium for their initial inventories of fuel than light water reactors.[10] This uranium requirement can be reduced somewhat by mixing in plutonium from reprocessed light water reactor fuel. Advanced converters require far less uranium ore over their operating lives than light water reactors.

The thorium-^{233}U fuel cycle can be used to greatest advantage in thermal advanced converters, and the uranium-plutonium fuel cycle can be used to greatest advantage in fast breeders. This suggests the possibility of using various integrated fuel cycles: combinations of fast breeders, advanced converters, and light water reactors.

These technical possibilities are unlikely to be realized unless nuclear power is publicly acceptable. Public opinion may show swings and trends in the future, as it has in the past. Public concern about nuclear power has centered on four issues: the safety of routine operation of the nuclear fuel cycle and of reactors; the possibility and effects of major nuclear accidents; the handling of radioactive wastes; and the production of nuclear bombs by nations or subnational groups using fissile materials obtained from nuclear-powered facilities.

At all stages of the nuclear fuel cycle, some radioactivity is released to the environment. The largest burden from these releases has come from the underground mining of uranium and from the milling process by which uranium is concentrated from its ores. The hazards of uranium mining have been estimated as resulting in about 15 deaths per year per 10,000 miners. The radioactivity in the mine increases the hazard of cancer, although the risk of accidental fatality in mining accidents is higher than the increased cancer risk.[11] Per miner-year, the hazards of uranium mining are comparable to those of coal mining, but because the same energy is recoverable from only about 1 percent as much material, the mortality of uranium mining is, per unit of power, far less serious than that of coal mining. (See chapter 9.)

Additional radioactive emissions come from the mill tailings—the residues from the uranium concentration process—which contain over 80 percent of the ore's original radioactivity. Past practices have been careless, resulting in exposure of the tailings to weathering, which releases some of the radioactivity to the environment, and in their incorporation into concrete and landfill for homes and schools, in extreme cases. Although the total morbidity from such handling has been quite small, these consequences have cast doubt on the seriousness with which the industry and the responsible federal agencies approach the job of protecting the public.*

Other routine sources of emission are the releases permitted from nuclear power plants (within set limits) of materials that have become radioactive, and potential releases of radioactive gases (such as krypton-85 (^{85}Kr), tritium, and carbon-14 dioxide) from reprocessing plants.

All these "normal" or routine releases of radioactivity are estimated to increase environmental radiation by a small fraction of the existing background, and on this basis, their effects per unit of power generated are small compared to the mining risks, or to the risks of other energy sources.†

More controversial is the possibility of reactor accidents. Much of the controversy has focused on the validity of risk assessments made in the Reactor Safety Study for the Nuclear Regulatory Commission (also known as the Rasmussen Report or WASH-1400). This report attempted to

*See statement 5-5, by E. J. Gornowski, Appendix A.

†Statement 5-6, by J. P. Holdren: The statement is too sweeping. NAS estimates prepared for CONAES imply 0.5–2.0 excess cancer deaths per GWe-year from routine exposures and emissions, excluding tailings.

estimate the probability (per reactor-year of operation) that accidents of varying severity would occur.[12] Its stated findings are that the actuarial risks (sums of the probabilities of consequences multiplied by the severity of consequences) are very small, and that the chances of severe accidents that would cause large numbers of casualties are extremely small—so small as to be within the range of risks we hardly deign to consider. Nevertheless, these findings have been challenged on several grounds: that the statistical treatment is in some respects incorrect and in others misleadingly presented; that casualty figures for the most severe types of accidents are underestimated; and that accident frequencies may have been overestimated (industry analysts typically arguing the latter, and nuclear critics, the former).[13]

The Risk and Impact Panel of this study examined the controversy, but could not reach more than qualitative conclusions. These conclusions are, briefly, that the statistical inferences of the report should be corrected upward, owing to the report's use of medians rather than means of certain probability distributions where the correct procedure would have been to use the mean values, and that in addition to this upward correction in the "best estimate" of the accident risk, the counterclaims of optimism and pessimism for accident frequencies and consequences ought at least to be interpreted as indicating that the uncertainties accompanying both probabilities and consequences are greater than the uncertainty factors stated in WASH-1400.

We would estimate higher average risks than WASH-1400—not so high as to be alarming, but with sufficient uncertainty that there remain legitimate grounds for controversy whether the risk of reactor accidents ought to be an important consideration in decisions about nuclear power. Thus on safety grounds alone, the expansion of nuclear power would be acceptable,* provided the rate of expansion were consistent with the rate of improvement of knowledge about accident risks, especially reductions in uncertainty.

The reactor accident at Three Mile Island occurred after most of CONAES's deliberations had been completed. That fact and the fact that several investigations of the accident are still in progress make it inappropriate for CONAES to discuss its implications at length, and impossible to do so with authority. The information so far released about the accident (and interpreted by nuclear specialists on the committee) seems consistent with CONAES's cautious, positive findings on reactor safety.

Another element of public concern is apprehension about the ability of

*Statement 5-7, by J. P. Holdren: Decisions on what is "acceptable" are the business of the political process, not of this or any other NAS committee.

institutions and industry to manage or dispose of radioactive wastes. The most acute concern is the fate of high-level wastes generated in reprocessing plants or contained in spent fuel, but the management or disposal of the much larger bulk of intermediate- and low-level waste generated throughout the nuclear industry also raises public apprehension. Most experts are of the opinion that no technological obstacles stand in the way of safe management of any of these wastes,[14] but governmental inaction, changes of program and emphasis, and the lack of approved facilities are not reassuring.

In the reprocessing and refabrication of fuel essential to making effective use of resources in advanced converters or breeders on either the thorium or the uranium fuel cycle, fissile material (either ^{233}U or ^{239}Pu) is separated from the spent fuel elements and is thus more readily subject to theft or illicit diversion than if it remained in the spent fuel elements. The appearance of pure plutonium or ^{233}U in some stages of the fuel cycle presents the troubling possibility that weapons-usable material could be stolen by terrorists. Proposals have been advanced for reprocessing methods that avoid separation of plutonium in pure form. These schemes are given the generic name "coprocessing" when the plutonium is chemically mixed with its parent uranium throughout the cycle, and "Civex" when it is given the additional protection of retaining some highly radioactive fission products. Such processes are not now available and would require development.

A graver possibility than illicit diversion is that countries installing reprocessing plants would thereby have the means to build up arsenals of nuclear weapons in short order. This concern is particularly acute for breeder reactors, which have little or no value without reprocessing, and it was this consideration that persuaded the Carter administration to defer both commercial reprocessing and commitment to the fast breeder.

A possible advantage of the thorium-^{233}U fuel cycle for fast breeders or advanced converters (it can be used in either) is that the ^{233}U or ^{235}U used to feed these reactors can be diluted with ^{238}U in a 4:1 ratio (for ^{235}U) or a 7:1 ratio (for ^{233}U), making either undesirable as weapons material without physical isotope separation as well as chemical reprocessing. This is the "denatured" thorium cycle. The efficacy of denaturing is now the subject of extensive debate. It is being studied in the United States and will be studied further in the ongoing program of the International Nuclear Fuel Cycle Evaluation (INFCE).

In spite of the unsettled state of the reactor-safety issue following the Three Mile Island incident (which occurred late in the committee's deliberations), the committee continued to regard proliferation and diversion as the most important—perhaps the overriding—issue in nuclear power. The degree to which the risks of national proliferation of nuclear armaments or subnational diversion of material for nuclear weapons could be controlled was discussed at length. The problem was acute: Subjective estimates of the magnitude of these risks were balanced against equally subjective estimates of the benefits that nuclear power might provide in easing the world's problems of energy supply.

There was general agreement that the greatest threat of nuclear technology lies in existing stockpiles of nuclear weapons and weapons material throughout the world. There was further agreement that to the extent that high enrichment of ^{235}U and isolation of ^{233}U and plutonium are needed for a civilian nuclear power industry, these steps of the fuel cycle should be conducted in secured plants, preferably under international control. However, some members of the committee believe that the economic importance of nuclear energy is not great enough to warrant accepting significantly increased risk of international proliferation or subnational use of nuclear weapons, and that such increased risk will attend the spread and growth of nuclear power if these should occur more rapidly than improvements can be made in existing safeguards and deterrents. Other members of the committee believe that the world's energy problems already pose a greater long-term threat than does proliferation, and that the benefits of the rapid spread of nuclear power in alleviating these problems outweigh any plausible increase in the risks of proliferation and diversion.* Divergent opinions on what steps to take follow from these beliefs.

Some argue that international solutions such as the Non-Proliferation Treaty, safeguards (monitoring by the International Atomic Energy Agency), and strengthened controls on fuel cycles can only be effected if the United States is an active participant, a reliable supplier of nuclear materials and know-how. These are arguments for carrying forward, and very probably exploiting, the development of reprocessing and breeder reactors, since both increase our ability to provide nuclear fuel.

Others argue that the current policy of the United States—staying the commercialization of reprocessing for the time being and limiting the development of breeders to technology-level studies—is essential as an example to others.† They maintain that this forbearance is necessary to avoid a situation in which countries that have legitimate domestic needs for major nuclear power enterprises are tempted to manufacture nuclear weapons. The argument is that the moral position of the United States is strengthened in international negotiations by what may be some self-sacrifice.

*See statement 5-8, by E. J. Gornowski, Appendix A.
†Statement 5-9, by E. J. Gornowski: The United States has lost this argument. Reprocessing is going ahead in other countries regardless of the U.S. position.

The issues of diversion and proliferation make the future of reprocessing and the breeder reactor uncertain. As a consequence, the future of nuclear power beyond the point of resource scarcity is also uncertain. The undecided future of reprocessing adds to uncertainty about the form of waste that must ultimately be banished from the environment. The committee cannot resolve these uncertainties, but in the recommendations that follow, suggests ways they might be reduced by improving the reliability of information, by narrowing and clarifying areas of dispute, and by instituting interim programs that preserve flexibility of response in anticipation of better information.

CONCLUSIONS

The committee draws the following conclusions about technical factors that should be considered in formulating nuclear policy.

- The rate of growth in the use of electricity is a primary factor affecting the strategy of nuclear power development. Low rates of growth allow the electric utilities sufficient flexibility to regard coal and nuclear capacity as interchangeable to a considerable degree. This becomes increasingly difficult for higher electricity growth rates; rapid expansion of both coal and nuclear capacity would be required. The highest growth rates in electricity use examined by the committee call for technically achievable rates of expansion of both new coal and nuclear capacity that many members of the committee regard as incompatible with environmental and political restrictions.
- The growth of conventional nuclear power (today's light water reactors) will be limited by the producibility of domestic uranium resources, probably before the year 2000. With today's once-through fuel cycle and no change in the prevailing policy against reprocessing, a maximum nuclear capacity of about 400 GWe could be reached by 2000, diminishing thereafter. This contribution could be extended to about 600 GWe with reprocessing and recycle of fuel in light water reactors. A more complete assessment is needed of domestic and world uranium resources, and of the rate at which they can be produced at various costs.
- A greater, or more sustained, contribution of nuclear power beyond 400 GWe and past the year 2000 could only be supported by the installation of advanced reactor systems, particularly those using recycle of nuclear fuel. Even if very extensive new uranium resources are identified before 1990, advanced converters would still be attractive because they could extend the uranium energy base appreciably. Nevertheless, only the

breeder can provide insurance of satisfying very high demand, or of abating a shortage of uranium.

• Several different breeder reactors could serve in principle as candidates for an indefinitely sustainable source of energy. Only the liquid-metal fast breeder reactor could be built and operated by the year 2000.

With regard to the major domestic issues that surround nuclear power, the committee draws the following conclusions.

• The short-term health risks from routine operation of the LWR nuclear fuel cycle appear to be far below the risks from the coal fuel cycle. This remains the case if reactor accidents are included, using the risk estimates of the Reactor Safety Study (WASH-1400). The accuracy of the WASH-1400 results and the validity of this type of comparison are disputed both inside and outside CONAES. Long-term risks are even more difficult to compare. The maximum estimates of nuclear power risks are within the range of risks for the coal cycle. An analysis of reactor safety such as WASH-1400 cannot be carried out for advanced reactors until specific commercial designs are available.

• No insurmountable technical obstacles are foreseen to preclude safe disposal of nuclear wastes in geological formations. All necessary process steps for immobilizing high- and low-level wastes have been developed, and there are no technical barriers to their implementation. Geological emplacement can be carried out with standard mining techniques. There is still some controversy about the assured integrity of the backfill.

• The main problems with geological waste disposal are site-specific: characterizing sites that exhibit a high degree of stability, transmit water only by pore flow, and offer no ready access to groundwater. Storage of waste at such sites would engender much smaller risk to the public than that of routine emissions from the rest of the fuel cycle. Routine emissions from the nuclear fuel cycle are generally recognized to present very small risks to health.

• Radiation has been released from stored nuclear waste, notably from the wastes of military production operations, but also from some wastes of civilian operations. These incidents have not so far resulted in public hazards. They do, however, illustrate the inadequacies of existing surveillance and regulatory practices, and they emphasize the need for permanent disposal facilities.

• Nuclear waste disposal has suffered in the past from decision making by the federal government that has been both dilatory and capricious. The time for decisions is upon us, but we have not yet arrived at a decision-making process that is both legitimate and authoritative.

Finally, with regard to international issues, we note the following conclusions.

• The United States, with relatively large reserves of both coal and uranium, is in a very favorable position compared to many countries of the world that have little or no indigenous fuels. In the absence of practical alternatives, these countries may well find nuclear power, especially breeder reactors, attractive as an energy source that greatly reduces reliance on fuel imports.

• The problem of diversion of nuclear materials by terrorist or criminal groups, and the related question of the vulnerability of nuclear facilities to sabotage, are serious matters. Domestic security measures, such as those practiced in laboratories and facilities handling enriched materials, can be effective. However, if our society moves in the direction of turbulence and polarization, questions might be raised about our ability to carry out domestic security measures properly.

• The proliferation threat must be viewed from the perspective that the overriding security problem is to avoid war; failing this, it is to avoid war between or among superpowers, and failing that, to avoid devastating nuclear exchanges among them. Nuclear power can reduce this threat by reducing the competition for scarce resources, one of the causes of war. Nuclear power can also increase this threat by facilitating the acquisition of nuclear weapons, particularly by countries whose possession or use of them might catalyze superpower war.

• Nuclear power is not the most likely route countries with the will to acquire nuclear armaments might follow, but it is not an impossible one. The most likely scenario by which nuclear power could contribute to nuclear armament is the appropriation of plutonium or ^{233}U from nuclear fuel cycles by a country that might not, in the absence of this opportunity, have made the decision to acquire nuclear weapons.

RECOMMENDATIONS

The committee's principal recommendations are listed below. More detailed recommendations appear in subsequent sections of this chapter.

• National policy should support the continued use of nuclear power for the next few decades.[15] The rationale for such support rests on the availability of nuclear power as a domestic energy resource whose risks are at worst comparable* to those of other energy sources, its competitive economics, and the undesirability of relying too heavily on coal or nuclear power, to the exclusion of the other, until the risks of each are better understood.

• Advanced reactor types should be developed to the point that they can be evaluated for possible introduction, if needed. Three advanced reactor types can be considered: the liquid-metal fast breeder reactor, the high-temperature gas-cooled reactor, and advanced versions of heavy water reactors. Of these, the LMFBR would be recommended if industrial economic factors were the only consideration. The LMFBR has the best chance of providing insurance that a nuclear industry could in the long term meet any electrical demands that might eventuate.† A major consideration is, of course, the question whether the fuel reprocessing necessary to operate LMFBR's (and eventually, other advanced converters) is compatible with national antiproliferation objectives. We recommend that development of proliferation-resistant reprocessing for LMFBR's proceed, pending a decision whether such methods as coprocessing, Civex, or radioactive spiking are sufficient or necessary to counter this policy objection to the LMFBR.

• High-temperature gas-cooled reactors or heavy water reactors (HWR's) (or both) are appropriate advanced reactors if (as in most CONAES scenarios) electrical demand grows at a moderate rate, or if appreciable uranium supplies can be produced at a price in the range of $100–$200/lb of U_3O_8. Uranium supplies in this price range are not believed to be abundant in the United States. These reactors could probably be installed with fewer siting restrictions than LMFBR's, and they would be compatible with LMFBR's in a mixed system, with the breeders operating as fuel factories. They might, for a considerable period, produce power more economically than breeders. Their development at some level of effort is therefore recommended, regardless of the decision on the LMFBR.

• Exploration for uranium resources and their specification must continue at a vigorous pace. This information is basic to the timing of expanding the existing light water reactor industry, and to the pace of commercializing advanced reactors.

• Technological designs and licensing requirements should be available "off the shelf" for the reprocessing plants that would handle fuel from the various reactor types to reduce the lead time needed if reprocessing is approved. This means that both research and development activities

*Statement 5-10, by J. P. Holdren: I disagree. Implied here is a kind of apples-plus-oranges comparison that cannot be done. The public might well decide some nuclear risks are intolerable.

†See statement 5-11, by L. F. Lischer, Appendix A.

through the pilot-plant stage must be undertaken, as well as licensing of these pilot plants. In view of the uncertain future for commercial reprocessing as a consequence of national policy, there is no alternative to government funding of this activity.

• Within the national nuclear energy program specifically, and as an integral part of the national technological research program, a much higher level of support is needed for fundamental engineering and materials sciences. These include basic studies in heat transfer, fluid flow, the mechanics of materials under dynamic stresses, corrosion, solid-state diffusion in technical materials, and many other similar fields. Such studies are necessary to resolve questions of reactor safety by means other than complicated and expensive overdesign, and should be the core of any nuclear safety research program.

• Ore tailings and low-level radioactive wastes from the nuclear industry need a sound program of environmental protection to ensure that they do not present significant health risks to the public. Since even the worst-case risks from these sources are quite low, the steps required should be based on the most probable values of costs and risks, rather than on the most conservative possible assumptions.*

• High-level and transuranic wastes from the nuclear industry should be disposed of by emplacement in a geological repository. The responsibility for site selection must ultimately rest with the federal government, as the benefits from a well-chosen repository accrue to the nation as a whole, and the risks to local populations are not high.† The target for placing a first repository in operation should be the mid-1990s. In the interim, spent-fuel storage should be practiced without assuming or precluding reprocessing.

• Radioactive wastes from weapons-material production facilities, military propulsion reactors, weapons fabrication facilities, and other military activities must be dealt with independently. These wastes exist in quantity and in less-than-ideal form for geological disposal. They should, at minimum, be immobilized and collected at a few isolated sites. After these steps, they can be prepared for geological disposal (probably at the same sites). Within the category of defense wastes, the transuranic waste and (possibly) the high-level waste from propulsion-reactor reprocessing are the only types that can be considered for disposal at the Waste Isolation Pilot Plant near Carlsbad, New Mexico.

*Statement 5-12, by J. P. Holdren: I disagree. Conservatism is particularly appropriate when the potential victims are in future generations, as is the case with tailings and low-level wastes.

†See statement 5-13, by J. P. Holdren, Appendix A.

Finally, what can we say about the problem of nuclear weapons proliferation that is not a homily or a statement of alternatives? It should be continued national policy to search for ways to avoid nuclear war, but it is by no means clear that this country is prepared to give up the option of nuclear weapons for use in extreme circumstances. In this quandary, any technical or industrial policy steps are, at best, supportive of larger national policy. All the antiproliferation measures we can conceive have an experimental character, including such relatively recent measures as the deferral of domestic plans for reprocessing and breeder reactors. If they promote a more peaceful and prosperous world (and thus, necessarily, a more peaceful and prosperous United States), such policies warrant continuation and can be used as the foundations for far-reaching policies. If they do not promote these ends, they should be firmly and quietly scrapped. In the committee's view, therefore, the country should recognize that any antiproliferation measures (Civex, no reprocessing, or whatever) are not ends in themselves, but only practical measures in the service of policy decisions.

NUCLEAR GROWTH RATES AND THE EFFECT OF REACTOR TYPES ON DEMAND FOR URANIUM

Nuclear power, as visualized now or for the intermediate period of this study, can contribute to our energy supply as a source of electricity. Thus, the contribution we can expect from nuclear power is tied to projections of the demand for electricity. A number of sources compete to satisfy this demand—coal, nuclear, hydroelectric, geothermal, and advanced systems. Nuclear power will only be a part of a mix. If a large part is desirable for nuclear power, it could be constrained by the availability of uranium, or by the availability of reactor types and fuel cycles that would make more efficient use of the available uranium.

This section takes up these questions within the context of a neutral policy. Policy decisions are rarely neutral. Even domestic decisions to select one reactor or fuel cycle over another cannot be simplified and isolated from their context, a context determined by the mutual interaction of technology, the environment, human values, and world affairs. The relative importance of various measures to achieve resistance to diversion and proliferation, the measure of risk and benefit the public assigns to nuclear power, the desirability of competing technologies, and the interaction between the United States and the rest of the world will figure as prominently in domestic decisions about nuclear power as considerations of demand and supply.

NUCLEAR GROWTH SCENARIOS, URANIUM REQUIREMENTS, AND REACTOR TYPES

The question examined here is, "What are the implications for uranium requirements of filling nuclear demands by various reactor and fuel cycle strategies?" The number of variables that affect the answer is large. Several studies have been conducted on the subject, using slightly different assumptions. The results, however, exhibit similar characteristics. The results presented here have their own set of detailed assumptions, but we consider them to be typical. The conclusions to be drawn are qualitative rather than quantitative. (The examples are taken from Zebroski and Sehgal.)[16]

Figure 5-2 shows introduction and installation schedules with corresponding requirements for uranium of two cases, both of which correspond to fairly low growth of demand for nuclear electricity. Although 200 GWe of installed electrical generating capacity from nuclear power plants in 1990 is slightly higher than the installed capacity now scheduled or expected by that date, the approximately 50 GWe assumed in the figure to be installed from 1990 to 2000, and the approximately 60 GWe to be installed between 2000 and 2010, are well below the expected figures. Case 1 assumes early introduction of advanced reactors (1987) and a rate of installation that results in about 80 GWe by 2000, or a third of total nuclear generating capacity. Case 2 assumes that advanced reactors are introduced in 1995* and installed at a rate resulting in 20 GWe by 2000 (about 8 percent of total nuclear generating capacity), and in 310 GWe by 2030 (62 percent of total nuclear generating capacity). In both cases, the total of the uranium consumption and forward commitments for fueling any of the combinations of reactor types that might be selected to achieve the postulated demand for nuclear generating capacity falls below 2 million tons, even by 2030. Indeed, with only light water reactors on a once-through cycle (the most resource-intensive system), cumulative ore requirements by 2010 would amount to just 1.2 million tons of U_3O_8. The use of more separative work in the enrichment process, leaving tails at an assay of 0.1 or 0.15 percent, would hold the resource requirement below 2 million tons until 2030, but not beyond.

Figure 5-3 illustrates three schedules of nuclear installation to meet moderately high growth of demand for nuclear-generated electricity. The starting point in 1990 is about 250 GWe rather than 200, but this difference does not affect the conclusions that can be drawn. The cases are (1) early introduction of advanced reactors and gradual installation, (2)

*See statement 5-14, by L. F. Lischer, Appendix A.

early introduction and rapid installation, and (3) late introduction and rapid installation. A resource base of 2 million tons of U_3O_8 is sufficient to meet any of these three schedules to 2010, but only those cases that assume the introduction of fast breeders or high utilization of ^{235}U from natural uranium (low tails assay) and fissile-isotope recycle in advanced converters (or both) manage to avoid outstripping 4 million tons of U_3O_8 in consumption and forward commitments by 2030.

Figure 5-4 depicts a very high rate of growth in nuclear generating capacity. In fact, Figure 5-4 is based on an unrealistic initial rate of installation—400 GWe by 1990—but as with Figure 5-3, the qualitative conclusions are not affected. In all cases, 2 million tons of uranium resources are exhausted soon after the year 2000, and 4 million tons are exhausted soon after 2010. Only those schedules that show early introduction and rapid installation of fast breeder reactors keep consumption and forward commitments of uranium oxide under 6 million tons by 2030. The simultaneous introduction and installation of advanced converters helps relieve some of the pressure on resources, but in all the cases illustrated, the demands on uranium, and possibly on thorium, exceed the rate of production the committee considers possible with present methods, as detailed in the section on uranium production. This scenario is therefore unrealistic unless a "heroic" program of breeder deployment is begun immediately.

The projections illustrated in Figures 5-2, 5-3, and 5-4 correspond roughly to the CONAES study scenarios, assuming that half or more of the electrical generating capacity required is supplied by nuclear power plants. Figure 5-2 represents projections compatible with the low- to medium-consumption scenarios I_3, II_2, and III_2; Figure 5-3, with the medium-high to high scenarios III_3, IV_2, and IV_3; and Figure 5-4, with the specific high-electrification variants of these latter scenarios. (See chapter 11 for details of the study scenarios.) One difference between the examples presented here and the CONAES study scenarios is that the former show demand for electricity continuing to rise at a signficant rate, while the latter show demand for electricity tending to saturate by 2010.

AVAILABILITY OF URANIUM[17]

In the previous section, the demands of various schedules for installation of nuclear electrical capacity, and of various strategies for developing and installing reactor systems, were expressed in terms of the amounts of uranium they might require. The availability of uranium could limit some of these strategies or schedules.

The federal government, for many years the only customer for uranium,

L = LWR
LP = LWR with PU Recycle
F = LWR + LMFBR
FP = LWR with PU Recycle + LMFBR
H = LWR + HWR
HP = LWR + HWR both with PU Recycle
FT = LWR with Thorium Cycles + LMFBR

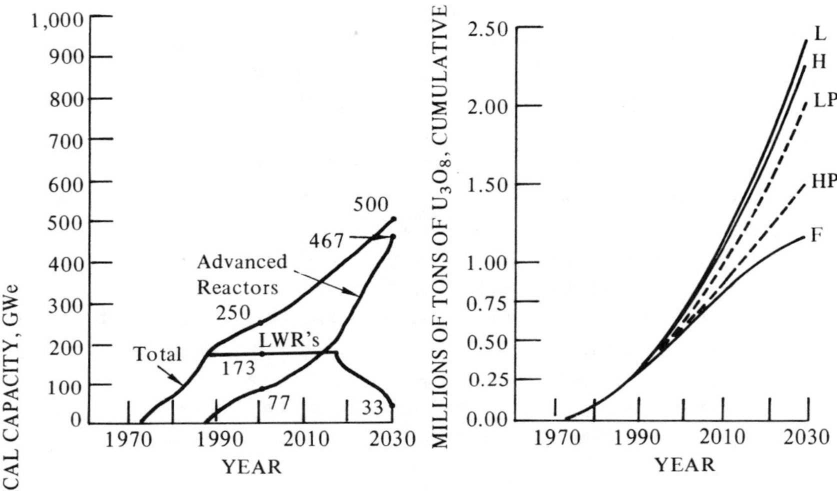

Case 1: Early advanced-reactor introduction

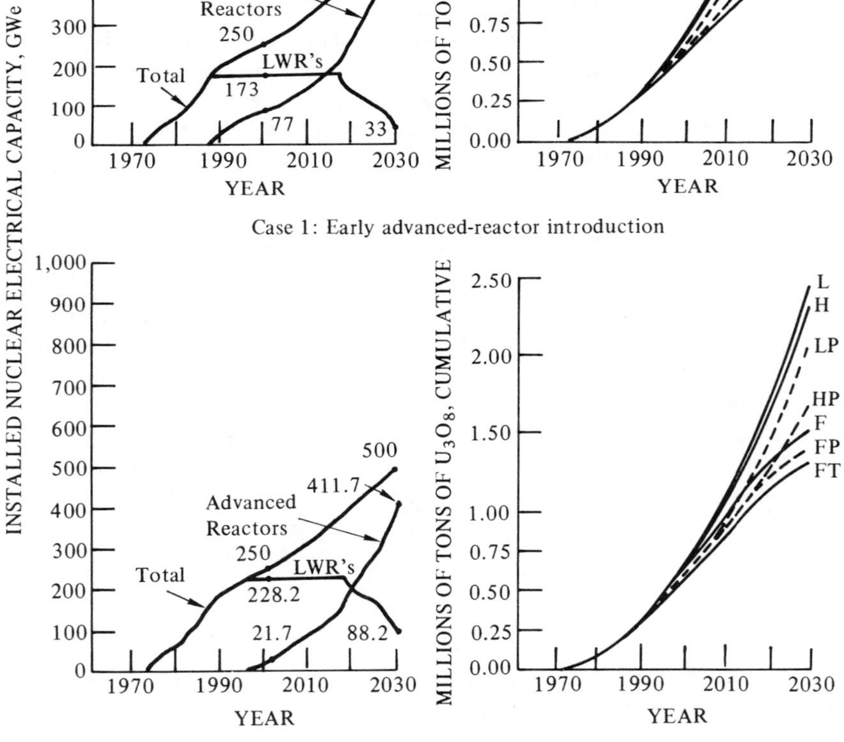

Case 2: Late advanced-reactor introduction

FIGURE 5-2 Low growth of installed nuclear capacity (350 GWe by 2010): comparison of reactor types and cumulative ore requirements from 1972 forward. In case 1, LWR's without recycle would approach an estimated 2.4 million tons of uranium supplies by 2030. Advanced converters (introduced in 1987 and reaching 80 GWe installed capacity by 2000) with reprocessing, along with reprocessing in LWR's, could significantly extend the life of nuclear power within the uranium supply estimate of the Uranium Resource Group (1.8 million tons) through 2030. In case 2, late introduction of advanced converters offers less extension time than case 1, but still more than 10 years. Advanced converters, introduced in 1995, reach an installed capacity of 20 GWe by 2000. Source: Adapted from E. L. Zebroski and B. Sehgal, "Advanced Reactor Development Goals and Near-Term and Mid-Term Opportunities for Development" (Paper presented to the American Nuclear Society, Washington, D.C., November 18, 1976).

---

publishes a systematic annual estimate of potential resources and reserves. The figures for reserves represent ore deposits that have been measured by drilling, and assayed within a 20 percent degree of accuracy, and material available under existing practices as by-products from mining or treating other materials. Reserves can be characterized as the supply of ore the mining industry is confident it can produce. Potential resources represent ore deposits inferred to exist from tacit knowledge and field judgments: the amount of ore that can reasonably be expected to occur in producing strata, or in nonproducing areas that display the characteristics of producing areas.

Figure 5-5 shows the annual estimates made in recent years. The estimates are set out by incremental or forward costs: the additional cost to recover uranium oxide over expenditures already incurred in exploring, filing claims, buying or leasing mineral rights, and determining the extent of deposits. "Forward costs" make no allowance for profit; they represent neither total costs to recover ores nor market prices. The reserves of uranium set out in various categories of forward cost should not be interpreted as representing the ores that can be economically recovered today. They do represent relative cost; for example, reserves in the $30/lb category will be about twice as expensive to recover as those in the $15/lb category. The estimates of potential resources are set out in three categories: probable, possible, and speculative.

The Uranium Resource Group of this study reviewed the 1976 estimate and compared it to information from geologists, experts advising the uranium industry, and others. The group concluded that the possible and speculative categories were not well enough established for resource planning. It judged that the estimate for probable resources, 1.06 million tons, was the closest approximation to a quantitative estimate for *all* potential resources[18] and estimated the domestic uranium producible by

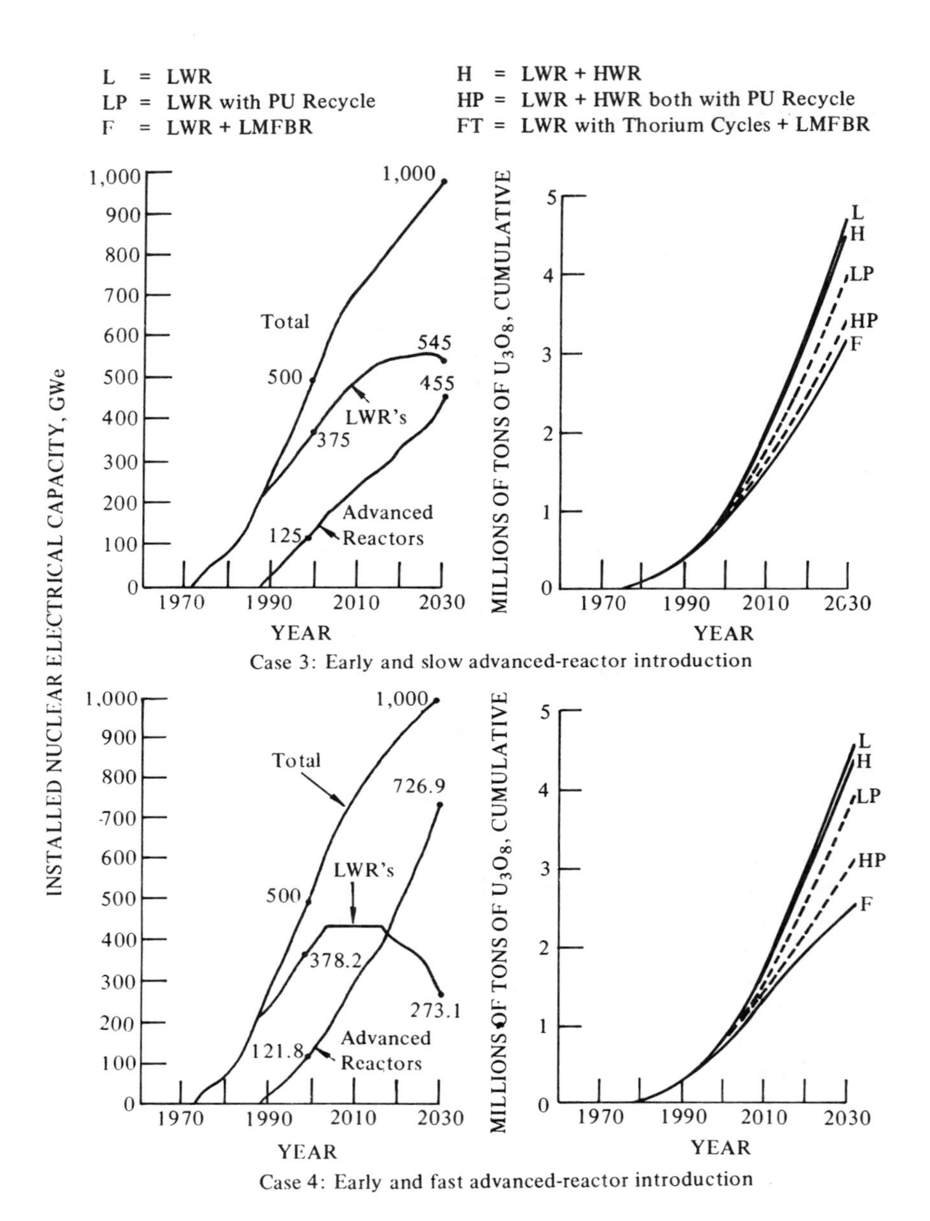

Case 3: Early and slow advanced-reactor introduction

Case 4: Early and fast advanced-reactor introduction

conventional mining at 1.7 million tons. The federal government has reported higher estimates since, partly as the result of better information on uranium in the \$30–\$50/lb forward-cost category. This uranium is not producible under prevailing market conditions, but its production will become economic as higher-grade deposits are depleted.

Table 5-1 shows the increase in estimates of reserves and probable potential resources since 1976. CONAES has adjusted its resource estimate, reflecting this increase, to 2.4 million tons of U_3O_8 (excluding by-product recovery).

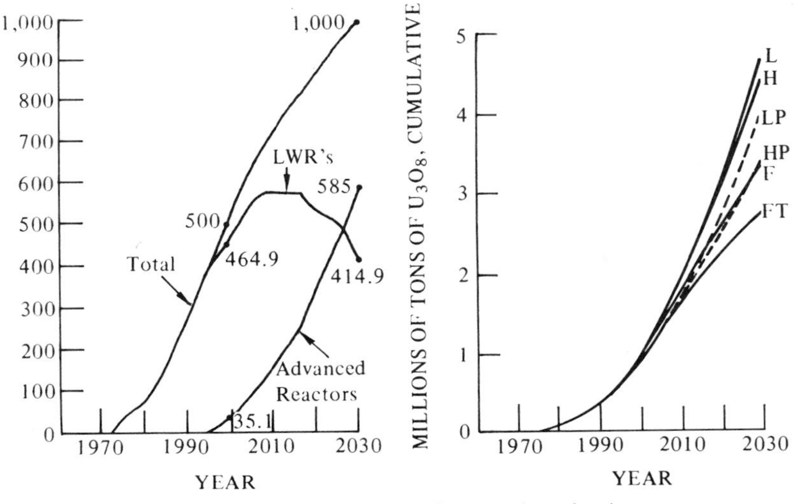

FIGURE 5-3 Moderately high growth of installed nuclear capacity (700 GWe by 2010): comparison of reactor types and cumulative ore requirements from 1972 forward. In all three cases, 2.4 million tons of uranium can sustain moderate rates of growth in installed nuclear capacity to 2010, with recycle of fissile isotopes in the selected mix of reactors. In case 5, however, the late introducton of advanced converters would realize significant savings in resource consumption only if the uranium resource base is more than 3 million tons. At the lower estimate, the late-arriving breeder would be necessary to sustain growth in installed nuclear capacity beyond 2020. In all three cases, the introduction of advanced converters or breeders, or both, is necessary to hold the consumption of uranium and forward commitments below 3 million tons to 2030. Source: Adapted from E. L. Zebroski and B. Sehgal, "Advanced Reactor Development Goals and Near-Term and Mid-Term Opportunities for Development" (Paper presented to the American Nuclear Society, Washington, D.C., November 18, 1976).

Experience with other minerals indicates that projections of potential resources usually fall far below the resources subsequently discovered and produced.[19] This is because increasing quantities of lower-grade ores are exploitable as the price rises. But uranium deposits that have been identified in the United States are either rather high grade ($>$ 700 ppm uranium by weight) or quite low grade ($<$ 100 ppm, or 0.01 percent) with no large intermediate range. Many geologists believe that this is an intrinsic and unique feature of uranium mineralization. Moreover, uranium deposits tend to be discrete and sharply bounded; the edges of the deposits, contrary to experience with other minerals, are barren. Finally, low-grade deposits (less than 0.01 percent uranium oxide), such as those in black shale, may not be producible in quantity because of the massive rock volumes to be moved, uncertain milling requirements, and vast quantities of mill tailings to be managed.

Uranium exploration is a chancy venture. Discovery of the Grants mineral belt in New Mexico (1956–1957), and of the Wyoming Basins deposit (1969–1970) added 60,000 tons/yr to uranium reserves, but

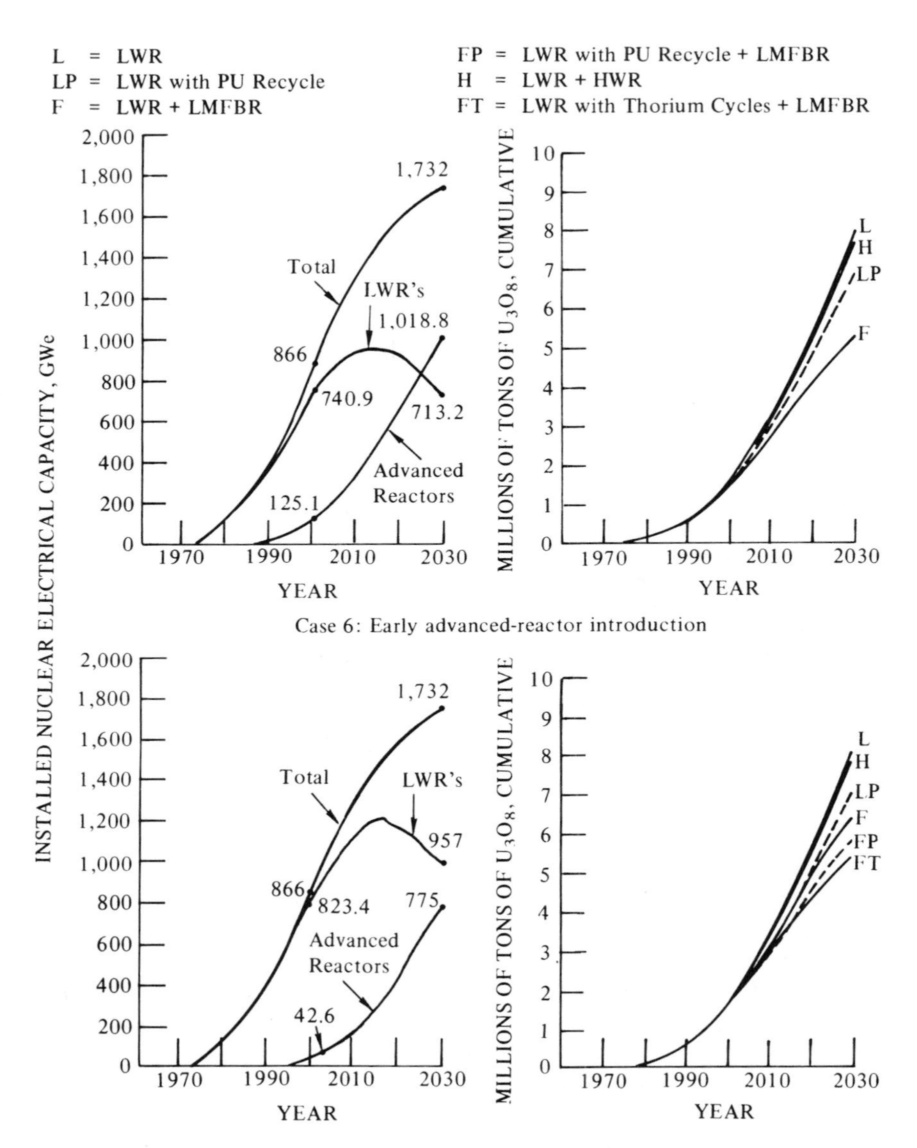

Case 6: Early advanced-reactor introduction

Case 7: Late advanced-reactor introduction

FIGURE 5-4 High growth of installed nuclear capacity (1300 GWe by 2010): comparison of reactor types and cumulative ore requirements from 1972 forward. Conditions can be met only by introducing fast breeder reactors and assuming abundant supplies of uranium—5 million tons for early introduction and rapid rates of installation. Source: Adapted from E. L. Zebroski and B. Sehgal, "Advanced Reactor Development Goals and Near-Term and Mid-Term Opportunities for Development" (Paper presented to the American Nuclear Society, Washington, D.C., November 18, 1976).

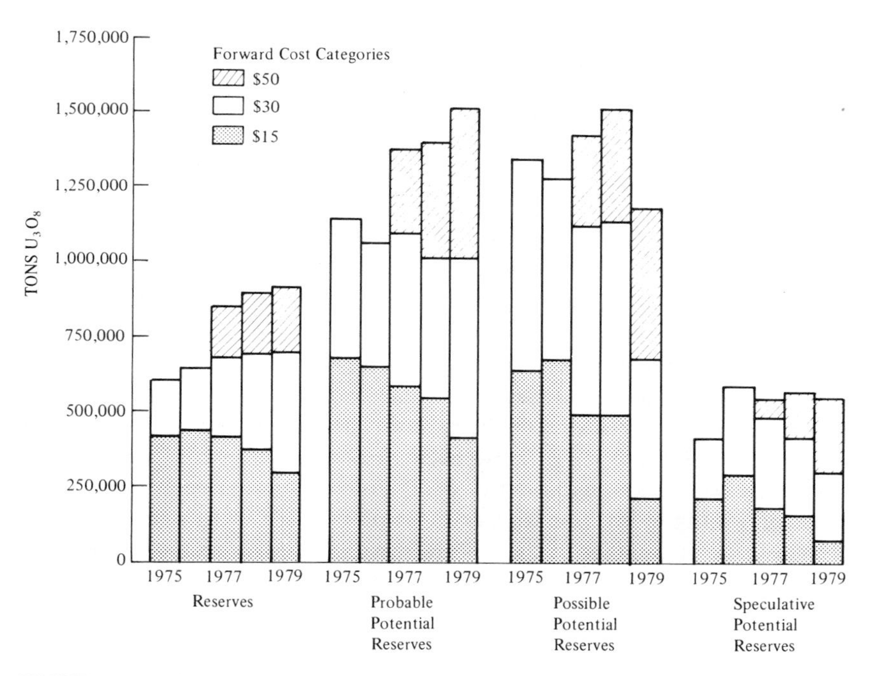

FIGURE 5-5 Changes in domestic uranium estimates, as of January 1 in each of the years 1975–1979. Source: Leon T. Silver, "Discussion of U.S. Uranium Supplies," testimony before the Subcommittee on Energy Research and Production, Committee on Science and Technology, U.S. House of Representatives, May 31, 1979.

between these two discovery periods, annual additions totaled only 10,000–30,000 tons/yr. Exploration for new uranium deposits concentrates on areas similar to those with known deposits. Only one new district has been discovered in more than 15 years. The odds that uranium exploration will lead to discovery can be improved by better understanding of the factors that govern deposition.

The Uranium Resource Group has estimated that under prevailing

TABLE 5-1 Estimates of Reserves and Probable Potential Resources of Uranium in the United States (tons of U_3O_8)

Year	Reserves	Probable Potential	Total
1976	640,000	1,060,000	1,700,000
1977	840,000	1,370,000	2,210,000
1978	890,000	1,395,000	2,285,000
1979	920,000	1,505,000	2,425,000

Source: U.S. Department of Energy, *Statistical Data of the Uranium Industry* (Grand Junction, Colo.: Department of Energy (GJO-100 [79]), 1979), pp. 21, 30–31.

conditions, uranium discovery will increase to a rate of 50,000 tons/yr in 1985 and decline thereafter. Under conditions favoring discovery, the rate would improve (for a resource base of 1.8 million tons) to 62,000 tons/yr by 1991, and under conditions favoring all-out discovery efforts, rates of 80,000 tons/yr might be reached by 1989.

To maintain adequate reserves—8–12 years' worth of production at prevailing rates—new ore deposits must be discovered and measured continuously. Production of these reserves normally lags behind discovery about 5 years. Under conditions of uncertainty in future demand similar to those prevailing today, this lag would lengthen to about 10 years. The 5-yr lag could be maintained if there were reasonable assurance of demand.[20]

Expansion of exploration and mining must be accompanied by expansion of milling[21] capacity. Discovery of new ore deposits takes 1–5 years after exploratory effort begins; evaluation by drilling and assay takes an additional 1.5–2 years; mine development, 1–3 years, and construction of a mill, 2–3 years. Some of these steps must be completed before others can be undertaken. The deposit must prove sufficient to justify a new mill, for example. Others may be undertaken simultaneously. Nevertheless, 10 years usually elapses between the decision to expand uranium exploration and the production of uranium oxide.

The nuclear power industry must plan for relatively distant futures. A light water reactor requires at least 10 years to license and build, and the utility must be assured fuel at reasonably predictable prices for at least 10 full-power years of its 30-yr life. In planning for future advanced converters or breeders, the nuclear power industry will be particularly concerned that fuel is both available and producible at rates that correspond to the planned rate of buildup of the industry. Here the uncertainties that plague uranium resource availability interact with other uncertainties to complicate the planning of both suppliers and consumers.

Under conditions that constitute business as usual for uranium producers, and under prevailing reactor fueling practices, the rate of uranium production in 2000 could fuel only 228 GWe from light water reactors. Accelerating the rates of discovery and increasing production could provide fuel for 310 GWe[22] around the year 2000. If the producible uranium oxide is 2.25 million tons, a 310 GWe capacity might then be sustained for 5–10 years before being restricted by a declining production rate as reserves decline.

URANIUM ENRICHMENT

Several enrichment methods were tested in pilot plants during World War II, including gaseous diffusion, thermal diffusion, and modified mass spectrography. Gaseous diffusion proved most effective and economical. In this process, the lighter isotope ^{235}U, in the form of gaseous uranium hexafluoride (UF_6), passes more readily than ^{238}U through porous barriers. By repeating the process in successive stages of a cascade, any degree of enrichment can be achieved. In the United States, three facilities (in Oak Ridge, Tennessee, Paducah, Kentucky, and Portsmouth, Ohio) are operated as a single enrichment complex. Enrichment plants using gaseous diffusion also exist in Great Britain, France, the U.S.S.R., and China.

Gaseous diffusion is itself energy intensive. On the average, a little over 5 percent of the electrical energy generated by a light water reactor is needed for the enrichment process.

The increasing cost of energy has made the alternative method of separation by gas centrifuge economically competitive. In this process, the heavier ^{238}U (again, as uranium hexafluoride) is spun to the outside of a centrifuge and the lighter ^{235}U withdrawn from the center. The process requires one twentieth or less of the electricity needed in gaseous diffusion. The savings in electricity justify a large capital cost, and the Department of Energy is now planning to add at least one gas centrifuge to its enrichment facilities, at a cost of $4.2–$4.5 billion.[23] Abroad, a tripartite consortium (URENCO), sponsored by Great Britain, the Federal Republic of Germany, and the Netherlands, is building one plant in England and planning others on the European continent.

Among other processes in the United States, laser separation of isotopes has been demonstrated in the laboratory. A chemical exchange method is being investigated in France, and separation by jet nozzles, a process developed in Germany, is being tried on a commercial scale in Brazil and (by a different process) in South Africa. Developers of the chemical exchange method claim it is practical for slightly enriched uranium, but not for the higher enrichments that would be needed for weapons. If this is confirmed, it would be preferred as a proliferation-resistant technology. The laser process has some promise as a method for almost entirely removing ^{235}U from the rejected material (tails) of existing enrichment plants, in which case it could increase effective uranium reserves by perhaps 20–30 percent (this is the fraction of ^{235}U in natural uranium that is now discarded). The nozzle separation method does not appear practical for the United States, as it is too energy intensive.

Both the laser separation method (which is not yet proved) and the gas-centrifuge method have the potential to produce weapons-grade ^{235}U in significant quantity at relatively small-scale installations.* Both processes employ sophisticated technology that few countries can use. However, technology always spreads. Over time, these processes might provide the easiest access to nuclear weapons: Because of the low level of radioactivity involved, enrichment plants could be built secretly with greater ease than reactors or reprocessing plants.

The capacity of enrichment plants is measured in separative work units (SWU's) per year, which have the dimensions of mass flow rate (e.g., kilograms of uranium per year). The amount of uranium and separative work required to deliver a given amount of reactor fuel at a given enrichment can be varied within the limits of enriching plants to operate at different tails assays. For the Department of Energy's existing enriching complex, the tails assay can be varied from 0.2 percent to 0.3 percent ^{235}U without loss of separative capacity. Table 5-2 illustrates the variation possible in feed and separative work per kilogram of light water reactor fuel within this range.

To a limited extent, separative work substitutes for natural-uranium feed. As illustrated in Table 5-2, light water reactor fuel enriched to 3 percent ^{235}U can be produced at 0.2 percent tails assay with about 20 percent less natural-uranium feed, but with 26 percent more separative work, than at 0.3 percent tails assay.

The three-plant complex operating in the United States has a capacity of 18 million SWU/yr, and as of 1977 was intended to reach 28 million SWU/yr in 1981.[24] (Requirements have since decreased, and expansion has been delayed accordingly.) The expanded plants are expected to reach full-capacity production by 1985.[25] This expanded capacity has been committed to domestic and foreign obligations (323 GWe of light water reactors—two thirds in the United States and one third in foreign countries). The new centrifuge plant is expected to operate at its full capacity of 8.8 million SWU/yr in 1988.[26]

*Statement 5-15, by E. J. Gornowski: The laser isotope separation procedure cannot readily produce weapons-grade ^{235}U. The technology is not easy, unsophisticated, or garage scale.

TABLE 5-2 Ratios of Uranium Feed and Separative Work Units to a Kilogram of Enriched Reactor Fuel, 3 Percent ^{235}U, for Selected Tails Assays

	Tails Assay (percent ^{235}U)		
	0.20	0.25	0.30
Natural uranium, kg	5.479	5.965	6.569
Separative work units	4.306	3.811	3.425

Source: A. de la Garza, "An Overview of U.S. Enriching Resources," report to the Supply and Delivery Panel, Committee on Nuclear and Alternative Energy Systems, National Research Council, Washington, D.C., 1976.

Figure 5-6 illustrates production expected from enrichment plants over the next decade and contract commitments for separative work. The apparent gap between commitments and production from 1981 to 1988 could be closed if operation at high tails assay—about 0.36 percent—were possible. But this inordinately high tails assay would require more feed at a time when uranium supplies may become scarce. Conversely, since enrichment plants are being added in Europe, and since the Soviet Union apparently has spare enrichment capacity, some relief might be available from these sources.

A recent report points out that utilities holding long-term fixed-commitment contracts are required to provide uranium feed to the enrichment complex in amounts that may not agree with their fuel requirements, and it suggests that most of the apparent gap between production capacity and commitments could be eliminated through case-by-case adjustments.[27] The substitution of fuel "enriched" by the addition of plutonium from reprocessed old fuel could also help prevent an "enrichment gap."[28]

Whether additional enrichment capacity will be needed beyond 1990, and if so when, depends on the number and type of reactors built and their particular fuel needs. Existing and planned enrichment capacity, for example, can supply the fuel for 215 GWe generated by today's light water reactors using a once-through cycle. Domestic capacity might approach this figure in the early 1990s, and in addition, it would be proper for the United States to supply its share of the enrichment needs of countries that must buy this service. The introduction of new reactors in the form of advanced converters would also affect projected demands for enrichment. Heavy water reactors using natural uranium, or uranium enriched to 1.2 percent ^{235}U, have no enrichment requirements in the first case, and

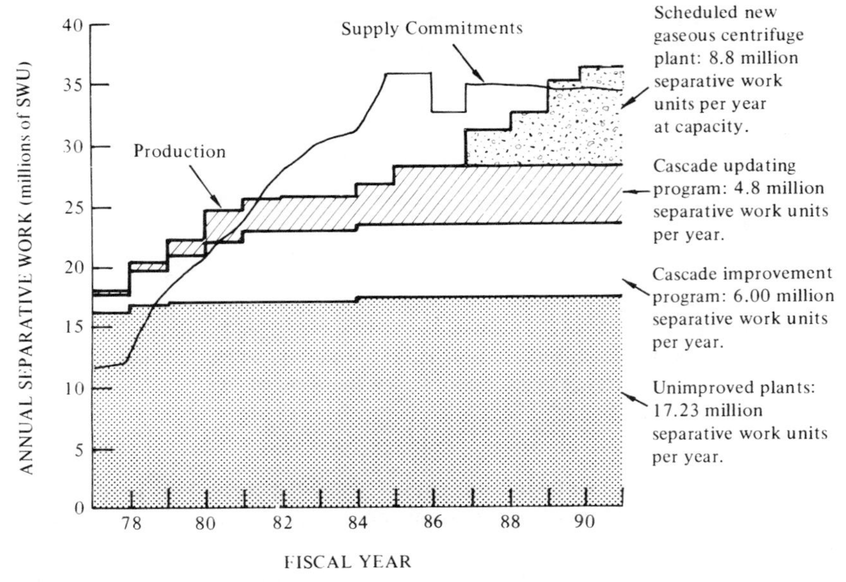

FIGURE 5-6 Separative capacity schedule for the Department of Energy complex of uranium enrichment plants (in Oak Ridge, Tennessee, Portsmouth, Ohio, and Paducah, Kentucky) and contract supply commitments to domestic and foreign utilities.

modest requirements in the second. Advanced converters on the thorium-uranium cycle (such as the HTGR or LWBR) with fuel recycle have heavy requirements for highly enriched uranium for their initial critical loadings, but after the first loading, they use less separative work than LWR's.[29] Breeder reactors started on plutonium have no separative work requirements and could indeed provide some fissile fuel (above and beyond what is needed to fuel new breeders) to converters.

However, in spite of the uncertainties in future demand for separative work, enrichment is not a bottleneck in nuclear expansion. An enrichment plant takes only as much lead time as a reactor to plan and build, so enrichment capacity can be scheduled to match reactor commitments.*

*See statement 5-16, by E. J. Gornowski, Appendix A.

NUCLEAR REACTORS AND FUEL CYCLES

Nuclear power is generated from the fission of heavy elements. The heat of fission is liberated in nuclear reactors. This heat is then transported by the reactor coolant to external equipment, where it is converted into electricity. The different types of reactors that have been developed for this purpose employ the major approaches that have been considered attractive. Table 5-3 lists these reactors and the status of their development.

Nuclear fuel is made of isotopes that are easily fissioned by slow neutrons. These are known as fissile isotopes. The only fissile isotope that exists in nature in usable abundance is ^{235}U. This makes up about 0.7 percent of natural uranium, and natural uranium is therefore the basic resource for nuclear power. There are two other isotopes from which fissile isotopes can be made in a reactor. These are ^{238}U, the more common isotope and the major constituent of natural uranium, and ^{232}Th, naturally occurring thorium. Fissile ^{239}Pu, the commonest form of plutonium, is made from ^{238}U, and fissile ^{233}U is made from thorium (^{238}U and ^{232}Th are called fertile isotopes).

As the fissile isotopes originally loaded in a reactor are destroyed, new fissile isotopes are formed from fertile isotopes. The relative rate of replacement is the conversion ratio. When the conversion ratio is greater than 1, new fissile isotopes are formed faster than the original fissile isotopes undergo fission, and the reactor is called a breeder. A reactor that operates at conversion ratios below 1 is known as a converter.

The fuel cycle for a reactor describes the way the fissile and fertile isotopes are used. If natural or slightly enriched uranium is the fuel, plutonium is formed and partly fissioned during reactor operation. Spent fuel still contains some ^{235}U and ^{239}Pu. It could be recycled, or the spent fuel could be considered waste material. Recycle reduces the amount of fresh ^{235}U that has to be supplied, and thus reduces the commitment of natural uranium needed to fuel the reactor.

More highly enriched uranium can also be mixed with thorium in reactor fuel, creating a Th-U fuel cycle. This is not generally economic unless the ^{235}U and ^{233}U left in spent fuel are recycled; however, some once-through cycles in some reactors are only slightly more expensive than their uranium counterparts. A version of the Th-U cycle, in which some ^{238}U is added to keep concentrations of fissile isotopes in uranium below weapons-usable amounts, is the denatured U-Th cycle. This cycle is discussed later in this chapter (under "Reprocessing Alternatives"). Similarly, plutonium could be extracted from spent reactor fuel and combined with ^{238}U to form new fuel. This would be done in recycling natural or slightly enriched uranium fuel, using plutonium at low concentrations. This is the fuel cycle for fast breeder reactors—at higher

TABLE 5-3 Nuclear Reactors and Fuel Cycles: Development Status

Reactor Type	Fuel Cycles	Development Status	Possible Commercial Introduction in the United States[a]
Light water reactor (LWR)	Slightly enriched U (~3 percent ^{235}U)	Commercial in United States	1960
Spectral-shift-control reactor (SSCR)	Th-U[b]	Conceptual designs, small experiment run; borrows LWR technology	1990; fuel cycle, 1995 or later[c]
Light water breeder reactor (LWBR)	Th-U[b]	Experiment running; borrows LWR technology; fuel cycle not developed	1990; fuel cycle, 1995 or later[c]
Heavy water reactor (CANDU or HWR)	Natural uranium	Commercial in Canada, some U.S. experience	1990
	Slightly enriched U (~1.2 percent ^{235}U)	Modification of existing designs	1995
	Th-U[b]	Modification of designs; fuel cycle not developed	1995
High-temperature gas-cooled reactor (HTGR)	Th-U[b]	Demonstration running; related development in Germany; fuel cycle partly developed	1985; fuel cycle, 1995 or later[c]
Molten-salt (breeder) reactor (MSR or MSBR)	Th-U[b]	Small experiment run; much more development needed	2005
Liquid-metal fast breeder reactor (LMFBR)	U-Pu[b]	Many demonstrations in the United States and abroad	1995
	Th-U[b]	Fuel cycle not developed	1995
Gas-cooled fast breeder reactor (GCFBR)	U-Pu[b] Th-U[b]	Concepts only; borrows LMFBR and HTGR technology	2000

[a] Based on the assumption of firm decisions in 1978 to proceed with commercialization. No institutional delays have been considered except those associated with adapting foreign technology. On the basis of light water reactor experience, it can be estimated that it would take about an additional 15 years after introduction to have significant capacity in place.
[b] Indicated fuel cycles demand reprocessing.
[c] Thorium-uranium fuel reprocessing is less developed than uranium-plutonium reprocessing. Indicated reactors could operate for several years before accumulating enough recyclable material for reprocessing.

concentrations of plutonium—although a Th-U cycle could also be used. The cycle with plutonium and uranium is designated the U-Pu cycle; as with the Th-U cycle, recycle is always assumed.

Table 5-4 presents fuel cycle characteristics of the reactors in Table 5-3. Some caution is needed in interpreting Table 5-4. There are continuous possibilities for varying enrichments, fuel concentrations, lattice spacings, and other reactor fueling parameters. How designers choose to operate a fuel cycle depends on finding a minimum fueling cost within broad technical limits, and only those parameters that seem interesting now are presented. Table 5-4 is schematic and highly simplified. In some cases, particularly those involving future developments, relatively crude estimates have been used.

Virtually all the nuclear power in the United States today, and all planned expansion of nuclear power, is in the form of light water reactors. As can be seen in Table 5-4, these reactors make large demands on supplies of natural uranium at economical prices. Such supplies (as pointed out under "Availability of Uranium") are limited; therefore, this type of reactor has a limited term of service. How long this term might be depends, of course, on the demand for nuclear power as well as on the supply of uranium. A very long-lasting nuclear industry could only persist by the use of some form of breeder reactor, whose ultimate source of fuel, the fertile isotopes, could probably be supplied at economical prices for hundreds of millennia.

All U-Pu recycle schemes and all Th-U schemes that do not use denatured uranium present the problem that pure fissile material could be chemically isolated during the recycle steps. In particular, breeder reactors are intrinsically fuel-recycle systems. As such, they present the possibility that nuclear materials usable in weapons could be diverted, or that the fuel cycle could be used in national proliferation of nuclear arsenals. On these grounds, the United States has deferred civilian nuclear fuel reprocessing and is attempting to persuade other countries to do the same. On the same grounds, the Administration has opposed proceeding with the latest demonstration breeder reactor project in the United States, the Clinch River breeder reactor (CRBR), a subject discussed later in this chapter.

Nevertheless, a vigorous and durable nuclear power industry could be a very important part of our future energy supply system. Therefore, in what follows, we explore the various reactor systems that might be used, even though many interesting systems rely on recycle of nuclear fuel.

ADVANCED REACTORS[30]

It has already been noted (Table 5-4) that the current generation of power reactors in the United States, consisting of light water reactors, is not very

TABLE 5-4 Fueling Characteristics of Various Reactors (each 1 GWe capacity)

Reactor Type	Fuel Cycle	Conversion Ratio	Natural U_3O_8 Committed to System Inventory (tons)[a]	Natural U_3O_8 Consumed over System Life (tons)[b]	Thorium Requirements (initial inventory plus tons per year)	Reference[c]
Light water reactor	~3 percent ^{235}U, once-through	0.5–0.65	700	5300	—	1
	~3 percent ^{235}U, U recycle	0.5–0.65	~1000	4200	—	
	~3 percent ^{235}U, U and Pu recycle	0.45–0.55	~1000	3200	.	
	Th-^{235}U, once-through	0.6	900	6200	120 + 30	
	Th-^{235}U, U recycle	0.65	~1200	2800	120 + 30	
Spectral-shift-control reactor	Th-^{235}U, U recycle	0.75	~1300	2300	120 + 10	2
Light water breeder reactor	Th-^{235}U, seed recycle	~0.75	700	5000	120 + 40	3
	Th-^{233}U, recycle	~1	~1000[d]	0	4[e]	
Heavy water reactor (CANDU)	Natural U, once-through	0.7	160	4300	—	4
	Natural U, Pu recycle	0.6	~350	1800	—	
	1.2 percent ^{235}U, once-through	0.7	310	3300	—	
	Th-^{235}U, once-through	0.75	700	2800	160 + 140	
	Th-^{235}U, U recycle	0.9	~1200	1200	160 + 140	
High-temperature gas-cooled reactor (pebble bed)	~8 percent ^{235}U, once-through	0.58	280	4400	—	5
	Th-^{235}U, once-through	0.58	300	4100	not available	

High-temperature gas-cooled reactor (prismatic)	Th-^{235}U, once-through Th-^{235}U, U recycle	0.6 0.7–0.9	280 ~600	4200 2300	150 + 10 150 + 10	6
Molten-salt breeder reactor	Th-^{233}U	0.9–1.07	~400[d]	0	Inventory not available; 3	7
Liquid-metal fast breeder reactor	U-Pu Th-^{233}U	>1.15 >1.1	1000–2000[d] 1000–2000[d]	Breeder Breeder	— Not available	7
Gas-cooled fast breeder reactor	U-Pu Th-^{233}U	>1.2 >1.15	1000–2000[d] 1000–2000[d]	Breeder Breeder	— Not available	7

[a] Estimate of reactor inventory plus out-of-reactor inventory for recycle systems.

[b] Does not include inventory, which is assumed to be recoverable at end of reactor life.

[c] Sources: As indicated on the table, the following sources were consulted for descriptions of reactor fueling characteristics. The conversion ratios and tonnages listed in the table (intended for comparative purposes) are the responsibility of CONAES.

1. P. R. Kasten *et al.*, *Assessment of the Thorium Fuel Cycle in Power Reactors* (Oak Ridge, Tenn.: Oak Ridge National Laboratory (ORNL/TM-5565), 1977), Appendix B.

2. Resources Planning Associates, *The Economics of Utilization of Thorium in Nuclear Reactors—Textual Annexes 1 and 2* (Oak Ridge, Tenn.: Oak Ridge National Laboratory (ORNL/TM-6332), n.d.), Table 2.13, p. 158.

3. Battelle Columbus Laboratories, *Study of Advanced Fission Power Reactor Development for the United States*, vol. 3 (Columbus, Ohio: Battelle Columbus Laboratories (BCL-NSF-C946-2), 1976), pp. C-80–C-120.

4. J. S. Foster and E. Critoph, "The Status of the Canadian Nuclear Power Program and Possible Future Strategies," *Annals of Nuclear Energy* 2 (1975):689–703.

5. E. Teuchert *et al.*, "Once-Through Cycles in the Pebble Bed HTR," *ANS Transactions* 27 (1977):460.

6. Battelle Columbus Laboratories, *Study of Advanced Fission Power Reactor Development for the United States*, vols. 2 and 3 (Columbus, Ohio: Battelle Columbus Laboratories (BCL-NSF-C946-2), 1976), vol. 2, pp. IV–II and vol. 3, pp. C-80–C-120.

7. National Research Council, *U.S. Energy Supply Prospects to 2010*, Committee on Nuclear and Alternative Energy Systems, Supply and Delivery Panel (Washington, D.C.: National Academy of Sciences, 1979).

[d] Estimates based on requirements for equivalent highly enriched ^{235}U to initiate fuel cycles.

[e] Published estimates of LWBR thorium requirements apparently assume refabrication.

efficient in its use of uranium. This is particularly significant if fuel is not recycled, as the fissile isotopes in spent fuel could replace some of the ^{235}U in freshly mined uranium. It is possible to use uranium more efficiently. The key is to use reactors with higher conversion ratios and long fuel lifetimes. Higher conversion ratios can substitute for recycling to the extent that the fissile atoms formed undergo fission in place. With recycle, the higher conversion ratio permits more fissile atoms to be substituted for natural ^{235}U.

Although light water reactors do not now have high conversion ratios, a great deal of the plutonium created in their operation undergoes fission in place. About one third of all the energy in LWR's is obtained from plutonium fission, and at the end of fuel life, more than 60 percent of the fissions occur in ^{239}Pu. LWR's could, in principle, be designed for higher conversion ratios and better use of natural uranium, a fact that should be remembered in comparing them to other reactors. Such designs would have lower enrichments and burnups than existing LWR cycles and could only achieve better use of natural uranium through plutonium recycle.[31] The required rate of reprocessing might be twice as high, per unit of electrical energy generated, as that for the standard LWR recycle mode estimated in Table 5-4, but lifetime uranium consumption would be less than 3000 tons. For all reactors, the conversion ratio varies with the composition of the fuel loaded and with fuel management. Differences among reactors often correspond to differences in the conversion ratio that can be readily achieved for fuel loading and management practices permitting economical power generation.

The reactors proposed to achieve greater efficiency in the use of fissile resources fall into two classes: advanced converters and breeders. Advanced converters can be designed to achieve conversion ratios ranging from 0.7 to slightly more than 1. (Light water reactors operate at conversion ratios of 0.6 or less.) Breeder reactors can be designed to achieve conversion significantly greater than 1, although they could obviously be designed and operated at lower conversion ratios. For some breeders, such as the molten-salt breeder reactor (MSBR), the reduction in fissile inventory could be sufficient for greater economy (i.e., the savings in charges against inventory could be greater than the loss of income from product sale and the extra cost of feed material).

Prototypes and designs for various types of advanced converters and breeder reactors have been developed in the United States and other countries. The functional and practical points that must be considered to evaluate the relative merits of these reactors and fuel cycles cannot all be assessed equally for the designs and prototypes. Some reactor designs are only conceptual, others have been tested through small pilot plants, and others are close to commercial status. A complete safety assessment, for example, requires a detailed commercial design. The probability of various accidents depends more on the details of design than on the reactor's generic characteristics. The only design that has been completely assessed for safety by the standards of the United States is the light water reactor.

ADVANCED CONVERTER DESIGNS

Light water reactors can be designed for conversion of thorium to ^{233}U. Although a light water reactor operated on a Th-U fuel cycle with reprocessing could achieve conversion ratios of about 0.7, the initial fissile inventory would require highly enriched uranium. (This highly enriched fuel may be ruled out by regulations to safeguard the fuel cycle.) The lifetime fuel requirements of a light water reactor on this fuel cycle could be 50–60 percent lower than those of an LWR on the once-through cycle, but the reactor would have to operate some time before enough ^{233}U accumulated for reprocessing. Preliminary studies suggest that Th-U fueling of light water reactors would be uneconomical;[32] however, the relatively modest changes required represent the most immediate opportunity to begin learning the engineering of Th-U fuel cycles. Spectral-shift-control reactors (SSCR's) are essentially similar to pressurized-water reactors (PWR's), one of the two light water reactors in use today. The coolant/moderator is changed during operation from heavy water to ordinary light water as the fissile content of the fuel in the core decreases. The development of this reactor consists mostly of conceptual studies, although a small pilot plant has been operated in Belgium.[33]

The light water breeder reactor, in spite of its name, is actually an advanced converter or break-even thermal breeder. Its design goal is to convert enough fertile material to fissile material to completely reload the core after accounting for fuel cycle losses. A demonstration LWBR achieved criticality in 1977, and experimenters anticipate that the reactor will be fully demonstrated by 1985.

The CANDU, an advanced converter designed in Canada to operate efficiently on natural (unenriched) uranium, supplies about 10 percent of the kilowatt-hours (kWh) generated by nuclear power in North America. This reactor employs heavy water as the reactor's moderator and coolant, and allows on-line refueling. The introduction and installation of heavy water reactors offers a relatively near-term opportunity in the United States to improve uranium efficiency on the once-through fuel cycle. The use of slightly enriched uranium oxide, perhaps 1.0–1.2 percent ^{235}U, would reduce the fuel requirements of a heavy water reactor 40 percent below the fuel requirements of a comparable light water reactor on a once-through fuel cycle. A CANDU-type reactor could also be designed to incorporate thorium fuel elements. This reactor could be operated either on a once-through fuel cycle or with reprocessing to recover the ^{233}U and thorium for recycle. A once-through cycle would require, however, considerably more uranium mining and enrichment. (See Table 5-4 for comparative uranium consumption.)

The high-temperature gas-cooled reactor loads fuel elements into graphite blocks that serve as the reactor's moderator. The coolant is high-temperature helium. A 330-megawatt (electric) (MWe) commercial plant is operating near Fort Saint Vrain, Colorado. HTGR's appear capable of operating at both higher thermal efficiency and higher conversion ratios than light water reactors, but their expanded use depends on successful development of economical reprocessing for graphite-based fuel. If the high-temperature gas used to cool the graphite core could be used to drive a gas turbine directly, the thermal efficiency of this reactor could be further improved and the reactor's operation would be freed of requirements for water. Moreover, HTGR's, or a version of the related German pebble-bed reactor (whose fuel is contained inside balls of graphite), could be used to supply process heat as well as electricity.

In the molten-salt reactors, solid fuel assemblies are replaced by uranium fluoride and thorium fluoride dissolved in a molten fluoride-salt mixture. The salt is circulated to a heat exchanger external to the core. The molten-salt breeder reactor adds chemical kidneys in the fuel's external circulation system for continuous reprocessing of the fuel/coolant. This feature reduces the total inventory of fissile material committed to the power plant and its fuel cycle, as compared to other breeder reactors. Volatile fission products are also continuously removed, a step that permits true breeding in the Th-U fuel cycle with a thermal-neutron spectrum. A small pilot molten-salt reactor (10 megawatt (thermal) (MWt)) was operated at Oak Ridge National Laboratories. The development of this concept is far behind that of other advanced converters and breeders. Success cannot be guaranteed because of formidable materials problems, but the advantages that might be realized in this type of system are considerable.

EVALUATION OF ADVANCED CONVERTERS

Several factors must be considered in evaluating advanced converters: ease of development, economic prospects, and compatibility with other policy objectives of the nuclear program. On this latter point, since policy objectives change, the important criterion is that a reactor type perform well under different, or variable, policy limitations.

Two advanced converter systems stand out as being clearly favored by these criteria: the prismatic HTGR (named for the shape of its fuel elements, which are hexagonal blocks) and the CANDU. A third, the spectral-shift-control reactor, is also worth mention. The prismatic HTGR has the lead in the United States by virtue of the licensed prototype unit (Fort Saint Vrain). Further development of the reactor requires only that this unit be scaled up to fully commercial size. Most generic problems have been identified. The CANDU coded by pressurized heavy water is fully commercial in Canada, and Canadian affiliates of companies in the United States are CANDU designers and vendors. Nevertheless, these companies are skeptical of the ability of the CANDU to meet domestic licensing requirements in its present design. Possible points of difficulty (under the industry's understanding of licensing philosophy) are the use of on-line refueling, thin-walled fuel cladding, and the design of the emergency core-cooling system.[34] If these design features must be changed, the CANDU would evolve in the direction of the British steam-generating heavy water reactor (SGHWR), a system that has been essentially abandoned in Great Britain as uneconomical. The SSCR is claimed to be a straightforward extension of LWR engineering, with its main point of development being an auxiliary unit: an in plant heavy water reconcentration unit that requires separate commercial development, but whose development may be considered independent of other reactor problems.

Two other systems seem to present more formidable development problems, but exhibit some development advantages. The organic-cooled CANDU has been developed in Canada to a point just short of prototype construction. One evaluation[35] suggests that its economic prospects are more favorable than those of the CANDU cooled by pressurized heavy water, but not by a margin sufficient to justify the cost of commercialization. However, if the United States were to undertake development of a CANDU for domestic use, there might be keen Canadian interest in joint development of the organic-cooled version.

The pebble-bed reactor is a version of the HTGR that has had successful prototype operation in Germany, and the vendor of the domestic HTGR (General Atomic) has access to its technology.

ECONOMIC PROSPECTS

Capital Costs Under present economic circumstances, none of the advanced converters appears to be competitive with LWR's. All types seem to involve appreciably higher capital costs. These capital costs must be counterbalanced by savings in fuel cycle costs.

The size of the capital-cost disadvantage for various systems is highly controversial; it is only natural that proponents of a concept present optimistic data, and it is extremely difficult to find economic evaluations for systems not yet built that are free of subjective judgments. With the

TABLE 5-5 Estimates of Nuclear Reactor Capital Costs (dollars[a] per kilowatt-electric)

Reactor Type	Cost
Light water reactor	625
Spectral-shift-control reactor	690[b]
High-temperature gas-cooled reactor	715
CANDU	915[c]
Fast breeder reactor	800

[a] 1977 dollars; includes interest during construction.
[b] Includes cost of heavy water.
[c] Fueled by natural uranium; includes cost of heavy water.

Source: Resources Planning Associates, *The Economics of Utilization of Thorium in Nuclear Reactors—Textual Annexes 1 and 2* (Oak Ridge, Tenn.: Oak Ridge National Laboratory (ORNL/TM-6332), n.d.), Table D.2, p. 15.

warning that the numbers are not firm, we offer Table 5-5 as a set of estimates for reactor capital costs.

As a rule of thumb, a capital cost of about $30/kWe can be translated, under today's charge rates for private capital, into an electrical cost of 1 mill/kWh. On this basis, the CANDU reactor would be ruled out. However, it should be noted that the cost cited for the CANDU is for the natural-uranium version, a larger machine that uses more heavy water than a CANDU designed for a slightly enriched uranium or thorium fuel cycle. Capital-cost estimates for these versions of CANDU are not available.

Once-Through Cycles The comparative cost of once-through cycles may be estimated as the differential cost of uranium enrichment and ore purchases. The arbitrary assumption is made here that the costs of fuel fabrication and storage (lesser factors in fuel cycle costs in any case) do not vary much per kilogram of fuel. To the extent that they do vary, small adjustments in comparative evaluations would have to be made. For this evaluation, we assign a separative work cost of $100/SWU, expressed as kilograms.

On a once-through uranium cycle, the spectral-shift-control reactor requires very little less uranium feed (about 10 tons/yr less of U_3O_8 per GWe) than its light water counterpart (a PWR)[36]; thus, this cycle is not listed in Table 5-4.

The natural-uranium CANDU requires no separative work and uses about 30 tons/yr less natural uranium per GWe than an LWR. Equivalent fuel cycle cost savings are about 2 mills/kWh on enrichment. At today's spot prices of $40/lb of U_3O_8, or $100/kg of uranium, the purchase of ore contributes about 4 mills/kWh to the cost of electricity. The savings in ore purchases are just under 1 mill/kWh. A natural-uranium CANDU could save 5 mills/kWh in fuel cycle costs (compared to an LWR), but only if uranium cost about $120/lb of U_3O_8. (This is, in fact, an underestimate of the break-even uranium price. As the cost of uranium rises relative to the cost of enrichment, the system shifts toward lower tails enrichment, and the fuel cycle cost rises less rapidly than this simple treatment indicates.)

The CANDU fueled with uranium enriched to 1.2 percent ^{235}U would require only about a fourth the separative work that an LWR fueled with uranium enriched to 3 percent ^{235}U would require for production of a given amount of electricity, and only about two thirds as much U_3O_8. In comparison to an LWR of the same size (1 GWe), operated at the same capacity factor (70 percent), a CANDU would use about 70 tons less U_3O_8 per year. At today's prices for ore and enrichment, the savings in the cost of ore purchases would be 1.5 mills/kWh, and in the cost of enrichment, 1.6 mills/kWh. The price of U_3O_8 would have to rise to $90/lb for these savings to balance the disadvantage of capital charges for the CANDU of 5 mills/kWh.

Finally, an HTGR on a once-through fuel cycle (using fuel enriched to 8 percent ^{235}U) would require essentially the same amount of separative work per kWh of electricity sold, and would save about 20 percent of the cost of ore purchases incurred by an LWR. The price of uranium would have to rise by a factor of about 6—to $240/lb of U_3O_8—before these savings on ore purchases counterbalanced a 5 mill/kWh capital-charge disadvantage. Fueling costs for prismatic and pebble-bed HTGR's on Th-U once-through cycles would be similar.

Table 5-6 lists these results for quick inspection.

Recycle Systems Recycle costs have been estimated, but not experienced. The estimated costs of fuel reprocessing and refabrication show little uniformity. A recent report[37] suggests that the cost (in constant dollars) to reprocess LWR fuel (recycling uranium) would be $200–$300/kg of heavy metal, whereas the cost of reprocessing HTGR fuel (Th-U) would range from $500–$900/kg, and the cost of reprocessing fuel from fast breeder reactors (U-Pu), from $300–$500/kg. These numbers are midrange among various estimates. Refabrication costs show an even larger reactor-dependent range of values. The fuel cycle cost savings that might be achieved by recycle in various reactors are difficult to estimate. Benchmark cases have been computed for LWR's using today's price schedules.[38] The results indicate that a slight saving could be achieved with recycle. The uncertainties in these calculations are large enough that slightly negative

TABLE 5-6 Cost of U_3O_8 for the Fuel Cost of Various Reactors on Once-Through Fuel Cycles to Fall 5 mills/kWh Below That of a Nominal Light Water Reactor (1978 dollars per pound)

Reactor	U_3O_8 Cost
CANDU (1.2 percent enriched uranium)	90
CANDU (natural uranium)	120
High-temperature gas-cooled reactor	240
Spectral-shift-control reactor	very high

results are also possible. The only definite conclusion that can be drawn is that economics is a minor factor (today) in the decision to reprocess. As uranium prices rise, however, recycle will become progressively more attractive—offering fuel cycle savings, for example, or the means to stay further increases in the prices of coal and uranium.

FLEXIBILITY

Both the CANDU and HTGR have the flexibility to operate on once-through uranium cycles and (with recycled uranium) on the Th-U cycle. In both cases, significant savings can be realized in the amount of uranium required (relative to the corresponding LWR case). The savings from heavy water reactors are generally greater than those from HTGR's (see Table 5-4). The SSCR, on the other hand, is only attractive as a recycle option: On a once-through cycle, it shows little better fuel economy than a standard LWR.

Again, both the CANDU and HTGR may be used in conjunction with a denatured Th-U cycle, in which fissile uranium is diluted below weapons grade with ^{238}U (see "Reprocessing Alternatives"). In both cases, the quantities of plutonium produced are far below those produced in a once-through uranium cycle. The SSCR would, on a denatured cycle, probably produce at least one third as much plutonium as the LWR it replaces.

The information just presented leads to the conclusion, drawn by other studies as well,[39] that the CANDU and HTGR are the best choices if an advanced converter is to be selected for the United States at this time.

In favor of the CANDU is that it consistently shows the smallest resource requirements (of natural uranium) for any given mode of fuel cycle operation (compare HWR and HTGR in Table 5-4). It has considerable flexibility of design, and a given reactor could accommodate a large range of fuel loadings, from natural uranium to denatured Th-U with recycle. (The reactor would have to be derated if a new fuel loading were significantly different from that for which the reactor was designed.) Besides making efficient use of resources, HWR's make efficient use of reactor fuel, as measured by fissions per initial fissile atom (FIFA). High FIFA translates into low frequency of reprocessing of fissile atoms, per unit of energy generated, and correspondingly small out-of-reactor inventories and process losses. Finally, HWR's can operate on a denatured Th-U fuel cycle with relatively small concomitant production of plutonium—about one tenth that of an equivalent LWR.

Among the disadvantages of heavy water reactors are that their capital costs appear high to evaluators in the United States, and that the plutonium produced in once-through operation is less contaminated with ^{238}Pu and ^{240}Pu, both isotopes that detract from the desirability of plutonium as a weapons material.

The HTGR displays significant advantages, irrespective of resource considerations. It permits operation at high temperature, giving a high efficiency for the conversion from thermal to electric power with less waste heat; it might achieve even higher thermal efficiencies with a gas-turbine topping cycle; its decay heat might be more easily dissipated in a loss-of-coolant accident, thereby reducing the probability of a major radioactive release; and evaluators rate its capital costs below those of HWR's. The HTGR also has considerable flexibility of operation and could be run on the denatured cycle without producing large amounts of plutonium. The very high burnups considered achievable in HTGR's make their plutonium product, even from a once-through cycle, relatively undesirable for weapons.[40]

Of the advanced reactors, the HTGR is considered the easiest to develop for application in the United States, but facilities for its fuel cycles require a longer development and commercialization period than those for a heterogeneous reactor such as the LWR or HWR. The fuel cycle operations are rated as correspondingly high in cost.

A balance of all of these considerations seems to favor the HTGR as the advanced converter whose development offers the best prospects of fuel economy, flexibility of operation, and economy in power generation. Nevertheless, it would be appropriate to undertake a careful cost comparison between an HTGR and an HWR optimized for domestic economic conditions and safety regulations before making a final choice between these two types.

Fast Breeder Reactors These reactors take their name from the fast neutrons (energy greater than 50 keV) that produce most fissions in their operation. The neutron energy spectrum in fast breeder reactors has several advantages.

- The number of neutrons produced in fission is slightly higher.
- The ratio of capture reactions to fissions is smaller.
- The fast neutron spectrum enables some ^{238}U to undergo fission, generating additional neutrons and raising the breeding gain beyond that possible with plutonium alone.
- Neutron absorption in structural materials, relative to fissions, is lower than in thermal reactors.

Fast breeder reactors obtain the fast-neutron spectrum by eliminating moderators such as graphite, water, or heavy water that slow down the neutrons emitted in fission to thermal energies before they produce additional fissions in the chain reaction.

The fuel for fast breeder reactors is considerably more concentrated than the fuel for thermal reactors. Fissile atoms may be 10–20 percent, or more, of the heavy-metal atoms in the reactor core (the region where most of the fissions occur and most of the power is generated).

Surrounding the core is a blanket of pure fertile material. In systems of reasonable size, a significant fraction of the neutrons produced escape from the core, and these must be caught in the blanket in order to achieve breeding.

The design of the liquid-metal fast breeder reactor seeks to conserve these gains by employing a coolant with good heat-transfer properties and low neutron absorption (liquid sodium).

The development of the LMFBR is significantly more advanced than other breeder concepts both in this country and abroad. In the United States, four reactors (EBR-I, EBR-II, Enrico Fermi-I, and SEFOR) have been operated, a test reactor (the Fast Test Reactor, or FTR) is under construction,[41] and a pilot commercial plant (the Clinch River breeder reactor, or CRBR) is in the design stage. Several large pieces of equipment have been fabricated, but further procurement is in abeyance as a result of the Administration's decision to terminate the pilot plant. Various test reactors and demonstration plants have already been operated abroad, notably pilot-size commercial plants in France, Great Britain, and the U.S.S.R. Other pilot-size plants are under construction in Germany and Japan. Commercial-size plants are under construction in France and the U.S.S.R. Preliminary results indicate that losses of fissile isotopes in reprocessing can be held to 1 percent—low enough to preserve the breeding gain in LMFBR's[42]—but full-scale reprocessing experiments with high-burnup fuel are just beginning.

The gas-cooled fast breeder reactor (GCFBR) is essentially similar to an

LMFBR. The principal design differences stem from the use of pressurized helium rather than sodium as the coolant. No test or pilot-plant GCFBR has been built, but much of the needed technology can be borrowed from HTGR's. Calculations on reference designs for GCFBR's indicate a higher breeding gain than that of the LMFBR, but the GCFBR would require a higher fissile inventory. The principal safety question is whether the core temperature can be held well below its melting point should the helium cooling system suddenly lose pressure.

Since this reactor has not been under intensive development, its program would lag at least 5 years behind that of the LMFBR in the United States.

Core and Blanket Cycles Fast breeders such as the LMFBR and GCFBR can be fueled with ^{239}Pu or ^{233}U, and can breed the same fissile materials by neutron capture in ^{238}U or ^{232}Th, respectively. In practice, the U-Pu cycle (^{238}U + ^{239}Pu) is preferred, as it yields a much better breeding ratio in the core. The use of thorium in LMFBR's leads to marginal breeding gain; the gain in GCFBR's could be significant.

The amount of plutonium produced in the core relative to the amount of whatever fissile product is bred in the blanket can be adjusted. Cores of smaller critical mass and with a higher ratio of plutonium to uranium will produce less plutonium in the core and leak more neutrons for breeding in the blanket. The breeding ratio of the system may actually increase with this adjustment. Either the LMFBR or the GCFBR could serve as a combination breeder-converter, with break-even (or less) plutonium production in the core, and with large production of ^{233}U in a thorium blanket, which could be used to fuel advanced converters. Up to now, this possibility has not received much attention, perhaps because the necessary design changes would result in fuel that could not remain so long in the reactor, and because more frequent reprocessing would be required. The possibility may be particularly suitable for the secured fuel cycle parks that have been proposed to safeguard the fuel cycle. In a fuel cycle park, the reprocessing, fuel fabrication, and some power generation would be colocated, eliminating the problems of transporting large volumes of fissile material outside a secure area.[43]

DEVELOPMENT STATUS OF VARIOUS CONCEPTS

The development status of the various reactor concepts was noted briefly in the reactor descriptions. The purpose of this section is to provide some insight into the effort that would be required to bring various reactor concepts to a state of readiness for introduction over the next few decades. The key considerations from the point of view of uranium-ore requirements are the date by which such new systems could be introduced, and the rate at which they might be installed in place of additional LWR's.

Experience with LWR's provides ample evidence for the time-consuming nature of developing and marketing large, technically sophisticated facilities. Generally, laboratory-scale experiments and component development are followed by a small-scale reactor experiment to test the technical feasibility of the concept. This is followed by one or more intermediate-size pilot or demonstration plants operated within utility grids. The purpose of these plants is to evaluate different design approaches for the concept (e.g., "loop" versus "pool" LMFBR), to begin developing the capacity to design and manufacture the special larger-scale components needed for a given concept, and to work out practical operating procedures in the context of producing power on a utility grid. Assuming that the demonstration is successful and efforts toward commercialization are warranted, the next step is to undertake construction and operation of one or more prototype plants—the first-of-a-kind, full-scale plants of the new concept. This step provides the kind of engineering data from which relatively firm cost estimates can be made. Bringing new reactor systems to this stage of development and commercial suitability may take 20–30 years after the decision has been made to proceed with development. The lengthy period includes time to budget, license, construct, and briefly operate the various plants, typically about 15 years. It also allows for some overlapping of steps; for example, proceeding with detailed designs of a prototype reactor while pilot or demonstration plants are in final construction phases. Some steps might be telescoped—pilot plant and intermediate-size demonstration plant—or bypassed altogether to save time and resources. Different levels of component testing conducted prior to building a plant can modify the technical and economic risks of that plant. Obviously, judgment must be exercised to prevent corner cutting that would increase the risk of failure.

Some concepts described above—such as LWR's with improved fuel utilization, spectral-shift-control reactors, and LWBR's—represent extensions or modifications of light water reactor technology. The development of these concepts could rely heavily on existing industrial capacity and experience to reduce developmental requirements; similarly, the GCFBR could make use of LMFBR and HTGR technologies, if these continue to evolve at a sufficiently rapid rate.

Significant development programs abroad have resulted in foreign capacity to construct nuclear power plants. Requirements for efforts in the United States could be reduced to the extent that the experience and capacity developed abroad can be transferred to domestic industry and regulatory organizations. The difficulty of useful transfer based on the trade secrets of industrial organizations, and differing regulatory criteria in various countries, should not be underestimated. In particular, differences in the treatment of proprietary information make government-to-government transfers to the United States impractical. Transfer can only be accomplished by the licensing of firms in the United States that agree to protect the information. The necessity of government involvement, however, makes it difficult to estimate how much the United States could benefit from CANDU technology in Canada or from LMFBR technology in France.

Developing the capacity to build and operate a large number of plants for a new concept is so costly (measured in billions of dollars) that no single private-sector entity is likely to make a conscious decision to proceed with the process on its own. Even with licenses from foreign sources, the domestic version of any new system would probably require a large pilot-size plant (about 300 MWe) to confirm satisfaction of regulatory requirements. There is a major risk that considerable redesign would be required for domestic application.

A cooperative program between government and the private sector would be required to bring these concepts to readiness, selecting those most attractive to vendors and buyers, and encouraging the development and commercialization of reactor types that have potential long-term economic benefits.

Our estimates for the earliest possible dates of commercial introduction for the principal breeder and advanced-converter designs, based on brisk efforts by government and industry, indicate that of the advanced converters, only the HTGR could have a commercial-size prototype operating before 1990. Of the breeders, the LMFBR could be readied for operation by the mid-1990s, and the GCFBR, 10 years later. Estimated schedules are included in Table 5-3. The schedule takes only technical problems into account; limitations implied by institutional or policy matters have not been (and cannot be) estimated. Any advanced-converter or breeder reactor system introduced by a date listed in Table 5-3 will need an additional 10–20 years to gain a significant share of the nuclear power market, and as noted in the table, any institutional delays (which must be expected if today's pattern of political, legal, regulatory, and financial uncertainties persists) would further lengthen the schedule.

One other aspect of the development problem has come to our attention. This is the matter of industry morale. In addition to the impetus from federal direction and funding, the LMFBR program has received considerable support from industry. This support includes industrial funding for the Clinch River breeder reactor program (about $250 million), as well as the assignment of vital personnel. The commitment was made under very strong federal pressure to do so from the Congress and the former Atomic Energy Commission, and industry personnel feel betrayed, both financially and psychologically, by the cancellation of the CRBR. Nevertheless, the LMFBR still commands industrial support because of this commitment.

Reprocessing followed the same course of strong federal encouragement, industrial commitment, and federal foreclosure. The industry is not likely to commit itself again to similar projects. Future reactor development and commercialization will probably require much more federal capital, with the industry assuming the role of developers, component vendors, and operators under contract. It is difficult to quantify the difference this might make, but it could tilt development schedules in favor of the LMFBR, relative to other reactors, by 2–5 years. This would contribute to the head start the LMFBR already has by virtue of a large array of developmental facilities: among others, the Fast Flux Test Facility at Richland, Washington, the Large Components Test Facility at Santa Susana, California, and the EBR-II and other major experimental reactors operated by Argonne National Laboratory at the National Reactor Test Station, Idaho. Exchange of noncommercial information between the United States and foreign countries operating breeders in the 300-MWe class (Great Britain and France) could also help accelerate the LMFBR's schedule.

BREEDERS VERSUS ADVANCED CONVERTERS

The most important point of comparison among reactor types and fuel cycles is likely to be contingent on their appropriateness under the conditions prevailing when a market appears for a new system. The conditions that dominate will be the accrued and projected growth of demand for electricity, the availability of uranium resources, the competitive economics of electrical generation, and the measures adopted to discourage or forestall diversion and proliferation.

The following conditions (not all of equal weight) would favor the use of fast breeder reactors over advanced converters in the United States.

- The demand for electricity in the United States grows steadily after the year 2000.
- Total domestic uranium resources are found to be at the low end of recent estimates.
- Very little intermediate-grade uranium ore is found that can be produced at costs in the range of $100–$200/lb.
- The world growth of nuclear capacity in conventional light water reactors exerts pressure on the United States to export some of its uranium or enriched fuel (or both) to offset the balance-of-payments deficit from oil imports, to discourage recycle of fissile isotopes or installation of breeder reactors elsewhere, or to meet other needs.

• Enrichment technologies reducing the cost of low enrichment tails do not become available early.

The following conditions would generally favor the use of advanced converters for nuclear-generated electricity.

• The demand for electricity in the United States grows slowly, especially after 2000.

• Sufficient uranium resources are found to fuel advanced converters at their projected rate of introduction and installation, particularly intermediate-grade ores producible at costs around $100-$200/lb.

• Capital costs of advanced converters turn out to be significantly less than those of breeders.

• The operation of advanced converters and their fuel cycles offers advantages for safeguarding against proliferation or diversion.

• New enrichment technologies that permit economic operation at low tails assays become available early.

Both lists of conditions require some qualification. As noted, economics and the measures adopted by the world to slow proliferation of nuclear weapons could dominate the choice. Both are highly uncertain factors. We can only estimate future costs qualitatively, and we can expect surprises in international decision making.

The conditions most favorable to a large role for advanced converters—low growth in demand for nuclear-generated electricity, and abundant uranium supplies—may also act against their development. Under these conditions, LWR's would also have a future. A slowly growing industry could not be expected to sponsor new and expensive development. More likely, it would continue to market a proved product—the LWR—with perhaps incremental improvements.

There are long-term reasons for developing breeders and advanced converters simultaneously. The availability of both could permit optimal mixes of the two types that would be superior to either type alone. For example, the fuel produced by breeders would compete economically with natural uranium purchases, holding down the fuel cycle costs of converters; advanced converters would provide customers for breeder operators. Breeders might fit well, along with reprocessing and fuel fabrication plants, into a system of secure fuel cycle parks, while advanced converters could be located near existing load centers. In short, these "symbiotic" systems might offer the best combination of economical power generation and security. But the strongest reason for parallel development is simply that one or both of the two types may be needed to permit substitution of nuclear power for coal if necessary, or to provide a desirable diversity of power sources.

CONCLUSIONS AND RECOMMENDATIONS: ADVANCED CONVERTERS AND BREEDERS

The relative roles advanced converters and breeders might play in the energy supply sector cannot easily be predicted. Conditions that have a reasonable chance of eventuating would be favorable to the installation of both types of reactors (steady growth in electrical demand; economic attractiveness of nuclear power relative to other sources of electricity; and satisfactory resolution of the political and social issues discussed later in this chapter).

Breeders are more flexible in their ability to respond to quite rapid growth in demand as well as to rather moderate growth. Thus, although the two types of reactors both serve, in a sense, as insurance that increased supplies of electricity could be provided if needed, breeders provide broader coverage. The probability that such coverage will be needed by, say, 2010 may not be very high. However, the risk of inadequate supply could be high, and insurance is of greatest value against high-risk, low-probability events.

Advanced converters offer insurance against moderate growth in demand for electricity, compared to past experience, and limited supplies of uranium, so long as they are not expensive. Advanced converters would also be a useful adjunct to breeders in a breeder economy. Thus, conditions favorable to *their* development are also flexible.*

CONAES concludes that these considerations lead to the recommendation that both types of reactors should be developed. If only one type can be developed, breeders should receive priority, as covering more contingencies. If, for whatever reasons, development of the breeder is so long deferred as to preclude the option of commercialization in the early twenty-first century, the commitment should be made to expeditious development of the advanced converter.

The committee recommends the following course of action.

- Development of the LMFBR should continue, but without immediate commitment to construction of prototype reactors. CONAES was divided on the issue of whether to recommend construction of the Clinch River breeder reactor as part of this development program.
- A majority of the committee considered the Clinch River breeder

*See statement 5-17, by L. F. Lischer, Appendix A.

reactor undesirable or unnecessary for reasons that varied within the majority, including: inappropriateness of its design as a developmental facility,† its incompatibility with President Carter's antiproliferation policies, and its possible contribution toward committing the United States to commercialization of the LMFBR. A minority considered it necessary, as a technological step that is well short of commitment to commercialization, but necessary if early commercialization turned out to be desirable.

- A reference design should be produced for a commercial-scale LMFBR to identify the problems that require solution in the research program.
- Commercial-scale experiments should be conducted to ensure that a workable fuel cycle for the LMFBR can be operated.* Proliferation-resistant schemes, such as the proposed Civex cycle or the denatured Th-U cycle, should receive particular attention.
- Development of the HTGR to full commercial scale should be encouraged (either the prismatic or pebble-bed version).
- A pioneer-scale reprocessing plant (a few hundred tons per year) for Th-U fuels should be built and operated. Further work on recycle of HTGR fuels in such a plant should be supported, and increased attention should be given to off-gas problems and to the special requirements of coated-particle composites.
- A joint program should be undertaken with Canada to explore and, if attractive, develop toward commercialization an advanced heavy water reactor design that can be adapted to the regulatory and economic climates of both countries. Such a design should be considered as the next major improvement of the heavy water line.

As is well known, many decisions about the domestic nuclear power program have been deferred pending the outcome of the International Nuclear Fuel Cycle Evaluation. The following actions should follow completion of that program, based on international agreements.

- The United States should act expeditiously to provide fuel cycle facilities of the types recommended. The best alternative breeder to the plutonium-fueled LMFBR should also be developed through a joint program with other supplier countries. This program of development should be carried out regardless of an INFCE recommendation for the LMFBR and U-Pu fuel cycle. We anticipate that world demand for nuclear fuel will lead to a breeder, rather than a break-even converter, as the most suitable alternative.
- After completion of INFCE and associated programs in the United

†See statement 5-18, by L. F. Lischer, Appendix A.

*Statement 5-19, by J. P. Holdren: The concept of a "commercial-scale experiment" is vague. If it means building a commercial-size plant for fast reactor fuel, I oppose it.

States, the schedule for breeder and advanced converter development or installation (or both) should be reevaluated and any required programs initiated.

DOMESTIC ISSUES IN THE FUTURE OF NUCLEAR POWER

PUBLIC APPRAISAL OF NUCLEAR POWER

Public opinion polls have repeatedly shown that the majority of people in the United States view nuclear power favorably.[44] Referenda introduced in seven states in 1976 that would have halted, postponed, or forestalled the expansion of nuclear power were all defeated. On April 7, 1979, just a week after the accident at the nuclear power plant near Harrisburg, Pennsylvania, citizens of Austin, Texas, voted to retain their 16 percent interest in a nuclear power plant under construction, and they extended the city council additional borrowing authority to cover anticipated and unanticipated costs.

Nevertheless, nuclear power is controversial, and is likely to remain so. The same polls cited indicate that there is a significant core of very strong opposition to nuclear power—opponents who will continue efforts to persuade the public to abandon this source of energy.

A factor that increases the effectiveness of the opponents of nuclear power is their development of a comprehensive information network. The bulk of the information circulated is, as might be expected, highly partisan, but it contains enough factual statements that the nuclear opposition is much better informed about nuclear issues than the general public.*

While it is no doubt important to understand the rational arguments and irrational appeals that may sway individual voters, the nuclear controversy can ultimately be explained only as a contest among *groups* in the society. The leadership of the antinuclear movement today appears to be in the hands of environmental organizations. The pronuclear forces are led by industries and professional associations within the nuclear power field. These groups are vying with one another to win public support.

For the foreseeable future, the scientific community will occupy a strategic position in this debate for at least two reasons. First, scientists themselves are found on both sides of the nuclear controversy. Second, other parties to the controversy are eager to claim scientific support for their views. This helps to account for the recent "proliferation of petitions, polls, and statements purporting to reveal what the nation's scientists and

*Statement 5-20, by J. P. Holdren: Completely symmetric statements could and should have been made about the information network of nuclear proponents.

engineers . . . really think about the controversial technology."[45] Moreover, scientists enjoy a great deal of public confidence—much more than any of the other main parties to the controversy and any of the other main sources of information, according to the Harris surveys. Indeed, Harris notes that many believe the pivotal factor in the California initiative was the widely shared impression that scientists support nuclear power.[46] But the public perception that scientists are not of one mind is itself an obstacle to acceptance of nuclear energy.

Public appraisal of nuclear power is difficult to analyze. Technical, political, and social issues flow together, change, and diverge. Public attitudes are influenced by technical information and opinion, and nuclear technology, in its continuing development, responds to political and social influences. For example, the publication in 1957 of the Atomic Energy Commission's report "Theoretical Possibilities and Consequences of Major Accidents in Large Nuclear Power Plants" (also known as WASH-740 or the Brookhaven Report) was an important factor in the emergence of public apprehension about reactor safety. The resulting demands for greater assurance of safety have profoundly influenced modern regulatory practice, leading to designs that are both higher in cost and protected against accidents that the industry would consider inconceivable.

For policy guidance, some principal concerns with the future and expansion of nuclear power are separated here into those that are primarily technical, institutional, or social.

Some public concerns about nuclear energy center on technical issues that have not been resolved to the satisfaction of a substantial number of scientific critics.

- The effectiveness of technical means to prevent or hinder diversion of weapons-usable material from the fuel cycle.
- The safety of nuclear reactors, including protection against sabotage.
- The long-term management of nuclear waste.
- Release of long-lived radioactive effluents from the nuclear fuel cycle.

Other concerns arise from public distrust of institutions responsible for the management of nuclear energy programs in the past. Primarily institutional issues include the following.

- Whether human institutions can be relied on to provide long-term management of radioactive waste, reactor safety, and secured weapons-usable material.
- Whether international institutions can be created and maintained to work effectively against the proliferation of nuclear weapons.

Other aspects of the public appraisal of nuclear power are principally social. They reflect differing perceptions of desirable future conditions.

- Nuclear power has become the most visible symbol of large-scale centralized technology for which many citizens feel they have surrendered control to experts who cannot be held accountable.
- A significant number of citizens dislike nuclear power, particularly the breeder, because if offers the continuation of a high-growth materialistic society that, in their view, will eventually prove disastrous to the physical and social environment of mankind.
- Some see nuclear power as competing for capital with other energy systems that are more nearly autonomous and under local control, and therefore, both in itself and as a symbol, as excluding social organizational patterns that are based on such autonomy.[47]
- On the other hand, some see nuclear power as essential if people are to have sufficient energy to live with dignity, achieve their aspirations, and improve their own lives and those of their children.[48]
- Many people feel that institutions, including utilities, government, and regulatory bodies, exist to provide services to citizens; that they can and should be economical (whether large or small); and that technologies, including nuclear power, can be controlled to serve man in a safe, environmentally acceptable way.

How should government and other institutions respond to these concerns? Even in controversies whose main content is technical, judgments are influenced by social and institutional preferences. On questions of fact or of likelihood, we can use existing institutions or create new ones. Ultimately, such questions are resolved in retrospect: by drawing inferences from experience.

On the social and institutional questions that have been raised, we can make no recommendation. They should be worked out through the political process. Each of us has opinions, but we agree that the only ethical way to act on them is through action outside the scope of this study.

It is obvious that a high level of confidence in nuclear power depends on consensus that the nuclear industry and the government have workable institutions to manage properly the whole enterprise, including the complete nuclear fuel cycle.

COSTS OF NUCLEAR POWER

Nuclear power plants began to be installed in quantity in the late 1960s in response to (1) the promise of more economical generation of electricity,

(2) features demanded of coal-fired generation, and (3) anticipated insecurity of oil supply. Nuclear power plants, in large sizes, appeared to be somewhat higher in capital costs than coal-fired plants, but the fuel cycle costs of nuclear plants were competitive with those of fossil fuels and had the added advantage that the price of uranium was considered much more stable than the price of fossil fuels.

Since that time, the costs of both nuclear and coal-fired power plants have escalated rapidly—nuclear somewhat more, both at a higher rate than overall inflation. The escalation was due partly, but not completely, to changes and additions to plant design required by regulation for improved safety or environmental protection. The costs of labor and construction materials have risen more rapidly than the general rate of inflation. The competition for new electrical generating plants is now between coal and nuclear power, with the two sources exhibiting the following characteristics.

- Coal plants take less time to plan, license, and build than nuclear plants (6–10 years versus 9–12 years).
- Coal plants have slightly lower capital costs than nuclear plants (0–25 percent cost differential).
- Nuclear plants have lower operating costs.
- Nuclear plants have lower fueling costs.
- Nuclear fuel is easily stored and does not have to be delivered frequently, making it less vulnerable to interruptions in mining and transportation.

One consequence of weighing these considerations has been that, for most private utilities in the United States, nuclear power is rated slightly to considerably less expensive than coal-fired electricity. However, there are uncertainties in future costs, both because there has been a history of downtime in nuclear plants arising from changes in safety regulations, and because escalations are intrinsically unpredictable (for both coal and nuclear plants). Therefore, utilities will usually order some mix of both types. This gives them flexibility: If one or the other type shows superior operational performance or lower cost, that type can be exploited more heavily.

The balance is more strongly in favor of nuclear power for publicly financed utilities, both in the United States and abroad. These utilities have lower carrying charges on capital, and the cost disadvantage of the initial nuclear investment is a less important consideration.

Although cost is not the only criterion for either utility or public decision making, it is an important one. Confusion is easily found in this area. For example, a period of rapid inflation, such as we are now experiencing, guarantees that future costs (in dollars of the future) will be greater than current costs (in today's dollars). There are also features of tax laws, of utility accounting practices, and of the regulatory requirements governing investment capital that compensate (or fail to compensate) in various ways for the effects of inflation and market fluctuations. Utility capitalization is based on the actual purchase costs, rather than on the replacement values of plants, and in times of rapid inflation, the plants' income-producing value diminishes correspondingly. To compensate, a very high capital charge rate is adopted. The result is that new investments with high capital costs are discouraged, including many that would be clearly economical in a noninflationary period. This result is paradoxical, since high capital-cost equipment is normally considered a prudent hedge against inflation if it reduces recurrent costs that are subject to inflationary pressures.

Because utility rates are subject to public regulation, they are affected by a variety of social issues. Should utilities be heavily taxed to support other regional services, or taxed only according to the services they use? Do they demand services that are not obvious, whose provision amounts to subsidy? What is a fair distribution of costs and economies between the utility owner (public or private) and the consumer? How much of future capital costs should be borne by the current consumer in anticipation of future benefits, and how much should be paid by the future consumer? Different answers to these questions, arising from the larger social debate, affect rates, the way those rates are evaluated, and thus the relative economies of nuclear power compared to other sources.

Capital Costs

A number of studies have estimated the capital costs of nuclear plants. In particular, the costs of nuclear power plants have been compared to the costs of competitive electrical plants; in the short and intermediate term, this means specifically coal plants.

The following expectations are typical.

1. Bechtel Corporation[49] projected typical costs in New England for fossil and nuclear plants. The most economical fossil fuel plant for this region consisted of three coal units burning high-Btu eastern coal and equipped with scrubbers. The plants were each of 700-MWe capacity. The cost in 1985 was projected as $850/kWe. A 5 percent inflation rate was implicit, and this cost becomes $600/kWe in 1978 dollars.

A comparable nuclear plant, consisting of two 1100-MWe units, was estimated to cost $1030/kWe in 1985, and with the same adjustment for inflation, $730/kWe in 1978 dollars. The larger total capacity of the nuclear plant is required because of adjustment for the reserve capacities of larger units.

2. Commonwealth Edison[50] cited comparable figures leading to a typical cost of about $530/kWe (1976 dollars) for coal units with scrubbers and to a virtually identical cost for nuclear units. Inflating to 1978 at a yearly rate of 6 percent, this cost would be about $595/kWe. The utility indicated an uncertainty band of 10 percent in the numbers. In 1978 dollars, their "high" estimates were $620/kWe and $630/kWe for coal and nuclear power, respectively. Part of the difference between the estimates of Bechtel and Commonwealth Edison is attributable to lower costs of construction in the Midwest.

3. Numbers cited by those who oppose nuclear power are often higher, but after correction for assumed inflation rates, are not very different. For example, Komanoff[51] indicates an expectation of $1200/kWe for nuclear plants and $950/kWe for coal plants in 1985. Deflated to 1978 dollars at 6 percent, these become $800/kWe and $630/kWe, respectively.

There seems to be general agreement, therefore, that the capital costs of nuclear power plants are between $600/kWe and $800/kWe in 1978 dollars, and those of coal plants are about the same to 20 percent less.

Cost Escalation

Estimates of future costs are colored by the estimators' expectations of cost escalations, over and above general inflation. Escalation of this kind makes future plants more expensive than present ones, even in terms of constant-value dollars. Brush, in testimony presented to a New England state utility commission,[52] introduced curves suggesting that escalation has added half again as much as general inflation to the costs of nuclear power plants over the last 10 years. This is an annual rate of roughly 3 percent. Brush's data imply that for fossil fuel plants, the annual rate has averaged about 2 percent. Bupp[53] agrees with the 2–3 percent figure for extra cost escalation of nuclear power up to 1975, but points out that there has been a much faster escalation of estimates since that date.

Arguments about future escalation center on the perceived causes. Industry representatives emphasize that past escalations have resulted either from general shifts in the costs of construction (i.e., field-labor rates, productivity decreases) or from changes in the scope of the jobs (thicker containment, more rigid quality assurance, extra engineered safety features for nuclear plants, scrubbers and other pollution-abatement systems for coal plants). The general cost shifts are felt in all large construction projects. As to changes in scope, the utilities argue that most of the regulatory tightening that increased the magnitude of nuclear projects has already occurred, whereas the standards for emissions from coal plants are still changing. This would lead to an expectation that the costs of coal plants will escalate faster than those of nuclear plants.

Another source of escalation is simply delay. Again, it is argued that nuclear plant schedules are now as protracted as can reasonably be expected, and further escalation from this cause is unlikely.

While it remains a matter of opinion, we find these arguments, based on attempts to identify the causes of escalation, more convincing than the arguments based on simple trends. Escalation is likely to make the future cost of electricity higher, but is not likely to make nuclear plants less competitive with coal.

Capacity Factors

Charges against capital continue no matter whether a plant is being used, whereas revenues are proportional to the product sold. Therefore, that part of the cost of the product that arises from capital charges is minimized if the plant works at maximum capacity. The measure of performance is the capacity factor, the ratio of electricity actually sent out to that which would have been sent out if the plant were in round-the-clock operation at maximum dependable capacity.

Capacity factors fall below 100 percent for three reasons: scheduled maintenance, unscheduled outage (for repair of malfunction or for other technical reasons), and lack of demand for the product electricity (economic downtime). Nuclear plants are designed for scheduled refueling and maintenance periods that vary between 7 percent and 15 percent of the time. Thus, their theoretical capacity factors for an average year are between 85 percent and 93 percent. Those of coal-fired plants are comparable. During refueling outages, turbine inspection and deferred repair work can be carried out.

Balancing this disadvantage of nuclear plants is an advantage in economic downtime. When demand is low, utilities operate at base load only, keeping in operation those plants whose incremental operating costs are minimal. For large plants, these incremental costs are almost exclusively fuel costs, and nuclear fuel is the cheapest source for thermal plants. This is offset to some extent by the difficulty of starting up and shutting down various plants, but the incentive remains to rule on nuclear plants more completely than on fossil plants for base loads. This effect suggested to early nuclear planners that capacity factors as high as 80 percent would be reasonable. More conservative estimators—realizing that the unscheduled outages commonly experienced in fossil fuel units would also occur in nuclear plants—tended to expect about a 70 percent capacity factor as a goal.

Experience has been disappointing. Although their physical condition is normally superior to that of fossil plants by virtue of the strict regulation imposed on nuclear plants, this strict regulation has also on occasion required shutdown for inspection or for repairs that in a fossil fuel plant would have been deferred to a scheduled maintenance period. Reserve margins have been high because of sharp reductions in demand growth in response to the events of recent years, and even some base-load plants have not experienced full-capacity demand. This latter fact has made the concept of availability—the fraction of time during which a plant could operate at full power *if called upon*—a measure of technical performance, while capacity factor remains the measure of economic performance.

Table 5-7 presents the histories of capacity factors and availability factors for large coal and nuclear units from 1970 to 1975. The chief feature of this table is year-to-year fluctuation. Komanoff[54] has interpreted the data as indicating a trend toward decreasing capacity and availability factors for nuclear plants, and has attributed this trend primarily to decreased availability of the larger sizes of nuclear units that entered service progressively during the period. Perl[55] has criticized both Komanoff's statistical inferences from his limited data base and some of Komanoff's data adjustments. A separate analysis by Commonwealth Edison[56] generally agrees with Perl, but is even more sanguine about the likelihood of high capacity factors for large units.

On this controversy, we find that the data base for nuclear units is too small for these analyses to be significant. For example, it may be noted that if entries in Table 5-7 had been dated from 1971 rather than 1970, no general trend in capacity factors would be apparent. Few large nuclear plants were in service in 1970. There *may* be a trend to lower availability for large nuclear units, but it is too early to tell. In any case, it seems that capacity factors for large coal and nuclear units will be similar, as they have been in the past, hovering between 55 and 60 percent. Capacity factors for both types of plants will increase if reserve margins decrease and peak loads are leveled. New pricing policies, such as off-peak cost reductions, would promote this outcome. Higher capacity factors for both coal-fired and nuclear plants would improve the relative economies of nuclear power by decreasing the fraction of power cost represented by capital charges.

Nuclear Fuel Cycle Costs

Throughout the discussion of costs, the point was repeatedly made that nuclear fueling is less expensive than fueling with coal. We now examine the data on this point.

Table 5-8 was prepared by CONAES to illustrate the main features of nuclear fuel cycle costs. The major variables are the price of uranium and

TABLE 5-7 1970–1977 Average Annual Capacity and Availability Factors for Coal Units over 400-MWe Capacity and Nuclear Units (percent)

	Capacity Factor		Equivalent Availability	
Year	Coal	Nuclear	Coal	Nuclear
1970	59	53	66	67
1971	61	58	68	72
1972	61	54	66	68
1973	63	57	69	71
1974	56	55	64	68
1975	58	59	65	72
1976	59	57	66	69
1977	58	63	61	75

Source: For coal, Edison Electric Institute, *Report on Equipment Availability for the Ten-year Period—1968–1977* (Washington, D.C.: Edison Electric Institute, 1979), and for nuclear, Division of Nuclear Power Development, *Operating History of U.S. Central Station Nuclear Power Plants* (Washington, D.C.: U.S. Department of Energy, April 1979).

the cost of working capital. The other factors were selected as typical by the following line of reasoning.

- Enrichment cost of $100/SWU is slightly above the present value. Including associated chemical conversion, present cost is a little over $90/SWU. In the past 3 years, enrichment charges have been tracking general inflation (after a threefold increase in 1975 to account for changes in government accounting of both capital charges and power costs). The present charges are very close to estimates of cost if the plants were privately owned and operated. (European enrichment plants are charging slightly higher prices for future services.) The new gas-centrifuge facility planned for the 1980s is projected to have similar costs and charges. Future price increases could arise from increases in the cost of electricity to run enrichment plants, but these electricity prices are not expected to rise faster than the rate of general inflation. The minimum combined cost of enrichment and uranium purchases is quite insensitive to the exact tails assay chosen. A conventional and representative value is 0.25 percent ^{235}U tails assay.
- Fabrication charges of $100/kg of uranium dioxide (UO_2) have been typical for more than 10 years. The cost in constant dollars has therefore

TABLE 5-8 Contributions to Electricity Cost of Nuclear Fuel Cycle Operations (mills per kilowatt-hour)[a]

Capital Charge Rate	Fuel Cycle Costs, Without Uranium Purchase				Uranium Purchase Cost, with UO_2 at:			Total Fuel Cycle Cost with UO_2 at:		
	Enrichment	Fabrication	Storage	Total	$50/kg	$100/kg	$200/kg	$50/kg	$100/kg	$200/kg
6	1.61	0.49	0.43	2.53	1.54	3.07	6.14	4.07	5.60	8.67
10	1.84	0.54	0.38	2.76	1.78	3.56	7.11	4.57	6.32	9.87
15	2.14	0.62	0.33	3.09	2.12	4.25	8.50	5.21[b]	7.34[b]	11.59
20	2.48	0.72	0.29	3.49	2.52	5.04	10.07	6.01[b]	8.53[b]	13.56

[a] Assumptions:

1. Once-through (throwaway or stowaway) cycle.
2. Enrichment to 3 percent at 0.25 percent ^{235}U tails assay.
3. Burnup at 3 MWd (heat) per kilogram of UO_2.
4. Conversion efficiency at 31 percent, heat to electricity.
5. Unit costs and payment schedules: Uranium—variable cost, mean payment 4 years before mean receipt of revenue; includes costs of mining, milling, exploration. Enrichment—$100 per separative work unit (uranium), paid $3\frac{1}{2}$ years before receipt of revenue; includes chemical conversion costs. Fabrication—$100/kg of UO_2, paid 3 years before mean receipt of revenue; includes chemical conversion costs. Storage fee—$125/kg of UO_2, paid 3 years after mean receipt of revenue; includes transportation costs.

[b] These values give the approximate range of present market conditions.

been declining. At worst, these charges can be expected to increase at the general inflation rate.

• Storage fee charges of \$125/kg of UO_2 in spent fuel are a guess. Numbers between \$25 and \$150 have been mentioned. The fuel cycle cost is not very sensitive to this charge, because the cost is incurred after the revenue has been produced and is therefore discounted.

• The schedule of payments is a crude approximation to conditions that would exist if well-developed markets for all the cost components existed. In fact, except for fabrication charges, actual payment schedules are a complicated collection of advance payments, interest credits, and (in the case of uranium purchasing) investment sharing and crediting. To that extent, the payment schedules are both approximate and arbitrary.

The capital charge rates used in Table 5-8 roughly correspond to the following economic circumstances: 6 percent, no inflation; 10 percent, mild (3–4 percent per year) inflation; 15 percent, mild inflation plus use of equity rather than borrowed capital; and 20 percent, strong inflation, equity capital. The rate is now between 15 and 20 percent for most utilities.

The price of uranium is generally quoted commercially in units of dollars per pound of uranium oxide. The uranium costs listed in Table 5-8 (\$50, \$100, and \$200/kg of UO_2) correspond, respectively, to prices of \$22, \$44, and \$88/lb of U_3O_8. Prices average about \$18/lb, but include deliveries made under existing purchase contracts, negotiated when uranium prices were very low. The present "spot" price—the price of immediate delivery of a new order—is about \$40/lb, having declined slightly from \$45/lb in 1977. The high value of \$88/lb is representative of a price that might be reached if low-grade deposits have to be mined.

Under present market conditions, then, the nuclear fuel cycle contributes between 5.2 and 8.6 mills/kWh to the cost of electricity (see footnote *b* in Table 5-8).

For comparison, Commonwealth Edison[57] lists nuclear fuel cycle costs in 1976 dollars of 6 mills/kWh (equal to 6.72 mills/kWh in 1978 dollars), and Perl[58] lists costs of 6.89 mills/kWh in 1985, presumably deflated to 1978 dollars.

Commonwealth Edison[59] suggested 10 mills/kWh as the fuel cost from high-sulfur coal (in 1976 dollars—closer to 11 mills/kWh in 1978 dollars) and 16 mills/kWh as the cost from low-sulfur coal. These are on the same basis as the nuclear fuel cycle costs: estimated costs of new fuel supplies.

Depending on the type of coal used, the cost advantage of the nuclear fuel cycle is about 4 mills/kWh (compared to high-sulfur coal, for a midwestern utility).

Add-On Costs

One contention of opponents to nuclear power is that the cost of nuclear power is not complete. They argue as follows that subsidies, either from the government or future electric ratepayers, will be required.

- Past research and development by the government has not been incorporated into the bill.
- The cost of current research and development by the government ought to be borne by the industry.
- Government services are provided at a loss to the taxpayer (for example, enrichment services, licensing charges, insurance, and waste disposal, when available).
- Costs of decommissioning nuclear power plants ought to be added into the bill.

While CONAES has not investigated these items in great detail, we are of the opinion that none should, or can, be a source of large increase in the cost of nuclear power, for the following reasons.

- It has not been the practice of the government to recover sunk research and development costs from industries that profit from the work. The rationale has been that the economic stimulus from new products yields a return to the government in general taxes. There is a certain amount of ideology involved in any contention of this nature. However, the existing practice serves the clear economic benefit of minimizing marginal costs for the benefits provided.
- The industry is supporting most of the research and development that it considers necessary for its own continued profit. A great deal of the ongoing government research and development (such as safety research) consists of projects intended to support the general welfare. Other areas receiving large government support are justified by future general economic benefits, which are not recoverable by the industries involved. Both the LMFBR and solar power benefit from this policy.
- We have not yet found a government "subsidy" by the accounting standards in force. Both licensing and Price-Anderson "insurance"[60] seem to be charging fair fees for the services offered.* Enrichment services have been continuously scrutinized and found to be without subsidy. The price has gone up because the capital costs of plants have been allocated to the users. The government was originally the main customer for separative

*Statement 5-21, by J. P. Holdren: I am not convinced that a fair fee is being charged for the limitation that Price-Anderson places on the total liability for an accident.

work, but the main customer now is the nuclear industry, and the industry pays base costs rather than marginal costs.

• Decommissioning costs are far in the future and can in no way be considered comparable to construction costs. The Atomic Industrial Forum, the only organization that has conducted a detailed study of these costs, concludes that they might, in constant-dollar terms, be as much as 10 percent of original costs.[61] Discounting this estimate at 5 percent over 50 years yields a present worth for this item of less than 1 percent of the original plant cost.

In summary, we consider that the costs of nuclear power, as computed now or projected into the future, represent a fair statement, and that no significant additions to these costs have been identified.*

Risk Costs

Nuclear power, as an industry subject to accidents and government regulation, may incur costs from uninsured risks. These arise from the excess costs of replacement power when plants are shut down following an accident or regulatory action. A related set of costs may result from delays in licensing that add to the capital costs of plants under construction.

These risks are subsumed under the capacity-factor projections and the contingencies included in construction schedules that are now part of the industry's standard accounting. The accident at the Three Mile Island nuclear power plant in 1979 raises the question whether the accounting is adequate. Are the capacity-factor projections and construction schedules that seemed reasonable before this incident still reasonable?

These questions cannot yet be answered. The rate of regulatory shutdown does not appear much greater than the rate prevailing before the accident at Three Mile Island. A licensing hold that has been in effect since then has delayed the schedules of several new reactors, but it may be lifted in the future.

The prolonged shutdown of the Three Mile Island plant represents a financial blow to its operating utility. The loss could be mitigated by an assessment against other nuclear units that would add less than 1 percent to nuclear generating costs. Institutions and arrangements to spread the risks in this or similar ways do not yet exist, but they are being explored. If there were many accidents, the costs would become significant, but in that case, nuclear power would no longer be considered a major energy option.

*Statement 5-22, by J. P. Holdren: A major uncertainty neglected here is whether large LWR's will in fact be able to operate at high capacity factor for the lifetimes advertised.

LMFBR Costs

Unlike those of existing coal-fired or light water reactor power plants, the costs of a commercial breeder reactor have only been estimated, not experienced. However, it is generally acknowledged that the capital costs of LMFBR's will be higher than those of today's light water reactors. This is explained partly by the need for an extra intermediate heat-transfer loop, partly by the extra complexity involved in using sodium as coolant, and partly by the more complex refueling and auxiliary systems. These additional costs are partly offset by the inherent advantages of eliminating the highly pressurized primary system of the LWR's, and by economies in the turbine-generator and condenser systems resulting from the higher thermal efficiency of the LMFBR power cycle.

For a light water reactor, the nuclear steam-supply system accounts for 10–20 percent of the total capital costs of the plant. The other 80–90 percent of the cost is for the so-called balance-of-plant, mostly conventional structures (piping, turbine, generator, condenser), cable, installation labor, engineering, and indirect costs. There seems to be no intrinsic reason why these costs should be higher for a developed LMFBR than for an LWR. On the other hand, the nuclear steam-supply system is expected to cost 2–3 times as much as the system for an LWR. Adding and subtracting these items, the capital cost of a developed LMFBR power plant is expected to be 10–40 percent higher than the cost of an equivalent LWR power plant.

An analysis conducted of cost estimates for the Clinch River breeder reactor, of the actual costs of the French LMFBR (Phenix), and of the economic improvements expected for a commercial plant suggests that capital costs of commercial LMFBR's should be about 40 percent greater than those of LWR's.[62]

The Clinch River breeder reactor is a first-of-a-kind demonstration plant, with costs much higher than those expected for commercial plants. Its construction cost is estimated as 3–5 times that of an LWR of equal capacity. Cost reductions of a magnitude sufficient to bring LMFBR costs down from this starting point to a target value within 40 percent of LWR costs are not uncommon in industrial development, but they are large enough that achievement of the target is uncertain.

Balancing the higher capital costs expected of LMFBR's is the expectation that their fueling costs will be lower than those of LWR's. There is no need for continuous fissile feed, for example, and the excess fissile material has by-product value. High fuel burnup is not limited by large reactivity losses. The reduction in fuel costs expected from these factors could compensate for the high fuel reprocessing and fabrication costs assumed for LMFBR's.

TABLE 5-9 Typical Nuclear Fuel Cycle Costs in the 1990s (1976 dollars)

	Light Water Reactor	Liquid-Metal Fast Breeder Reactor
Assumptions		
U_3O_8 ($/lb)	60	N.A.
Enrichment ($/SWU)	100	N.A.
Fabrication ($/kg)	100	800
Reprocessing ($/kg)	200	350
Capacity factor (percent)	70	70
Fuel burnup (percent heavy metal atoms)	3	6
Fissile plutonium value ($/g)	24	24
Waste management[a]	—	—
Contributions to generating costs (mills/kWh)		
U_3O_8, net	4.33	0
Enrichment	2.36	0
Fabrication	0.55	1.92
Reprocessing	0.74	0.63
Plutonium sale	(0.43)	(0.43)
Plutonium inventory	N.A.	1.03
TOTAL	7.6	3.2

[a]Not included, but assumed identical.

Table 5-9 compares projected fuel cycle costs for an LMFBR and LWR (with uranium and plutonium recycle), at a U_3O_8 price of $60/lb in 1977 dollars. At this uranium price, and with the warning that the unit costs for such items as fabrication and reprocessing are estimates rather than firm values, the LMFBR fuel cycle cost would be 4.4 mills/kWh less than the fuel cycle costs of an LWR.

Anticipating a result obtained later in this section (that capital charges now contribute about 20 mills/kWh to the price of electricity), an LMFBR that was 40 percent more expensive would contribute 8 additional mills to that cost. From Table 5-9, we may infer that the LMFBR will save 8 mills in fueling costs when the price of U_3O_8 reaches $110/lb. Uranium concentrates might cost $60/lb (constant value) in the late 1980s, but are unlikely to cost $110/lb until well past 2000 (unless the demand for nuclear power and the parallel demand for uranium accelerate in the intervening period). On the other hand, if the LMFBR cost targets are met, these results also indicate that the breeder could place a cost ceiling on electric power at a level no higher than 30 percent above present prices.

To the extent that breeders replace LWR's, they enable more economic operation of the remaining LWR's by slowing the demand for uranium and the escalation of its cost. They would also exert downward pressure on the costs of coal. A system of breeders and LWR's would be economically more attractive than a system of either by itself. A system of breeders and advanced converters would show even greater mutual cost benefits. With reprocessing and recycling, various types of breeders and converters could compete for the market.

Summary of Costs

Table 5-10 represents an appraisal of the costs of LWR and coal power. These are planning figures from a midwestern utility, based on replacement-cost accounting as of 1976. They are therefore useful for comparing expected costs. Table 5-10 gives nuclear power plants an expected 18 percent cost advantage over coal plants with scrubbers.

Another recent comparison is presented in Figure 5-7. According to these data, if both coal and nuclear plants are run at 70 percent capacity factor, and particularly if the best available emission control technology is required for coal plants, nuclear power is cheaper in most regions of the United States. However, if nuclear plants are run at capacity factors around 55 percent and coal-fired plants at around 70 percent, with no new emission control technology required, coal-generated electricity would be cheaper. (As all new coal-burning plants are now required to install scrubbers, this latter comparison may be of little interest.)

There is a strong incentive to minimize investment risk in the utility sector. This will lead to decisions for a mix of coal and nuclear plants, since future costs for both sources of power have large uncertainties and each is a hedge against the other. If at some time the costs become more reliably predictable, the mix can be adjusted.

We have not found any costs within the nuclear estimates that can be identified as sources of differential cost escalation (relative to the costs of coal-generated electricity), nor are there any new charges against nuclear power that would increase its relative cost. There are potential requirements for further emission control devices on coal-burning plants that could significantly increase the cost of coal power. Therefore, if we were forced to make a prediction, we would guess that nuclear power would dominate the electrical generation market if cost were the only consideration. Fortunately, this guess is unnecessary, as the decision will be made on an investment-by-investment basis, and cost is not the only consideration.

TABLE 5-10 Comparison of Estimated Total Busbar Generating Costs (1976 dollars) for a Midwestern Utility (mills per kilowatt-hour)[a]

Costs	Nuclear	Low-Sulfur Coal Without Scrubbers	High-Sulfur Coal With Scrubbers	Oil
Carrying charges	20	14	20	14[b]
Fuel (replacement costs)	6	16	10[c]	23[d]
Other	2	2	4[e]	1
TOTAL	28	32	34	38
Nuclear advantage		4	6	10
		(13%)	(18%)	(26%)

[a]*Nuclear Fuel Assumptions*:

1. \$35/lb of U_3O_8.
2. \$75 per separative work units for enrichment—0.20 percent tails assay.
3. \$100/kg for fabrication.
4. Burnup in megawatt-days per metric ton: 33,000 for pressurized-water reactors and 29,000 for boiling-water reactors.
5. Net salvage cost (cost of reprocessing and waste disposal, less salvage recoveries) equivalent to about 0.5 mill/kWh.

Fossil Fuel Assumptions (*delivered cost to Chicago area*):

1. High-sulfur coal, \$1.00 per million Btu.
2. Low-sulfur coal, \$1.60 per million Btu.
3. No. 6 oil, \$2.30 per million Btu.

[b]Roughly the same as coal without flue-gas scrubbers.

[c]Includes 0.5 mill/kWh for fuel required to generate power for scrubber operation.

[d]Based on \$2.30 per million Btu oil—roughly equivalent to \$15 per barrel for no. 6 oil and 10,000 Btu/kWh.

[e]Includes 2 mills/kWh for flue-gas scrubber operation and maintenance expense other than fuel. (Actual expenses have been costlier so far.)

Source: Gordon R. Corey, testimony before the Environment, Energy, and Natural Resources Subcommittee, Committee on Government Operations, U.S. House of Representatives, "Nuclear Power Costs (Part I)," 95th Cong., 1st Sess., Sept. 12–19, 1977, p. 883.

The following points should be noted for breeders and advanced converters.

- It is very unlikely that more than a few pioneer commercial-scale units of either type can be put on line this century.
- The economics of a single unit or a few units will not have a significant effect. The economies of new types of reactors will only be realized after they are fully commercial.
- Advanced reactors, and particularly breeders, offer the prospect of a durable cost ceiling on the price of electricity. Projections of the cost of

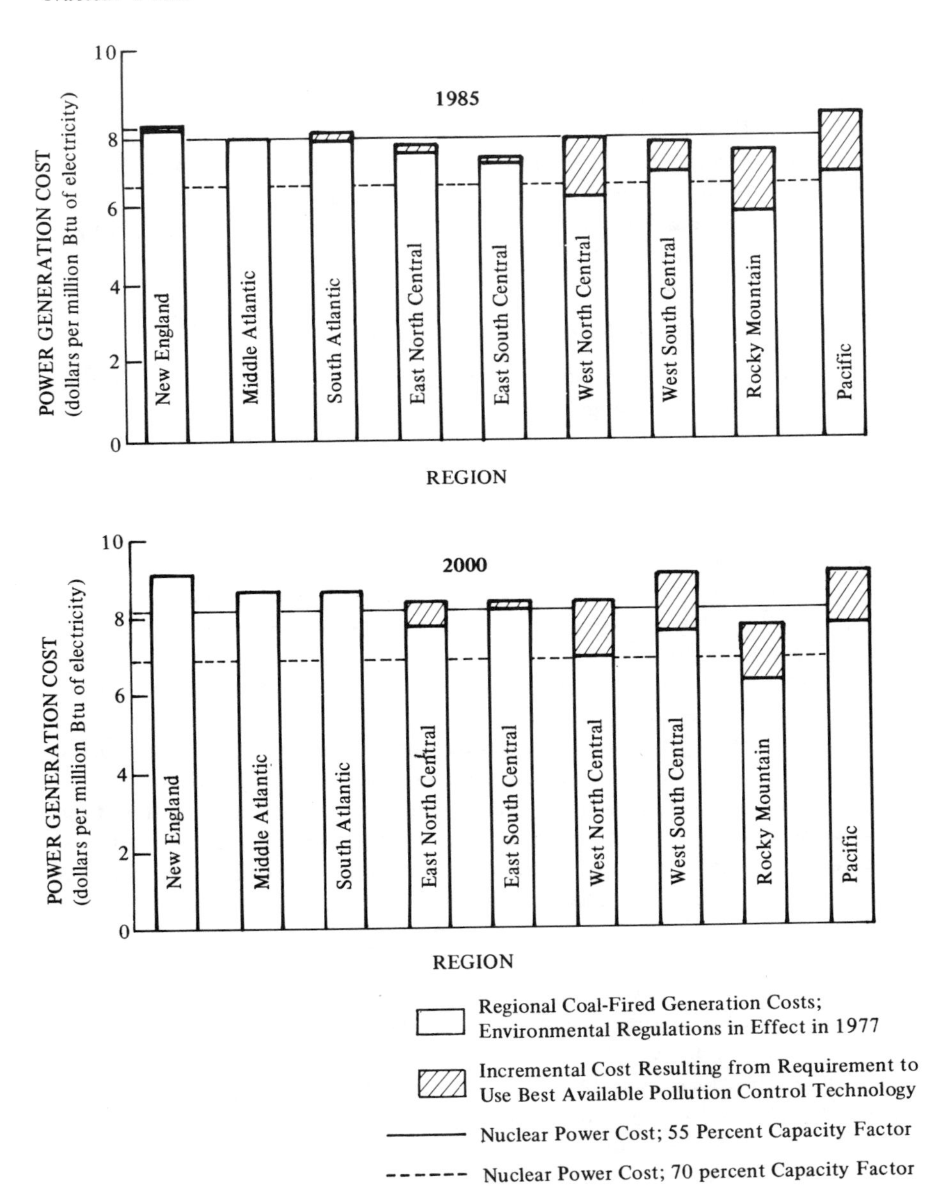

FIGURE 5-7 Comparison of coal-fired and nuclear power costs under existing and proposed environmental regulations. Source: D. Gunwaldsen, N. Bhagat, and M. Beller, *A Study of Potential Coal Utilization, 1985–2000* (Upton, N.Y.: Brookhaven National Laboratory (BNL 50771), 1977).

electricity from commercial breeders range from 10 percent to 30 percent above today's cost; similar costs can be expected for advanced converters.

- The economics of a mixed system of breeders and converters is likely to be more favorable than the economics of either reactor type by itself.
- The decision for or against the ultimate possibility of a breeder economy will have a profound effect on decisions about other reactors—LWR's in particular—but advanced converters as well, since breeders would help hold down future uranium demand and cost.

REGULATION OF NUCLEAR POWER PLANTS

The regulatory process affects the industry by lengthening the time between planning a new generating facility and placing it in operation, by retroactive changes in plant design arising from the unique surveillance responsibility of the Nuclear Regulatory Commission, and by providing a special forum for public opposition to nuclear plants.

Fossil fuel plants typically require 8–10 years from the start of planning to the completion of construction, while nuclear plants require 10–12 years or more. A large part of this extra time is claimed by the extensive reviews required for nuclear plants prior to construction. The extra time costs money and adds to the uncertainty that a project's cost targets will be met. The nuclear industry has a large stake in shortening the period taken up with regulatory processes.

Many rules and regulations engender large capital and operating costs, not all of which (in the industry's view) can be easily justified by cost-benefit analysis or improved public safety. Contentions of this type between regulators and regulated are fairly standard in our political system. A more vexatious matter is the changes in regulations imposed during the plant's construction or operation. Such changes are more expensive to implement than changes during the design stage—often an order of magnitude greater.[63] Justification for these retroactive changes is always at issue, particularly since each change in regulations increases the opportunities for further legal and administrative interventions.

Interventions have been used to build up opposition to nuclear power, and in some cases, have forced postponement or cancellation of nuclear power plants, either by generating resistance in the region affected, or by delays that bring the economics of the plant into question. The regulatory process, reinforced by judicial interpretations, serves the clear function of guaranteeing that all proper points at issue are raised and judged. However, very little is accomplished if the same points are debated again and again.

Although frequently frustrating to the industry, the close attention paid to nuclear power through the regulatory process has contributed to the safety record of this energy technology. Moreover, a competent and independent regulatory agency is necessary for public acceptance of nuclear power. The involvement of the Nuclear Regulatory Commission in events following the accident at Three Mile Island is widely credited as decisive in maintaining confidence that a nuclear accident need not lead to public catastrophe. In streamlining regulation of the nuclear power industry, therefore, close attention to the objective requirements of protecting the public must be maintained, as well as attention to the requirements of public confidence: legitimate participation in technical decisions and observance of due process. Three complicated aspects of the regulatory process could be simplified without affecting its integrity.

1. Responsibility is shared but not clearly partitioned among federal and state agencies for safeguarding fuels, overseeing the storage of nuclear wastes, monitoring the health and safety of plant employees, regulating financing and insurance, and protecting the environment. This responsibility is in a state of flux. A division of responsibility should be worked out that is clear, reasonable, and stable.
2. Issues in nuclear power should be settled as much as possible on generic bases. Given generic findings based on comprehensive hearings, regulatory agencies can conduct hearings on individual systems, taking up specific items of contention that have not been previously heard and judged. The records of hearings on generic issues provide a sound basis for determining whether new data can affect the findings, and if so, whether such data should be sought and heard.
3. The regulatory process too often degenerates into an adversary process when, in fact, all parties share an interest in safety, economy, and service. The regulatory process should be scrutinized and corrected so as not to discourage conciliation and mediation.

SAFETY OF NUCLEAR REACTORS AND THEIR FUEL CYCLES

This section examines technological aspects of nuclear power safety. Chapter 9 discusses the health effects of the nuclear energy cycle and compares them with those of fossil fuel use. A recent report of the National Academy of Sciences presents an analysis of these and other safety issues, drawing on a review of the literature to lay an ample base and set legitimate bounds for dicussion.[64]

Radioactivity is an inherent feature of nuclear energy. Nuclear fuels and fertile materials are radioactive. The chain reaction produces neutrons and gamma rays from fission, as well as new radioactive materials. These latter are the products of fission and of nonfissioning neutron capture. It has long been known that radiation causes damage to living tissues. The safety of nuclear power depends on isolating these sources of radiation from the biosphere.

Isolation cannot, and need not, be absolute. Radiation is a natural part of our lives. By comparing the radioactivity added by nuclear power to natural radiation backgrounds, or to the variation in background among locations generally considered to be healthy living environments, exposure standards can be set at some fraction of "normal" background. (See chapter 9.) At some level of addition, small increases of radioactivity may be judged to be inconsequential, as a practical matter, simply because the consequences are very much less than those from risks already accepted in exchange for similar benefits. This is a judgment that must be made by society ("how safe is safe enough?"). In making this judgment, the most appropriate standards (in the case of nuclear power) will be based on comparisons with the risks of alternative sources, or with the risks of not having this source of power.

At least two characteristics of radioactive emissions must be supplied to inform this judgment: their frequency (probability per unit time of emissions of given magnitude), and their consequences. Of these characteristics, the frequency is the more difficult number to ascertain. The health effects of radiation have been studied more completely than the effects of any other type of emission. There is still some controversy (as detailed in chapter 9), but models of radiation effects that most authorities believe to be conservative (i.e., that predict more severe consequences than are considered likely to result) are used by regulatory agencies to set exposure limits and to estimate the consequences of very low levels of irradiation.

Such studies, however, do not address the probability of radioactive emissions—the frequency variable. They serve to dramatize the principle that releases of large magnitudes must not be allowed. This is then a design criterion for nuclear systems. The success of design must be judged against this criterion by a numerical estimate of the frequency of a release. If the frequency is sufficiently small, a case can be made for the practical impossibility of the event. However, at smaller frequencies the background of experience disappears. Estimates must be made on the basis of event models, and the uncertainties of the estimates become very large.

"Risk" is the measure of the expected average value of consequences from some unit value of operation of a system. For example, "fatalities per passenger-mile of commercial aircraft," "person-days lost through accidents per person-year of employment in industry X," or "corrosion damage costs (in dollars) per unit of release of chemical Y" are all measures of specific risks. We have adopted the GWe-year of reactor operation as the unit of system operation, since this gives some insight into the risks associated with a single large nuclear power plant. Actuarial risk, of course, says nothing about the distribution of risk (for example, among different models of commercial aircraft), or about whether the risk arises from many small accidents or a few large ones.

Safety of Normal Operations Although greater public concern surrounds reactor accidents, controversy surrounds the safety of the normal operations of nuclear power systems. Three separate areas of operation have figured in these controversies.

1. The operation of reactors involves some discharge of radioactivity to the environment. Attempts have been made to implicate these discharges as significant causes of public morbidity.
2. During the course of reactor operations, and specifically during maintenance periods, workers at nuclear power plants are exposed to radiation from contaminated equipment. It has been charged that this represents undue and uncompensated risk to the workers.
3. At various stages in the nuclear fuel cycle, radioactive effluents are discharged. This is another potential source of public morbidity.

These points are discussed at greater length in chapter 9 of this report. The last point will be treated in the subsequent discussion of nuclear waste management and disposal. It is worth noting here that continued review has indicated that normal reactor discharges, within existing regulatory limits, are *not* significant causes of public morbidity. The evidence on which the first charge is based has not stood the test of scientific scrutiny. With regard to occupational exposures of workers in nuclear installations, on the other hand, it follows from the linear dose-response hypothesis that existing limits on occupational exposure to radiation present a marginal risk to workers of slightly increased risk of cancer. (This hypothesis is still in dispute, as discussed in detail in chapter 9.)

Reactor Accidents More than 98 percent of the radioactive atoms made in a reactor, and an even greater fraction of those that remain radioactive after a few seconds, are generated in the fuel. The two types of new radioactive atoms are fission products and actinides. Fission products are a congeries of nuclides of medium atomic mass number (66–172 amu, with the largest quantities found near mass numbers 95 and 140). Actinides are the elements beyond actinium, and include both natural and manufactured isotopes of thorium, protactinium, uranium, neptunium, plutonium, and heavier elements. The half-lives of fission products vary from fractions of a second to very large values, and different fission products become important at different times after they are formed. The artificial actinides of consequence have longer half-lives than the more important fission products, and correspondingly less radioactivity (for a given number of atoms, radioactivity is inversely proportional to half-life). The radioactivity of the actinides in nuclear fuel just after reactor operation is inconsequential compared to that of fission products, and it is not until after several centuries of fission-product decay that the two sources of radioactivity become about equal.

Reactors are designed to contain the radiation they produce. Thick biological shields of concrete, or deep pools of water, are used to attenuate neutrons and gamma rays. In addition, their fuel elements (with the exception of the molten-salt reactor) are designed to contain the fission products and actinides produced. If individual fuel elements fail, the primary reactor system can contain the radioactivity they release, and as a backup, the whole system is placed inside a stout, pressure-type containment building.

The efficacy of these multiple containments is the subject of reactor accident analysis. Accident conditions are postulated, and the consequences of the accident are modeled analytically. From this analysis, further engineered safeguards may then be indicated, which either reduce the probability (frequency) of the accident or mitigate its consequences (chiefly, exposure of people to radiation).

No reactor considered for nuclear power generation can explode like an atomic bomb; nevertheless, certain of the postulated types of accidents might have severe consequences.

The efficacy of containment is also tested by reactor incidents and accidents. In the accident at Three Mile Island, for example, the contaminated water in the reactor building was not contained there.* This is now recognized as a flaw in the plant's design.

In the case of light water reactors, there is general agreement that the worst type of accident would be one that led to melting of a large part of the reactor core. Such an accident might conceivably be caused by a large power increase, beyond the capability of the coolant to remove the energy generated, or by an interruption of coolant flow. Of these potential mechanisms, the power increase can be ruled out because it would immediately result in a decrease in water density, which would act to decrease the power again. In this respect, light water reactors are inherently self-regulating. An interruption of coolant flow could, however, be brought about by a break in the coolant-flow line—a loss-of-coolant accident (LOCA).

Since in a LOCA the loss of water would quickly quench the fission reaction, only the decay power of the radioactive products would remain. This is about 6 percent of reactor thermal power immediately after

*Statement 5-23, by L. F. Lischer: The reactor building structure was not breached; design of controls and operator action permitted the pumping of contaminated water to the auxiliary building.

shutdown, and it decreases with time.[65] One hour after shutdown, a 1200-MWe (3800-MWt) LWR would still be generating about 50 MWt of heat from radioactive decay. (This is about the time after shutdown that the most serious event in the Three Mile Island accident occurred. We infer from the information published so far that the decay heat from the reactor core had raised the fuel temperature to the point that its cladding began to react chemically with the reactor's water.)

To remove this heat under the assumed conditions that the normal water circulation would have been lost in the "blowdown" following the LOCA, an emergency core-cooling system (ECCS) is an engineered safeguard feature of LWR's. This is a system of several subsystems, each capable of removing the heat from a reactor after a LOCA and designed to function independently.

If the ECCS were to fail to cool the reactor fuel, the fuel could melt, forming a large glob. Further cooling would be extremely difficult, and the molten fuel could melt through the reactor vessel. Throughout this period, it would be releasing fission products, whose radioactive decay would heat the containment building, as well as the structures and the air within it. The steam escaping from the reactor through the ruptured pipe could further heat the building. In the absence of a cooling system for the building, or controlled venting, the containment might conceivably rupture from the internal pressure. Alternatively, the containment might be faulty from some other cause, or the molten fuel might itself melt through the floor of the building and be released to the earth below.*

While engineering safety features, such as the containment building, the ECCS, containment cooling devices, and filtered vent lines, can reduce the likelihood and consequences of serious accidents in light water reactors, the probability of the core's melting followed by release of a large amount of radioactivity cannot be reduced to zero.

Using an analytical technique called fault-tree analysis, the Reactor Safety Study (also referred to as the Rasmussen Report or WASH-1400)[66] has estimated the expected median frequency and consequences of various accidents in light water reactors of contemporary design.[67] The method consists of analyzing failure rates of various components in the operating reactor system (including operator failure where appropriate), determining what further events must occur to lead to significant accidents, determining what failure rates are appropriate to these further events, and so on. The result is statistical assignment of a very large number of events into a frequency distribution relating magnitude of release (of radioactivity) to frequency of release.

*See statement 5-24, by L. F. Lischer, Appendix A.

There are so many components and potential pathways for release that the process can never claim to have evaluated all possible accidents exhaustively; however, the authors of the study maintain that they have analyzed enough cases (which, in truth, comprise a very large number) that their statistical treatment is likely to be of a valid sample. Moreover, they have corrected their frequencies upward, as is proper, to account for unsampled accident chains. A major effort of the study was to conceive of events that have the potential for very large releases. Therefore, the authors believe (and we consider it reasonable) that the frequency of large releases is not likely to be low by virtue of "missed" accident sequences.*

Any fault-tree analysis depends on its input data. Many criticisms of WASH-1400 from the nuclear reactor industry are based on the contention that the fault frequencies of individual components were consistently overestimated. This is a consequence of two separate factors. First, failure data from eclectic sources were used in the absence of statistically valid samples from the nuclear industry itself (for example, with regard to pipe breaks, losses of motive or signal power, and so on). Second, the translation of these data into failure analysis in nuclear systems requires engineering judgment. The engineering protocol of basing such judgment on "conservative" or worst-case analysis is almost automatic, and it is claimed that much of the translation was on this basis.

A degree of conservatism that has been documented since the time most critiques were filed can be found in the assumptions made about the rate of heating of uncooled reactor fuel[68] and about the release of fission products from melted fuel.[69] It is now believed that a "best value" assumption of the heat input would imply delayed onset of meltdown and a lengthier period over which meltdown might occur, significantly improving the likelihood of corrective action. A "best value" assumption of fission-product release would significantly reduce the estimate of fission-product escape from a core melt.†

Criticisms have been directed against the treatment of "common-mode" failures in WASH-1400. A common-mode failure is an accident in which a single initiating event dislocates protective sequences designed to deal with the consequences of that event. For example, protective circuits are usually arranged so that a failure of one is backed up by the operation of another; if the event that caused one protective circuit to fail also caused the others to fail, that would be a common-mode failure. Data on common-mode failures are very difficult to validate; nevertheless, some WASH-1400 assumptions on this point are questionable.[70] The Risk and Impact Panel, after investigating this point, concluded that the criticisms were valid, but

*Statement 5-25, by J. P. Holdren: The history of attempts to identify *a priori* the ways that complex systems could fail warrants more skepticism than is expressed here.

†Statement 5-26, by J. P. Holdren: See my dissenting view, statement 5-27, Appendix A.

could not specify what the effects of the WASH-1400 assumptions might be. Other critics contend that the assumptions are optimistic (i.e., low) in their effect on the predicted frequency of large release.

It has also been noted that the statistical presentation of risk in WASH-1400 is not appropriate for actuarial purposes. The report presents median values for frequencies of accidents of varying severity, whereas for actuarial purposes, mean values are more appropriate. Means are "expectation values." For the sort of frequency distribution assumed (log normal), and in the case of a large spread (standard deviation) of such a distribution—as in the WASH-1400 results—the mean may be many times greater than the median.

A number of organizations, including the Atomic Energy Commission, the American Physical Society, the Environmental Protection Agency, and the Union of Concerned Scientists,[71] identified a number of omissions, errors, and additional sources of uncertainty in the report. The Nuclear Regulatory Commission responded to these criticisms by commissioning the Risk Assessment Review Group. The group's recently issued report[72] states essentially the same conclusions presented here. It is notable that there was little consultation between members of CONAES and members of the review group. The conclusions—both positive and cautionary—may represent a growing consensus on the status of reactor safety.

At the time of our review, detailed evaluations of the recent incident at the Three Mile Island nuclear power plant near Harrisburg, Pennsylvania, were not available. It is possible that it may be an example of the class of accidents—potentially catastrophic, but in the event, controllable—from which improved safety practices would follow. If improved practices do result, then the outlook for nuclear safety might be favorable in a technical sense, regardless of the justifiably negative effect of the accident on public appraisal of nuclear power.

The Risk and Impact Panel also considered WASH-1400 and concluded that there did not seem to be any consistent biases in the study, but concurred in the qualitative judgment that the uncertainties in accident frequencies and consequences should be larger than reported.[73]

An important critique of WASH-1400 was presented by the Nuclear Energy Policy Study Group. Their conclusion was that considering all the uncertainties (those highlighted above, and others), LWR's are unlikely to have an actuarial risk more than 500 times greater than that inferred from the median value points presented in the report. The group notes that at this upper limit, the risk from an LWR is not higher than the upper limit of risk from coal power, and therefore concludes that reactor safety against accidents ought not to be a factor inhibiting nuclear expansion, at least as compared with coal.[74]*

WASH-1400 is a monumental piece of work, one that can be used to define those safety problems and system designs for which further work would be most significant, but it cannot prove either that reactors are safe or that they are dangerous.

With all the qualifications just presented, it is nevertheless useful to examine the Reactor Safety Study's conclusions (set out in Table 5-11). It should be noted that risk is the product of frequency (chance per reactor-year) and consequences. Table 5-12 presents the data in this form. The frequency (probability per year) of loss of coolant is estimated as 1 in 2000 per reactor-year. The probability that emergency systems designed to prevent meltdowns in loss-of-coolant accidents will fail is estimated as 1 per 10 accidents, leading to a meltdown frequency of 1 in 20,000 per reactor-year. Further, WASH-1400 has estimated that only 1 meltdown in 100 would release large enough amounts of radioactivity to cause 10 or more deaths among members of the general public. The product of this 1-in-1000 figure and the estimated probability for meltdown of 1 in 20,000 per reactor-year gives the much-quoted WASH-1400 estimate of the probability of severe accidents in light water reactors: 1 in 2 million per reactor per year.

The Reactor Safety Study can be used to draw certain inferences, in spite of the large uncertainties that must be attached to the frequencies at which accidents of various consequences might occur.

The shape of the risk curve is a good deal less sensitive to uncertainty than is the actual magnitude of risk. This is because the larger releases tend to be consequent to the same initiating events. The entire risk curve will go up or down as the frequencies of the initiating events are changed, but the relative frequencies of the larger-consequence events will not change. From this, we may draw other inferences.

For example, accidents with lesser consequences will be far more prevalent than those with greater consequences. What this implies is that a vigorous program to diagnose and correct flaws in reactor safety systems as they become apparent through small accidents will not fail to decrease the expected frequency of severe accidents.

Another inference that is relatively firm concerns the nature of the dominant risk. This is the type of accident from which (over the long term) the greatest damages are expected to accrue. Table 5-12 exhibits maximum values among the entries for a particular column in the region of dominant risk. It can be noted that this is roughly the "one-in-a-million-reactor-

*See statement 5-27, by J. P. Holdren, Appendix A.

TABLE 5-11 Consequences of Reactor Accidents as a Function of Median Frequency of Their Occurrence

Frequency (chance per reactor-year)[a]	Early Fatalities	Cases of Early Illness	Total Property Damage (billions of dollars)	Decontamination Area (square miles)	Relocation Area (square miles)	Latent Cancers[b]	Thyroid Nodules[c]	Genetic Effects[d]
One in 20,000[e]	—	1	—	—	—	—	—	—
One in one million	—	300	0.9	2,000	130	170	1,400	25
One in 10 million	110	3,000	3	3,200	250	460	3,500	60
One in 100 million	900	14,000	8	8,000[f]	290	860	6,000	110
One in one billion	3,300	45,000	14	14,000[f]	300	1,500	8,000	170
Uncertainty[g]	0.25–4	0.25–4	0.20–2	0.20–2	0.20–2	0.17–3	0.33–3	0.33–6

[a]These are median probabilities, with uncertainties of a factor of 5 in either direction.

[b]Fatal cancer cases per year over an assumed 30-year latency period, resulting from the postulated release.

[c]Nodules per year in an assumed 11- to 40-year period following the postulated release. The thyroid nodules counted here are benign or successfully treatable. Thyroid cancers resulting in death are included in the latent cancers.

[d]Induced effects per year over the span of one human generation following the postulated release. Later generations would have fewer cases from that release.

[e]One in 20,000 years is the estimated median frequency of meltdown, per reactor.

[f]There is no risk measure for these quantities; they are ways of characterizing consequences.

[g]Factors within which there is 95 percent confidence that consequences have been accurately predicted.

Source: U.S. Nuclear Regulatory Commission, *Reactor Safety Study: An Assessment of Accident Risks in U.S. Commercial Nuclear Power Plants* (Washington, D.C.: U.S. Nuclear Regulatory Commission (WASH-1400), 1975).

TABLE 5-12 Risks of Reactor Accidents as a Function of Median Frequency of Their Occurrence (per reactor-year)[a]

Frequency (chance per reactor-year)	Early Fatalities	Cases of Early Illness	Total Property Damage (thousands of dollars)	Decontamination Area (square miles)[b]	Relocation Area (square miles)[b]	Latent Cancers[c]	Thyroid Nodules[d]	Genetic Effects[e]
One in 20,000[f]	—	0.00005	—	—	—	—	—	—
One in one million	—	0.0003	0.9	—	—	0.0051	0.042	0.0025
One in 10 million	0.000011	0.0003	0.3	—	—	0.0014	0.011	0.0006
One in 100 million	0.000009	0.00014	0.08	—	—	0.00026	0.0018	0.00011
One in one billion	0.0000033	0.000045	0.014	—	—	0.000045	0.00024	0.000017

[a]These are median probabilities, with uncertainties of a factor of 5 in either direction.
[b]There is no risk measure for these quantities; they are ways of characterizing consequences.
[c]Fatal cancer cases per reactor-year.
[d]Nodules per reactor-year. The thyroid nodules counted here are benign or successfully treatable. Thyroid cancers resulting in death are included in the latent cancers.
[e]Genetic effects per reactor-year if the total number of cases is 100 times the annual rate exhibited in the first generation.
[f]One in 20,000 years is the estimated median frequency of meltdown, per reactor.

Source: U.S. Nuclear Regulatory Commission, *Reactor Safety Study*: *An Assessment of Accident Risks in U.S. Commercial Nuclear Power Plants* (Washington, D.C.: U.S. Nuclear Regulatory Commission (WASH-1400), 1975).

year" category, characterized by few immediate casualties of any type, by tens to hundreds of delayed health effects, and by hundreds of millions of dollars in property damage. Therefore, such cases should be considered as characteristic of "catastrophic" nuclear accidents. We infer that a similar type of reasoning may have guided the Nuclear Energy Policy Study Group in concluding that "the consequences of an extremely serious accident are not out of line with other peacetime catastrophes that our society has been able to handle . . . ."[75] Damages in the same range as those from dominant nuclear accidents have, after all, been experienced in other industries: refinery and chemical plant fires and explosions, airplane crashes, shipwrecks, and toxic chemical and metal releases.

The assumption implicit (if not explicit) in the reactor safety studies conducted so far is that the equipment and the people operating it and regulating its use behave approximately according to the conditions specified. Nevertheless, there may be shortcomings in the people and equipment. Mistakes, laxity, and incompetence can overcome technological barriers. In the nuclear power industry, as in any industry in which mistakes can have expensive consequences, human errors and inadequacies constitute a significant source of risk that is difficult to quantify. It would seem that the uncertainties in estimations of risk have themselves been underestimated by failing to take these factors into account.

On the other hand, human ingenuity eventually brought the two most serious nuclear power accidents (Brown's Ferry and Three Mile Island) under control, and this quality has evidenced itself in the prevention and mitigation of many other incidents.

Thus, we find reason to assign an uncertainty to the possibilities calculated for nuclear power accidents, ranging higher or lower than those published, and note as a consequence the great value of maintaining as well trained a work force as possible for the design, construction, operation, maintenance, inspection, and supervision of nuclear power plants.

Safety of Other Reactors Fault-tree, event-tree analysis can also be applied to compare the risks of various reactor systems against one another. The analyses require information that has not yet been assembled for advanced converters or fast breeders: specific designs, recognized design criteria, and results of accident analyses.

The only document produced in the United States on the safety of LMFBR's and available for study is the draft environmental statement for the Clinch River breeder reactor. Events that might lead to large-scale release of radioactivity to the public are the class-8 and class-9 accidents listed in Table 5-13.

Accidents that form the basis for the plant's design are grouped in class

TABLE 5-13 Postulated Liquid-Metal Fast Breeder Reactor Accidents That Could Result in Radioactive Release to Public, with Postulated Light Water Reactor Accidents Shown for Comparison

Accident Classification	Description	Light Water Reactor	Liquid-Metal Fast Breeder Reactor
8	Accident-initiating events considered in design-basis safety evaluation	Transients in reactivity; rupture in primary piping system; steam-line breaks	Leaks in steam generator; steam-line breaks, failures in primary sodium storage tank; leaks cold trap; large rupture in primary piping system[a]; events leading to core disruption[b]
9	Hypothetical sequences leading to accidents more severe than those in class 8	Successive failures of multiple barriers provided and maintained to prevent the escape of large amounts of radioactive material	Successive failures of multiple barriers provided and maintained to prevent the escape of large amounts of radioactive material[b]

[a]The Clinch River breeder reactor design has a closed-cycle secondary heat-transport system that separates the primary coolant from the power-conversion system. Class-4 failures (events that release radioactivity into the primary cooling system) and coincident heat-exchanger leaks would not result in release of significant amounts of radioactive material to the environment.

[b]While the Nuclear Regulatory Commission does not consider these events as design-basis accidents for the plant, the commission has asked that mitigating features be provided.

Source: National Research Council, *Risks and Impacts of Alternative Energy Systems,* Committee on Nuclear and Alternative Energy Systems, Risk and Impact Panel (Washington, D.C.: National Academy of Sciences, in preparation).

8. The plant must be designed to withstand such accidents without failure of containment. These postulated accidents begin with the failure of major components or piping and threaten the release of significant amounts of radioactive material from the reactor's primary system. The design must incorporate special features to mitigate the consequences of class-8 accidents: sealed equipment cells, a double-level containment, and two independent, diverse systems to shut the reactor down. Siting regulations require the offsite doses calculated for the full range of class-8 accidents to fall within guideline values.

In the LMFBR, the sodium coolant circulates at atmospheric pressure and at temperatures well below the boiling point. Thus, loss of the coolant by sudden evaporation (blowdown) is impossible. This represents a significant safety advantage over light water reactors.

Potentially severe accidents that are physically possible, but so extremely improbable that it is not considered reasonable to counter them by expensive engineered safeguards or consider them in siting decisions, fall into class 9. Class-9 accidents require that major failures in the reactor system be accompanied by independent and concurrent failure of safety systems and barriers to the escape of radioactive material. The consequences of these accidents could in some cases exceed the consequences calculated for the worst accidents considered in the safety report on light water reactors. It seems intuitively reasonable to assume that the likelihood of such accidents would be substantially lower than even the very low values predicted by the Reactor Safety Study for high-consequence LWR accidents, but this intuition must be confirmed by a probabilistic fault-tree analysis.

The compact core of a liquid-metal fast breeder reactor displays both high power density and high plutonium content. This core is not designed for maximum reactivity. The principal concern over a possible disruption of the core is that it might take on a more reactive shape, or that shifting pieces of the core might create areas of high reactivity. The energy released in these cases could exert enough pressure to disassemble the core, terminating the chain reaction. But core disruption or subsequent recriticality might conceivably release sufficient energy to generate mechanical forces that threaten the containment. The Department of Energy has sponsored considerable research on this topic because of its crucial importance, and it is now believed that a containment-threatening, core-disruptive accident is precluded by proper design.[76]

The Nuclear Regulatory Commission takes the position that designs for prototype liquid-metal fast breeder reactors must attempt to reduce the probability of class-9 accidents leading to large-scale core melting to 10^{-6} per year. In addition, the Nuclear Regulatory Commission recommends that features be provided in the design to protect against the effects of mechanical forces that might be generated by core melting, and to contain the bulk of fission products in a class-9 accident. This would be a stricter criterion than those applied to light water reactors.

A number of commentators point out that reducing the probability of core melting 10–50 times below the probability of core melting in LWR's can be set as an objective for LMFBR designs, but that it is unrealistic to expect conclusive demonstration that the objective has been met.[77]

The unanswered questions for LMFBR safety, then, are: whether inherent or engineered safety features eliminate or greatly reduce the probability of core melting; whether, if this probability cannot be reduced to desirable unlikelihood, engineered features can contain the consequences; and by what mechanisms reasonable consensus can be reached that these objectives have or have not been met.

A preliminary analysis of HTGR safety has been conducted by the vendor, and there has been some assessment of its features in design reviews of the Fort Saint Vrain reactor and other proposed HTGR installations. Salient features of this analysis are discussed in the report of the Risk and Impact Panel. The Working Group on HTGR Safety of that panel concluded tentatively that the HTGR may be less susceptible to large radioactive releases than the LWR. The HTGR has demonstrably better tolerance for storing decay heat without releasing fission products and may be less subject to a large LOCA, but it has the extra mechanism of graphite oxidation for potential release of fission products and heat. In case of loss of the helium coolant, for example, air could not be used for emergency cooling because it would burn up the graphite.[78]

Sabotage of Nuclear Facilities As already noted, deliberate sabotage has not been included in the discussion of nuclear accidents, as it is not usually included in accident analysis of other systems. Nevertheless, the question has been raised whether the existence of nuclear facilities presents an "attractive nuisance" to would-be terrorists, who might use the threat of sabotage to extract concessions from society, or to people bent on destruction.

This discussion is limited to a general review. Details can be found in the recently issued report of the National Academy of Sciences on the safety of nuclear power.[79] That report (like this discussion) is necessarily limited to information in unclassified literature.

Three points must be considered: the degree of vulnerability of nuclear installations, consequences that might credibly ensue, and comparison of vulnerability and consequences with those of other energy sources.

Nuclear systems are probably less accessible, and harder to sabotage, than many competing energy systems. Access is more carefully controlled than to other thermal power plants, and all thermal plants are inherently easier to protect (because of their relative compactness) than more dispersed sources, such as dams or solar or wind installations. Thus, with regard to vulnerability against loss of generating capacity, nuclear plants must be rated as highly secure.

With regard to vulnerability against attacks or threats aimed at endangering the public, nuclear plants have considerable intrinsic protection. The main reason is that the plants themselves are complex. A limited number of individuals have knowledge sufficient to initiate a severe accident, and the steps that must be taken to ensure that an accident leads to large release are numerous. The multiplicity of systems available to the defense-in-depth strategy of nuclear plants would have to be disabled, and the degree of planning required would seem to demand the collusion of a great many "insiders." Preparations would be necessarily time consuming, and no threat could be voiced until they were complete, since shutting a reactor down would quickly decrease the severity of consequences from delayed releases of radioactivity.

Nevertheless, the consequences of sabotage against nuclear plants must be rated as potentially severe. In a hierarchy of risk, nuclear plant sabotage could lead to consequences of the same order of severity as those following the breach of a major dam or sabotage of a natural gas storage facility. Oil refineries would present a medial level of risk, and coal plants and solar and wind facilities would be at the bottom.

The range of consequences that can be produced by sabotage of a reactor is probably rather similar to that for reactor accidents. The aim of a saboteur, presumably, would be maximum release of radioactivity at a time when the weather conditions would be most conducive to directing the radioactivity to populated areas nearby. This aim would complicate the saboteur's task by introducing a further element of timing. Not only would the sabotage have to be prepared and set, but also perpetrated at the most damaging time.

We might also consider the additional threat to the public that might ensue from bombing a nuclear plant (in the course of war, for example). For anything but direct hits on the reactor building with very heavy conventional bombs or penetrating missiles, the reactor containment is an effective barrier. Although a direct hit on a spent-fuel storage pool would disperse the radioactivity contained in it, experience with reactor destructive tests, such as BORAX,[80] and accidents, such as that of the SL-1, indicate that little of the radioactivity is transferred a significant distance, and that the overwhelming bulk of the contamination would be confined to the plant site. However, a direct hit with a nuclear bomb would very much enhance (by orders of magnitude) the subsequent damage due to fallout.

Three points can be made about the sabotage potential of nuclear systems.

1. It is extremely difficult to make quantitative estimates of the expected frequency of effective attempts to sabatoge nuclear plants.
2. Nuclear systems are easy to sabotage into a state of inoperability, if penetrated by saboteurs, but they are very difficult to sabotage into a public hazard because of the redundancy of safety features.
3. Many other systems in our society offer more dangerous combinations of vulnerability to sabotage and likelihood of causing major damage: large aircraft, tanks of liquefied natural gas, major dams, and chemical plants.*

As with many other large industrial installations, it would appear that the greatest degree of defense against sabotage should be concentrated at sites near large population centers. The Nuclear Regulatory Commission is responsible for plant protection standards and appears to have given the matter of sabotage adequate emphasis.

Conclusions and Recommendations

It is important to recognize three quite separate issues in reactor safety that many published discussions of the subject fail to distinguish clearly.

1. The set of "best estimates" of the probabilities and consequences of various kinds of accidents, and the ranges of uncertainty that bracket these values.
2. Interpretation of these values and uncertainties to yield some understanding of expected values of consequences.
3. What "mean" values of probabilities and consequences of accidents in nuclear power plants, and what degrees of uncertainty about these values, are acceptable in exchange for the benefits of nuclear power.

The first two issues are essentially technical; the third is essentially social and political.

The probabilities of very low frequency accidents are difficult to estimate with precision. One expert suggests that the probabilities and consequences of catastrophic nuclear accidents can never be estimated more accurately than within an order of magnitude.[81] Thus, the safety of nuclear reactors will continue to be a matter of judgment. Perhaps the most valuable feature of fault-tree, event-tree analysis is that it points out where design improvements could be most effective. Those improvements that make a

*Statement 5-28, by J. P. Holdren: I find the existence of other points of vulnerability only modest consolation. Nor am I convinced (paragraph following) that the NRC's program against sabotage is "adequate."

significant contribution to further reducing the likelihood of accidents can be identified and should be incorporated in new plants. Similarly, "safety" systems that do not make such a contribution might be dropped from licensing requirements, particularly if significant cost savings result.

There are three legitimate ways by which judgments of risk may be better specified for policy purposes. Improved precision could result in part from more reactor-years of experience and from improved methods of analysis. It might also be possible to improve the data base of the models used in reactor safety studies by conducting better tests and analyses of nuclear-system components. The third way would be to improve the design of nuclear power plants and to make stringent inspections at critical stages of construction.

These considerations permit CONAES to reach a conclusion on the question of reactor safety. It is consonant with that of many other review groups, but is both more optimistic and more cautionary. We believe that the expansion of nuclear power can proceed without untoward public risk from reactor accidents, but only under certain conditions.

1. Institutions to review experience and enforce improvement of safety design must be vigorous and independent. In the case of the United States, this institution is the Nuclear Regulatory Commission. For international concerns, such bodies as the International Atomic Energy Agency (IAEA), which has safety consultation and review authority, should be strengthened for maximum effect of its recommendations.

2. Both the nuclear industry and regulatory authorities must be more receptive to design modifications that would enhance safety. The design and licensing process is now so lengthy that there are strong economic disincentives to consideration of any design changes. It must be recognized that improvements in safety by design are difficult to prove; under these circumstances, there is a natural tendency to continue with existing practices. Nevertheless, it is only by change that improvements can be made. Evaluations of proposed improvements (with regard to their effect on the results of the Reactor Safety Study's estimates of accident frequencies and consequences) should be a continuous process, and the proposed improvements that receive favorable evaluation should be instituted.

3. It goes without saying that research on the safety of nuclear power should receive continued support. There is need to reconsider the sort of work to be emphasized. Significant advances in our understanding of reactor safety come from improved knowledge in the engineering sciences. Such topics as two-phase flow, mechanics of materials, metal-water chemical reaction processes, steam explosion theory, fission-product decay heat, fission-product chemistry and volatility, and others are basic to

understanding the phenomena that might occur during accident conditions, and to quantitative specifications of the consequences. These deserve strong research support, preferably as independent studies, but if necessary, under efforts to improve safety. (Their level of support as research topics has been grossly inadequate for more than a decade.) Conversely, integral experiments to "verify" conclusions should be used sparingly; too often, they are demonstrations of the already known. For example, measurements of the physics or heat-transfer behavior of systems that have already been simulated by critical experiments or electrically heated loops should only be performed if there is some question about the feedbacks among various types of phenomena in the reactor.

4. The Reactor Safety Study gives valuable guidance for decisions the public must make about expansion of nuclear power.* The committee makes the following recommendations.

- For existing reactor types, such as LWR's now operating in the United States, studies should be updated every 10 years. The purpose of the update should be to quantify the rate of improvement in knowledge pertinent to safety and safety records.
- The safety of new reactor types, such as LMFBR's or advanced converters, should be compared with the safety of existing reactors. This should be done before commissioning a large number of new reactors. An appropriate time to conduct such studies might be after about the first 10 such reactors have been granted construction permits. Before this point, there is likely to be too much design variation to permit generic comparision. If the study is delayed too long, the same sort of conflicts may develop that trouble the country today about LWR safety.

MANAGEMENT OF RADIOACTIVE WASTE

Radioactive wastes consist of a variety of natural uranium and thorium decay products, fission products, products of neutron activation, and transuranic isotopes or actinides with intermediate to very long half-lives. These wastes must be sequestered from the biosphere for as long as their radiation represents a hazard. Burial or "geological isolation" is the reference method for doing this.

The principal technical consideration in assessing modes of geological isolation is the transport of radionuclides by groundwater. For longer-lived wastes, this requires selection of geological formations that themselves would be proof against the failure of containers after one or two hundred

*Statement 5-29, by J. P. Holdren: The Reactor Safety Study's "guidance" would be misleading unless accompanied by full awareness of its understatement of uncertainties.

years, in the sense that migration of the waste nuclides must be slow enough, or accompanied by so much dilution, that the radioactivity of the water when it reaches the biosphere is a small fraction of natural radioactivity.

Public concern with the management of radioactive waste centers on society's ability to achieve and maintain the necessary isolation. Both technical and social aspects of this topic have received lively attention. The technical aspects will be taken up first; the social considerations, while not wholly separable, will be discussed briefly near the end of this section.

What Is the Required Isolation?

The degree of isolation required of radioactive materials varies from one form to another. Some materials are easily mobilized and transported (e.g., by groundwater), some are not. Some materials are concentrated by living tissue, some are not. Radioactive materials decay, losing their radioactivity, at different rates.

Over and above these considerations there is the question of what level of concentration or release is dangerous. Standards of comparison can be natural radioactivity (background) or projected health effects. Some types of radioactivity released during the nuclear cycle can occur naturally: uranium, thorium, and the decay daughters of uranium and thorium, which are the nuclides at issue in mining and milling wastes, and ^{14}C and tritium, which can be formed during reactor irradiation and released either at that time or during the course of reprocessing. For such materials, comparisons of the contribution from the nuclear cycle with their natural abundance can provide insight into the safety factors needed to render them effectively harmless. (See chapter 9.) Other materials—fission products, higher actinides, and activation products—have no natural source of any consequence. For these, projected health-effect calculations are used to determine the standards for release. Complicating these simple criteria is the fact that they are not necessarily consistent with one another. Projected health-effect calculations (that attempt to err always on the side of predicting higher mortality and morbidity than a best estimate) can predict effects from natural radiation that are well beyond observed values; in the other direction, predictions of health effects have led to hypotheses that some natural sources of radiation are significant causes of morbidity.

The differences due to concentration effects are also important. Low concentrations of radioactivity have a statistically low probability of harming any given individual. As with toxic heavy metals, society tends to accept "low enough" concentrations in the environment. The rationale seems to be that at some level, the risk posed can be ignored in comparison to risks that are orders of magnitude larger. Yet, it is (properly) considered poor practice to manage most wastes by dilution to an acceptable concentration level, even when that is feasible. We have (again, properly) confidence in our ability to sequester or destroy wastes if we concentrate and package them to delay or prevent their return to the biosphere. Yet, paradoxically, the existence of concentrated wastes, specifically radioactive wastes, is viewed with more apprehension than the existence of radioactivity in very dilute form already mobilized within the biosphere (such as radioactivity from fallout of weapons tests). It therefore appears that the problem of disposing of radioactive wastes has two parts. The first is to package and isolate the wastes as well as possible, and the second is to arrange for sufficient delay and dilution, in case the isolation fails, to ensure that the concentration possibly returning to the biosphere is not a major source of risk by prevailing standards.

The most radioactive materials (e.g., most fission products and most products of neutron activation) have relatively short half-lives; indeed, specific radioactivity is inversely proportional to half-life for a given number of atoms. It is reasonable to demand a high degree of assurance that the isolation of these products will continue until natural decay has reduced their radioactivity to low levels. As half-life increases (and radioactivity decreases), such assurance becomes both more difficult to provide and less necessary. The difficulty is, of course, a direct result of the long time required for decay, but the radioactivity is subsequently less intense and the consequences of release would be less severe.

The result of these considerations is a waste management philosophy that incorporates and combines two separate types of waste isolation: physical and chemical isolation, and disposal in a geological setting expected to delay and hinder the return of radioactivity to the biosphere, or at worst to dilute it considerably during the course of such return. The physical and chemical isolation is more of a backup precaution, forming an additional barrier to the geological isolation that constitutes the main safety factor for long-lived waste. For the short-lived waste, physical and chemical isolation is designed to be an effective barrier in its own right.

Types of Radioactive Waste

Uranium Ore Tailings About 80 percent of the original radioactivity in uranium ore remains in uranium ore tailings. Processing (milling) removes only the uranium, leaving behind about 7 percent of the uranium (as processing loss), the daughters ^{234}Th, ^{231}Th, and ^{230}Th, their decay products, and any natural ^{232}Th in the ore. Among these radioactive materials, the most troublesome are ^{230}Th (half-life: 77,000 years), radium-226 (^{226}Ra) (half-life: 1600 years), and their daughters. These radioactive isotopes are widely dispersed in nature through the natural weathering of uranium-containing rock. It is therefore the surface concentration, rather than the quantity of these nuclides in the ore tailings, that must be dealt with. The total quantity of tailings accumulated since 1948 totals 123 million tons, or approximately 70 million yd^3. The quantity is large because virtually all the uranium ore mined appears in the tailings: Typical uranium concentrations are well under 1 percent.

A standard light water reactor requires the mining of about 150 tons/yr of U_3O_8, and at the typical concentration of 0.1 percent in the ore, this amounts to about 150,000 tons of rock, or about 40,000 yd^3 (30,000 m^3). The contained radioactivity is about 500 times greater than that of ordinary soil. The parent nuclide of most of the radioactivity is ^{230}Th, and all the lighter radioactive nuclides in the uranium radioactive series are "fed" by it. At any given time, there are about 10 other radioactive disintegrations for each disintegration of ^{230}Th, the whole chain decaying with the thorium half-life.

The most important members of the disintegration series from the point of view of radiation hazard are ^{226}Ra and radon-222 (^{222}Rn). Radium is chemically a member of the alkaline earth family (along with calcium, strontium, and barium) and shares with other members a relatively high leachability and mobility in the presence of groundwater. Radon is a noble gas that, unless trapped in a crystalline medium or in sub-surface pores, diffuses into the air. Radon is the daughter of radium, so that if the radium has migrated, radon is released at the point of radium disintegration.

Abandoned mines and active piles of ore tailings are the sources of this radiation. Protective measures should meet the criterion of reducing this source to the same order of magnitude as ordinary soil. As noted, this means reduction by a factor of about 500. Achieving such reduction is not a technically difficult matter, although it represents a small incremental cost to the mine operator. Filling in mines, burying tailings, and avoiding massive invasion of the tailing piles by water are recommended procedures. A few feet of earth above the pile is an effective seal against the escapes of radon, because the gas has only a 3.85-day half-life and diffuses slowly. Asphalt seals further delay the release of radon, but serve the more important function of preventing or slowing seepage of surface water through the pile. Soil fillers or conditioners could also be used to inhibit seepage.

Over the course of time, wind or water erosion could conceivably reexpose the pile. However, filling and covering are equally likely results of both wind and water action.* In the United States, uranium is mined mostly in arid regions where surface water is not common, but care must

*Statement 5-30, by J. P. Holdren: I cannot agree. The average net effect of wind and water is to uncover and displace, as the whole operation of the sedimentary cycle reveals.

be taken not to disturb subsurface aquifers. Further exposure is a possibility, but the level of exposure would not be significantly above background and would not result in serious consequences to any human generation. If our descendants are sophisticated and "radiation conscious," they would find detection of the source to be very simple, and correction of the condition equally simple. If they are not, it is likely to be because their technology has so regressed that they are subject to much more pressing dangers.

Wastes of Reprocessing Four major types of waste are produced in fuel reprocessing: the aqueous raffinates from solvent extraction, the metal hulls sheared from the fuel rods, the scrubbing solutions or solids generated by reaction with the gases containing radioactive iodine-129 (^{129}I), ^{131}I, tritium, and ^{14}C, and the gases released during dissolution of the fuel.

The aqueous raffinates are by far the most radioactive of these wastes. They contain almost all the fission products generated, together with a small fraction of the actinide elements uranium, plutonium, neptunium, and americium. Regarding the fission products, strontium-90 (^{90}Sr) and cesium-137 (^{137}Cs) are the most notorious components. Each has a half-life of about 30 years, and the reduction of their activities by a factor of 1000 in 300 years, 1 million in 600 years, and 1 billion in 900 years establishes the period 500–1000 years for social concern with their custody. The ratio of atoms of actinides to fission products is less than 1 percent. Their radioactivity is at first negligible compared to that of fresh fission products, but their radioactivity lasts longer. In conventional reprocessing, the raffinates appear as concentrated nitric acid solutions, amounting to about 56 gal/ton of spent fuel. These raffinates are commonly considered to be the crux of the radioactive waste management program.

Other radioactive wastes from reprocessing—insoluble residues, raffinates from product purification steps, hulls and off-gases—are treated separately. In particular, off-gases are of short half-life or low radioactivity. The most troublesome is ^{85}Kr, which is to be collected and stored for about a century when the amounts become significant.

Tables 5-14 and 5-15 set out, respectively, concentrations of chemical elements in spent light water reactor fuel and the radioactivity of important nuclides.

Table 5-14 is primarily of interest in considering source concentrations of elements that would be subject to both total and isotopic dilution during migration. Table 5-15 presents the radioactivities of spent fuel up to 10 years after reactor discharge. Of interest is the very high radioactivity after 10 years of cooling, and the increasing contribution of the actinides. Thirty days after discharge, less than 2 percent of the activity is from actinides, but after 10 years of cooling, the actinides produce almost 20 percent of the activity.

Alpha-Active Wastes[82] Throughout the nuclear industry, uranium is converted from one chemical form to another: from uranium oxide to uranium hexafluoride for enrichment, from uranium hexafluoride to uranium dioxide for fuel-material preparation, and so on. There are radioactive wastes from these processes. They are not as radioactive as mill tailings, and ought probably to be mingled with them for disposal,* but institutional arrangements have not been made for this step.

Laboratory operations involving plutonium, ^{233}U, and other actinides also produce wastes that contain long-lived alpha activity. In the past, the concentration of alpha activity governed the method of disposal: Low concentrations were considered low-level wastes, to be buried or dispersed. However, the increased concentrations and quantities of this material being disposed of, and public fear of the consequences of its dispersal, stimulated a change in policy, and this material is now considered in the same disposal category as high-level waste.

Looking to the future, we can expect to see a large increase in the generation of alpha-active waste if ^{233}U and plutonium are recycled. A considerable amount of waste is generated during nuclear fuel fabrication: dusts from grinding operations, contaminated fabrics from filters, contaminated crucibles and tools, contaminated metal pieces from rejected fuel elements, and so on. (For a given number of atoms, radioactivity is inversely proportional to half-life, and the half-lives of possible contaminants are 700 million years for ^{235}U, 160,000 years for ^{233}U, 23,000 years for ^{239}Pu, and 6500 years for ^{240}Pu.)

A further consideration in recycled fuel is the contamination introduced by other actinide nuclides. These materials (^{232}U, ^{238}Pu, ^{241}Pu, americium-241 (^{241}Am), and others) have even shorter half-lives and higher activities than those just mentioned, and for that reason, they have been suggested as radioactive "spikes" to safeguard nuclear fuel, discussed in this chapter under "Safeguarding the Domestic Fuel Cycle."

Reprocessing Wastes from Military Production Although not within the responsibility of a civilian industry, military production wastes are the major focus of current concerns about waste management. The important wastes are the aqueous raffinates from reprocessing of fuel to recover plutonium for weapons. Three different processes have been used in the

*Statement 5-31, by J. P. Holdren: No presently agreed-to plan for managing tailings justifies either the term "disposal" or the addition of other wastes to the piles.

TABLE 5-14 Element Concentrations in Spent Light Water Reactor Fuel (grams per metric ton of heavy metal)[a]

Element	Concentrations					
	After 30 days	After 90 days	After 150 days	After 1 year	After 3 years	After 10 years
^{3}H	0.075	0.074	0.074	0.071	0.064	0.043
Kr	383	383	382	381	378	369
Xe	5,580	5,590	5,590	5,590	5,590	5,590
Rb	341	341	341	342	346	355
Cs	2,830	2,810	2,800	2,750	2,630	2,380
Sr	932	921	914	903	877	794
Ba	1,410	1,420	1,440	1,490	1,610	1,850
Y	486	482	480	477	477	477
La	1,300	1,300	1,300	1,300	1,300	1,300
Ce	2,890	2,830	2,790	2,690	2,570	2,550
Pb	1,210	1,220	1,230	1,230	1,230	1,230
Nd	3,910	3,950	3,990	4,090	4,200	4,230
Pm	113	109	104	88.8	52.3	8.2
Sm	824	829	834	849	885	926
Eu	194	192	191	189	184	172
Gd	111	113	113	116	122	136
Te	1.9	1.9	1.9	1.9	1.9	1.9
Dy	1.1	1.1	1.1	1.2	1.2	1.2

U	954,000	954,000	954,000	954,000	954,000	954,000
Np	500	500	500	500	501	504
Pu	9,090	9,090	9,080	9,050	8,960	8,700
Am	137	145	153	182	274	532
Cm	47.2	44.8	42.9	38.7	33.9	26.5
Zr	3,770	3,760	3,760	3,760	3,790	3,870
Nb	32.0	21.7	12.9	1.5	0.002	
Mo	3,480	3,520	3,540	3,560	3,560	3,560
Tc	863	863	863	863	863	863
Ru	2,400	2,360	2,340	2,300	2,240	2,220
Rh	371	386	391	394	394	394
Pd	1,320	1,340	1,350	1,390	1,460	1,480
Ag	62.6	62.4	62.3	62.1	61.8	61.8
Cd	88.2	88.4	88.4	88.7	88.9	89.0
In	1.2	1.2	1.2	1.2	1.3	1.3
Sn	53.9	53.7	53.5	53.2	53.1	53.1
Sb	17.9	17.8	17.7	16.9	14.4	11.1
Se	53.3	53.3	53.3	53.3	53.3	53.3
Te	583	582	582	583	585	589
Br	15.7	15.7	15.7	15.7	15.7	15.7
I	276	277	277	278	278	278

[a]Isotopic mixtures, including radioactive and stable nuclides. Assumptions: 3.3 percent enriched uranium fuel; burnup, 34,000 MWd/metric ton of heavy metal; specific power, 29.5 MWe/metric ton of U_3O_8.

Source: H. O. Haug, *Calculations and Complications of Composition, Radioactivity, Thermal Power, Gamma and Neutron Release Rates of Fission Products, and Actinides of Spent Power Reactors' Fuels* (Karlsruhe, Federal Republic of Germany: Reactor Research Institute, 1974).

TABLE 5-15 Radioactivity of Selected Nuclides in Spent Light Water Reactor Fuel (curies per metric ton of heavy metal)[a]

Nuclide	Half-life	Radioactivity					
		After 30 days	After 90 days	After 150 days	After 1 year	After 3 years	After 10 years
Fission Products							
^{3}H	12.3 years	727	720	713	690	616	415
^{85}Kr	10.8 years	11,400	11,300	11,200	10,800	9,490	6,060
$^{131}Xe^{b}$	12.0 days	2,600	104	3.2	—	—	—
^{133}Xe	5.3 days	37,300	14	—	—	—	—
^{134}Cs	2.1 years	250,000	237,000	224,000	184,000	93,300	8,750
^{136}Cs	13.0 days	12,800	522	21	—	—	—
^{137}Cs	30.0 years	111,000	110,000	110,000	108,000	103,000	87,900
$^{137}Ba^{b}$	2.6 min	103,000	103,000	103,000	101,000	96,600	82,200
^{89}Sr	52.1 days	464,000	209,000	93,800	5,340	0.3	—
^{90}Sr	28.1 years	78,900	78,600	78,300	77,200	73,500	61,800
^{140}Ba	12.8 days	277,000	10,800	417	—	—	—
^{90}Y	64.0 hours	78,900	78,700	78,300	77,200	73,500	61,800
^{91}Y	59.0 days	642,000	316,000	156,000	12,400	2.2	—
^{140}La	40.2 hours	319,000	12,400	480	—	—	—
^{141}Ce	32.3 days	716,000	198,000	55,000	553	—	—
^{144}Ce	284 days	1,020,000	880,000	760,000	450,000	75,500	150
$^{144}Pr^{b}$	17.3 min	1,020,000	880,000	760,000	450,000	75,500	150
^{143}Pr	13.7 days	287,000	13,800	663	—	—	—
^{147}Nd	11.1 days	87,900	2,070	49	—	—	—
^{147}Pm	2.6 years	104,000	101,000	96,400	82,500	48,600	7,630
^{93}Zr	1.5×10^{6} years	1.9	1.9	1.9	1.9	1.9	1.9
$^{93}Nb^{b}$	13.6 years	0.2	0.2	0.2	0.2	0.4	0.9
^{95}Zr	65.2 days	973,000	513,000	271,000	27,300	11	—
$^{95}Nb^{b}$	90.0 hours	20,700	10,900	5,750	580	0.2	—

^{95}Nb	35.0 days	1,250,000	852,000	508,000	58,100	24	—
^{99}Tc	2.1×10^5 years	15	15	15	15	15	15
^{103}Ru	39.5 days	710,000	249,000	86,900	2,020	—	—
$^{103}Rh^b$	57 min	711,000	249,000	86,900	2,020	—	—
$^{106}Ru^b$	1.0 year	524,000	468,000	418,000	278,000	70,000	560
^{106}Rh	30.0 s	524,000	468,000	418,000	278,000	70,000	560
^{129}I	1.7×10^7 years	0.1	0.1	0.1	0.1	0.1	0.1
^{131}I	8.0 days	65,600	375	2	—	—	—
Actinides							
^{234}U	2.5×10^5 years	0.7	0.7	0.7	0.7	0.8	0.8
^{236}U	2.4×10^7 years	0.3	0.3	0.3	0.3	0.3	0.3
^{237}U	6.7 days	39,500	86	2.7	2.5	2.3	1.6
^{238}U	4.5×10^9 years	0.3	0.3	0.3	0.3	0.3	0.3
^{236}Pu	2.8 years	0.4	0.4	0.3	0.3	0.2	—
^{238}Pu	88.9 years	2,970	3,010	3,030	3,070	3,060	2,900
^{239}Pu	24,400 years	323	323	323	323	323	323
^{240}Pu	6,760 years	485	485	485	485	486	487
^{241}Pu	14.6 years	108,000	107,000	106,000	103,000	94,000	67,400
^{242}Pu	3.8×10^5 years	1.5	1.5	1.5	1.5	1.5	1.5
^{241}Am	433 years	105	134	162	260	575	1,460
^{243}Am	7,650 years	20.2	20.2	20.2	20.2	20.2	20.2
^{242}Cm	163 days	32,000	24,800	19,200	7,710	352	7.7
^{244}Cm	18.1 years	2,820	2,810	2,790	2,730	2,520	1,930
Sum, Fission Products		1.06×10^7	6.14×10^6	4.38×10^6	2.24×10^6	806,000	325,000
Sum, Actinides		1.89×10^5	1.39×10^5	1.32×10^5	1.18×10^5	101,000	74,600
TOTAL		1.08×10^7	6.28×10^6	4.51×10^6	2.36×10^6	907,000	400,000

[a] Assumptions: 3.3 percent enriched uranium fuel; burnup, 34,000 MWd/metric ton of heavy metal; specific power, 29.5 MWe/metric ton of U_3O_8.
[b] Nuclides in metastable states that can decay into more stable form by emission of a gamma ray.

Source: H. O. Haug, *Calculations and Complications of Composition, Radioactivity, Thermal Power, Gamma and Neutron Release Rates of Fission Products, and Actinides of Spent Power Reactors' Fuels* (Karlsruhe, Federal Republic of Germany: Reactor Research Institute, 1974).

past (bismuth coprecipitation, Redox, and Purex), and they differ in the quantities and types of process chemicals that accompany the waste to liquid storage. However, the main features of these wastes are similar, and they are different from those of civilian Purex wastes.

- The actinide content is much lower. Almost no higher actinides are formed during irradiation to make weapons-grade plutonium, and process yields of plutonium are more nearly total, leaving only a small fraction in the waste.
- The waste solution is neutralized. This produces wastes, most of which have precipitated readily from solution (as hydroxides or carbonates), loading the solution with sodium salts. After prolonged liquid storage, the storage tanks contain both solid "salt cake" and a salt solution. Unfortunately, almost all these wastes from the weapons programs were stored in carbon-steel tanks, an expediency adopted during World War II and regrettably continued until recent times. The salt cake and salt solution have corroded some of these tanks and are expected to corrode more. The wastes cannot be redissolved and pumped out without also dissolving the tanks. Thus, these weapons wastes present a unique one-time problem not to be experienced in any civilian program, where the use of acid solutions in stainless steel tanks eliminates the difficulty.

The carbon-steel tanks at the Hanford reservation have leaked many times, and severely at least once. The leaks have been without hazard to the public[83] because of the low rate of ion migration of radioisotopes through the soil at Hanford, and the low concentration of long-lived actinides indicates that they will never be a danger. The radioactivity will have decayed to innocuous levels by the time the material reappears in the biosphere.

However, the peculiar nature of this waste has made it very difficult to find a disposal method that is more appropriate than letting the salt cake evaporate to dryness and entombing the tank in concrete.

Chemically similar wastes are stored in carbon-steel tanks of better design at the Savannah River plant in South Carolina, and in relatively small amounts at West Valley, New York. Leaks at Savannah River have been minute, and zero at West Valley, but the local hydrology is by no means so favorable in those two places as in Hanford.

Spent Fuel as Waste President Carter announced in 1977 that the United States would defer reprocessing of spent reactor fuel indefinitely, to avoid potential diversion of reprocessed plutonium for weapons. Many have interpreted this policy as tantamount to declaring spent fuel to be waste. CONAES has not accepted this interpretation, and prefers to consider the announcement to be one of reserving judgment. Even those members who oppose nuclear power consider it prudent for now to preserve spent fuel as a potential resource. For the short term, the stowaway fuel cycle (guarded storage of spent fuel) protects against subnational diversion because the spent fuel is literally too hot (radioactive) to handle.

In the future, a decision *could* be made to treat spent fuel as waste. The fuel is in a ceramic form and is not expected to be easily leachable. However, it is at least as concentrated a heat source as solidified reprocessing wastes, and contains 50–100 times more actinide radioactivity per GWe-plant-year than reprocessed waste.

Treatment of Radioactive Waste for Disposal

The recommended treatment of mill tailings has already been discussed. The same considerations ought to govern the treatment of small quantities of alpha-active waste, and indeed we consider it reasonable to define the term "small" to be that quantity that could be blended with mill tailings without noticeably increasing their radioactivity. In fact, a considerable amount of low-level wastes might actually be disposed of with mill tailings.

The more concentrated wastes require separate treatment. Cladding hulls, reprocessing-plant solids, and medium-level solid wastes from reactor operations are all representative of materials that can be (and have been) treated by encapsulating them in drums and filling the drums with asphalt, cement, concrete, or thermosetting plastics. The key to the success of this technique is the low heat generated by these wastes, which permits them to be handled without concern for cooling or ventilation.

Cladding hulls, which are voluminous, are generally compressed and stored in water-filled drums. To reduce their volume further and with an eye to eventual recovery of zirconium, which is a valuable metal, various forms of chemical reprocessing of the zirconium are under development. The aim today is simply to get more zirconium into a drum.

It is usually assumed that these miscellaneous materials, in their drums, will be buried underground. The Federal Republic of Germany has already committed a salt formation (Asse) to this use. The same facilities used for alpha-active waste disposal should also serve here. In the case of both alpha-active and medium- to high-level activation products, disposal in a high-level waste repository appears to be a straightforward matter and should not increase the problems of handling the high-level waste.

Radioactive off-gases from reprocessing get variable treatment. At present, only the radioiodines are permanently fixed, as silver iodides. There are few problems with this technique, since most of the radioactivity is in the form of 8-day ^{131}I, which decays into stable xenon, and since ^{129}I, which has a very long half-life (16 million years), is formed at such low yield that it can be allowed to build up indefinitely in the filter bed, until the plant itself is decommissioned.

The gas ^{85}Kr is not trapped. It contributes a small amount to our radioactive background—of the order of 0.01 person-rem/yr. However, if large-scale reprocessing for a large nuclear industry (> 1000 GWe) is instituted, it will be desirable to isolate this gas. A thorough analysis of the problem has been conducted at Karlsruhe, in Germany.[84] The preferred handling method is isolation by cryogenic distillation, and storage in pressurized steel spheres. If small spheres are used and stored in an underground cavern, even an occasional leak or rupture would not lead to significant radiological hazard.

Tritium and ^{14}C in off-gases represent small additions to natural radioactivity and are rapidly dispersed, mostly in the atmosphere and the oceans, which provide enormous dilution. In consequence, their level of hazard is small, though spread worldwide. They also emit relatively low-energy beta radiation, which reduces their biological consequences, and they have relatively short biological half-lives (mean residence time *in vivo*), which somewhat mitigates the severity of local exposures.

The environmental inventory of ^{14}C comes from nuclear reactions initiated by cosmic radiation in the atmosphere. It is estimated to be 280 megacuries (MCi)[85] and to deliver a whole-body dose of 0.7 person-rem/yr to the average individual. To this, an LWR reprocessing plant of 1500 tons/yr uranium throughput (serving about 50 GWe of reactors) would, if all the ^{14}C in the fuel were released, add 0.000006 person-rem/yr for each year of operation.[86] After 500 years of operating such a plant, the dose from this source would be 0.003 person-rem/yr, and if 40 such plants were operating worldwide for that period (a 2000-GWe world industry), the dose from ^{14}C from that source would be 0.12 person-rem/yr. The maturing of a large HTGR industry could, however, increase this dose appreciably. In HTGR's, ^{14}C is formed from the following reactions (the latter from residual nitrogen in graphite):

$$^{13}C + {}^{1}n \rightarrow {}^{14}C$$
$$^{14}N + {}^{1}n \rightarrow {}^{14}C + {}^{1}H$$

Similarly, if all the tritium produced in LWR fuel from a 2000-GWe world industry were released, the dose to each individual would reach a constant value of about 0.03 person-rem/yr after about 20 years; further additions to the environmental inventory would be balanced by its decay. The widespread use of heavy water could significantly increase the quantity of tritium formed, due to the following reaction.

$$^{2}H \text{ (deuterium)} + {}^{1}n \rightarrow {}^{3}H \text{ (tritium)}$$

Since the nuclear industry is well below the 2000-GWe level used for the examples above, and since the incremental dose is small compared to background ^{14}C (0.7 person-rem/yr) and the total background (100 person-rem/yr), reprocessing that is now being practiced releases both ^{14}C and tritium (as well as ^{85}Kr, from which the dose is about the same as from tritium). Research is in progress on methods for trapping even this small amount of radioactivity,[87] and since the chemistries of hydrogen and carbon are very well known, processes for trapping at least 90 percent of these effluents can almost certainly be developed. It is expected that these processes will be used ultimately, in accordance with the philosophy of ALARA (as low as reasonably achievable) used by regulatory commissions for effluents from the nuclear industry.

The technology for handling wastes from the reprocessing of spent LWR fuel is relatively well developed in both the United States and Europe. The typical waste is a nitric acid solution of fission products and actinides partially evaporated immediately after discharge from the reprocessing plant to reduce the volume to be stored and to recover nitric acid for recycle. After evaporation, the concentrated wastes—amounting to about 56 gal/ton of spent fuel—are routed to large double-walled underground storage tanks cooled by water and made of stainless steel, where they are held for up to 5 years from the time of reactor discharge.

When the waste has aged to the point that its radioactivity no longer requires strong cooling (less than 5 years after reprocessing), it can be solidified. For example, the Waste Calcination Facility at the Idaho Reprocessing Plant (which reprocesses highly enriched fuels from propulsion and research reactors) reduces the nitric acid solution to a frit (the partly fused state necessary for glass making, or for introduction into ceramics) of oxide granules by fluid-bed calcination. The French reprocessing plants at Marcoule and La Hague carry the process a step further by continuously incorporating the calcined solids into glass cylinders encased in steel cans. The radioactive solids are an integral part of the glass, which has the appearance of an opaque, smoothly glazed ceramic. It is expected that all reprocessing plants will ultimately use either glass or a metal or ceramic matrix as the vehicle for incorporating solidified high-level wastes.

If spent fuel is not reprocessed, it must be stored. To relieve utilities of the responsibility for storing increasing amounts of spent fuel in their temporary cooling ponds, the government proposed in 1977 to accept title and transfer of spent reactor fuel on payment of a one-time storage fee.[88] At least for the time being, the stowaway fuel cycle will prevail, and the high-level waste process will involve early storage of discharaged spent fuel in water-filled canals at the reactor site (to provide gamma-ray shielding and a medium for heat dissipation), later encapsulation of the unprocessed assemblies in sealed containers, and delivery of canned assemblies to the government for storage. The waste forms and storage facilities will have to be designed for safe isolation over two or three decades, and for economical recovery of spent fuel.

Status of Technology for Ultimate Disposal in Underground Repositories Several methods have been considered for the ultimate disposal of encapsulated waste, including ejection into extraterrestrial orbit by rockets, disposal on or under the seabed, nuclear transmutation of long-lived actinides, and isolation at depth in suitable continental geological formations. Only the last of these options, using conventional underground mining technology, is believed to be practical in the near future.

Desirable geological properties for an underground repository include absence of groundwater, low permeability, high plasticity, freedom from joints and faults, good ion-exchange capability, and location in an area of low seismic activity. Rock types that exhibit some or all of these desired qualities include bedded evaporites such as salt and potash, marine shales, unjointed and unfaulted crystalline rocks (igneous and metamorphic), and limestone in arid regions.

Since percolation of groundwater is the only significant mechanism for releasing waste forms from their matrix, evaluation of the suitability of an area depends substantially on properly modeling the transport of radioactive atoms once dissolved. The highly active fission-product waste is hardly at issue here. The matrix in which it is incorporated is expected to be at least mostly resistant to leaching over the period (300–1000 years) during which the waste—principally ^{90}Sr and ^{137}Cs—decays away. (But see below for the U.S. Geological Survey's reservations on this point.) The actinides of medium half-life, such as ^{239}Pu (24,000 years), ^{240}Pu (6500 years), and ^{241}Am (450 years), are the critical nuclides. This underscores the significance of dealing properly with alpha-active waste. There is a good chance that *if* groundwater were to intrude upon the repository, the matrix *could* be leached out before these nuclides decayed.

Most actinide transport studies that have been conducted indicate that migration in groundwater will be very slow, being governed by absorption-desorption equilibria with solids in the aquifer, rather than solution transport mechanisms.[89] This could be the controlling process in return to the biosphere, and indicates that the ion-exchange behavior of the disposal environment for actinide ions is the most important selection criterion. The resistance to leaching of the glass and the absence of groundwater serve only as "insurance" factors.

This question has recently been reviewed by the American Physical Society (APS) and the U.S. Geological Survey (USGS).[90] Although the tones of the reports are different, the APS being generally optimistic and the USGS emphasizing reservations, our study of these documents indicates that their findings are similar. In both cases, there is confidence, primarily based on past experience with radium as a natural tracer of mineral migration in groundwater, that a site chosen with reasonable care will provide the necessary holdup of waste radioactive nuclides. Both reports emphasize the importance of improved characterization of the geochemistry and chemical hydrology of the chosen site. The USGS suggests that the geochemical reactions that might occur in the presence of the heat sources from high-level waste might rapidly alter the form of that waste. By inference, the authors are more optimistic about the chances that properly encapsulated alpha-active waste (which produces far less heat than high-level waste) would remain intact. Both sources, finally, believe that underground emplacement in mined salt cavities is not necessarily the best method of geological isolation, and they particularly recommend that emplacement in deep drill holes, and in such conventional rocks as granites and basalts, be reevaluated. Interestingly, a very thorough study by the Swedish commission appointed to study the question[91] has selected temporary storage (for decay of the heat production), corrosion-resistant jacketing (to hinder geochemical reactions), and ultimate emplacement in granite as the preferred method.

The drilling and mining operations required for waste emplacement are considered to be conventional technologies.

Three other problems, which we believe to be quite minor, have also received attention. One of these concerns the absolute dedication of disposal sites to that purpose, foreclosing further exploitation of subsurface minerals, geothermal energy, and so forth. We consider that any reasonable characterization of a site would confirm the presence or absence of mineral deposits of high value, that the foreclosure of mineral exploration at an unpromising site would be readily accepted, and that the monitoring of a site to confirm adherence to regulations against drilling would be easy.

Another problem concerns the protection of the underground emplacement from surface-water seepage, and the parallel problem of protecting the surface against escape of radioactivity through the bore holes. As we understand these problems, they are within the capabilities of sound practice in mining engineering, and the sort of leakage that could be expected would be minor.

A third problem is that of designing the underground repository in such a way that it does not disrupt its geological milieu. The problem seems similar to that involved in drilling deep tunnels, a technology in which, again, standards of good practice are well established. In some settings, however, this problem may cause deep drilling to be a preferred technique over excavation of a mined cavity; for example, when the integrity of the rock above the cavity is uncertain.

The Nature of the Waste Disposal Hazard An informed public response to the hazard presented by stored radioactive waste must begin with a qualitative understanding of its nature. Because hazards from high-level reprocessing wastes are potentially much larger than those from other parts of the nuclear industry, we use them for illustration.

A basis for conceptualization can be found in an estimate of the consequences of simply abandoning the accumulated high-level wastes at the Savannah River plant.[92] This estimate was prepared to quantify the base case against which various waste treatment and disposal techniques could be evaluated for incremental costs and benefits. Such abandonment, with no precautions taken against tank corrosion and seepage of the waste, is estimated to result in the eventual delivery of 620,000 person-rem to a surrounding area population of 70,000 people. Criticisms of this report have shown that the risk is underestimated, particularly with regard to persistent radiation in the water and soil. It is nevertheless true that such cavalier treatment of the waste should be a wild upper limit to the hazards. This particular limit is not much greater than the lifetime dose to the same population from natural background, and the expected consequences (most importantly, 100 extra cases of cancer, compared to a normal incidence of about 10,000 cases) in the population affected would be tragic but not catastrophic.

The Savannah River plant has turned out a quantity of high-level radioactive waste over its lifetime that is large in absolute terms, but small compared to the product of a civilian reprocessing plant. The plant size usually considered to be of full commercial scale would handle 1500 tons of spent fuel per year, the annual throughput from about 50 GWe of LWR's. The fission products handled would be many times greater than at Savannah River, and there would be orders of magnitude more plutonium and higher actinides in the waste. However, the waste would be relatively quickly converted to a solid, rather than stored indefinitely as a liquid, and we are concerned here with its hazards after solids have been emplaced underground. These extra steps offer orders-of-magnitude reductions in public exposure that counterbalance the increases in scale, so that the results for Savannah River remain an easily improvable upper limit.

The APS study[93] considered a large number of ways in which this extra level of protection might be negated. The natural forces of tectonic activity (earthquakes), volcanism, erosion, and meteoritic impact can be ruled out in any well-chosen site, at least to the extent that such events might bring wastes back to the biosphere by physical movement. Similarly, the effects of random anthropogenic activity, such as drilling, surface blasting, and so on, would not be a credible threat to the integrity of emplacement, and sabotage after emplacement would be virtually impossible. The only event that needs to be considered seriously is transport of radioactive materials by groundwater. Groundwater could exist in the repository as a result of poor choice, or could enter the repository as a secondary consequence of other catastrophes. The chain of events would be this: intrusion of the water; contact of water with waste bodies; dissolution of radioactive material; transport by the water through the host rock, and finally delivery (still by water) to the surface in springs or seepage outlets. The modeling of the various steps is an imperfect art. There is a very wide spread in estimates of transport times and amounts and concentrations of the radioactivity delivered.

Nevertheless, some limiting considerations can be applied. First, the process of leaching and transport does not produce highly concentrated solutions of radioactive material; rather, the conditions that favor rapid water transport also favor dilution of the material. These conditions include high permeabilities of the rock, and they are associated with rapid flow rates. Thus, any rapid delivery of radioactive waste to the biosphere would be at low concentration, and the consequences would be expected to be measurable as small increases in background radiation over a large area.

The conditions just outlined are quite rare. More common is a condition of naturally low flow rates of water through small pores. The water percolates, rather than flows. This condition favors an ionic absorption-desorption mechanism as the detailed transport phenomenon. A limiting case would be the chromographic process, in which such mechanisms permit certain ionic species to migrate in a concentrated "band" at a lower rate than the water. These processes are many times slower than ordinary convection—slow enough that they are only significant for longer-lived radioactivities (if applicable to shorter-lived species, the transport time becomes many half-lives, and the radioactivity decays en route). Thus, only the longer-lived radioactivities, primarily actinides and a limited number of fission products such as technetium-99 (^{99}Tc) or ^{129}I, present any possibility of reappearing in the biosphere in concentrated form, and such reappearance can be expected to occur in the very far future (millennia to eons).

Thus, the nature of the public hazard lies between two extremes: a relatively widespread, highly diluted reappearance of fission-product activities, and a much delayed, relatively more concentrated reappearance of actinides and a few fission products of long half-life. The consequences of this reappearance would in both cases mean an increase of environmental radioactivity, either in the form of an increase in background radiation over a large area, or in the form of pockets of radioactive materials. In the former case, the results of this increased radioactivity might be seen as similar to those from the fallout of distant nuclear weapons tests. In the latter case, they would be similar to those arising from natural concentrations of radiaoctive ores. A widespread increase in background radiation is not necessarily negligible in its public health implications, but is nevertheless a very small risk to any individual. An addition to surface concentrations of long-lived radioactivity would not affect a considerable number of people in any generation. It is only by adding these "occasional" effects over many generations that a large total effect could be inferred, and this would persist only as long as the radioactivity remains unidentified and therefore unabated.

To estimate the effects of this latter possibility, one study[94] considered "abandoned" wastes from the Idaho Chemical Processing Plant. The wastes are stored as solids in steel bins contained in concrete bunkers just below ground level. The worst exposures to individuals from a variety of pathways, including radon emanations from the bins, intruders digging superficially at the site, or settlements and farms appearing on the site, amounted to less than 5 rads per individual, or a doubling of the lifetime dose exposure to natural background radiation. The frequencies of exposure to such doses were estimated to be of the order of a few individuals per century, or less.

Whatever risk exists from release of radioactive waste stored in repositories can therefore be characterized qualitatively as one that is very small to any individual, but in the worst cases, as one that may be quite persistent or broadly diffused. It is definitely not a catastrophic risk in the usual meaning of that term.

Some Social and Institutional Considerations Understanding the problems of nuclear waste management requires discussion of more than the technology. The decisions to be made are principally of a social nature, such as how safe is safe enough? How should questions of equity be settled when the benefits are widely distributed but the wastes are disposed of in one (or a few) places?

The whole matter has received a great deal of public attention since about 1971, but little before. Present difficulties have three main causes: the generally negligent and uncommunicative attitude of the Atomic Energy Commission (AEC), even up to its dissolution; continuing federal indecision about the acceptability of nuclear power; and an unfortunate mistaking of goals for working policy. The management of radioactive waste has been a political issue, which itself has several features.

Regarding the historical role of the AEC, we have already discussed briefly how the handling of nuclear waste in the weapons program led to severe difficulties which are, however, of a nonrecurring type. The unfortunate technical choices were at the same time accompanied by the practice of holding information about the weapons waste program to a minimum, sometimes justified on grounds of national security, sometimes not. The AEC was at the same time declaring the nuclear waste problem to be tractable and straightforward, and was also deluding itself that an abandoned salt mine in Lyons, Kansas, originally selected only for nonradioactive experiments, was ideally suited as a permanent repository, despite the presence of solution mining in the same salt bed less than 2 miles away, and despite the many unrecorded drill holes throughout the strata. When all these matters came to light, the AEC and its successor agencies lost credibility that they have not yet regained.

Regarding the second cause, federal indecision about the long-term acceptability of nuclear power seems to be amplified in continuing indecision and changes in goals for handling nuclear wastes. First, the distinction between the problems posed by weapons wastes and those of commercial wastes have never been adequately emphasized publicly, nor has an adequate working policy been laid out for either. On the contrary, goals have been stated (e.g., a demonstrated repository by 1985), as if they themselves were the policy being carried out, and the distinction has been ignored.

Regarding the third point, the issue is naturally—almost ideally—political. Storing the wastes anywhere involves questions of equity, of local political jurisdictions, and of local acceptance. The difficulty is acute, considering that hazards with much larger social costs are routinely accepted merely because they are distributed. Many other much larger social costs are inequitably borne (e.g., coal-mining hazards) mainly because the problems are less recent and fashionable.

The quality of the information circulated to the public about the nuclear waste problem can only be described as abysmal. Indeed, misinformation is rife. At one extreme, many nuclear proponents have claimed that the waste is less radioactive than the ore after 500 years, a statement that is simply not true. At the other extreme, opponents have raised spectres of 250,000-yr hazards without a hint of their minute magnitude, or of the minute consequences of such hazards.*

Discussion

Before presenting our conclusions and recommendations, a few further points of discussion are necessary.

Until about 2 years ago, all plans for waste disposal anticipated that the fuel from LWR's would be reprocessed and that the resulting high-level wastes from the reprocessing plants would be solidified and encapsulated in glass or ceramic matrices for geological storage, most likely in salt formations. For many years, waste disposal was treated by government and industry as a problem that could safely be postponed, since it appeared to have a number of reasonable technical solutions. In the meantime, spent fuel from existing nuclear plants was temporarily stored in water basins

*See statement 5-32, by L. F. Lischer, Appendix A.

adjacent to reactors, pending reprocessing. The amount of waste to be handled was modest and did not present a problem that was viewed as urgent by the responsible technicians. Public distrust of this approach, however, was exacerbated by difficulties with the weapons wastes, already described, and was further increased when official statements continued to change in response to criticism.

Regardless, it is important to emphasize that waste management, unlike reactor safety, diversion, and proliferation, does not present hazards of large consequence. The risk is of slow leakage due to unforeseen contact with water and consequent exposure of people to low levels of radiation.

Finally, we must point out that the most general sorts of geological and geochemical knowledge, on which predictions of geological repository performance are based, cannot be expected to improve in the foreseeable future. To expect confirmation by experiment of expectations for integrity beyond 1000 years is simply impossible.

Our own conclusions and recommendations are essentially identical with those reached by the American Physical Society's study group on nuclear fuel cycles and waste management, with regard to the feasibility of radioactive waste isolation.[95] Among other points, the study group notes that waste isolation is feasible in salt and other media; that detailed technology for waste solidification, encapsulation, transport, and emplacement in mixed salt caverns is within the scope of existing knowledge; that confidence in geological isolation arises primarily from limitations on the rate of ion migration in underground formations; that continued investigation of geological and geochemical transport modeling is the most important current research topic; and that unreprocessed spent fuel should not be considered as waste, at least at this time.

Our own comments follow up these conclusions. While it appears that adequate technical solutions to radioactive waste disposal exist (e.g., geological disposal), the implementation of a program will require overcoming several political and institutional barriers. The foremost of these barriers is misunderstanding by the public of the nature of the problem. As evidenced by local hostility in many places to *investigation* of sites, it appears that the public is under the misapprehension that waste management poses local, high-intensity risks, rather than (at worst) widespread, low-intensity risks.

A second barrier is the failure of the government to agree which agency is responsible for setting standards—the Department of Energy or the Nuclear Regulatory Commission. Thus, standards (which can be set) are still pending.

A third barrier is the widespread opinion (at least in government) that a demonstration facility is necessary before a working repository is engineered. This seems to have led the public to conclude that there remains some significant doubt that waste disposal can be engineered on the basis of existing knowledge and site-specific investigations. This leads to the paradox of requiring proved demonstration in advance of establishing the facility. Arguments over what is meant by "demonstration" could be used to delay action for many years. Finally, there is an unnecessary emphasis on waste retrievability, which leads (again) to an impression of uncertainty and also raises costs.

Even if civilian nuclear power were to disappear, there would be a substantial radioactive waste inventory from military programs. Efforts should be instituted as soon as possible to develop and act on a workable program to manage military wastes. At Hanford alone there are about 200,000 m^3 (47 million gal) of high-level wastes, containing well over 200,000 tons (mostly sodium salts) of solids and over 200 Mci of radioactivity, primarily longer-lived fission products.[96]

Solving the political and institutional problems connected with management of this waste would be an important step forward. The chief problem seems to be devising a method of balancing federal and state interests in such a way that legitimate concerns are adequately addressed. The current policy, which gives localities and states arbitrary veto power, seems to be unworkable because local opinion has proved particularly vulnerable to scare tactics.

The potential for the future development of improved methods of dealing with radioactive wastes should not be ignored. Further research and development could be very valuable, while not impugning the validity of whatever short-term decisions are made. (An analogy may be made with enrichment. While the gas centrifuge process appears to be superseding the gaseous diffusion process, this does not mean that the latter is not a good process.)

Conclusions and Recommendations

Our recommendations follow from these observations and address primarily the institutional problems.

- The nature of the risks from geological disposal of nuclear waste must be clearly spelled out and vigorously publicized. The risks are those of chronic, dispersed, low-level radiation and are not comparable, for example, with risks from catastrophic reactor accidents.
- The federal government should immediately proceed to set criteria for geological waste disposal. These should be performance criteria (leach rates, heat rates) for waste forms in categories that recognize the risks from different types of wastes, and site criteria (groundwater standards,

seismic stability, resource and mining restrictions). The Nuclear Regulatory Commission has this responsibility, which is appropriate.

• The problem of disposal must be separated from the problem of spent fuel storage.

• The weapons waste problem must be settled, and the issue separated from that of commercial wastes. It may well be that long-term entombment is appropriate. If so, it should be effected. We note again that this waste is mostly fission products, and that its period of high risk is therefore relatively short.

• The federal government should accept full responsibility for any waste in existence, leaving the question of joint state-federal responsibility to be resolved for wastes generated in the future.

• Standards must be set and enforced for the treatment of abandoned mines and of tailings from mines and mills. These standards should permit disposal of low-level alpha-active wastes (i.e., alpha-active wastes that, if blended with the tailings, would not significantly increase the risk from tailings) in tailings piles.* This will require collaborative effort between the federal government and the uranium-mining states.

• While retrievability of waste forms after emplacement is a desirable feature of a test facility, and such a facility would be useful for a research program, retrievability ought not to be a consideration in designing a repository for actual waste disposal.

SAFEGUARDING THE DOMESTIC FUEL CYCLE

Atomic bombs are made from fissile material, and fissile material is the fuel of nuclear power. The term "safeguards" is the rubric under which we collect all the measures by which the manufacture of bombs from nuclear fuel materials can be prevented.

An intrinsic safeguard is isotopic dilution. Natural or slightly enriched uranium contains fissile ^{235}U in low concentration, with most of the uranium being ^{238}U. Enrichment is necessary for bomb material: A bomb requires a high concentration of ^{235}U. Since enrichment is a laborious, high-technology operation, now carried out in national or international facilities, stealing low-enrichment uranium would not be of much use for making bombs.

Another intrinsic safeguard is radioactivity. Spent reactor fuel is so radioactive that it must be handled under water or behind thick shielding. Most such fuel contains plutonium or ^{233}U, both of which are also bomb

*Statement 5-33, by J. P. Holdren: See my statement 5-31 and my longer dissenting view, statement 1-36, Appendix A.

materials, but to separate these fissile isotopes, a high-technology reprocessing operation must be carried out. While technically simpler, this process is even harder to carry out secretly than enrichment. Reprocessing releases radioactive gases that are readily detected at a considerable distance.

Spent nuclear fuel is a material of potential value. Recovery of its fertile material—thorium or low-enrichment uranium—permits substitution of recycled chemical elements for material that would otherwise have to be mined. Of greater value is the fissile material it contains: ^{233}U when thorium is the fertile material, ^{239}Pu when uranium is the fertile material. The recycle of this material in breeders or advanced converters could markedly reduce the need for ^{235}U from nature. However, both ^{233}U and ^{239}Pu can be made into bombs. Therefore, if they are to be recycled, they must be safeguarded from the time they are recovered from the spent fuel to the time they are inserted as recycled fuels into a reactor.

REPROCESSING

Reprocessing is the key operation that triggers concern about safeguards. The reference process for uranium-rich fuels is known as Purex, and for thorium-rich fuels, Thorex. They have similar characteristics. In both cases, fuel pieces from power reactors are chopped up or chemically declad, to expose the fuel material, and the fuel material is dissolved in acid. Uranium and plutonium (or thorium and uranium, in Thorex) are extracted by a solution of tributyl phosphate (TBP) in dodecane (a light paraffin oil). The TBP forms a complex chemical compound with the heavy elements and remains dissolved in dodecane, which is immiscible with the original acid solution. The fission products remain behind in the acid, which now is high-level waste. After separating the solution of dodecane and TBP, the heavy elements are washed (stripped).

The result of reprocessing can be either separated uranium and plutonium (or uranium and thorium) or mixed fuel materials.

The plants in which reprocessing takes place are very heavily shielded, and until the purified product is isolated, the physical barriers of the shielding and the intense radioactivity of the material make diversion essentially impossible. The type of construction needed also makes it convenient to provide very secure vaults for storage of the product material. Therefore, concern about safeguarding the reprocessing operation centers on prevention of "inside jobs" that might lead to diversion of the final product. The methods used for control of this product include materials accounting (i.e., checking that fissile material brought into the plant is either in process or has been shipped to a legitimate user) and personnel screening. However, the chief safeguard is security: limitation and control of access (as with vaults for precious materials); multiple checking of shipment authorizations and deliveries to shipping; and, since the product fissile material is still quite radioactive, multiple radiometric monitoring of access points and the personnel using them.

Radioactivity of Reprocessed Fuel

Recovered uranium from slightly enriched reactor fuel has more ^{236}U than natural uranium, but after purification from fission products and plutonium, it can still be handled essentially as virgin material. Recovered thorium is highly contaminated with ^{228}Th and its radioactive daughters. It must be stored for 10–20 years before it can be reused. By that time, the activity from ^{228}Th will have decayed to its low natural level.

The plutonium produced from ^{238}U and the uranium produced from ^{232}Th are much more radioactive. Typical plutonium recovered from spent LWR fuel contains 3 percent ^{238}Pu, 57 percent ^{239}Pu, 23 percent ^{240}Pu, 12 percent ^{241}Pu, and 5 percent ^{242}Pu. After several recycles, concentrations might be 5 percent ^{238}Pu, 31 percent ^{239}Pu, 27 percent ^{240}Pu, 17 percent ^{241}Pu, and 20 percent ^{242}Pu.[97] Representative numbers from thorium cycles are less firm, since there is far less experience with these fuels. The main radioactive constituent of ^{233}U, however, is ^{232}U, and it has been calculated that ^{233}U from HTGR irradiation could have 300–1000 ppm (i.e., 0.03–0.1 percent) of ^{232}U.[98]

Each kilogram of the mixture of plutonium isotopes coming from spent reactor fuel has typically over 600 curies (Ci) of alpha activity and more than 10,000 Ci of beta activity. (One curie is 3.7×10^{10} disintegrations per second, the radioactivity of 1 g of purified radium.) While these radiations are readily shielded by relatively thin containers, they are accompanied by penetrating gamma rays and neutrons. A kilogram of reactor plutonium emits 3×10^8 gamma rays per second with energy greater than 300 keV, considered "hard" gammas, and 3×10^5 fast neutrons per second from spontaneous fission of ^{240}Pu. These radiations deliver a dose of about 140 rads/hr, on contact (as in a pocket), that is dangerous and easily detectable, even with shielding. Only a minute quantity could escape detection unless lead or other heavy shielding is used in its removal, and these materials are readily sensed by other means.

Freshly reprocessed ^{233}U has less radiation; its alpha activity, mostly from ^{232}U, is only about 30 Ci/kg, and the penetrating gamma radiation that accompanies it is one tenth that of plutonium. This is probably detectable in large quantities. However, it may be desirable to "age" ^{233}U in order to increase its detectability, as illustrated below.

If the fissile material is not used after reprocessing, its short-lived isotopes (^{238}Pu, ^{241}Pu, and ^{232}U) begin to decay. From plutonium, the principal new radioactive constituent is ^{241}Am, the product of ^{241}Pu beta decay. After 3 months, gamma radiation from americium starts to contribute to the gamma-ray field surrounding plutonium, and after about a year, it almost doubles the emission of the most penetrating gamma rays (over 600 keV). Thus, aged plutonium is slightly easier to detect than freshly reprocessed material.

The situation with ^{233}U is more dramatic. Decay of the accompanying ^{232}U produces, in turn, ^{228}Th (1.9-yr half-life) and all the short-lived daughter nuclides, of this isotope, which quickly come into radioactive equilibrium. These daughters include some *very* hard gamma-ray emitters, such as thallium-208 (^{208}Tl), lead-122 (^{122}Pb), and lead-212 (^{212}Pb). After about 1 year, the alpha activity from these decay products is several hundred curies per kilogram of ^{233}U, and the penetrating gamma radiation is several hundred to several thousand times greater than that from plutonium. Aged ^{233}U is radiologically self-protected against diversion.

REPROCESSING ALTERNATIVES: DENATURING URANIUM-233, CIVEX FOR PLUTONIUM, AND OTHER DEVICES

Since reprocessing is the fuel cycle step in which potentially weapons-usable material is isolated, ways of achieving reprocessing while supplying extra technological safeguards would be very desirable. The principal approach for ^{233}U is "denaturing": blending in enough ^{238}U to make isotopic separation necessary to recover weapons-usable material. The principal approach for plutonium is to increase its radioactivity so that further, highly shielded reprocessing would be necessary to convert it to weapons-usable material.

Denaturing of ^{233}U is readily performed by mixing a quantity of depleted uranium solution in the fuel-dissolution stage of thorium fuel-element processing. The isotopes of uranium behave identically in the chemical process. After separation of the uranium from the thorium and isotopic analysis of the separated uranium, further blending of the product would finally yield 10–12 percent concentration of ^{233}U. This is a concentration at which a nuclear weapon would be so large and awkward as to be essentially impractical.

The chief drawback with denatured ^{233}U is the effect of depleted uranium (^{238}U) on nuclear fuel cycles. The ^{238}U substitutes for thorium, so that denatured fuel is, chemically, largely uranium. For example, the nuclear equivalent of 3 percent ^{235}U in slightly enriched uranium, such as that used in LWR's, is about 6 percent ^{233}U in thorium. If ^{233}U is used in denatured (say, 12 percent) concentration, almost 40 percent of fuel material would be uranium, and only 60 percent would be thorium. Almost 30 percent of the new fissile material produced would be plutonium, which is a less desirable fissile material in thermal reactors. A heavy water reactor would be markedly superior in this regard, since it would require less fissile material in its fuel, and a feasible loading might be of the order of 15 percent uranium at 12 percent ^{233}U concentration, and 85 percent thorium. Only 10 percent, or less, of the new fissile material would be plutonium.

A second effect is that the concentration of ^{233}U in spent fuel is less than 12 percent of the uranium. In order to recycle the fissile material, either more highly enriched uranium must be added, or the reactor must be reloaded even more heavily with uranium in subsequent cycles. The preferred method would be to "re-enrich" with 20 percent ^{235}U in ^{238}U, which is also undesirable for weapons (the critical concentrations of ^{235}U are higher than those of ^{233}U in assemblies of the same dimensions). Ultimately, however, unless the reactor is virtually a breeder, this will also lead to a "mostly uranium" fuel.

Denatured fuel cycles require further study before definitive conclusions can be drawn. The heavy water reactors, HTGR's, and LWBR's appear the most likely candidates for denatured fuel application.[99] The denatured Th-U cycle does not appear attractive for conventional LWR's.

For plutonium, a recently proposed alternative is the Civex cycle.[100] In this cycle, the uranium and plutonium are extracted together, and enough fission products are intentionally carried along with them that the product is still highly radioactive. Only a small sidestream of separated uranium is permitted, to bring the plutonium concentration of the product to reactor-usable levels. Coextracting some of the fission products along with the fuel materials actually simplifies the flow sheet of the Purex process. In principle, it can be accomplished using less careful purification of the tributyl phosphate. TBP is partially decomposed by the intense radiation of the reprocessing environment, and its decomposition products extract trivalent elements such as zirconium, lanthanides (rare earths), and higher actinides.

These elements include several fission products with sufficiently high activities to make the product material "hot" enough to be disabling, yet with half-lives long enough to persist. However, the lanthanides include a number of nuclides that strongly absorb low-energy neutrons. As a result, the Civex cycle has only been proposed for recycle of plutonium in fast reactors. Other flow sheets, rejecting lanthanides but retaining zirconium, other transition elements, and (preferably) strontium and cesium, would be needed for "activity-protected" thermal fuel recycle.

A related idea is to denature plutonium with ^{238}Pu. Because of its high rate of radioactive decay (half-life: 88 years) through alpha particle emission, this nuclide produces a great deal of heat (for which reason it is used as a heat source for remote power systems and, in very small amounts, for implanted cardiac pacemakers). If the concentration of ^{238}Pu (in plutonium) is above about 5 percent, any large pieces of weapons material would have to be cooled, or they would at least deform, and (if metal) possibly melt. Such material would be very unattractive, even for use in a national nuclear weapons arsenal, and would be virtually forbidding as weapons material for a subnational diverter. It could, however, be used as reactor fuel.

The ^{238}Pu would be manufactured by mixing a substantial quantity of neptunium-237 (^{237}Np) with the first recycle of plutonium fuel from LWR's and subsequently coprocessing neptunium. The nuclear reaction is the following.

$$^{237}Np + n \rightarrow {}^{238}Np \rightarrow {}^{238}Pu + e^{-}$$

(In the discussion that follows we subsume this concept under Civex for brevity, although its originators are different from the originators of the Civex concept.)

Many in the nuclear industry question Civex on the grounds that it would require remote fuel fabrication. Today's fuel elements are manufactured by processes that include a variety of direct manipulations, but this would not be possible with highly radioactive fuel material. However, we consider that the development of remote fabrication processes is inevitable. Recycled plutonium already demands some shielding, and aged ^{233}U requires either reprocessing (to remove ^{228}Th) or remote handling. Development of remote, automated inspection equipment is already quite advanced, and its use would obviate the direct inspection steps that figure among the main reasons for human intervention in the fabrication process.

Moreover, remote fabrication would, in our opinion, eliminate one more diversion route; the fewer the hands in the pie, the fewer the possible number of sticky fingers. Finally, the Civex cycle requires colocation of reprocessing and fabrication facilities to minimize shipment of radioactive material, and this is a highly desirable antidiversion practice in itself.

However, it must also be noted that wastes from the fabrication of Civex fuel (in contrast to the alpha-active wastes from fabrication of ordinary nuclear fuels) would be both alpha active and high-level. Minimizing population exposure from such wastes would require meticulous process design and operation.*

Finally, to complete the picture of technical measures that have been proposed to discourage diversion, we might mention the suggestion (from Manson Benedict) that conventionally recycled fuel elements be irradiated

*Statement 5-34, by L. F. Lischer: The reference to population exposure is misleading. *All* wastes require comparable safe handling.

in a research-type reactor before shipment. A relatively short irradiation period, of the order of 1 week, would be required. This would build up, in situ, a fission-product inventory with sufficient radioactivity of a persistent nature to obviate diversion of fresh fuel in shipment.

All of these measures require further research and development: for denatured ^{233}U, mostly analytical studies to follow the process in specific reactors, but also further investigation of Thorex reprocessing with heavy uranium loadings; and for Civex, technological development of both the reprocessing and fabrication steps. Even pre-irradiation would require some special reactor development, albeit of a straightforward nature. Moreover, standard security measures may be sufficient to guard adequately against diversion in the United States; if so, the added security of such technical safeguards as denaturing and Civex might be very costly and of small incremental value.

HIJACKING AND THEFT IN STORAGE AND SHIPMENT

If separated plutonium or fissile uranium is the end product of reprocessing, transport of this material between reprocessing and fabricating plants presents a major period of vulnerability to theft. For security, these two plants could be colocated in a guarded fuel cycle park. Colocation would be convenient, since rejected material from the fabrication plant must sometimes be returned to the processing plant for recovery of fissile isotopes.

Shipment of fabricated reactor fuel to the reactor plant and management of the fuel before loading into the reactor pose the greatest diversion hazards. The fuel must be loaded onto a vehicle for shipment; the vehicle could be hijacked, misplaced in railroad systems, or waylaid.

Some maintain that shipments of nuclear fuel must be accompanied by unusual security measures. These measures include lead and follow vehicles with guards, air surveillance of the shipment, and super safes to contain the material. Safes might be tagged with radioactivity for easy location by air surveillance. Aerial radiometric surveillance has been used to locate lost gamma-ray sources. Proponents of reprocessing point out that similar security measures have been successful in the far more dangerous transportation of nuclear weapons, and that the attendant military presence has been unobtrusive. The expense of these measures would contribute little to the cost of nuclear power.

Fortunately, reactor fuel comes in large pieces that cannot be moved without the cooperative interaction of several people to operate cranes, hoists, and fueling machines. In view of this, sabotage is generally considered a greater threat to nuclear power plants than diversion.

Security measures are generally acknowledged to be effective deterrents to theft under two conditions. First, the protective force available must be sufficient in numbers and equipment so as to resist effectively the maximum attacking force that could be expected in the case of overt theft. There is some disagreement as to how large such an attacking force might be, but a report of the Nuclear Regulatory Commission suggests that an on-duty force of about a dozen guards, properly equipped and with reinforcements available, would be sufficient to protect a nuclear facility such as a reprocessing plant.[101] This is judged to be a competent group to defeat a small attacking force or to delay a large one until reinforcements could be brought in.

The other condition is that the security force and key operating personnel (particularly, managers and professional employees) be loyal to security. Indeed, preventing "inside jobs," particularly those of a covert nature, appears to be a major concern of much of our existing policy. For example, there is considerable emphasis on materials accounting as a safeguards measure. This is a tool specifically to detect covert theft. Security against dissident personnel depends on personnel clearance, separation of function control over detailed responsibilities, and management integrity to enforce controls.

Some argue that these measures might or might not work. There has been insufficient critical discussion of their efficacy. Even in the domestic weapons program, there have been incidents of lax security. Security measures tend to deteriorate with time. In a turbulent society, the rate of attempts might be quite high—higher, for example, than the rate of attempts on payroll trucks—in view of the enormous possibilities for blackmail presented by nuclear weapons. Concern has been expressed that the measures necessary for dealing with such a threat might erode civil liberties. Others find these expressed concerns to be contrived and exaggerated.

CONCLUSIONS

The subject of nuclear power security is evolving, and it is difficult to reach a firmly based conclusion on the adequacy of measures proposed or the threat they might pose to civil liberties.[102]

Levels of security now used or proposed are not unusual in our society. Some degree of personnel clearance for employment is normal for operations involving access to precious or dangerous materials, and voluntary citizen cooperation with police activities is high enough on matters of great public risk that civil liberties are rarely compromised. The point at issue is therefore whether measures required in the nuclear power area, particularly when fissile materials are recycled, will reach a magnitude so great as to convert acceptable security into a choice between insufficient security and unacceptable intrusion.*

It therefore seems premature to us to commit the United States to expensive, additional technical safeguards, such as denatured fuel cycles, Civex, pre-irradiation, and so forth. They deserve further development, because such development might reduce their expense or demonstrate further benefits that would counterbalance their cost. Among the benefits of such development is that of having such safeguards available in the case that other security measurements are either insufficient or socially unacceptable.

Colocation of fuel cycle facilities is an example of a measure likely to lead to both improved economy and increased security. Improved methods of materials accounting also seem to be in this desirable category.

Finally, our conclusions pertain only to the United States. Other countries have different national and social institutions, including different standards of citizen rights against government intrusion. We must therefore expect that different countries will evolve different internal safeguards, and we should avoid judging these safeguards by too detailed comparisons with our own.

NUCLEAR POWER AND PROLIFERATION OF NUCLEAR WEAPONS

The view is increasingly expressed that the most serious liability of commercial nuclear power is the link between this technology and the international proliferation of nuclear weapons. This was the conclusion, for example, of the report issued in 1976 by the United Kingdom's Royal Commission on Environmental Pollution[103] and of a study sponsored by the Australian government to assess the effects of uranium mining in that country.[104] In the United States, the concern permeating recent reports on nuclear power was bluntly summed up as follows in a 1977 report sponsored by the Ford Foundation.[105]

> In our view, the most serious risk associated with nuclear power is the attendant increase in the number of countries that have access to technology, materials, and facilities leading to a nuclear weapons capability . . . . If widespread proliferation actually occurs, it will prove an extremely serious danger to U.S. security and to world peace and stability in general.

*Statement 5-35, by L. F. Lischer: The conclusions of two major documents are that sufficient measures can be found (see note 97) and that their social effects would not be large (see note 102).

The broad spectrum of views held by informed analysts results from the complexity of the issue. Proliferation is not one problem, but the intersection of many. The motivations of nations to acquire both nuclear reactors and nuclear weapons are as much an issue as the technical means, and those motivations arise from concerns for the security and independence of energy supplies, trustworthiness of military alliances, regional antagonisms, disparities in arsenals of conventional armaments, aspirations to greater status in the community of nations, and perceptions of rich-poor, big-small, and north-south inequities. It is this complexity and diversity on the motivational side that led one analyst to insist, "There are no simple solutions that are feasible, no feasible solutions that are simple, and no solutions at all that are applicable across the board."[106]

How should concern for proliferation influence the use of nuclear technology in the United States and shape the action this country takes to help or hinder the use of nuclear power technologies abroad? Differences of opinion on this issue can be usefully classified by the question that first elicits disagreement in the following hierarchy.

1. Is proliferation very important?
2. If so, is the link between nuclear power and proliferation very important?
3. If so, is the influence of the position taken by the United States on nuclear power very important?
4. If so, what should the United States do?

In what follows we try to state concisely the principal positions held on each of these questions by various parties to the proliferation debate, in the hope that this will illuminate policy alternatives.

IS PROLIFERATION VERY IMPORTANT?

The "yes" position holds that proliferation is important because the more nations that possess nuclear weapons, the greater the likelihood of nuclear war. Countries that do not have nuclear arms cannot use them in whatever conflicts they enter. For a given number of conflicts, the greater the fraction in which one or more parties have nuclear arms, the greater the chance that these weapons will be used.

The "no" position* holds that proliferation may as easily be a stabilizing as a destabilizing force. It would diminish one form of inequality, where inequality is correlated with instability. In a world where many countries

*Statement 5-36, by L. F. Lischer: The single "yes" and "no" positions are perhaps an oversimplification. I know very few advocates of nuclear power who think that proliferation is not important.

have nuclear weapons, all countries will tread more carefully and there will be fewer conflicts altogether.

The antiproliferation position replies that this view relies too heavily on the universality of rational decision making. What of revolutions in which national stockpiles of nuclear weapons fall into the hands of fanatic groups or unstable dictators? What of miscalculations in the face of perceived threats to national survival? The pro-proliferation or neutralist position responds that there is only one historical example of the use of nuclear weapons in hostilities, and in that instance, only one of the adversaries had them. The U.S.-Soviet nuclear "balance" has been stable for almost 30 years.

An intermediate position of sorts has been stated by Greenwood[107]: a world of many nuclear weapons states may not be *intrinsically* less stable than today's world of a few, but there is danger in the possibility that the *rate* of spread of weapons will be too fast for political systems and international institutions to make the appropriate adjustments.

There are irreducible uncertainties in speculating about the future behavior of nations. This study takes proliferation to be important and dangerous, not necessarily by virtue of a general relation between the number of weapons states and the probability that nuclear weapons will be used, but in considering that proliferation may occur in particularly fragile regions or at rates too high for institutions to accommodate.

IS THE LINK BETWEEN NUCLEAR POWER AND PROLIFERATION VERY IMPORTANT?

The spectrum of opinion on this question is encompassed by three positions: first, that the spread of nuclear power has little to do with the real prospects for proliferation; second, that the spread of nuclear power is likely to alleviate pressures for proliferation; and third, that the spread of nuclear power is likely to aggravate the proliferation problem.

There is general agreement that the spread of nuclear weapons among nations has been limited up until now by various combinations of the following four factors: (1) good intentions, as manifested by the Non-Proliferation Treaty (NPT) signed by more than a hundred nations; (2) lack of the technical skills required to design and fabricate a reliable nuclear weapon; (3) reluctance to commit the necessary technical and financial resources to this particular task; and (4) lack of the means to acquire the necessary fissile materials, which until recently only a few nations have had the technical wherewithal to obtain. The category "good intentions" includes the case of national policy that open acquisition of nuclear weapons is not in the country's interest, and cases in which fear of detection by the International Atomic Energy Agency deters a clandestine nuclear weapons program.

In this context, the link between nuclear power and proliferation rests on the fact that commercial nuclear power programs can spread both bomb-relevant technical skills and bomb-quality fissile materials into countries that formerly possessed neither, undermining two of the four obstacles listed. The spread of pertinent knowledge has already proceeded to the point that access to fissile material almost certainly poses the greater threat.

The various types of nuclear power technologies differ in the ease with which they can be made to yield bomb-quality fissile material. The most sensitive technologies in this respect are enrichment and reprocessing. As indicated in the section on enrichment (in "Nuclear Reactors and Fuel Cycles"), low concentrations of ^{235}U can be "cascaded" to produce the much higher enrichment needed for nuclear weapons. In the reprocessing of spent fuel, fissile plutonium (or ^{233}U) is separated from the highly radioactive fission products in a form that requires no further isotopic separation for use in nuclear bombs.

One argument against a strong link between nuclear power and proliferation holds that nuclear weapons can be manufactured by a number of methods that are far less costly and troublesome than trying to use nuclear fuel cycles. For example, almost any country that sets out to acquire nuclear weapons can expend a few years of effort and a few tens of millions of dollars to build a natural-uranium, graphite-moderated, air-cooled reactor capable of producing a few bombs' worth of plutonium each year. More ambitious countries could build bigger production reactors with much higher outputs of plutonium, at a cost below that of commercial power reactors with comparable plutonium production. Fuel for reactors dedicated to the production of plutonium is easier to reprocess. Moreover, the isotopic *quality* of plutonium from reactors dedicated to the production of weapons material can be used more readily to make efficient bombs than can the plutonium from most kinds of commercial reactors dedicated to the economic production of power. Moreover, several different enrichment processes now at various stages of development—centrifuges, nozzles, laser schemes—might become widely accessible in the future, even to countries with no commercial nuclear power programs.

Some counterarguments can be advanced. All enrichment schemes known today are, as far as can be deduced from the open literature, high-technology enterprises. If not spread through nuclear power programs, they may remain out of the reach of many developing countries for years—perhaps long enough to slow proliferation by enrichment. The use of dedicated production or research reactors and modest indigenous reprocessing capacity is probably a more troublesome possibility, considering the case of India, but at least two arguments are put forward by those contending that this possibility does little to alleviate legitimate concern about proliferation through commercial nuclear power.

First, production reactors and reprocessing plants of the size readily constructed by most of the countries in question dribble out plutonium, whereas commercial-size reactors pour it out, on the order of 20 Nagasaki bombs per year for each large light water reactor. With a commercial nuclear power program operating, particularly if it includes reprocessing, a country can go from no bombs to many bombs in short order.

Without a nuclear power program, a nation must commit itself to the construction of a reactor and reprocessing plant, or set up uranium enrichment for the sole purpose of making nuclear weapons. Reluctance to make such a commitment or to risk being caught in a clandestine operation has served as a barrier to proliferation. Commercial nuclear power, or more explicitly, national reprocessing or enrichment facilities, lowers the barrier. It supplies the option without the commitment, and by shortening the time between decision and bombs if a country's intentions change, it enhances the option. The time between commitment and possession of nuclear weapons, the argument goes, is likely the period when a country's decision makers feel most vulnerable to censure or intervention from their own people, other nations, or international organizations. The shorter the period, the less the risk. In a world of well over a hundred nations there must be some that would not initiate a weapons program from scratch but might succumb to the temptation of having most of the needed ingredients at hand in the form of commercial nuclear facilities and materials.

The principal argument for the view that nuclear power is of less importance than international diplomatic measures rests on the relative success to date of measures to enforce international agreements to limit the spread of nuclear weapons. Since the period immediately after World War II, only three nations have actually been added to the "weapons club." At the same time, nuclear technology has become common knowledge throughout the advanced and developing world. The spread of nuclear technology has been accompanied by general acceptance—with some outstanding exceptions—of IAEA safeguards and the Non-Proliferation Treaty. The terms of the treaty and safeguards explicitly offer favorable status for nuclear energy development as a reward for nonproliferation, an apparently effective bargain.

The most optimistic view, of course, is that the spread of commercial nuclear power will exercise an influence against the proliferation of weapons greater than the temptation it represents. The argument underlying this position is that poverty in general and shortage of energy in particular are major causes of tension among nations. By providing a source of energy needed for economic development, the spread of nuclear power reduces this tension. Poor and developing nations cannot be expected to compete with the United States and other industrial nations in bidding for Middle Eastern oil. Widespread intensive use of nuclear power reduces the possibility of a war erupting over access to Middle Eastern oil.

A counterargument on this point is that nuclear power is rather badly matched in scale, energy quality, and other characteristics to the needs of many developing countries. Thus, the spread of nuclear power may narrow the rich-poor gap only slightly—and, if it contributes mostly to the well-being of urban areas, may even widen that gap *within* some countries—all at the cost of making nuclear weapons more accessible. On the other hand, it can make a major contribution if the availability of power expedites such uses as water pumping to increase productivity of small agricultural holdings, or if power is used to improve rural life, as exemplified by the Rural Electric Association in this country.

As in other questions where perceptions, motivations, and future actions of sovereign states play a role, the effect of the spread of nuclear power on the spread of nuclear weapons defies completely persuasive argumentation for any position. On one point, however, the members of CONAES agree: Stopping the spread of nuclear power, or limiting its evolution to forms considered proliferation resistant, cannot stop proliferation. Countries with sufficient determination will get bombs by other routes. The issue is more a quantitative than a qualitative one: whether the spread or evolution of nuclear power would speed proliferation, either in the number of nations that decide to take the plunge or in the number of bombs they have at any given time. Accordingly, the argument that there is a potentially malign link between nuclear power and nuclear weapons proliferation (which, by implication, should be manipulated if possible to reduce the threat) is most plausibly an argument for buying time, for slowing a process that almost all analysts agree can no longer be completely stopped. And, naturally, buying time will avail nothing if statesmen do not use it to fashion institutions capable of deterring the use of nuclear weapons by however many nations possess them.

IS THE INFLUENCE OF THE POSITION OF THE UNITED STATES ON NUCLEAR POWER VERY IMPORTANT?

Again the spectrum of opinion can be illuminated by considering three answers to this question: yes; yes, but only if the position of the United States is one of continued active participation in the world nuclear power market; and no, regardless.

The "no" viewpoint is perhaps the easiest to state simply. It is basically that "the genie is out of the bottle." Nuclear power is going to spread whether the United States likes it or not, since so many nations besides the United States can supply it, and since so many others have valid and legitimate reasons for wanting it. And since the main driving force behind proliferation of weapons is the motivation of countries to possess them, not technology itself, posturing about technology by the United States can only distract attention from the more important issues and thus diminish this country's influence over the course of events.

The "yes, but . . . " position argues that only by remaining an active participant in the international nuclear market can the U.S. hope to retain any influence over the kinds of nuclear technologies that are supplied to other countries, and over the kinds of safeguards against proliferation that are exercised. If this country backs out of the export market, no one will listen to advice from the United States about how that enterprise should be managed, a situation that could well worsen if the United States begins to phase out nuclear power at home.

An important component of the "yes, but . . . " position pertains to the Non-Proliferation Treaty. The NPT is, in the view of many, the most important single instrument available for discouraging proliferation. Accordingly, any position taken by the United States on proliferation must be at pains not to undermine the treaty. Since the treaty requires that weapons states cooperate with non-weapons states in making available the technology for peaceful applications of fission, the United States could not withdraw from the international nuclear market, and perhaps cannot even limit selectively the export of particularly proliferation-prone technologies without weakening support for the treaty among non-weapons states.

The unequivocal "yes" viewpoint—that the United States can exercise influence against proliferation through its position on exports of nuclear technology, on the one hand, and through the character of its own nuclear energy supply, on the other—rests on three propositions. The first is that the substantial share of the world nuclear export market now controlled by the United States is in itself sufficient to give this country considerable leverage in governing what is available for other countries to buy. Other nations could even totally fill orders the United States refuses, but the time new suppliers require is time gained against proliferation. Second, the United States has some influence over other nations supplying nuclear power, in the form of persuasion, in the form of the direct dependence of their nuclear programs on our parts and enrichment services, and in the form of other political and economic incentives and disincentives this country might be prepared to wield in pursuit of nonproliferation. Third, the nuclear power policies of the United States might serve as an example. The United States is still a model that influences the behavior and aspirations of many other countries. Even where the direct force of the example on other governments is minimal, a strong position taken by the United States against proliferation may undermine the public support in other countries for proliferation-prone policies pursued by their governments, or strengthen the hand of antinuclear movements within those countries.

It cannot be doubted in any case, the argument continues, that if the United States does *nothing* against proliferation, other countries are likely to do nothing as well. Business as usual in the country with the biggest nuclear industry in the world can only be taken as a clear signal for business as usual everywhere. The more the United States pushes nuclear energy at home, particularly plutonium recycle and plutonium breeders (which are claimed by some to be more subject to proliferation than other technologies), the more convinced other nations will become that they cannot do without these technologies. The counterargument is that heavy-handed use of persuasion and political pressure by the United States to dissuade other countries from deploying certain nuclear technologies may backfire by hardening their resolve and even accelerating their programs.*

The "yes" viewpoint—that positions taken by the United States restricting the export or use of nuclear technologies can slow proliferation—must confront the liability of potential damage to the NPT. Supporters of this view often do so by suggesting that the treaty is not doing as much good as is often supposed, thus the possibility of undermining it should not stand in the way of measures likely to be much more effective. Many nations have not signed it, including some thought to be particularly interested in proliferation, or having signed, have not ratified it. Those that have ratified it can withdraw on 3 months' notice. The IAEA safeguards provided by the treaty can at best detect diversion, not prevent it, and some have asserted that the IAEA inspectorate is too overtaxed to provide even reasonable assurance of detection. Moreover, it is argued, the sanctions likely to be exercised if a country happens to be caught in violation are too feeble to be effective. Finally, the weapons-states parties to the treaty have themselves already undermined it by not meeting their own obligations, either in making significant progress toward nuclear disarmament, or in giving preferential treatment to other parties in transfers of nuclear power technology.

*Statement 5-37, by L. F. Lischer: This has actually been happening in several countries.

WHAT SHOULD THE UNITED STATES DO?

The policy options open to the United States in pursuit of nonproliferation can be organized for purposes of discussion in terms of the following hierarchy.

1. Approaches seeking to reduce motivations toward proliferation, versus approaches dealing with the potentially proliferation-related characteristics of nuclear power itself.
2. Among those approaches dealing with nuclear power, seeking increased resistance to proliferation in the characteristics of reactors and fuel cycles, versus developing management techniques and institutional arrangements for nuclear power that act against proliferation.
3. Among the measures in the previous category, elements that are part of purely domestic policies on nuclear power, versus policies on U.S. exports of nuclear technologies, versus other kinds of policies intended to influence the behavior of other nations.

Naturally, the elements in this hierarchy are not mutually exclusive. Many could be pursued at once. In the following subsections, we give brief attention to approaches seeking to reduce motivations, followed by more detailed treatment of proliferation-resistant management.

Motivations

Some of the reasonably obvious ways to try to reduce the motivations driving nations toward acquisition of nuclear weapons are the following.

- Maintaining, strengthening, or extending security guarantees provided by the United States to certain non-weapons states.
- Working to resolve or stabilize regional disputes.
- Reducing the prestige and symbolic importance of nuclear weapons in world politics, including working vigorously for reduction of the nuclear arsenals of the superpowers.
- Seeking to satisfy some of the economic and political ambitions of certain potential weapons states.

While these approaches seem attractive in the perspective of nonproliferation, many of them have significant economic, military, and political costs. Investigation of these matters was well beyond the scope of this study.

Proliferation-Resistant Management and Institutions It should be apparent from the preceding sections of this chapter that there is no technical antidote to the possible use of nuclear power for the proliferation of nuclear weapons. We list here the major kinds of management measures and institutional changes that have received serious attention in the proliferation debate, emphasizing the role the United States could play. The following list moves from moderate to drastic measures.

- Ways to strengthen the NPT.
- Bilateral (U.S.-recipient country) agreements stronger than the NPT.
- International control of sensitive parts of the fuel cycle.
- Barriers to the spread of fission technology or to its evolution in particularly proliferation-prone directions.

For each category, we consider the role of domestic policy, policy on nuclear exports, and other policies.

In principle, the Non-Proliferation Treaty could be strengthened by the following: tightening the safeguards incorporated into agreements concluded under the treaty between the IAEA and non-weapons-state parties; funding more inspectors to enforce the safeguards; adding sanctions to be imposed on violators; bringing the behavior of weapons-state parties into line with the letter and spirit of the treaty. Some of these measures could be accomplished without rewriting the treaty, a tedious and risk-laden procedure.

The United States could contribute expertise and money to achieve other measures, refuse expertise and materials to non-party nations, stop withholding assistance and materials from party nations, use its influence to encourage other supplier nations to do the same, and upgrade efforts to make real progress toward nuclear disarmament. A particularly dramatic—some think dangerous—measure in this last category would be to take unilateral steps to reduce the domestic stockpile of nuclear weapons. Modifying the treaty to add sanctions would probably be very difficult, and the likelihood that the United States would find the benefits worth the required political investment seems low.

The United States could choose, in its own relations with recipient nations as a supplier, to reach safeguards agreements more stringent than those enforced under the treaty. One such possibility is to lease nuclear fuel to non-weapons states rather than selling it, requiring return of the fuel, when spent, for reprocessing or storage without reprocessing in this country. A recipient country could break such an agreement and reprocess a batch of fuel to extract the plutonium, but detection would be assured and it would only work once—perhaps a small consolation. Some domestic opposition to taking back spent fuel from other countries might be expected, on grounds that this would burden us with environmental liabilities from the energy use of other countries. Furthermore, as a practical matter, the United States (despite positive statements by the Administration and members of Congress) has not acted on any plan to provide adequate spent-fuel storage—even for anticipated domestic needs.

A politically much more difficult, but also more promising, approach to proliferation control is to place the most sensitive parts of the fuel cycle under international control, including enrichment plants, reprocessing plants, plants for the fabricating of plutonium fuel, and shipping links wherein plutonium flows unprotected by accompanying fission products. Breeder reactors using undenatured fuel would perhaps also come under international control. Colocation of many of these facilities in international fuel cycle centers would reduce the problem of surveillance. Dispersed reactors supplied with denatured fuel from these centers and returning their spent fuel to the centers could be under national control. The political and organizational problems connected with this degree of international cooperation are generally considered to be formidable obstacles. Nevertheless, it is possible that a really vigorous effort by the United States to muster support for the approach, also underway in the IAEA, could lead to several pilot examples.

The most drastic category of measures the U.S. might entertain consists of erecting and maintaining barriers to the spread of fission technology and to its evolution in particularly proliferation-prone directions. For any or all of the sensitive technologies—enrichment plants, reprocessing plants, breeder reactors—the United States could refuse export, restrict their domestic use or development, and influence other supplier nations to do the same. Carrying this approach to its limit would require refusing to export reactors or fuel cycle facilities of any kind, and sharply limiting or phasing out domestic nuclear power in an attempt to force the world away from the nuclear option.

Any approach in this category has the liability of undermining to some degree the Non-Proliferation Treaty, as discussed above. These approaches also open the United States to the accusation that this country is insensitive to the needs of countries poorer and less well endowed with energy alternatives than itself, and they run the risk of aggravating non-weapons states into developing unsafeguarded nuclear facilities. Proponents of the "barrier" approaches believe that the risk of aggravating others is one the United States must take, and that necessary accompaniments to the barriers are that the United States must supply a plausible substitute for what has been denied and also begin to exercise some real leadership in nuclear disarmament.

The Administration's proposal is to guarantee supply of enrichment services for converter reactors as a compensation for the denial of enrichment and reprocessing plants. Countries capable of developing plutonium recycle and marketing plutonium breeders are urged to defer these steps. The United States would defer as an example. These proposals are seen by some nations as self-serving, inasmuch as they promote what the United States has to offer (enrichment services) while disparaging alternatives (reprocessing and breeders) being promoted by competitors in the world market. The offer to provide enrichment services is questioned on grounds of this country's record on nuclear policy, and on grounds that the United States is doing little to stretch its own uranium supply.

Certainly it is a liability of these proposals that non-weapons states are not likely to be much more inclined to rely on the United States for enriched uranium than they are to rely on OPEC for oil. This problem may diminish with the development of a strong international market in enrichment services, characterized by a number of independent suppliers. But critics who think the proposals too mild assert that continuing to export reactors themselves will encourage the owners to complete a measure of energy independence by seeking, as quickly as possible, domestic enrichment or reprocessing capacity, or both, and critics who think the proposals too severe believe that resentment of moralizing from the United States on these matters will diminish any influence we might have had in securing better international safeguards for the full range of nuclear facilities sure to be demanded almost everywhere.

If the United States were to go further than the Administration's proposals by trying to erect barriers against the spread of all nuclear technologies, it seems clear that, for consistency, domestic policy would have to phase out nuclear power, and foreign policy would have to assist other countries with a variety of alternative energy supplies. This assistance might well emphasize "income" energy sources available in particular abundance in some of the poorest regions. But the sorts of trades that would have to be considered in this situation would likely include diverting Alaskan oil and gas to Japan and exporting a good deal of coal mined in the United States. Many Americans might think this too high a price to pay.

CONCLUSIONS

The nature of the proliferation problem is such that even stopping nuclear power completely could not stop proliferation completely. Countries can acquire nuclear weapons by means independent of commercial nuclear power. It is reasonable to suppose if a country is strongly motivated to acquire nuclear weapons, it will have them by 2010, or soon thereafter, no matter how nuclear power is managed in the meantime. Unilateral and international diplomatic measures to reduce the motivations that lead to proliferation should be high on the foreign policy agenda of the United States. Nevertheless, the potential links between nuclear power and proliferation of nuclear weapons must be taken seriously, both because the *rate* of proliferation may be increased by the availability of commercial nuclear power facilities, and because the "attractive nuisance" that indigenous stocks of weapons-usable material constitute may nudge some nations, not sufficiently motivated to develop a separate weapons program from scratch, to topple into weapons status.

A minimum antiproliferation prescription for the management of nuclear power is to try to raise the *political* barriers against proliferation through misuse of nuclear power by strengthening the Non-Proliferation Treaty, and to seek to raise the *technological* barriers by placing fuel cycle operations involving weapons-usable material under international control. Any such measures should be considered tactics to *slow* the spread of nuclear weapons and thus earn time for the exercise of statesmanship. It is essential that statesmen use this earned time to find ways to end proliferation, to begin to shrink the weapons stockpiles, and to reduce the probability that nuclear weapons will ever again be used.

The question remains whether measures more comprehensive and more disruptive of the nuclear enterprise are warranted by antiproliferation goals. Weighing the often counteracting political and technological considerations outlined in the body of this section, different members of the committee reach different answers.

So many of the political factors, particularly, contain unpredictable elements, that no completely convincing analysis of the likely outcomes of given measures is possible. It is hardly surprising, therefore, that individuals with different perceptions of the likely future behavior of governments, of the incremental dangers of risk reduction associated with given technological changes, and of the likelihood and jeopardy of energy shortages, do not agree whether the United States should try to accelerate or decelerate the use and spread of nuclear power in general and breeder reactors in particular.

We do agree that any proposed policy should recognize the possibility that it is based on wrong judgments, and accordingly, should incorporate escape routes—ways to pull back from a policy decision if evidence accumulates that the consequences run counter to its aims.

NOTES

1. Atomic Industrial Forum, *Electricity from Nuclear Power* (Washington, D.C.: Atomic Industrial Forum, 1979), p. 1.

2. National Research Council, *U.S. Energy Supply Prospects to 2010,* Committee on Nuclear and Alternative Energy Systems, Supply and Delivery Panel (Washington, D.C.: National Academy of Sciences, 1979), chap. 2.

3. National Research Council, *Alternative Energy Demand Futures to 2010,* Committee on Nuclear and Alternative Energy Systems, Demand and Conservation Panel (Washington, D.C.: National Academy of Sciences, 1979); and National Research Council, *Supporting Paper 2: Energy Modeling for an Uncertain Future,* Committee on Nuclear and Alternative Energy Systems, Synthesis Panel, Modeling Resource Group (Washington, D.C.: National Academy of Sciences, 1978).

4. See chapter 11 for display and discussion of study scenarios.

5. National Research Council, *Supporting Paper 1: Problems of U.S. Uranium Resources and Supply to the Year 2010,* Committee on Nuclear and Alternative Energy Systems, Supply and Delivery Panel, Uranium Resource Group (Washington, D.C.: National Academy of Sciences, 1978).

6. Atomic Industrial Forum, *op. cit.*, p. 1. This figure includes some reactors not scheduled to operate until as late as 1995.

7. Supply and Delivery Panel, *U.S. Energy Supply Prospects to 2010, op. cit.*, chap. 5.

8. Assuming that the available resource base is sufficient to provide initial fuel requirements. See, for example P. R. Kasten *et al., Assessment of the Thorium Fuel Cycle in Power Reactors* (Oak Ridge, Tenn.: Oak Ridge National Laboratory (ORNL/TM-5565), January 1977), p. xi.

9. *Ibid.*; and Alfred M. Perry, "Thermal Breeders in Today's Context" (Paper presented at the International Scientific Forum on an Acceptable Nuclear Energy Future of the World, Fort Lauderdale, Fla., November 7–11, 1977).

10. Kasten, *op. cit.*; and Perry, *op. cit.*

11. U.S. Atomic Energy Commission, *Comparative Risk-Cost-Benefit Study of Alternative Sources of Electrical Energy* (Washington, D.C.: U.S. Atomic Energy Commission, (WASH-1224), 1974); L. A. Sagan, "Human Costs of Nuclear Power," *Science* 177 (August 1972): 487–493; E. E. Pochin, *Estimated Population Exposure from Nuclear Power Production and Other Radiation Sources,* Organization for Economic Cooperation and Development (Paris, France: Nuclear Energy Agency, 1976); U.S. Nuclear Regulatory Commission, *Final Generic Environmental Statement on the Use of Recycle Plutonium in Mixed Oxide Fuel in Light Water Cooled Reactors,* vol. 4 (Washington, D.C.: U.S. Nuclear Regulatory Commission (NUREG-0002, or GESMO), 1976); Nuclear Energy Policy Study Group, Spurgeon M. Keeny, Jr., Chairman, *Nuclear Power: Issues and Choices* (Cambridge, Mass.: Ballinger Publishing Co., 1977), pp. 173–174.

12. U.S. Nuclear Regulatory Commission, *Reactor Safety Study: An Assessment of Accident Risks in U.S. Commercial Nuclear Power Plants* (Washington, D.C.: U.S. Nuclear Regulatory Commission (WASH-1400), 1975).

13. For a complete review of the literature on reactor safety, see National Academy of Sciences, *Risks Associated with Nuclear Power: A Critical Review of the Literature,* Committee on Science and Public Policy, Committee on Literature Survey of Risks Associated with Nuclear Power (Washington, D.C.: National Academy of Sciences, 1979).

14. See, for example, "Report to the American Physical Society by the Study Group on Nuclear Fuel Cycles and Waste Management," *Reviews of Modern Physics* 50 (January 1978): S-1–S-185; and National Research Council, *Radioactive Wastes at the Hanford Reservation: A Technical Review,* Commission on Natural Resources, Committee on Radioactive Waste Management (Washington, D.C.: National Academy of Sciences, 1978).

15. It must be recognized that this statement reflects the least common denominator of a wide range of views in CONAES, and is accordingly ambiguous. Some members of CONAES, for example, believe that nuclear energy is the lowest-cost and least environmentally risky technology available for the generation of electricity and should be encouraged to expand as rapidly as warranted by the electricity demand projections of the industry. Others believe that nuclear power is the technology of last resort and its expansion should be restricted to the

maximum extent compatible with no harm to the economy. Specifically, these members believe that plants operating or under construction should be allowed to live out their useful lives, and that additional LWR's on the "planned" list should perhaps be allowed to be built and operated if carefully sited, but that more than the 150–200 GWe implied by these categories is likely to be unnecessary and should be avoided if possible.

16. E. L. Zebroski, and B. Sehgal, "Advanced Reactor Development Goals and Near-Term and Mid-Term Opportunities for Development" (Paper presented to the American Nuclear Society, Washington, D.C., November 18, 1976).

17. These estimates assume neither export nor import of uranium.

18. National Research Council, *Problems of U.S. Uranium Resources and Supplies to the Year 2010,* Committee on Nuclear and Alternative Energy Systems, Supply and Delivery Panel, Uranium Resource Group (Washington, D.C.: National Academy of Sciences, 1979), p. 17.

19. The Ford Foundation cites this experience, with qualifications appropriate to the particular situation of uranium discovery and production, in concluding that government figures "substantially underestimate the amounts of uranium that will be available at competitive costs." Nuclear Energy Policy Study Group, *op. cit.*, pp. 9, 71–94.

20. Supply and Delivery Panel, Uranium Resource Group, *Problems of U.S. Uranium Resources and Supplies to the Year 2010, op. cit.*, pp. 47–63.

21. Milling is the process by which uranium is extracted from its ores. It is only economical if done near the mine.

22. Assuming that each 1000-MWe (1-GWe) reactor requires 5750 tons of U_3O_8 over its 30-yr operating life—an average figure for today's two versions of the light water reactor operating on a once-through fuel cycle.

23. U.S. Department of Energy, "Proposals Requested for Centrifuge Facility," *Weekly Announcements* 1 (December 23, 1977): 4.

24. The Cascade Uprating Program and Cascade Improvement Program, in addition to plans for ensuring supply of full power to enrichment plants.

25. A. de la Garza, "An Overview of U.S. Enriching Resources," unpublished working paper for the Supply and Delivery Panel, August 27, 1976, p. 2.

26. U.S. Department of Energy, "Proposals Requested for Centrifuge Facility," *op. cit.*, p. 4.

27. *Nuclear Fuel,* October 31, 1977.

28. W. R. Voight, Jr., "Enrichment Supply" (Paper presented at the Fuel Cycle Conference, Atomic Industrial Forum, Phoenix, Ariz., 1976).

29. At some sacrifice in net conversion and with some plutonium as an unavoidable by-product, uranium enriched to 20 percent ^{235}U could be substituted for highly enriched uranium, if this is considered a useful safeguard. The effect on separative work requirements would be very small.

30. The reactors mentioned are all well described in the nuclear literature; only the briefest list of characteristics is given here. In addition to the references cited for this section, the interested reader is referred to the files of the journal *Nuclear Engineering International* for descriptive articles on most of the important reactor types and prototypes.

31. There is a strong program now investigating "improved" light water reactor fuel cycles. The principal incentive of this program is to decrease the volume of spent fuel to be stored, and the program emphasizes longer reactor lifetime of the fuel. This, in turn, is conventionally approached by raising the enrichment of the fuel, making higher conversion efficiency increasingly difficult to achieve. The reader is cautioned that long fuel life (high burnup) does not by itself indicate improved use of uranium and may be counterproductive in that regard.

32. Kasten *et al., op. cit.*

33. The Vulcain. See "High Burnup Irradiation Experience in Vulcain," *Nuclear Engineering International* 15 (1970): 93–99.

34. Supply and Delivery Panel, *U.S. Energy Supply Prospects to 2010, op. cit.*, chap. 5.

35. J. S. Foster and E. Critoph, "The Status of the Canadian Nuclear Power Program and Possible Future Strategies" (Paper presented at the Wingspread Conference on Advanced Nuclear Converters and Near Breeders, Racine, Wisc., May 1976).

36. Resources Planning Associates, *The Economics of Utilization of Thorium in Nuclear Reactors—Textual Annexes 1 and 2* (Oak Ridge, Tenn.: Oak Ridge National Laboratory (ORNL/TM-6332), n.d.), p. 158, Table 2.13.

37. *Ibid.*, p. 20.

38. Energy Research and Development Administration, *Benefit Analysis of Reprocessing Light-Water Reactor Fuel,* (Washington, D. C., Energy Research and Development Administration (ERDA-76/121), December 1976).

39. For example, two such reports are "Report to the American Physical Society by the Study Group on Nuclear Fuel Cycles and Waste Management," *op. cit.*, and Kasten *et al., op. cit.* The former favors the CANDU , and the latter, the HTGR. Both urge further research and evaluation of several advanced-converter concepts.

40. ^{238}Pu in high-burnup plutonium fuel makes the plutonium thermally and radioactively hot, creating considerable difficulty in fabricating weapons safely and reliably. ^{240}Pu in high-burnup plutonium fuel significantly decreases the reliability and yield of a weapon, unless it is very sophisticated.

41. The Fast Test Reactor (FTR) is also known as the Fast Flux Test Facility (FFTF), as the latter refers to the reactor and laboratory complex of which the FTR is the main component.

42. For the French Phenix reactor, reprocessing losses (for fuel irradiated to 130,000 MWd/metric ton) were about 1 percent; for the United Kingdom's prototype fast reactor (PFR), reprocessing losses (average burnup: 7.5 percent) were less than 0.1 percent; for the EBR-II in the United States, 1 percent or less. Phenix: George Vendryes, Commissariat a l'Energie Atomique (CEA), personal communication to W. Kenneth Davis, Chairman, Supply and Delivery Panel, Feb. 9, 1978; PFR: R. H. Allardice, C. Buck, and J. Williams, "Fast Reactor Fuel Reprocessing in the U.K." (Paper presented at the International Conference on Nuclear Power and Its Fuel Cycle, Salzburg, Austria, May 2–13, 1977); EBR-II: P. Murray, Westinghouse Electric Corporation, personal communication to W. Kenneth Davis, Chairman, Supply and Delivery Panel, February 7, 1978.

43. Among others, Peter Fortescue, "Sustaining an Adequately Safeguarded Nuclear Energy Supply" (Paper presented at the International Scientific Forum on an Acceptable Nuclear Energy Future for the World, Fort Lauderdale, Fla., November 7–11, 1977).

44. See, for example, Louis Harris and Associates, *A Survey of Public and Leadership Attitudes Toward Nuclear Power Development in the United States* (New York: Ebasco Services, Inc., 1975); Louis Harris and Associates, *A Second Survey of Public and Leadership Attitudes Toward Nuclear Power Development in the United States* (New York: Ebasco Services, Inc., 1976); and results of successive surveys by Edison Electric Institute in *The Electric Utility Industry Today, 1971–1976* (not for general circulation).

45. Luther J. Carter, "Nuclear Initiatives: Two Sides Disagree on Meaning of Defeat," *Science* 194 (1976): 811–812.

46. Louis Harris and Associates, *A Second Survey of Public and Leadership Attitudes, op. cit.*, p. 116.

47. See, for example, the report of the Consumption, Location, and Occupational Patterns Resource Group of this study; and Amory B. Lovins, *Soft Energy Paths: Toward a Durable Peace* (New York: Ballantine Books, 1977).

48. National Association for the Advancement of Colored People (NAACP), *Report of the NAACP National Energy Conference,* (Washington, D.C.: NAACP, December 21, 1977).

49. Harvey Brush, testimony before the Connecticut Public Utilities Control Authority, January 21, 1976.

50. Gordon R. Corey, testimony before the Environment, Energy, and Natural Resources Subcommittee, Committee on Government Operations, U.S. House of Representatives, "Nuclear Power Costs (Part I)," 95th Cong., 1st Sess., September 12–19, 1977.

51. Charles Komanoff, testimony before the Environment, Energy, and Natural Resources Subcommittee, Committee on Government Operations, U.S. House of Representatives, "Nuclear Power Costs (Part II)," 95th Cong., 1st Sess., September 20–22, 1977, p. 1186.

52. Harvey Brush, testimony before the Connecticut Public Utilities Control Authority, January 21, 1976.

53. Irvin Bupp, testimony before the Environment, Energy, and Natural Resources Subcommittee, Committee on Government Operations, U.S. House of Representatives, "Nuclear Power Costs (Part II)," 95th Cong., 1st Sess., September 20–22, 1977, p. 1401.

54. Komanoff, *op. cit.*, p. 1187.

55. Lewis J. Perl, testimony before the Environment, Energy, and Natural Resources Subcommittee, Committee on Government Operations, U.S. House of Representatives, "Nuclear Power Costs (Part I)," 95th Cong., 1st Sess., September 12–19, 1977, p. 661 *et seq.*

56. A. David Rossin, paper prepared for the Commonwealth Edison Company, submitted with testimony of Gordon R. Corey before the Environment, Energy, and Natural Resources Subcommittee, Committee on Government Operations, U.S. House of Representatives, "Nuclear Power Costs (Part I)," 95th Cong., 1st Sess., September 12–19, 1977, pp. 843–884.

57. *Ibid.*

58. Perl, *op. cit.*, p. 694.

59. Rossin, *op. cit.*, p. 881.

60. The Price-Anderson Act (Public Law 85-256, as amended) provides for insurance and partial indemnification of the civilian suppliers and users of nuclear power equipment.

61. Atomic Industrial Forum, *Engineering Evaluation of Nuclear Power Reactor Decommissioning Alternatives* (Washington, D.C.: Atomic Industrial Forum (NESP), 1976).

62. M. Levenson and C. P. L. Zaleski, "Economic Perspective of the LMFBR," unpublished monograph, 1976.

63. *The Future Development and Acceptance of Light-Water Reactors in the U.S.*, report prepared by the Energy Lab in collaboration with the Department of Nuclear Engineering, Massachusetts Institute of Technology (Cambridge, Mass.: MIT (MIT-EL-78-035), 1978).

64. Committee on Science and Public Policy, Committee on Literature Survey of Risks Associated with Nuclear Power, *op. cit.*, summary and synthesis chapter.

65. *Decay Heat Power in Light Water Reactors,* ANS-5.1 Proposed Standard, American Nuclear Society, June 1978.

66. U.S. Nuclear Regulatory Commission, *Reactor Safety Study, op. cit.*

67. WASH-1400 specifically excludes reactor sabotage as a cause of release of radioactivity. This omission has been criticized. We nevertheless concur that the exclusion of sabotage and its consequences is proper. Other energy sources, to which nuclear power must be compared, are not normally evaluated on the basis of sabotage risks.

68. The ANS-5.1 Proposed Standard (see note 65) is 20 percent below the previous standard.

69. *Fission Product Behavior in LWR's,* quarterly report (Oak Ridge, Tenn.: Oak Ridge National Laboratory (ORNL/NUREG/TM-186), 1978).

70. National Research Council, *Risks and Impacts of Alternative Energy Systems,* Committee on Nuclear and Alternative Energy Systems, Risk and Impact Panel (Washington, D.C.: National Academy of Sciences, in preparation), chap. 4.

71. Critiques, WASH-1400.

72. Risk Assessment Review Group, H. W. Lewis, Chairman, *The Risk Assessment Review Group Report to the U.S. Nuclear Regulatory Commission* (Washington, D. C.: U.S. Nuclear Regulatory Commission (NUREG/CR/0400), September 1978).

73. Risk and Impact Panel, *Risks and Impacts of Alternative Energy Systems, op. cit.*, chap. 4.

74. Nuclear Energy Policy Study Group, *op. cit.*

75. *Ibid.*, p. 232.

76. J. D. Griffith and F. X. Gavigan, "Reactors—Safe at Any Speed—1979 Update," *Proceedings of the International Meeting on Fast Reactor Safety Technology,* Seattle, Washington, August 19–23, 1979.

77. L. Cave *et al.*, "Designing for Safety in Fast Reactors in the Presence of Uncertainties," and D. Okrent, "Some Thoughts on Reactor Safety," both in *Proceedings of the International Meeting on Fast Reactor Safety and Related Physics,* U.S. Department of Commerce, October 5–8, 1975. Available from the National Technical Information Service, Springfield, Va. (Report no. CONF-761001).

78. Risk and Impact Panel, *Risks and Impacts of Alternative Energy Systems, op. cit.*, chap. 4.

79. Committee on Science and Public Policy, Committee on Literature Survey of Risks Associated with Nuclear Power, *op. cit.*, summary and synthesis chapter.

80. U.S. Nuclear Regulatory Commission, *NRC Plans for Research Directed Towards Improving Safety of Light-Water Nuclear Power Plants,* Report to Congress, April 1977.

81. F. R. Farmer, "Risk Quantification and Acceptability," *Nuclear Safety* 17 (1976): 418–421.

82. Also known as transuranic waste (TRU) if the alpha activity is due to neptunium, plutonium, or heavier elements.

83. National Research Council, *Radioactive Wastes at the Hanford Reservation: A Technical Review,* Committee on Radioactive Waste Management, Panel on Hanford Wastes (Washington, D.C.: National Academy of Sciences, 1977).

84. R. D. Penzhorn, *Alternativerfahren zur Kr-85-Endlagerung* (Karlsruhe, Federal Republic of Germany: The Reactor Research Institute (KFK-2482), 1977).

85. "Report to the American Physical Society by the Study Group on Nuclear Fuel Cycles and Waste Management," *op. cit.*, Table 5-B-3.

86. *Ibid.*, Table 5-D-7.

87. See, for example, studies conducted on veloxidation by Oak Ridge National Laboratory, Chemical Technology Division, for the Nuclear Regulatory Commission.

88. U.S. Department of Energy, "DOE Announces New Spent Nuclear Fuel Policy," *Weekly Announcements* 1, no. 2 (October 21, 1977).

89. G. de Marsily *et al.*, "Nuclear Waste Disposal: Can the Geologist Guarantee Isolation?" *Science* 197 (1977): 519.

90. "Report to the American Physical Society by the Study Group on Nuclear Fuel Cycles and Waste Management," *op. cit.*; and J. D. Breckhoeft *et al., Geologic Disposal of High-Level Radioactive Wastes,* U.S. Geological Survey Circular 779 (Washington, D.C.: U.S. Geological Survey, 1978).

91. Thomas B. Johansson, and Peter Steen, *Radioactive Waste from Nuclear Power Plants: Facing the Ringhals-3 Decision*, 1978.

92. Energy Research and Development Administration, *Alternatives for Long-Term Management of Defense High-Level Radioactive Waste, Savannah River Plant, Aiken, South Carolina,* (Washington, D.C.: Energy Research and Development Administration (ERDA77-42), 1977).

93. "Report to the American Physical Society by the Study Group on Nuclear Fuel Cycles and Waste Management," *op. cit.*

94. H. Lawroski *et al.*, "What Would Happen if High-Level Nuclear Wastes Were Stored Near the Surface of the Earth?" (Paper prepared for the Tucson Waste Symposium, March 1979).

95. "Report to the American Physical Society by the Study Group on Nuclear Fuel Cycles and Waste Management," *op. cit.*

96. Committee on Radioactive Waste Management, Panel on Hanford Wastes, *op. cit.*

97. U.S. Nuclear Regulatory Commission, *Safeguarding a Mixed-Oxide Industry: A Technical Report to Assist in Understanding Safeguarding* (Washington, D.C.: Nuclear Regulatory Commission (NUREG-0414), 1978), Table 3.5, pp. 3–17.

98. J. W. Roddy *et al., Correlation of Radioactive Waste-Treatment Cost and Environmental Impact of Waste Effluents in the Nuclear Fuel Cycle—Fabrication of High-Temperature Gas-Cooled Reactor Fuel Containing Uranium-233 and Thorium* (Oak Ridge, Tenn.: Oak Ridge National Laboratory (ORNL/NUREG/TM-5), 1977).

99. The use of denatured cycles in the light water breeder reactor and high-temperature gas-cooled reactor has the advantage that the uranium and thorium are physically separated. On recycle, most of the uranium can be coprocessed with the thorium to yield a desirably denatured product, and only a fraction of the core material requires "salting" with ^{235}U. Also, the higher the conversion ratio, the less feed of ^{235}U is needed.

100. M. Levenson, and E. Zebroski, "A Fast Breeder System Concept—A Diversion-Resistant Fuel Cycle" (Paper presented at the Fifth Energy Technology Conference, Washington, D.C., February 27, 1978).

101. U.S. Nuclear Regulatory Commission, *Safeguarding a Mixed-Oxide Industry, op. cit.*, chap. 5.

102. See, for example, Theodore B. Taylor and Mason Willrich, *Nuclear Theft: Risks and Safeguards* (Cambridge, Mass.: Ballinger Publishing Co., 1974).

103. Royal Commission on Environmental Pollution, Sir Brian Flowers, Chairman, *Nuclear Power and the Environment*, 1976.

104. R. W. Fox, Chairman, *Ranger Uranium Environmental Inquiry*, October 1976.

105. Nuclear Energy Policy Group, *op. cit.*

106. Richard K. Betts, "Paranoids, Pygmies, Pariahs, and Non-Proliferation," *Foreign Policy* (1977): 157–193.

107. Ted Greenwood, George W. Rathjens, and Jack Ruina, "Nuclear Power Technology and Nuclear Weapons Proliferation," unpublished monograph, July 1976.

6 Solar Energy

The comprehensive term "solar energy" embraces a wide variety of processes for converting the sun's energy into high- and low-temperature heat, electric power, and liquid and gaseous fuels, on scales that can vary from small household systems to large centralized power plants. Their common feature is that they use the sun's radiation as a source of energy—either directly (as in the increasingly familiar rooftop solar heating systems) or indirectly (as in wind power, ocean thermal energy conversion, wood burning, or those prospective technologies designed to turn plant matter, or "biomass," into liquid and gaseous fuels).

Obviously, these technologies are a disparate lot. Some are essentially fully developed technically and under certain circumstances are competitive with the replacement costs of other forms of energy. Some are very far from economic and technical practicality. With appropriate subsidies many of the former could become valuable conservation measures over the next two or three decades; others can be considered as important energy sources only for the twenty-first century and beyond.

For convenience in this report, the many different solar options are grouped, on the basis of end-use, into the following three categories.

• *Direct use of collected solar heat for heating and cooling.* This group of solar technologies is characterized by low to medium temperatures. Some of the designs and processes are fairly simple and lend themselves to such applications as domestic space heating, domestic hot water heating, and production of hot water or low-pressure steam for industrial and agricultural processes. "Active" solar systems transfer heat by working fluids circulated mechanically; "passive" systems make use of natural forces such as convection, conduction, and radiation. Efficient and economical solar cooling remains a difficult problem. This group of solar technologies is, in general, the most nearly economical today. Some of the methods can be considered fairly well developed and among the most probable candidates for widespread commercialization in the intermediate term.

• *Solar electric technologies.* Solar radiation can be used in many different ways to generate electricity. Photovoltaic solar cells are one promising technology, though high cost and the lack of an economic electric storage technology are barriers to widespread commercial use today. Wind-powered generators are another means of converting solar energy to electricity. In favored locations they are technically practical already; however, their ability to displace utility generating capacity is limited by the intermittent nature of wind. In another solar electric method the rays of the sun are focused on large boilers with large arrays of tracking mirrors (heliostats); such an installation might cover thousands of acres and have a generating capacity of 10–100 megawatts (electric) (MWe). This approach is referred to as solar thermal central station energy conversion. The Department of Energy is building a 10-MWe pilot plant for this concept. More speculative is the ocean thermal energy conversion (OTEC) concept, which involves using the temperature differential between the tropical ocean's surface and subsurface waters to run large heat engines, thereby generating electricity. This last alternative is only in the research stage. An attractive feature of OTEC is that it could be used for base-load generation without storage devices, since the ocean itself would act as the storage medium.

• *Solar production of fuels.* Solar radiation can be used in a number of ways to produce solid, liquid, and gaseous fuels. Solar heat can be used, though with low efficiency, to decompose water molecules and produce hydrogen for use as a fuel or as a chemical raw material in place of hydrocarbon fuels. Photochemical processes can in principle be much more efficient but are in the early stages of research. Solar energy can also be used in the form of biomass for direct burning or conversion to synthetic fuels. Given the coming decade's anticipated fluid fuel supply problems (chapters 1 and 3), this group of technologies could become valuable in the intermediate term.

In dealing with these options, government energy planners face two critical issues: (1) the extent to which solar technologies can save oil and gas in the intermediate term (1985–2010) and (2) how best to assure that the appropriate solar technologies will be available to help meet the nation's long-term energy needs.

The answer to the first question depends largely on how much the public is willing to pay for the fuel conservation benefits of solar technologies over the next few decades. At present, the relatively high capital costs of even the most economical solar technologies form a barrier to wider use. The required initial investments are very high in comparison with those of conventional systems, which spread their main costs—for fuel—over many years. However, solar energy has attractive advantages. First, of course, are the potential economic and political benefits of conserving fossil fuels and thus in some measure reducing the need for oil imports. Also of great importance are the environmental benefits; solar energy technologies can be among the most benign of all energy sources in this regard. Furthermore, we are convinced of the necessity for diversity in the nation's energy system; adding a solar component to the system would reduce the danger inherent in relying on one or two primary energy sources. Finally, most solar systems perform well in small-scale installations, so that the time lag between the decision to build and actual operation can be short in comparison with the lag involved in siting, building, and licensing very large coal-fired or nuclear power plants. This means that once the economic and technical availability of a particular solar technology is established, wide deployment can begin relatively quickly.

Today's market for energy does not reflect these advantages very well. Conventional energy sources, with their lower first costs, average (rather than replacement) fuel costs, and established industries, will be favored for some time to come unless the barrier to solar energy posed by high initial investments can be surmounted. Nonetheless, if national energy policy mandates it for social and environmental reasons, the more nearly economic solar applications can be rendered competitive with conventional alternatives by some form of subsidy. For example, California has adopted a 55 percent tax credit on solar heating systems intended to promote significant market acceptance at the current prices of competing fossil fuels.

Subsidies for energy technologies are, of course, not unknown. The various incentives that have been provided the oil and natural gas industry in the form of depletion allowances, intangible drilling benefits, and the like are examples. The nuclear industry has received support in the form of federal research and development support, limits on utility liability (under the Price-Anderson Act*), and federal uranium enrichment facilities. Other energy and nonenergy minerals are accorded similar subsidies. However, to ensure the market penetration of most solar technologies at present costs, subsidies for solar energy would have to be considerably greater than those that have been accorded other energy forms. The

*See statement 6-1, by B. I. Spinrad, Appendix A.

subsidies required for at least some of the solar technologies would, however, be expected to decrease greatly or become unnecessary in the future as a result of advances from research and development and commercial experience.

SOLAR ENERGY SCENARIOS

To illustrate the potential contributions that solar energy could make in the intermediate term, this study developed two special scenarios of future use of solar energy. The first scenario embodies estimates of solar energy use under the assumption that the prices of competing energy sources remain near present levels and that solar energy is given no special incentives. The second represents the potential solar contribution under a policy of vigorous government intervention in the market for energy. More detailed information on how these scenarios were developed is given in the report of the Supply and Delivery Panel's Solar Resource Group.[1]

LOW-SOLAR-ENERGY SCENARIO

This scenario is based on the assumption that no policy other than the current federal tax credits is implemented to assist the entry of solar energy into the energy market, that the costs of other energy sources increase only slowly, and that the costs of solar energy technologies remain high (i.e., follow the estimates of the Solar Resource Group without breakthroughs in the costs of advanced solar technologies). This yields a very small market penetration of solar energy by 2010 (Table 6-1). Solar heating systems do not achieve significant market penetration until after 2000, at which time it is assumed that the prices of electricity and natural gas have risen greatly. No solar electric technology becomes economically competitive before 2010, and municipal waste is converted to fuels only in major urban regions.

HIGH-SOLAR-ENERGY SCENARIO

This scenario is based on the assumption that by 1985 a national policy decision mandates vigorous incentives to bring about the use of solar energy. This scenario is driven not by economic forces, but by government intervention in the energy market. It is assumed that this government policy mandates, independent of cost, adoption after 1990 of solar energy for heating all new buildings and for all industrial process heat where feasible; sets in force a mandatory schedule for deploying several solar electric technologies; and requires rapid adoption of technologies for

TABLE 6-1 Solar Contributions in Low-Solar-Energy Scenario (primary energy displaced, quads)

Solar Application	1985	1990	2000	2010
Heating and cooling	0	0	0	0.3
Solar electricity	0	0	0	0
Fuels	0	0	0.1	0.3
TOTAL	0	0	0.1	0.6

converting municipal and agricultural wastes to fuels. This scenario leads to a solar energy contribution about equal to President Carter's announced goal of 20 percent solar energy in the year 2000. The 20 percent figure includes present use of hydro and biomass, estimated to be about 5 quadrillion Btu (quads), to which about 13 quads of solar energy would be added to form a total of 18 quads (or nearly 20 percent of the Administration's assumed total).*

Heating and Cooling by Solar Energy

It is assumed in the high-solar scenario that all new construction after 1990 uses solar space heating and water heating and that half the solar heat collected for this purpose is used also for solar total energy systems, which cogenerate[2] heat and electricity. An accelerated schedule for use of industrial process heat is assumed,[3] and it is again assumed that half this solar heat is used in total energy systems. With these assumptions, the schedule of energy contributions from direct use of solar energy for heating and cooling is as shown in Table 6-2 (not including the electricity produced in total energy systems).

Solar Electricity

A number of solar electric technologies are under active research and development, and it is impossible to predict which would be technically and economically most suitable under the policies assumed in this solar-intensive scenario. The high-solar scenario is based on mandated deployment of three technologies—solar thermal central station energy conversion, solar thermal total energy systems, and dispersed wind energy systems—without implying that these would necessarily be the specific

*See statement 6-2, by H. Brooks, Appendix A.

TABLE 6-2 Heating and Cooling in High-Solar-Energy Scenario (primary energy displaced, quads)

Application	1985	1990	2000	2010
Domestic water heating	0.2	0.4	1.2	1.3
Passive space heating	0.1	0.2	0.3	0.4
Active space heating	0.02	0.06	0.64	1.21
Nonresidential air conditioning	—	0.06	0.36	1.50
Industrial process heat	0.2	0.4	1.6	6.6
TOTAL	0.52	1.12	4.1	11.01

technologies used. Photovoltaic or ocean thermal conversion, for example, might serve instead.

For the purpose of defining this scenario it is assumed that public policy requires the use of specified amounts of solar electric capacity for new generating capacity, regardless of cost. The estimates for central station solar thermal conversion are based on an assumption that in 1990 three full-scale production facilities for heliostats (sun-tracking concentrating mirrors) are placed in operation and that the production of heliostats is doubled every 5 years thereafter. This is a 15 percent annual growth rate. The estimates for solar thermal total energy systems are based on the estimates given above for solar heat delivered for space heating, air conditioning, and industrial process heat. It is assumed that 50 percent of the solar heat delivered for space heating and industrial processes (Table 6-2) is used for total energy conversion, and 70 percent of the solar heat delivered for nonresidential air conditioning is so used. The heat rate of the total energy systems is assumed to be 17,000 Btu per kilowatt-hour (kWh). Table 6-3 gives estimates of solar electric contributions under these conditions. The estimates in the table correspond to an installed capacity in the year 2010 of 250 GWe of central station solar thermal plants (with storage for a load factor[4] of 0.4), 74 GWe of total energy generation (load factor of 0.3), and 50 GWe of wind turbines (load factor of 0.4). (Total U.S. generating capacity in 1976 was 494 GWe.)[5]

Use of Biomass for Fuels

We assume in the high-solar scenario (Table 6-4) that a federal decision is made to mandate recovery of the energy content of municipal wastes and agricultural residues. By the year 2000, 95 percent of municipal waste is processed for its energy, and 35 percent of the energy content of

TABLE 6-3 Solar Electricity in High-Solar-Energy Scenario (primary energy displaced, quads)

Solar Electric Technology	1985	1990	2000	2010
Central station	—	—	1.7	8.7
Total energy	—	0.07	0.5	1.9
Wind	0.1	0.5	1.4	1.8
TOTAL	0.1	0.57	3.6	12.4

agricultural residues is recovered as methane, yielding a total of 5.4 quads annually in addition to the 2.7 quads now used annually by the forest products industry.[6] We do not include here any contributions of energy farms. Such farms could possibly provide another 3.4 quads in the year 2010, albeit at substantial ecological cost.

Combining all these estimates yields a total of about 29 quads of primary energy displaced in 2010 by solar energy in this high-solar scenario (Table 6-5).

The high-solar scenario was not intended to represent a recommended national strategy for solar energy development, but rather to indicate the upper bound of what is technically feasible. The total cost of implementing such a scenario, at today's costs for solar technologies, might be around 3 trillion dollars, perhaps 2–3 times the cost of obtaining equivalent energy from conventional nonrenewable sources. The increment, if provided as a subsidy by government, would be much larger than the total amount provided to date by government to stimulate energy production by conventional means (including nuclear), which was estimated recently at about one tenth of a trillion dollars.[7]

TABLE 6-4 Use of Biomass for Fuels in High-Solar-Energy Scenario (primary energy displaced, quads)

Biomass Source	1985	1990	2000	2010
Municipal wastes	0.5	0.8	1.9	1.9
Agricultural residues	0.5	0.9	3.5	3.5
Energy farms	—	—	—	—
TOTAL	1.0	1.7	5.4	5.4

TABLE 6-5 Total Solar Contributions in High-Solar-Energy Scenario (primary energy displaced, quads)

Solar Application	1985	1990	2000	2010
Heating and cooling	0.5	1.1	4.1	11.0
Solar electricity	0.1	0.6	3.6	12.4
Fuels	1.0	1.6	5.4	5.4
TOTAL	1.6	3.3	13.1	28.8

THE ROLE OF RESEARCH AND DEVELOPMENT

Preparing solar energy to take its place in the long-term energy system depends even more on government action than does planning for nearer-term contributions. At issue is the total amount of research and development effort and its allocation to various solar technologies now in their infancies. It would be unfortunate if national decisions about long-term energy options were to settle on a single option, such as the breeder reactor, merely because it is relatively more developed today. The government's role should be to bring the state of knowledge of all long-term technologies to the point at which some appropriate combination of energy sources can be chosen on the basis of realistic comparisons. Decisions that restrict the variety of our long-term energy options should be deferred as long as possible.

Government solar energy research and development policy therefore should provide for exploring many long-term solar options, without concentrating too heavily on large-scale demonstrations of one or a few, far in advance of their expected deployment. Indeed, premature demonstrations could well prove counterproductive, since their results might well be taken as evidence that the technologies could never become economic. The government's present concentration of funds on demonstrations of solar thermal central station power plants may turn out to be an example of such premature demonstration programs.

DIRECT USE OF SOLAR HEAT

Heating buildings and domestic water and providing industrial and agricultural process heat and low-pressure steam are by far the simplest and most economical applications of solar energy. (Solar space cooling, while it can be done, is not now efficient, reliable, or economical.) For these reasons, this group of technologies is the most suitable for deployment in the intermediate term as part of the nation's energy conservation program.

However, these applications are in general more costly than conventional alternatives, Btu for Btu, and even more costly in terms of the initial investments in complete heating or cooling systems. Solar applications are likely to develop a large market before the end of this century only if the prices of alternative fuels rise quite sharply (either from market forces or from taxation) or if solar installations are subsidized heavily.*

The cost of heat from solar systems cannot be computed with certainty, and it is hard to generalize about the economic standing of solar systems as compared with alternatives. Their economic success depends on a number of variable and uncertain considerations, including the following.

- The initial cost of the system, including storage.
- The amount of sunshine available at the site.
- The amount of the collected solar heat that is actually used. (Solar hot water or process heat systems, for example, operate all year long and thus yield fairly constant returns on the initial investments. Solar space heating systems normally do not, and this lowers their value.)
- Local prices of conventional energy sources.
- The total cost of providing backup energy.
- Interest rates, taxes, and subsidies.
- The performance, reliability, and maintenance and repair costs of the particular system.

Obviously, these factors vary from place to place and time to time, as well as from system to system. Most, moreover, are difficult to project, especially over the expected system lifetime of about 20 years. Judgments about the economics of installing a solar system rather than a conventional source of heat, therefore, depend on assumptions about these uncertain quantities. It can, however, be stated with confidence that most forms of solar heating are not economic now except in special circumstances, and that either fairly rapid changes in the prices of alternatives or a government incentive program larger than the one now in place will be necessary if solar heating is to attain much importance before the end of this century.

A consumer is more likely to decide in favor of purchasing a solar heating system if he considers the life cycle cost of the system, including estimated fuel costs, and is not deterred by the high initial cost. The decision is essentially a trade-off of present capital investments against future costs of fuel. In fact, consumer decisions are not often made in this

*See statement 6-3, by H. Brooks, Appendix A.

way. Because people tend to value the present more than the future, most near-term decisions would favor nonsolar approaches because of their lower first costs, even if the relative life cycle cost per million Btu of solar energy and the alternatives were equal. Consumers may also be deterred by expectations of technological improvements (hence lower costs) in solar systems.

However, if public policy so dictates, because of the social and environmental benefits from solar heating, it is possible to bring decisions on new heating systems more into line with minimum life cycle cost and thus to render consumers more likely to select solar systems on a Btu-for-Btu comparison with conventional alternatives. This can be done by means of additional solar tax credits, low-interest or interest-free loans, thermal performance standards for buildings, or additional taxes on nonrenewable fuels. Such measures would help solar heating gain a significant market share earlier than it would otherwise. (There is now in place a federal tax credit of up to $2200, as a variable percentage of the system's cost, for installing solar, wind, or geothermal equipment in the home.)

WATER HEATING

Domestic water heating is easily and efficiently performed by present solar technology. The water temperature desired is about 140°F, which is well within the capacity of simple flat plate collectors. Systems being marketed today are superior in performance to the thermosyphon systems once widely used in Florida.

The cost of a solar water heating system for an average single-family residence has stood at about $1500 or more in recent years, increasing with inflation. This compares to about $200–$300 for a conventional system. The life cycle cost of delivered heat from such a solar water heater might be as low as $7.50 per million Btu, although it is more likely to be at least $10 per million Btu. The latter cost is comparable to that of heat from electricity at its present national average residential cost (but about half the replacement cost of electricity). It is about twice the average residential cost of heat from natural gas or fuel oil. The cost advantage of natural gas and fuel oil would be increased if the efficiency of fuel-fired water heaters, now about 50 percent, is improved in the future. Such improvements in efficiency might offset fuel price increases for a decade or more and maintain the present relative cost relationship of solar and fossil fuel water heaters. Solar water heating is a mature technology, and no great reductions in the cost of units are anticipated, though experience with solar heaters should lead to more efficient use of labor for installation, slightly reducing the first cost of installed units.

SPACE HEATING AND COOLING

Solar space heating and cooling technologies fall into two groups. In so-called active systems, water, air, or some other working fluid is forced through heat exchangers heated by the sun; the heated fluid is then moved by pumps or fans to where heat is required or to a heat storage volume, where the heat can be retained until needed. Passive systems involve careful design of buildings to take advantage of solar radiation, natural convection, shading, and the like to heat or cool without the need for special heat-transfer systems. Active systems are more versatile and allow better temperature control. They are also much more expensive in general.

Active Solar Space Heating

Active solar energy systems provide space heating for several hundred buildings in the United States, and the number will soon be in the thousands. Such systems collect solar heat in water, antifreeze, or air and move it through pipes or ducts to heat the building space directly or to serve as a source of low-temperature heat for further boosting by heat pumps. Some of these systems have been wholly or partly supported by governments or other organizations as demonstrations, but others have been paid for entirely by the building owners. Experience with these systems has shown the need for further performance and reliability.[8]

Is the existence of these solar-heated buildings proof of the adequacy of present technology for solar space heating? To answer this question, it is necessary first to define the objective of such systems. Such active systems are intended to serve as thermostat-controlled heat sources to replace some of the energy used by conventional heating units. Most of the buildings in which they are installed are in most respects of normal design and construction. In such buildings active solar space heating systems are in direct competition with conventional systems.

To make economic comparisons of solar space heating systems with conventional systems, it is obviously necessary to know the costs of installed solar systems. There is considerable uncertainty about these costs even at present. Following are assessments from several sources.

- Energy Research and Development Administration/Mitre: A current installed system cost of $20/ft^2 of collector is assumed.[9]
- Department of Energy/CS: Current installed system costs range from $25 to $40/ft^2 of collector.[10]
- Solar Energy Research Institute: Surveyed costs of actual installed systems are $39–$43/ft^2 of collector.[11,12]
- Lovins: Intelligently designed and installed systems in today's market are $10–$15/ft^2 of collector.[13]

For the following calculations, we use a value of $20/ft^2 of collector. The collectors themselves account for about $8 of this cost. The chances of large cost reductions in commercial, contractor-installed systems have been widely discussed in the past few years, and it appears unlikely that the cost of collectors can be brought down much below $8/ft^2, though future collectors at this price may be more efficient than those available now. The costs of components other than collectors (e.g., pipes and valves) have been rising faster than the rate of general inflation. On-site labor hours can be reduced by the use of prefabricated units, but it can probably be anticipated that the hourly cost of construction labor will rise at or above the rate of general inflation. Thus, the installed cost of solar space heating systems will probably remain close to what it is today, though homeowners willing to invest their own labor can lower these costs greatly. Similarly, very large systems, built to provide heat to large buildings or entire neighborhoods, could take advantage of some economies of scale.

Given an assumed system cost of $20/ft^2, the cost of energy from such a system can be calculated. For example, each square foot of collector, of average efficiency and at an average location, will provide annually about 0.12 million Btu of useful heat (i.e., delivered when heat is actually required; this quantity can vary ± 50 percent with climate and design factors). If the interest rate on a 20-yr mortgage is 11 percent, then that square foot costs $2.50 annually in mortgage payments, and the cost of delivered energy is $2.50/0.12, or about $20 per million Btu. This is about 4 times the current national average price of heat from natural gas with a well-designed (70 percent efficient) furnace and about twice the present price of electric resistance heat (or about equal to the replacement cost). Note that the range of system costs given earlier would result in costs of delivered energy from half to twice this value.

The true cost to the consumer, of course, will also vary because of factors other than system costs. For example, the interest on the mortgage would represent an income tax deduction. Further, the owner of a solar heating system now receives a federal income tax credit (and in many states, a state income tax credit or rebate). Property taxes might be higher because of the added value represented by the system, though some states have passed laws exempting solar systems from property taxes. In addition, the yearly output of useful heat per square foot will vary from place to place; systems installed in very sunny, relatively cool climates where heating is necessary for longer periods each year would cost less per million Btu than those in less appropriate climates.

These uncertainties are compounded by the need to weigh the costs of the solar system against the expectation that costs of more conventional alternatives will rise more rapidly than general inflation over the assumed 20-yr life of the system. The reliability of the system and the cost of maintenance and repairs (and the prospect of technological advances that improve performance) must also be estimated. Expectations about its resale value also play an important part.

Each potential buyer of a solar system will, for individual reasons, make certain assumptions about these cost considerations. Some may choose to compare the cost of solar heat with that of using electricity rather than natural gas, even though gas is now cheaper, because they expect gas to be unavailable or extremely expensive in 20 years. Some may assume that the price of alternatives will rise fast enough over the 20 years to make solar heating economic. Some may install solar systems for noneconomic or personal reasons, just as consumers often prefer higher-cost electricity over cheaper fuels because of its perceived advantages in convenience or cleanliness. We can state no universal conclusion as to the relative attractiveness to consumers of solar space heating and conventional sources. However, because of the expected long-term economic, social, and environmental benefits, it is appropriate for the government to grant tax advantages or other subsidies that alter market economics, thereby accelerating the introduction of this technology.

Solar Space Cooling

No technically and economically adequate solar space cooling technology is now available for small residential applications. A number of approaches that are being explored are described in the report of the Solar Resource Group.[14] None of these approaches yet meets the requirements of simplicity, low cost, efficiency, and effectiveness of waste heat rejection that will likely be necessary for consumer acceptance.

A fundamental problem is that it is very difficult to operate an air conditioner from a flat-plate solar collector when the waste heat must be rejected by a dry cooling unit (the type used by most residential air conditioners). The efficiency of such a unit is very low, and at other than design conditions, the unit may fail to perform at all. To solve this problem at present, one must either (1) operate the collector at higher temperatures (which requires more sophisticated collectors and some form of energy storage other than hot water) or (2) use an evaporative, or "wet," cooling tower (which consumes significant amounts of water and requires frequent maintenance). Neither solution is satisfactory, and a new approach to solar cooling for individual residential buildings may be required.

Solar cooling of larger buildings or clusters of residential buildings is less difficult. This is because economies of scale allow use of fairly complex units and because such facilities already have maintenance arrangements for their heating and ventilation systems. A number of systems have been demonstrated successfully, and a commercial 25-ton solar cooling unit is now on the market. However, such systems (including collectors) cost too much at present to compete with gas-fired absorption air conditioners in regions where natural gas is available (gas-fired, rather than electrical, units dominate this market).

The use of solar energy to supplement conventional fuels in nonresidential cooling will be made generally feasible if low-cost solar collectors that provide low-pressure steam become available, since large cooling units are now generally run on low-pressure steam (usually obtained from combustion of natural gas). Such collectors are now under development for industrial process heat applications. If this development effort is successful in meeting its cost goals (see below), it will greatly improve the feasibility of solar space cooling in large buildings.

Impact on Utilities

If electricity is used to provide backup energy for a solar heating and cooling system, there is potential for a significant and adverse impact on the utility. It would be inefficient and uneconomic to use electrical generating capacity only to provide occasional supplemental energy, whether to solar energy systems or to other sources of intermittent demand such as all-electric homes. The Electric Power Research Institute investigated this question from the perspective of minimizing the total cost of the system including the building and the utility and concluded that solar heating and cooling systems are not inherently less efficient in their use of electrical generating capacity than conventional all-electric homes.[15]

INDUSTRIAL PROCESS HEAT

Another potentially important direct use of solar energy is to provide industrial process heat for use in small- and medium-size applications, as in laundries, food processing operations, and crop drying. At present about half of industrial process heat in the United States is supplied by natural gas, which is increasing in price and becoming less readily available. The temperature requirements of industrial processes range from warm water to very high-temperature (more than 1000°F) gases, but the heat is used mostly in the form of low-pressure steam.

Lower-temperature (less than 200°F) processes can be served by flat-plate collectors, so here the performance of present solar technology for fuel saving is adequate. However, for most industrial hot water applications the cost of present solar systems is not competitive with that of fossil fuels at current prices. If natural gas and fuel oil become much more expensive or government regulations limit their availability to industries, and at the same time environmental restrictions preclude on-site combustion of coal, then solar hot water systems may be selected over the principal alternative, electricity, on the basis of lower life cycle cost.

No adequate technology is now available for producing low-pressure steam with solar collectors. Significant contributions by solar energy to industrial processes will require development of collectors for producing steam at about $5 per million Btu. Steam-producing collectors are under development and should soon be available, but it is not known if or when this cost goal will be met.

There is also no available solar technology for producing very high temperatures (greater than 600°F) for industrial processes. Such technologies will be difficult to develop, because of requirements for high concentration of sunlight and for high temperature. They may therefore be more costly than systems using the relatively low temperatures provided by stationary collectors.

ACTIVE AND PASSIVE SOLAR ENERGY SYSTEMS AND ENERGY-CONSERVING BUILDING DESIGN

In the past few years most of the efforts to conserve energy in buildings have focused on alterations that can be made in existing buildings to reduce their conduction or infiltration heat losses in winter. However, over the time period considered by this study, much of this stock of existing buildings will be replaced. Thus, a more important question is how new buildings can be designed to use less energy.

The combination of energy-conserving design and passive solar design can reduce energy use per square foot of building area to half its present value, and in favorable circumstances perhaps to a third or less. This would reduce the energy requirements for building space conditioning from a major part of the nation's energy demand (22 percent today) to a minor one.*

If buildings are to be properly designed, all factors determining their external environment must be considered. The external environment is usually thought of simply as causing part of the energy demand within the building, but it may also be used to displace some of the demand. Of particular interest are the possibilities of using the sunshine falling on the surfaces of the building to provide some of the functions of space conditioning, by so-called passive solar design.

In a passive solar energy system the transfer of solar heat takes place by natural convection, radiation, or conduction without the use of special mechanical pumps or blowers. Active solar systems generally provide

*See statement 6-4, by B. I. Spinrad, Appendix A.

better control of the use of solar energy in a building than do passive systems, but at a substantial increase in complexity and first cost. In some cases, the use of passive systems may lead to greater excursions in temperature and humidity than the more expensive active systems.

Active and passive solar energy systems are not mutually exclusive; that is, a building can be designed with passive solar concepts to minimize its energy demands for space conditioning and can use an active solar system to satisfy part of the remaining demand. Further, some systems that use air for collection and transfer of solar heat cannot be clearly categorized as either active or passive but represent a hybrid.

Passive solar energy systems can do more than simply provide heat. Those aspects of a building that constitute a solar passive heating system will also alter the other energy requirements of the building. For example, a carefully conceived passive system may decrease the cooling load in summer by using thermal convection exhausts to help draw breezes, or air from a cool storage volume, through the building in summer. The passive system may also improve internal lighting of the building by "daylighting" and thus reduce the air conditioning load normally due to waste heat from lamps. In designing a passive solar system, it is necessary to consider simultaneously all the energy requirements of the building, as well as the ultimate objective of creating pleasing and functional space in the building. This is not a trivial task, and the processes involved in the operation of a passive system are subtle. While a passive system is physically less complex than an active one and has fewer or no moving parts, it requires a more sophisticated match of building design to the external environment and the needs of the inhabitants. Active systems have heavy first costs in components and installation; the first costs of passive systems are in careful thought and delicate balance of design.

A number of design concepts related to solar active and passive design do not involve direct use of the sun. For example, cooling can be done by radiating heat to the night sky or blowing cool night air through crushed rock to provide storage for daytime cooling. Whether or not these concepts should be labeled as solar energy is irrelevant, since they are part of the repertoire available to the designer of a solar-heated building, and they work together with the solar part of the design in determining the overall performance of the building.

How effectively can a passive solar system provide the space conditioning of a building? Experience is still limited to only a few building types and climates, but the results are very promising. One example of the use of passive design in a large nonresidential building is a California state government office building in Sacramento. Some residential passive solar buildings constructed in moderate climate areas, such as the Skytherm House in Atascadero, California, require no other source of energy for space conditioning. Others use only small stoves (often wood burning) as backup heat sources. In climates with harsh winters, more backup is needed. In climates with hot, humid summers, passive design will not always provide sufficient comfort, and air conditioning will be required. However, on a national scale, the widespread use of solar passive design concepts can drastically reduce the amount of energy required for space conditioning.[16–18]

How expensive is the energy provided by passive solar systems? Although a passive system provides energy in a way that cannot be directly compared with the workings of a normal heating and cooling system, the energy cost of a passive system can in principle be estimated by comparing the total initial cost of the building and its total energy requirements for all purposes with those of an otherwise similar building without passive solar design. Convincing analyses of this type have yet to be done. The type, layout, and quality of the internal space in a passive building depend on the special features of its passive design. For example, if the passive design includes a greenhouse placed on the south side of the building to serve both as a solar collector and for aesthetic purposes, what fraction of the cost of the greenhouse should be included in cost comparisons?

What needs to be done to bring about the use of solar passive designs? A number of things—some technical, others institutional—are required. Much knowledge is required concerning how various elements of building structures and heating, ventilation, and air conditioning systems interact with the building's environment. This will require data and methods of analysis, design, and performance prediction that are not yet available. Research on these topics is essential. In addition, some new passive system components will need to be developed, though most passive concepts require only proper use of existing components.

Building standards that properly regulate the use of passive concepts in design need to be adopted. Some proposed changes in existing standards, intended to achieve energy conservation, might actually forbid or restrict passive designs, for example, by imposing arbitrary limits on amounts of window space. All changes in building standards intended to aid energy conservation should be properly related to the ultimate performance of the building, so as not to exclude passive solar design innovations that prove effective in certain situations. Training and education of architects, mechanical engineers, and contractors will be necessary to bring about well-informed use of passive concepts.

Finally—and this problem is shared with active solar systems and other methods of energy conservation in buildings—the obstacle posed by the present emphasis on first cost of a building must be resolved. Financial or regulatory arrangements to bring purchase decisions into line with life cycle costs are necessary to speed implementation of energy conservation in this and other ways, with the social and environmental benefits discussed elsewhere in this report.

Extensive future application of passive solar techniques and energy-conserving building design may have adverse implications for the economics of active solar heating and cooling. Successful use of passive solar and conservation techniques reduces energy requirements, which substantially limits the further savings obtainable by active solar systems. Such small savings could make the high capital costs of active systems unattractive. This might be true even though a smaller collector surface area would be required, since the cost of the noncollector parts of active solar heating systems do not decrease in proportion to collection area as size is reduced. This intimate technical and economic relationship among active and passive solar systems and energy-conserving building design makes it important that government-sponsored research and development and demonstration programs in these areas be integrated.

RETROFITTING FOR SOLAR HEATING

An important determinant of the rate at which solar energy can supplement the nation's energy supplies is the extent to which retrofitting of existing buildings can contribute to the use of solar heating and cooling.

That existing buildings tend to be so energy inefficient is a major barrier to retrofitting them with solar energy. It would be economic nonsense to install a solar system to provide heat that leaks rapidly out of the building, just as it is to use increasingly expensive fossil fuels for the same purpose. Thus the first step in retrofitting should be proper insulation and leak tightening. Furthermore, when a building is thermally inefficient, a much larger area of solar collectors is required to heat it. This much free area is often hard to obtain on the roofs of existing buildings, generally cluttered with vents, pipes, and other obstructions. Thus a retrofit solar system will often provide only a small fraction of the total heating load.

However, the greatest barrier to retrofitting buildings with active solar energy systems is the high cost of installation in existing structures. It is expensive to add collector fittings and supports on roofs, to run pipes or ducts through walls, and to fit heat storage tanks into buildings. Further, existing roofs are not always properly exposed to sunlight. For these reasons, active system retrofits are sometimes placed externally to buildings, with the collectors on separate, specially constructed structures and the storage tanks buried in yards rather than in basements. The few demonstrations of retrofit of active space heating systems have cost in the range of $40–$84 in total system cost per square foot of collector area, which corresponds to paying roughly $40 per million Btu of space heat.[19]

Another barrier to solar heating retrofit is the imposing level of understanding of solar design principles, architecture, and construction practice necessary to custom-design a retrofit system for an existing structure. Buildings vary so greatly that there are literally millions of special cases. To make a retrofit at minimum cost, all aspects of the existing structure must be used to advantage. Thus, designing a good retrofit is more difficult than designing a new house with a common type of active solar system. A further barrier to solar retrofit is encountered in financing the construction. When a solar system is included in the initial construction of a building, the cost can be included in the mortgage. However, for a solar retrofit a special loan must be arranged, generally with a higher interest rate than that typical for a mortgage on a new building.

It is worthwhile as part of the nation's energy conservation effort to seek techniques for solar heating retrofit of buildings at reasonable cost, and ways to lower the barriers inhibiting their use such as by encouraging and financing such retrofits. Inability to establish a market for solar retrofits will greatly retard the potential market penetration of solar heating. Attaining significant contributions from solar heating and cooling by 1990, for example, would require retrofitting a substantial fraction of existing residences, because only about 25 percent replacement is expected from new construction during that period.

It is much easier to retrofit for domestic water heating, since the cost is almost as low as installation in a new building. The solar heater can be tied into the existing water heating system and needs little additional plumbing and no heat storage. Most roofs have room for the small collectors. Since in many regions as much energy is used for water heating as for domestic space heating, such retrofits could have significant impacts on energy consumption. A good solar water heating system can provide 75 percent of the energy used for water heating in a household, or on the order of 20 million Btu/yr in favorable circumstances.[20]

Some promising approaches to solar retrofit for space heating have begun to appear. These are generally passive, rather than active, design approaches. At present the most successful is the add-on greenhouse. These are light, frame structures with double glazing, which are added to the south sides of existing buildings. To temper the swings of temperature between day and night, thermal mass is added to the greenhouse in the form of crushed rock, water-filled drums, or the like. Warm air is exchanged with the building by natural convection or by fans and ducts. The result is a passive solar heating system that may not be as effective as it would be if the entire building had been designed initially for this, but it does provide significant energy savings. This approach is being rapidly adopted in northern New Mexico, where the clear, cold winters allow this system to work very well. Such units should also be able to contribute some useful space heating in other regions.

The widespread use of solar water heater retrofits and passive add-ons for space heating could make significant energy contributions by 1990. However, solar and conservation measures must be designed to complement each other; such retrofits should always be done in conjunction with other measures for energy conservation in the building. With the use of economic retrofits for conservation and solar heating (20 percent or better payback on investment),[21] the average energy requirement per house can be cut by half or more; were 50 percent of U.S. residences so retrofitted, a savings of 3 quads/yr of oil and gas would be realized.* There are significant challenges to governments in designing building codes, incentives, and education programs to aid the adoption of such approaches to solar retrofit.

SOLAR ELECTRICITY GENERATION

Four concepts for generating electric power from solar radiation are under active development. Today the one that is receiving the most attention in government research programs is solar thermal conversion, which involves concentrating sunshine to achieve high-temperature heat. Next are photovoltaic generation with solar cells; wind power; and ocean thermal energy conversion, which uses floating power stations that exploit the temperature difference between the ocean's surface and subsurface waters to run heat engines.

Each of these concepts has important potential for the long term as an inexhaustible source of power. Some wind power applications are technically well developed and fairly economical and could play significant roles in the intermediate term as fuel savers. However, solar-generated electricity, today costing several times average electricity costs, is unlikely to penetrate the market on a substantial scale much before 2010 unless the prices of competing fuels continue to rise rapidly or obstacles are met in installing more conventional power plants at the rate desired. In general, these solar technologies are so much less economic than other sources of power that their entry into the market would need to be massively supported by government, a policy that could be justified only if the environmental advantages over other forms of electricity generation or the need to conserve fossil fuels becomes of overriding national importance.

*Statement 6-5, by B. I. Spinrad: My statement 6-4, Appendix A, also applies here.

electric technologies with utility grids have been examined by a number of studies.[22] However, general conclusions have not yet been attained because of the complexity of the issues. The actual situation depends on the region and type of resource (e.g., wind resources vary greatly between the Great Plains and New York),[23] on the load profile of the utility, and on the generating fuel mix. In favorable circumstances, solar electric generation will have significant value as a fuel saver, even without capacity credit.[24]

Solar electric research and development has been neglected until recently. Although the federal research and development program has been growing rapidly, actual research and development projects, apart from analyses and design studies, have been under way generally only a few years, and it is too early to expect significant results. The next 5 years should provide the beginning of a strong technology base for solar electric conversion and should allow a much more realistic estimate of the extent to which solar technologies may be competitive with other sources of electricity. Present technology, lacking the benefits of extensive previous efforts in research and development, nonetheless provides some basis for estimating future capabilities and costs.

SOLAR THERMAL CONVERSION

Solar thermal electric conversion dates back to the early years of the twentieth century; several technologically successful plants have been operated over the past 70 years. The solar thermal conversion approach is typified by the central receiver concept now being developed by the U.S. Department of Energy and the Electric Power Research Institute, in which a large field of mirrors, built to follow the sun, concentrates heat onto a boiler atop a tower. Such a combination of mirror field and tower would be designed to provide from 10 to 100 MWe of peak electrical output and would occupy up to 1 square mile. This system could probably be designed to provide intermediate-load power to a utility grid. Because of the intermittent nature of sunshine and the lack of a practical energy storage technique, however, it would be unable to displace much generating capacity.

Solar thermal conversion systems for smaller-scale on-site generation would be roughly similar, but smaller. Such systems would be sized to match the requirements of local loads and might range from 1 to 10 MWe in electrical capacity. Note that 1 MWe requires about 6 acres; this amount of land may be difficult to find for local loads.

The present state of central receiver solar thermal generation is indicated by the design for a federally funded 10-MWe pilot plant at

Barstow, California, and by design studies for solar thermal repowering of existing oil- and gas-fired plants in the Southwest. In repowering, a solar thermal field provides supplementary power, using existing turbines and generators. Construction of the Barstow pilot plant is expected to cost $123 million, or over $10,000 per kilowatt (electric) (kWe). (This is about $1 kWh, assuming a load factor of 0.2 and an annual fixed charge rate of 15 percent.) A commercial solar thermal electric plant built with present technology and operational in 1985 would probably cost about one fourth this much and have generating costs about 5 times the current average. Such plants would probably be designed with more energy storage than Barstow. Increasing the plant load factor, which would reduce the cost per kilowatt-hour, depends on the development of a practical technology for storing high-temperature heat or electricity. Early experiments conducted on storage systems for the 10-MWe pilot plant are encouraging, but much needs to be done before high load factors can be achieved. An alternative to storage is the use of oil and gas for backup generating capacity, as in repowering.

One promising extension of solar thermal electric conversion designs is the so-called solar total energy system, in which the waste heat from the generator is used near the power plant. No commercializable technology for this has been demonstrated, but an experimental system is being built at Shenandoah, Georgia. The electricity from these early demonstrations will probably have costs of the same order of magnitude as the first electricity from central station solar thermal plants, or several hundred mills per kilowatt-hour. However, until the first solar thermal total energy system demonstrations are completed, it will be difficult to estimate costs precisely.

Environmental, Health, and Social Considerations

These impacts have been considered by the Risk and Impact Panel,[25] and the following material makes use of that analysis. An environmental assessment of solar energy is also being performed by the Department of Energy.[26,27] The most significant environmental impacts of large solar thermal conversion plants would be their land and material requirements. Studies suggest that such a plant, with enough storage to generate 100 MWe at a 40 percent load factor, if located optimally (e.g., in the southwestern desert) would take up at least 1 square mile. This is about as much land as would be taken during the entire expected life of an equivalent coal plant (including underground mining) and an order of magnitude greater than that for the equivalent nuclear plant using high-grade ore (including mining, milling, etc.). Depending on assumptions about the efficacy of reclamation, a coal-fired plant based on strip mining may require more or less land than the solar plant.

The materials requirements of this approach are also great. A 100-MWe solar electric plant would use, for the arrays of heliostats alone, 30,000–40,000 tons of steel, 5000 tons of glass, and 200,000 tons of concrete, as compared with about 5000 tons of steel and 50,000 tons of concrete for the construction of equivalent capacity with nuclear power. Equivalent capacity in a coal plant would require considerably less steel and concrete.[28] Many of the air pollutants produced in mining and manufacturing the steel, glass, and cement for such a solar thermal plant—notably sulfur and nitrogen oxides, carbon monoxide, and particulates—would be comparable in kind and amount with 1 year's effluents of an equivalent coal-fired plant using current control technology, except for the particulates, which would be an order of magnitude greater for the solar plant. Thus, over the (30-yr) lifetimes of the two systems, the solar plant would be an order of magnitude more benign in most pollutants. The solar plant would be somewhat worse in effluents over the lifetime of the plant than a natural-gas-fired electric generating station of equivalent capacity (with the exception of nitrogen oxide emissions and the long-term carbon dioxide hazard) and much worse than the equivalent nuclear generating capacity.

Ecological and environmental effects of this technology in a desert location would be considerable. Burrowing animals and their habitats would be destroyed during construction, and sites would have to be chosen to avoid dense wildlife populations or endangered species. The desert surface would be altered by construction, road building, off-road vehicle traffic, building of transmission lines, and so on. This would affect erosion in the region. Wind erosion would increase because the protective desert crust, or pavement, would be broken, and water erosion and runoff would also increase, especially along roads. The hydrological cycle would be affected by this and by modification of evaporation rates due to the heliostat canopy. Evaporative losses resulting from the use of wet-tower cooling or storage reservoirs would also significantly affect this cycle. Availability of cooling water (chapters 4 and 9) may restrict the deployment of this or any other electric generating option in the Southwest and elsewhere.

Central receiver electric generating plants could also alter local and regional climates by modifying the radiation balance of the natural desert. Considerable amounts of dust are likely to be introduced into the atmosphere by construction activity. The longer-term climatic effects of this would probably not be as significant as those of modifications of the radiation balance, but the temporary potential for distant effects might be greater, since dust clouds can travel long distances. Some fogging could occur in the vicinity of the plants if wet-tower cooling were used.

The social impacts of building central receiver plants in large numbers in the Southwest could be substantial. There might be significant population shifts to that region. Eventually, the availability of more energy in this region might attract major industries. Some people of the Southwest may object to added stresses on the environment, such as new demands on water supplies; others may welcome new industries.

PHOTOVOLTAIC CONVERSION

Photovoltaic cells are semiconductor devices that generate small electric currents when exposed to light. Power can be produced by wiring the cells together in arrays, either exposed directly to the sun or equipped with concentrating mirrors or lenses that track the sun. Such arrays have been used for a number of years in the U.S. space program and for power generation on earth in remote places. They are much too expensive at present to compete with conventional means of producing electricity. Because of the complex semiconductor technology by which the cells are made, and their relatively low efficiencies, silicon and cadmium sulfide cells now sell for $10,000–$30,000/kWe of peak output (depending mainly on quantity purchased) when used without concentrating mirrors or lenses.[29] This corresponds to costs of about 1000–3000 mills/kWh, more than 20 times the prevailing cost of residential electricity. The Department of Energy has identified and is funding development of processes that, if integrated, could produce silicon cell arrays for $2000/kWe of peak output. These costs might be further reduced by technical breakthroughs; the federal photovoltaic program has a goal of reducing prices of photovoltaic arrays (not including associated electronics and storage) to $500 per peak kilowatt by 1986. Whether this is possible is the subject of some controversy, though most experts agree that these array costs could be brought down to $1000–$2000. The major thrust of this cost reduction effort is presently directed toward incremental advances in the technology for automated and mass production of silicon photovoltaic cells. However, it is argued by some that achieving the cost goals for large markets will require a new type of photovoltaic device which will come only from research on advanced device concepts, perhaps thin films or amorphous (rather than crystalline) materials. If so, the creation of a major industry based on silicon cells may be wasteful.[30]

Photovoltaic systems may be used in any size, from single households to large central generation stations. Photovoltaic conversion for on-site generation of electricity may also be used in a total energy system. The land requirements for large photovoltaic systems would be perhaps twice those of solar thermal systems unless highly efficient photovoltaic devices are developed.

Environmental and Health Considerations

Large-scale photovoltaic generators would probably be located in arid regions. They might be installed with concentrating heliostat fields or in multiple mass-produced arrays. Greater land areas would be required than for solar thermal plants unless efficiencies can be increased, but local ecological disruption and habitat destruction would be similar. The capsules of the solar cells might degrade in time, and their decomposition under normal conditions or during a fire could produce hazardous reaction products, depending on the materials of which they are made. These should be studied carefully. Some proposed solar cells contain environmentally dangerous substances, and the mining, manufacturing, and distribution of these must be handled with care. These hazards are roughly similar to those of other products using semiconductor components, though much greater in scale. With attention to these hazards during the development of advanced photovoltaic technology, it will be possible to ensure adequate safety.

WIND ENERGY CONVERSION

Wind turbines are fairly simple mechanical devices, and the problems of connecting wind generators to an electric grid have been successfully resolved by experience in Europe and the United States. The greatest use of wind-generated electricity was in fuel-short Denmark during World War II, when wind turbines generated 18 million kWh for local networks over the course of more than 7 years (an average power of about 300 kWe).

Wind energy conversion may be accomplished by horizontal-axis propellers, by vertical-axis turbines, or by various other types of devices. The resulting mechanical energy is easily converted to electricity. Units can range in size from half a kilowatt to several megawatts. Wind energy is more limited in choice of sites than other solar technologies, because available wind energy density varies greatly with average wind speed. To accommodate variable wind energy to the regular cyclic needs of a utility system, large amounts of electrical storage capacity are needed unless the delivered energy can be economically used when the wind happens to be blowing or if the wind energy is used in a supplemental mode. It has been proposed that some energy-intensive industries, such as nitrogen fertilizer production, may be able to use wind energy in an intermittent mode. Wind turbines are also useful for pumping water or compressing air. The promise of wind energy for operation in power grids is thus significantly enhanced where pumped hydroelectric storage or other energy storage capacity is available.

Small wind generators are now sold for several thousand dollars per kilowatt of rated output. Dominion Aluminum, Ltd., of Canada offered in 1977 to sell for $175,000 copies of a 200-kWe prototype machine built in Canada. This cost corresponds to about 37 mills/kWh on a good site. The value of such a machine as a fuel saver in a utility grid will generally be less than this cost would imply, because of the need either to supply additional energy storage or to maintain backup generating capacity for times when peak electricity demand and peak winds do not coincide.

The most immediate prospect for wind technology would be the development of a diversified design and manufacturing effort directed generally at machines with generating capacities of about 1 MWe. The market potential is likely to be highly differentiated and, relative to total domestic energy demand, modest. The Solar Resource Group[31] estimated, in its high-solar scenario, that about 50 GWe of capacity from wind energy could be installed in the United States.

Environmental Considerations

The main environmental impacts of wind power are its use of land, local ecological disruption, and the aesthetic considerations of noise and cluttered landscapes. Because the devices are rather small in comparison with conventional power plants, it takes a large number of them to produce the power equivalent of a typical 1000-MWe fossil or nuclear power plant. These must be spread over a large area so that the individual devices do not interfere with one another, and the land area required would be several times that of a solar thermal power plant of equivalent output. The land required for 1000 MWe of installed wind electric capacity has been estimated to be between 200 and 500 square miles.[32] (This land could still be used for other purposes, such as agriculture, so it need not be considered lost.) Spinning turbine blades could injure birds, and the noise could disrupt wildlife. A potentially serious problem is that spinning metal turbine blades cause interference with television video reception, especially in the UHF bands. Access roads for construction and maintenance, and the electrical interconnection of many units, would pose a severe environmental problem, especially in relatively primitive areas.

OCEAN THERMAL ENERGY CONVERSION

Ocean thermal energy conversion (OTEC) uses the approximately 40°F temperature difference between the surface and deep waters of tropical oceans to drive heat engines. A single plant might produce 100 MWe or more of electrical output. The energy could be transported to shore as electricity or as fuel generated by electrolysis of water, or it might be used on the ocean site for energy-intensive industrial processes. This is the only earth-based solar electric technology that is naturally suited for base-load generation of electricity; all others require special storage devices or other backup systems.

A small ocean thermal conversion experiment was operated briefly in 1929. However, no plant of the closed-cycle type now under consideration by the Department of Energy was built and operated until 1977, when the Mini-OTEC plant was demonstrated in Hawaii. This plant generated 10 kWe from a gross output of 50 kWe. (The balance was consumed by water pumping and other plant functions.) If present experiments on heat exchangers are successful, a multimegawatt pilot plant may be built in the 1980s.

There is now no basis for cost estimates except conceptual designs that assume various structures and operations that have not been used in the marine environment. Such estimates correspond to a busbar cost of 70 mills/kWh for electricity delivered to shore. There is great uncertainty and controversy about the cost estimates for ocean thermal conversion. Resolution must await systems tests under realistic conditions. There is also technical uncertainty about the potentially serious problem of fouling of the heat-transfer surfaces in the plant by organisms.[33] This problem is being addressed by current Department of Energy research.

Some attention is being given to deriving power from ocean waves, tides, and currents. Research and development on wave energy converters is under way in England, and some limited use has been made of energy from tides, but these resources are too small to contribute significantly to U.S. energy needs.

Environmental Considerations

Two OTEC designs, differing radically in their environmental implications, are being considered. In one, cold water is pumped from a depth of more than 1000 ft and released at the surface. This could bring to the surface nutrients that could support a plankton bloom. It could also lower the surface temperatures, possibly altering local climate and commercial fish populations in the region. Evaporation rates would be reduced in the ocean near the plant; if many plants were built, large-scale precipitation patterns could be altered. Biocides might be used to prevent biofouling, but their release could be harmful to marine life. Working fluids could also be harmful if released accidentally. It has been suggested that the upwelling of cool water from the ocean depths might release substantial amounts of carbon dioxide into the atmosphere (at a rate about one third that of an equivalent fossil fuel plant). Because of the complexity of the ocean carbonate chemistry involved, this is only speculation.

The second design would pump the cold water back down to about the depth from which it was taken. This would avoid some of the problems mentioned above but would be more expensive.

SOLAR ENERGY SYSTEMS IN SPACE

Proposals have been advanced for development of space-based solar energy systems designed to generate terrestrial electricity with microwaves beamed from space. A considerable amount of ingenuity and analysis has been devoted to design concepts and performance criteria for such systems, which might be constructed from earth-based materials transported aloft by space shuttles, or perhaps from lunar or asteroid material. In the distant future such a system might become the most suitable energy source for large electricity grids if energy sources like breeder reactors, nuclear fusion, and earthbound solar electric technologies should prove to be unattractive. The attraction derives from a longer solar day, freedom from weather, and a solar flux about twice that of the southwestern desert on a clear day. In our judgment this technology, at huge estimated cost, cannot possibly become a substantial energy source until major earth-based competitors are shown to be unattractive or insufficient. Funds spent on satellite power for some decades, except for limited research and development, would be funds spent prematurely.

Space solar systems would require much initial space shuttle traffic, which would introduce significant amounts of energy and effluents into the atmosphere at all levels, though the effluents might be largely water vapor. The environmental consequences of this are not known. Microwave beams appear to be the most practical way to transmit the energy to earth, and the effects of this on the upper atmosphere might be significant. The effects of microwaves on living things are cause for concern also, although present plans call for fairly diffuse beams.

SOLAR FUEL PRODUCTION

As the availability of natural oil and gas declines, the need for liquid and gaseous substitutes will become critical. Large quantities can be provided by coal conversion and perhaps oil shale conversion, as discussed in chapters 3 and 4, but at rather substantial environmental costs. There are a number of solar technologies, most of them using energy trapped in plant matter (biomass), that could supply useful amounts of these fuels in the intermediate term. Other solar technologies can produce fuels over the long term.

At present the federal solar research and development program places too little emphasis on these technologies; only 6 percent of the program's funds go to support biomass technologies.[34] Some research on generating fluid fuels from organic municipal and agricultural wastes is supported, but this is given a rather low priority. The potential, however, is large; this study has estimated that by the year 2010 an annual total of 5.4 quads of energy could be drawn from municipal and agricultural wastes alone.

Production of plants especially for their energy content, on so-called "energy farms," could conceivably contribute another 3.4 quads, but the potential is limited by competition with conventional agriculture for land, nutrients, and water and by the severe ecological impacts involved. Growth of algae or other water plants in ponds or in the open ocean may be a more practical alternative.

For the long term, the most attractive potential solar alternative for fluid fuel production is direct photochemical decomposition of water to produce hydrogen. The hydrogen could be used either directly as a fuel or to synthesize hydrocarbon fuels from various sources of carbon, including carbon dioxide from the atmosphere. The approaches to photochemical conversion can be divided into three categories: biological, biochemical, and synthetic. In the biological approach, microorganisms are used to separate the oxygen and hydrogen in water. In the biochemical approach, enzyme systems are extracted from such organisms and used in reactions to decompose the water molecules. In the synthetic approach, chemical systems would be devised without using components from living organisms. Each of these approaches is under development, but none can be considered commercializable before the end of this century.

In addition, there are other techniques that are not uniquely suited to solar energy, but could be driven also by other energy sources, such as nuclear fission or fusion. These include thermochemical processes and electrolysis. Neither approach is very promising at this time, for reasons given below.

ENERGY FROM BIOMASS

The solar energy stored in biomass, whether obtained from municipal and agricultural wastes or grown to be used as an energy crop, can be made available either directly through combustion (at 85–90 percent efficiency) or indirectly. (When the feedstocks are waste materials, efficiency refers to net yield of energy. However, when biomass is grown specifically for energy, the energy inputs for production must be deducted to give the net energy yield.) Indirect methods that have been proposed and studied are pyrolysis to produce mixed liquid and gaseous fuels, anaerobic fermentation to produce gaseous fuels, and hydrolysis of carbohydrates to produce glucose followed by fermentation of the glucose to produce alcohol. The maximum possible energy recovery efficiencies of these conversion processes are about 50–60 percent for anaerobic fermentation, about 35 percent for pyrolysis, and about 55 percent for hydrolysis and fermentation. Many of these processes are technically well developed.

Municipal Wastes

The use of municipal wastes in such processes holds great promise for the near-term future. Such wastes are already being collected and carried to central locations, often near municipal power plants that could use them or their conversion products as fuel. In addition, disposing of these wastes is becoming increasingly difficult and expensive in many cities, and the energy conversion projects would have value as means of disposal in addition to the value of the fuel produced. Furthermore, such wastes are increasingly being sorted to recover various nonfuel materials, and this operation would at least partially offset the expense of energy conversion.

The economics of recovering energy from municipal wastes thus depend strongly on the value of the nonenergy materials recovered and on the credit given to the operator for disposal of the wastes, which would otherwise be a cost to the municipality. A recent economic analysis of methane production from municipal solid wastes cited a total cost of $5.93 per million Btu for the facility and operating expenses. The credits to the operator correspond to $3.84, so that the net cost of the methane produced is only $2.09, a quite competitive cost for gas. While this pattern varies from place to place and from process to process, it is a reasonable estimate of the cost of converting municipal wastes for their energy content. A variety of processes are now being developed and tested, and it is yet unclear which will be the best in practice.[35]

Agricultural Wastes

Agricultural wastes include the inedible portions of food crops, animal manures, and unused portions of trees harvested for paper or lumber. The total amount of energy contained in such wastes is much more uncertain than the energy content of available municipal wastes. Neither the gross availability of such materials (the total amount produced by U.S. agriculture and forestry) nor their real availability (the amount that could be economically collected, transported, and fed into a conversion process) can be precisely determined. The Solar Resource Group estimates that the gross annual availability of these wastes is about 8 quads and that their real availability is about half that value.

The energy available as fuel from agricultural wastes depends on the conversion process. The decision on which process to use in a given situation will be based on the character of the wastes, the ease with which they can be brought to large central conversion plants or converted in smaller plants on site, and the local prices of various fuels and residues from the processes. The decision will sometimes, but not always, be made in favor of the process with the highest net energy yield. The Solar Resource Group estimates that in about a decade collection and conversion of agricultural wastes (with a recovery factor of 25 percent, to take account of both real availability and conversion losses) could yield about 2.2 quads of methane or some equivalent fuel. This value sets the order of magnitude for the following discussion.

The U.S. food system consumes significant amounts of energy, only a fraction of which is used in the actual production of food. Most is used in transporting and processing food after production. The amount of energy used on farms for producing food has been estimated to be about 2.1 quads/yr.[36] Of this amount, about half is actually consumed as fuel on the farm, and about half is used in producing fertilizer, pesticides, and other farm chemicals. The amount of energy used in U.S. agriculture has increased rapidly over the past few decades. This increase has been associated with a decrease in the labor inputs to agriculture but reflects also a trend to centralized activities. Cattle feedlots and "chicken factories" are but two examples of this centralization.

This trend has had two major consequences. One is increased consumption of energy; the other is a growing problem of waste disposal from large central facilities. For example, the control of water pollution from chicken factories has been a major goal of the Environmental Protection Agency for a number of years. Disposal of these wastes is costly, and the wastes contain both energy and nonenergy (nutrient) values. Thus, the economic situation is similar to that of municipal waste conversion, except that colocation of the waste with a major energy conversion facility is less likely.

Most of the wastes from the existing U.S. agricultural system occur not in central facilities of the type just described but in widely dispersed locations. Their disposal is not seen as a costly process. In fact, the wastes produced at dispersed sites often are used to provide nutrients and other factors important for soil quality (e.g., crop residues are plowed under, and the nutrients are returned from manure on sparsely grazed ranges). While it is possible (though costly) to collect wastes from such dispersed locations and convert them to fuel, it is not clear that there will always be a net energy gain from doing so. If these wastes are not used to maintain soil quality, they must be replaced by other inputs, such as fertilizers. The result may be a net energy loss.

It is interesting, though perhaps not in itself important, that the energy estimated to be available from all agricultural wastes (about 2.2 quads) is about equal to the present energy inputs for agriculture. This fact does gain some importance when it is considered in light of the trade-offs possible in agriculture between energy recovery from wastes and reuse of wastes at the site of production to reduce energy inputs. The degree to which the use of wastes in the field can replace the use of fertilizers or other farm chemicals is not really known, even though it is widely debated.[37]

It is possible that in future years, through some combination of reuse of wastes and conversion of wastes to fuels, the agricultural sector will be made essentially self-sufficient in energy. However this balance works out in detail, the agricultural sector will not be a major producer of fuels for other U.S. energy needs. The resource is simply too small to have significant impact on a national scale. Conversion of waste materials, however, may reduce or eliminate agricultural energy demands and thus eventually make a net contribution of a few quads to the U.S. energy system.

Growth of Biomass for Energy

Existing farming methods could be used in "energy farms" to produce significant amounts of biomass. Since in an energy farm the goal is to produce the maximum amount of biomass, crops such as eucalyptus trees, rubber plants, or sunflowers might be used because of their rapid growth and high energy content. The energy efficiency of land agriculture is quite low; typically less than 0.5 percent of the solar energy is stored as biomass. However, the photosynthetic process is capable of higher efficiencies, up to 12 percent being noted in some experiments. The efficiency of energy farms could be greatly improved if crops can be developed to perform photosynthesis efficiently in full sunlight and not lose excessive amounts of energy by respiration. Furthermore, in the United States competition with food production for land, water, and nutrients is a major consideration. Energy farming would, like other agricultural operations, require substantial amounts of water. The availability of vital minerals such as phosphorus may also be a limitation. Crops now being raised command higher prices as food or fiber than as fuel, and this will probably not change in the near future. Those studying the world food situation do not foresee a period of general crop surpluses such as would be required to provide low-cost biomass from food crops for conversion to fuel.

Competition with conventional agriculture might be avoided by producing biomass in the open ocean. The surface waters there have low natural productivity, but if nutrients could be provided (for example, by pumping nutrient-rich deep ocean water to the surface), a biomass crop like giant kelp could be grown. An experimental test of the concept is under way, but technical and economic feasibility have yet to be proved.

A variety of cost estimates for energy farming have been published. They range from about $0.50 to $2.00 per million Btu of "raw biomass" energy. (If the same land, water, nutrients, etc., could be used with the same efficiency for producing food crops, sale of the crops would provide a return of several times this estimate.) If this biomass is to be converted to either liquid or gaseous fuels, the conversion cost must be included. Estimates for these vary widely, depending on the technologies, plant capacities, and conversion efficiencies assumed. The range of conversion cost estimates is about $0.70 to $3.50 per million Btu. This means a gaseous fuel (methane) from biomass production could cost as much as $5.50 per million Btu, or as little as $1.20 per million Btu (less than the present wellhead price of natural gas).[38] It should be emphasized, of course, that these estimates are not based on proven and tested applications of existing technology.

The exemption of "gasohol" motor fuel (90 percent gasoline, 10 percent ethyl alcohol) from federal taxes, and in some cases state taxes, has given a great incentive to the fermentation of grain to fuel alcohol. However, the total amount of fuel available from grain is small, and such crops have high value in international trade. It is unclear whether this use of grain is a net producer of liquid fuels in the United States.[39]

Deployment of a biomass production technology in marginal U.S. lands would cause drastic environmental alterations. The natural ecosystems of the regions would be replaced by plant monocultures. The impacts on insect and animal life would also be drastic. If the plant type used grows extremely rapidly (as it must for high efficiency) there should be no trouble maintaining it in competition with native plants. However, plant monocultures are very susceptible to infestations and disease. The methods used to control infestations and disease are major sources of the pollutants released by food-producing agriculture, and these pollutants should be expected to be a significant feature of biomass production. One possible problem is escape of the biomass plant species from the confines of the energy farm. A plant that grows extremely rapidly under marginal conditions, with no extra water or nutrients, has little to prevent it from becoming widely dispersed and displacing native plant species.

Another question would be raised in the event that a very fast growing plant that prospers in marginal conditions can be developed. There would then be great incentive to adapt this plant to produce food for human consumption, or at least feed for livestock. We assume that efforts in this direction would be made, and that if successful they would put biomass production and food crops again in direct competition for even marginal land.

Another approach to biomass growth is to use the nutrients in liquid wastes from homes, industry, and agriculture to support the growth of algae in ponds. Such systems use bacteria to oxidize waste materials and produce nutrients to feed algae, which in turn collect solar energy to produce biomass. The advantage of this method is that it can be done with efficiencies of better than 5 percent and that it does not compete with agriculture for nutrients. In fact, it consumes waste nutrients that are responsible for eutrophying bodies of water. However, algae produced in such ponds might have a higher value as animal feed than as either a source of energy or a chemical feedstock.

PHOTOCHEMICAL, THERMOCHEMICAL, AND ELECTROLYTIC CONVERSION

In the long term, solar energy can be used with a number of chemical processes that supply fuels without the need for producing or gathering biomass. All must be considered speculative for one reason or another; some await a great deal of technical progress, while others are barred from near-term use by their high costs. The methods being investigated hold the promise of supplying fuels in virtually unlimited amounts, however, and much greater attention should be paid them in federal research and development programs. Commercialization is not likely in this century.

Photochemical Conversion

One attractive long-term alternative to the growth and conversion of biomass is direct solar fuel production by photochemical conversion. The processes now being most widely investigated involve photolysis, the decomposition of water to produce hydrogen by means of radiant energy.

This can be done by a number of biological, biochemical, and synthetic methods.

In a biological approach, living organisms are used to decompose water into hydrogen and oxygen; thus the term biophotolysis. However, the liberation of molecular oxygen generally inhibits the activity of hydrogenases, the biological enzymes that produce molecular hydrogen. Thus, an important task for research in this area is to find species or mutations with hydrogenases that are effective in the presence of oxygen. One option is to produce the hydrogen and oxygen separately, as has been done with cultures of a blue-green alga. However, in all known cases of biophotolysis the rate of hydrogen production is extremely small, and great progress will be required before practical conversion schemes can be designed.

In a biochemical approach, enzyme systems would be obtained from biological organisms and then combined in an appropriate reaction cell to perform all the steps involved in collecting energy and driving the water-decomposing reactions. Production of hydrogen, at least at low rates for short periods of time, has been demonstrated, but more basic research on the biochemical mechanisms of photosynthesis, and inventive ideas for incorporating molecular components into systems, will be required for this technique ever to become practical.

In a synthetic approach, a complete chemical or electrochemical system for photolysis would be designed and synthesized without using any components taken from plants or algae. This has a great potential advantage in that problems of instability of biological components would be avoided. There are some promising ideas about the form such chemical systems might take, and some electrochemical systems have been demonstrated in the laboratory, but this must be considered a long-term research problem.

The achievement of a practical technology for photochemical conversion to produce fuels will depend on significant advances in our fundamental understanding of primary photochemical processes and the subsequent processes involved in the transfer of the energy of electronic excitation to stable chemical products. Experiments performed thus far on biochemical systems have suffered from very low efficiencies (often less than 0.1 percent) and from instability of the reactant systems. Some electrochemical systems have shown higher efficiencies, at least for short periods of operation. The difficulties to be overcome in the development of a practical technology of photochemical conversion are formidable. However, the potential rewards are great.

Theoretical considerations indicate that efficient photochemical processes should be possible. The attainable conversion efficiency might be on the order of 20–30 percent, based on incident solar energy.[40,41] With this efficiency, which is more than 10 times the efficiency probably to be attained through biological conversion in energy farms, the quantity of fuels required by a technological society could be obtained from a relatively small amount of land. For example, production of fuels with energy content equal to present U.S. consumption of petroleum and natural gas (55 quads) would require only about 20,000 square miles (about 0.6 percent of the coterminous U.S. land area) at a conversion efficiency of 25 percent. Furthermore, photochemical conversion would not require the great amounts of water and nutrients required by energy farming. However, attainment of a photochemical technology will demand a long-term commitment to both basic and applied research in this area; even then, ultimate success is uncertain.

Thermochemical Conversion

Processes for thermochemical decomposition of water to produce hydrogen are being actively investigated because of their potential for use with high-temperature nuclear reactors. Hundreds of possible processes that use various reactants in closed cycles have been investigated by computer simulation. Concentrating solar collectors could provide temperatures high enough (above 1350°F) to drive these cycles. However, these thermochemical processes are complex and might be impractical. Large amounts of reactants; high temperatures and perhaps high pressures; and extensive mixing, reaction, and separation steps are required. The design of a practical process will be a challenging task.

Electrolytic Conversion

An alternative to thermochemical production of hydrogen is to use any solar electric technology to generate electricity initially and then use the electricity to produce hydrogen by electrolysis of water. Electrolysis is a proven technology that has been used commercially in both small and large applications. Efficiencies are good, with only about 115 kWh of electricity required to produce 1000 ft^3 of hydrogen, which corresponds to an energy efficiency of 83 percent. Once the hydrogen is produced, it can be converted to other fuel forms, such as methanol. However the economics are poor: If the cost of solar-produced electricity is 70 mills/kWh, this contributes $25 per million Btu to the cost of the hydrogen.

CONCLUSIONS

1. The aim of the government's solar energy program should be to place

the nation in the best possible position to make realistic choices among solar and other possible long-term options when choices become necessary. This requires continuing support of research and development of many solar technologies. Comparisons of the present costs of various solar technologies and other long-term technologies should not be regarded as critical at the present stage of development. Of more importance is the potential for significant technical advances.

2. In the intermediate-term future, the direct use of solar heat can contribute significantly to the nation's energy system. Solar heating technologies should be viewed, along with many conservation measures, as means of reducing domestic use of exhaustible resources. The role of the government program should be to support the development and assist the implementation of the most cost-effective solar techniques, used wisely in combination with energy conservation. In particular, the government should stimulate the integration of solar heating into energy-conserving architectural design in both residential and commercial construction through support and incentives for passive solar design. Since all solar energy technologies are capital intensive, uses that are distributed throughout the year, such as domestic water heating and low-temperature industrial process heating, are likely to be economically competitive earlier than uses for which there are large seasonal variations in demand.

3. Under present market conditions, solar heating systems are usually not competitive with other available technologies, and therefore market forces alone will bring about little use of solar energy by 2010—probably less than 6 quads even if average energy prices quadruple.[42] Nevertheless, important social benefits would accrue from the early implementation of these systems: They would contribute to the nation's conservation program, they are environmentally fairly benign, and they would increase the diversity of the domestic energy supply system and its resilience against interruption. National policy should stimulate the early use of solar energy by intervening in the energy market with subsidies and other incentives.

4. Many solar energy applications require long-term development, and these technologies should properly be compared with breeder reactors or fusion. It would be unfortunate if alternatives to the breeder were rejected because too little is known about them today to count on them. It would also be wrong to assume that the choice will or should fall on a single long-term option. Diversity in the nation's long-term sources can provide valuable resilience in the face of interruptions in the supply of a single fuel or technology. Decisions that restrict the variety of our long-term options should be deferred as long as possible.

5. The cost picture for a number of solar technologies is likely to change radically in the future, with successes and failures in development.

Competing technologies will display parallel trends. The costs of many factors of production are likely to change, affecting various technologies differently. In most cases, the economics of solar energy depend critically on advances in ancillary technologies, such as energy storage. It is important that the benefits of these ancillary developments be assessed for other energy technologies on the same basis as for solar, however. For example, cheap energy storage systems would benefit the economics of all systems containing capital-intensive generating technologies.

Large-scale government demonstrations of long-term solar technologies, such as the planned demonstration of a solar thermal central station power plant, could be counterproductive if undertaken prematurely. Such projects may suggest (possibly incorrectly) that the technologies could never become economically competitive, whereas waiting for additional technical developments could result in a considerably more favorable outlook.

6. An imbalance exists in the federal solar energy program in favor of technologies to produce electricity at the expense of those to produce fuels. The program overemphasizes technologies for the production of electricity, yet electricity accounts for only about 11 percent of energy end-use. Much more attention should be given to the development of long-term solar technologies for fuels production, although there is at present no prime candidate besides biomass production (which is limited by ecological considerations).

7. The diversity of solar technologies is so great that it is difficult to make decisions among alternatives in a centralized way. To a great extent, the actual choice of which solar technologies to deploy should be made in as decentralized a manner as possible. In other words, the decisions should be left to private industry and individual consumers. The government's role should be development of a broad scientific and technological base in support of solar energy (much as it did for nuclear energy prior to 1960 and for aeronautics after World War I) and provision of economic incentives that favor solar alternatives.

NOTES

1. National Research Council, *Domestic Potential of Solar and Other Renewable Energy Sources,* Committee on Nuclear and Alternative Energy Systems, Supply and Delivery Panel, Solar Resource Group (Washington, D.C.: National Academy of Sciences, 1979).

2. See chapter 2 for a description of cogeneration and its advantages.

3. Supply and Delivery Panel, Solar Resource Group, *op. cit.*

4. The load factor of a power plant can be thought of as the proportion of the time that the plant is actually producing power at its rated capacity. It is thus the ratio of its actual output

to its potential output. There are 8760 hours in a year, so that a plant operating at a load factor of 0.4 runs about 3500 hours/yr.

5. National Electric Reliability Council, *Fossil and Nuclear Fuel for Electric Utility Generation: Requirements and Constraints* (Princeton, N.J.: National Electric Reliability Council, August 1977).

6. U.S. Department of Energy, *Report of the President's Domestic Policy Review of Solar Energy* (Washington, D.C.: U.S. Department of Energy, 1979).

7. *An Analysis of Federal Incentives Used to Stimulate Energy Production* (Richland, Wash.: Pacific Northwest Laboratory (PNL-2410), March 1978).

8. *Ibid.*

9. Energy Research and Development Corporation/Mitre Corp., *An Economic Analysis of Solar Water and Space Heating* (McLean, Va.: Mitre Corp. (M76-79), November 1976).

10. U.S. Department of Energy, *An Analysis of the Current Economic Feasibility of Solar Water and Space Heating,* Assistant Secretary for Conservation and Solar Applications (Washington, D.C.: U.S. Department of Energy, November 1977).

11. Solar Energy Research Institute, *Economic Feasibility and Market Readiness of Eight Solar Technologies* (Golden, Colo.: Solar Energy Research Institute (SERI-34), June 1978).

12. D. S. Ward, *Solar Heating and Cooling Systems Operational Results Conference, Summary* (Golden, Colo.: Solar Energy Research Institute (SERI/TP-49-209), 1979).

13. Amory Lovins, "Soft Energy Technologies," in *Annual Review of Energy,* ed. Jack M. Hollander, vol. 3 (Palo Alto, Calif.: Annual Reviews, Inc., 1978), pp. 477–517.

14. Supply and Delivery Panel, Solar Resource Group, *op. cit.*

15. Arthur D. Little Co., *Individual Load Center, Solar Heating and Cooling Project* (Palo Alto, Calif.: Electric Power Research Institute (ER-594), December 1977).

16. A. H. Rosenfeld, *Building Energy Compilation and Analysis* (Berkeley, Calif.: Lawrence Berkeley Laboratory, 1979).

17. B. Anderson and M. Riordan, *The Solar Home Book* (Harrisville, N.H.: Chesire Books, 1976).

18. E. Mazria, *The Passive Solar Energy Book* (Emmaus, Pa.: Rodale Press, 1979).

19. Ward, *op. cit.*

20. *Ibid.*

21. Rosenfeld, *op. cit.*

22. E. Kahn, "The Compatibility of Wind and Solar Technology with Conventional Energy Systems," in *Annual Reviews of Energy,* ed. Jack M. Hollander, vol. 4 (Palo Alto, Calif.: Annual Reviews, Inc., 1979), pp. 313–352.

23. W. D. Marsh, *Requirements Assessment of Wind Power Plants in Electric Utility Systems,* vol. II (Palo Alto, Calif.: Electric Power Research Institute (ER-978 V.2), 1979).

24. J. W. Doane *et al., A Government Role in Solar Thermal Repowering* (Golden, Colo.: Solar Energy Research Institute (SERI/TP-51-340), 1979).

25. National Research Council, *Risks and Impacts of Alternative Energy Systems,* Committee on Nuclear and Alternative Energy Systems, Risk and Impact Panel (Washington, D.C.: National Academy of Sciences, in preparation).

26. U.S. Department of Energy, *Technology Assessment of Solar Energy* (Washington, D.C.: U.S. Department of Energy, 1979).

27. M. Yokell *et al., Environmental Benefits and Costs of Solar Energy* (Golden, Colo.: Solar Energy Research Institute (SERI/TR-52-074), 1979).

28. M. Davidson, D. Grether, and K. Wilcox, *Ecological Considerations of the Solar Alternative* (Berkeley, Calif.: Lawrence Berkeley Laboratory (LBL-5927), February 1977).

29. D. Costello *et al., Photovoltaic Venture Analysis* (Golden, Colo.: Solar Energy Research Institute (SERI/TR-52-040), 1978).

30. H. Ehrenreich, *Solar Photovoltaic Energy Conversion* (New York: American Physical Society, 1979).

31. Supply and Delivery Panel, Solar Resource Group, *op. cit.*

32. U.S. Energy Research and Development Administration, *Wind Energy Mission Analysis* (Washington, D.C.: U.S. Energy Research and Development Administration (COO/2578-1/2), February 1977).

33. National Research Council, *Selected Issues of the Ocean Thermal Energy Conversion Program,* Assembly of Engineering, Marine Board, Panel on Ocean Thermal Energy Conversion (Washington, D.C.: National Academy of Sciences, 1977).

34. U.S. Department of Energy, *Solar Energy Research and Development Program Balance* (Washington, D.C.: U.S. Department of Energy (DOE/IR-0004), February 1978).

35. D. L. Klass, *Energy from Biomass and Wastes* (Chicago, Ill.: Institute for Gas Technology, 1978).

36. J. S. Steinhart and C. E. Steinhart, "Energy Use in the U.S. Food Supply System," *Science* 184 (1974):307–316.

37. S. Flaim, *Fertility and Soil Loss Constraints on Crop Residue Removal for Energy Production* (Golden, Colo.: Solar Energy Research Institute (SERI/RR-52-324), 1979).

38. Supply and Delivery Panel, Solar Resource Group, *op. cit.*

39. D. I. Hertzmark, *A Preliminary Report on the Agricultural Sector Impacts of Obtaining Ethanol from Grain* (Golden, Colo.: Solar Energy Research Institute (SERI/RR-51-292), 1979).

40. J. R. Bolton and D. O. Hall, "Photochemical Conversion and Storage of Solar Energy," in *Annual Review of Energy,* ed. Jack M. Hollander, vol. 4 (Palo Alto, Calif.: Annual Reviews, Inc., 1979), pp. 353–402.

41. G. Porter and M. D. Archer, "In Vitro Photosynthesis," *Interdisciplinary Science Reviews* 1, no. 2 (1976): 119–143.

42. Supply and Delivery Panel, Solar Resource Group, *op. cit.*

7 Controlled Nuclear Fusion

The nuclear fusion program of the United States should seek to develop this technology sufficiently for comparison with fast breeder reactors, solar power, and other long-term sources of energy.[1] The research and development program has proceeded nearly to the point of demonstrating the scientific feasibility of nuclear fusion. The technology must still be tested for engineering achievability, environmental characteristics, and reasonable cost. We recommend that the program pursue research on the principal physical concepts that have been advanced for nuclear fusion, and on the materials and design problems of fusion reactor technology, with attention to environmental characteristics and engineering and economic practicality whenever possible. The aim is to develop the most attractive forms of this technology on a timely schedule.

The following general circumstances establish the context for assessment of nuclear fusion and the program of research and development to test its possibilities.

- Only four long-term sources of energy have been projected for the future: nuclear fusion, nuclear fission breeders, solar energy, and geothermal energy. The latter three are not so free of problems that fusion can be disregarded or assigned secondary importance.
- Controlled fusion will require scientific and technological development for at least two decades, perhaps much longer. Its prospects as a source of power cannot be judged with confidence in advance of this development.
- The safety and environmental liabilities attending the use of fusion for

power cannot be analyzed in satisfying detail at the present stage of development. Some may be far more tractable than those of fast breeders.

• The complexity of problems in nuclear fusion may tend to isolate the scientific and technical community engaged in various aspects of this area of research and development. Care should be taken to ensure that this isolation does not lead to wrong decisions.

• Research and development in nuclear fusion is expensive. To demonstrate the technology sufficiently for decisions about its use as a source of energy will cost $10–$20 billion. Each new possibility explored in the effort to optimize the fusion option will add to the cost.[2] (The cost of developing any new long-term source of major significance, however, probably falls in the same range.)

THE FUSION REACTION

Large amounts of energy are released in the union or fusion of light nuclei. Nuclear fusion requires that two charged nuclei approach one another closely. They must approach with high enough energy to overcome their mutual electrostatic repulsion. High energy can be achieved through high temperature. The sun, for example, fuses hydrogen nuclei into helium at interior temperatures of about 20 million degrees Celsius. For the most promising earth-bound possibility—fusing deuterium and tritium into helium—temperatures about 10 times that of the sun's core must be achieved and maintained long enough to allow a significant fraction of the fuel to react. At the high temperatures of fusion reactions, matter has decomposed into atoms whose electrons are stripped away. The result is an ionized gas, or plasma, that conducts electricity and responds readily to magnetic and electric forces.

The practical use of fusion as a source of energy, then, depends on the simultaneous achievement of high temperatures and effective containment of the plasma. Heating and containing the plasma, and operating the ancillary equipment that may be used to drive the reaction, represent a large investment in energy that the net production of energy from fusion must pay back with interest.

FUSION FUEL CYCLES AND THE ENVIRONMENT

The fusion reaction that requires the lowest temperature and the least effective containment, and offers the prospect of the highest power densities, is one of deuterium and tritium.

$$\text{D(deuterium)} + \text{T (tritium)} \rightarrow \text{He (helium)} + \text{n (neutron)} + 17.6 \text{ MeV.}$$

Most of the energy released (14.1 MeV) is carried by the escaping neutron. By being slowed down in solid or liquid materials surrounding the plasma, these neutrons have their kinetic energy converted into heat, which is then converted into electricity in a conventional heat engine such as a steam turbine.

Deuterium makes up 1/7000 of the hydrogen in water and, at a present cost per unit of energy 1/10,000 that of coal, can be considered free. Tritium is weakly radioactive and decays into helium-3 (^3He) with a half-life of 12.3 years. It is almost nonexistent in nature and must be regenerated in the fusion fuel cycle after use in deuterium-tritium (D-T) reactions. The following two reactions of lithium (Li) with neutrons permit tritium breeding at a gain theoretically much greater than 1.

$$n + {}^7Li \rightarrow n' + T + {}^4He - 2.7 \text{ MeV}$$
$$n + {}^6Li \rightarrow T + {}^4He + 4.8 \text{ MeV}.$$

Here n′ represents a neutron that has lost 2.7 MeV of its kinetic energy in an inelastic collision with lithium, whose resulting excited state dissociates into tritium and ^4He as indicated by the equation.

The breeding reactions can be arranged to occur by absorption of some of the neutrons released by fusion in a surrounding lithium blanket, at temperatures conventional in power plants. Thus, a fusion reactor on this fuel cycle is a "breeder" reactor fed by deuterium and lithium.

Reasonably assured terrestrial resources of lithium represent a potential source of energy for fusion comparable to world resources of coal, at a fuel cost 1/1000 that of coal. Lithium in seawater represents an energy resource 10,000 times larger; it could probably be extracted at a cost low enough to make little contribution to the cost of electricity from fusion.

The prolific release of neutrons in the D-T fusion reaction will induce radioactivity in structural and other materials in the reactor. These activation products, together with the substantial inventory of tritium, could pose a significant radiation hazard to the public in the event of a major accident, and in any case will complicate routine operation and maintenance. Tritium will be subject to some routine release from the fusion plant. The magnitude of such releases (and the corresponding public exposure) will be accurately known only after specific designs emerge and control technologies are tested at commercial scale. Based on today's knowledge, it appears possible, at a cost, to hold routine public exposures from fusion reactors to the same low levels as those from routine operation of light water fission reactors. The neutron-induced radioactivity in fusion reactor materials can be reduced, in principle, by judicious selection of the materials used. (The radioactivity of the fission products produced in a fission reactor, by contrast, cannot be appreciably altered through reactor design.) How much of this apparent potential to reduce neutron activation in fusion can be realized in practice remains to be seen.

It depends in part on which of the low-activation materials meet the other demanding conditions of operation in a fusion reactor.

Besides the D-T cycle, other fuel cycles have been proposed; they rely on deuterium alone, on deuterium and ^{3}He, or on ordinary hydrogen plus heavier elements such as lithium, beryllium, or boron in the plasma itself. These fuel cycles could offer various advantages. For example, more energy would be released in the kinetic energy of charged particles (and thus the potential for high efficiency would be enhanced) and less would be released in the form of neutrons. There might even be no radioactivity either induced in structural materials or in the fuel or "ashes." These fuel cycles may eventually find application in the production of chemicals or in materials processing and perhaps ultimately in the production of electricity. They now seem unlikely to compete with the D-T-Li fuel cycle for the early generation of electricity from fusion, since they require higher temperatures and much more effective confinement.[3] If an exception materializes, it will likely be the D-D reaction, catalyzed by reinjection of product tritium and ^{3}He, which is next in difficulty after D-T and produces more neutrons than the other "advanced" reactions.

Hybrid fusion-fission fuel cycles, in which the neutrons emitted in D-T or D-D reactions are used to induce fissions or to create fissile isotopes (uranium-233 or plutonium-239) from fertile thorium-232 or uranium-238 in a blanket surrounding a fusion core, are also possible. The fissile material produced can be fissioned in place or removed to fuel nonbreeder fission reactors elsewhere. The result, in effect, is to multiply the energy release per fusion reaction by about an order of magnitude. This makes the conditions that must be achieved in the fusion core less demanding than those for a pure fusion reactor, but the additional engineering complexity of combining fusion and fission technologies in a single device will at least partly offset this advantage and may overwhelm it. The environmental and safety characteristics of hybrid devices would be substantially those of fission reactors, compounded by the addition of fusion's tritium and activation products. The chances of some kinds of accidents might be reduced; the chances of others increased. Most proponents of hybrids argue that these fuel cycles could best be applied in the hybrid device optimized to produce fuel for fission reactors located elsewhere. The safety or antiproliferation characteristics (or both) of such a fusion-fission nuclear energy system, they argue, might be superior to those of a system consisting primarily of fission breeder reactors.

The hybrid possibility calls attention to a link between *any* neutron-producing fusion energy system and the potential for proliferation of fission weapons; excess fusion neutrons can be "diverted" by a reactor's operators to producing fissile materials for bombs. The practical importance of this link may be small, however, given the difficulties of fusion energy technology. Any group or country capable in fusion could acquire fissile materials by a number of easier means. More troublesome, perhaps, is the possibility that *knowledge* derived from certain aspects of research on inertial confinement approaches to fusion could be applied to the development of fusion weapons. Firm conclusions on the exact nature of this link, or its importance, cannot be reached without access to classified information.

CREATING THE CONDITIONS FOR FUSION: HEAT AND CONFINEMENT

The two main classes of schemes that have been proposed to heat and hold thermonuclear fuel for fusion are magnetic and inertial confinement. Experiments with magnetic confinement were begun in the early 1950s; serious investigation of inertial confinement began in the 1960s.

MAGNETIC CONFINEMENT

The reacting fuel to be contained by magnetic fields is a tenuous deuterium-tritium gas at a millionth of solid density or less. The plasma is initially formed and heated electrically. Further heating may be accomplished by a variety of methods, of which the injection of high-energy neutral beams is the preferred method. Thereafter, in many concepts (particularly the Tokamak), the fusion reaction itself keeps the plasma hot. In its hot, reacting state, the gas is typically at a pressure of 10 atm. It must be held together in isolation from material that could cool or contaminate it for at least several seconds to allow an appreciable fraction of the fuel to react. (Machines operating at somewhat higher fuel densities—up to 1/100,000 that of normal solids—need correspondingly shorter confinement times.)

Magnetic fields, coupled to electric currents in the hot gas, generate forces that confine the fusion gas to a prepared vacuum space. The particular forms of microturbulence that limit plasma lifetimes and density depend on temperature and details of the configuration. Understanding the physics of this complex matter and searching for the optimum configuration have absorbed most of the worldwide research effort, on which some $4 billion has been spent.[4] Several geometrical arrangements have been advanced. So far, the leading contenders in design are all large and lead to large power systems (at least 500 megawatts (electric)*

*Statement 7-1, by H. Brooks: Although a 500-MWe design might be built, an economically optimized design would probably be even larger than that for a fission breeder.

(MWe)), principally because the vacuum region where fusion occurs must be designed with various inlets and outlets, must be sheathed in a 1-m-thick solid blanket for slowing the high-energy neutrons, breeding tritium, and carrying away the heat in cooling channels, and must be surrounded by superconducting coils and other electrical equipment. A few possibilities (much less thoroughly investigated) might work at the 100-MWe scale.

Details of the specific arrangements need not concern us here: Two leading examples are magnetic mirrors, analogous to the magnetic field in outer space that confines the earth's "radiation belts" of trapped particles, and toroidal (doughnut-shaped) devices, such as stellarators or Tokamaks. The science of plasma physics and magnetic confinement has matured rapidly, aided by an unhindered program of international cooperation and exchange coordinated by the International Fusion Research Council (under the International Atomic Energy Agency). The United States, the U.S.S.R., England, France, the Federal Republic of Germany, the Netherlands, Italy, and Japan participate. Toroidal Tokamaks (illustrated in Figure 7-1), conceived in the U.S.S.R., have been the most heavily funded research instruments in the United States since the early 1970s. Of magnetic fusion experiments, those in Tokamaks are closest to achieving the combination of confinement conditions needed for a reactor. Magnetic mirror experiments are being conducted in the U.S.S.R., the United States, and elsewhere, at a substantial level of funding, but well below that of Tokamaks.

At this point, the nature of uncertainty about the prospects for fusion can be understood. The Tokamaks are by virtue of their fundamental principles complicated in design, difficult to service, and probably operate on long pulses—minutes—rather than steadily. Mirror machines are simpler geometrically and can operate in steady state but seem likely for rather basic physical reasons to have poorer confinement than Tokamaks. Does something better than either exist—that is, an approach more likely to evolve into an attractive reactor—but because the research and development field is so complex it has yet to be recognized? The whole field of study is insufficiently advanced to guide selection, although the great variety of possibilities seems to give considerable grounds for optimism.

INERTIAL CONFINEMENT

In inertial confinement schemes, small pellets of deuterium-tritium fuel are dropped one at a time into a vacuum and irradiated by high-energy beams that cause the outer layers to evaporate explosively. The resulting forces compress and heat the remaining pellet core, generating fusion

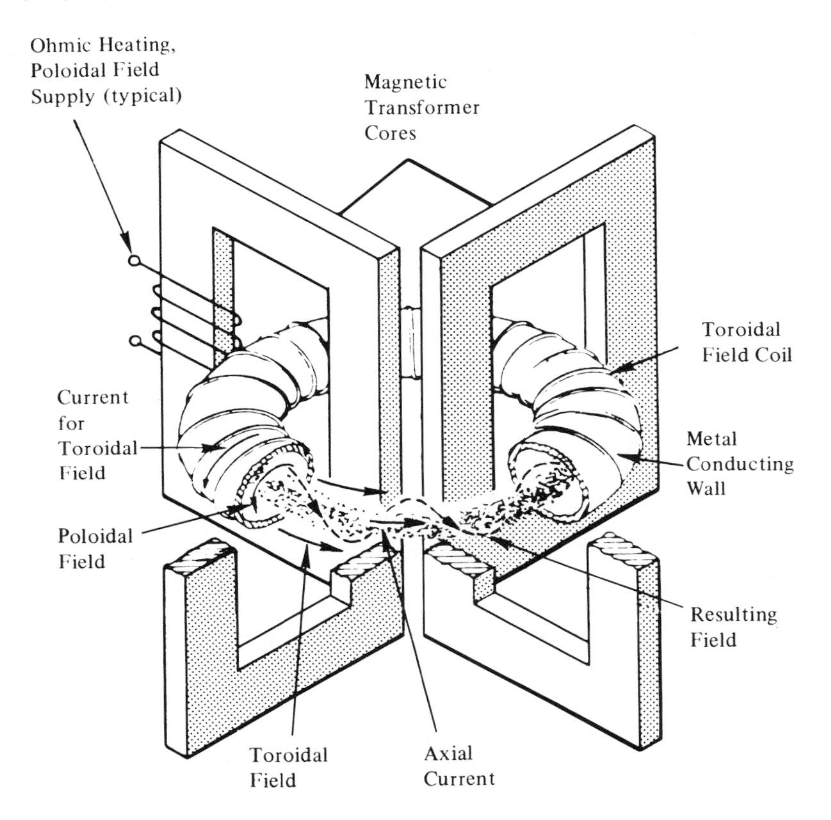

FIGURE 7-1 A common type of Tokamak. The current in the field coils generates a toroidal magnetic field; the plasma current induced by the current in the coils produces a poloidal magnetic field, and these two fields combine to form a spiral magnetic field, stabilizing the plasma.

reactions. The pellets used in experiments are 1 mm or less in diameter, and the energy release is controllable. The principal difficulty is that of achieving the necessary compression—100 billion atm—while preserving the stability of the pellet's core and preventing its premature heating. The first source of energy trained on the pellets was the laser (inertial confinement is thus often labeled "laser fusion"). More recently, energetic electron or ion beams that are focused and pulsed have been proposed to drive the reaction. An inertial confinement scheme is illustrated in Figure 7-2.

Development of inertial confinement reactors depends on the beam system: In about one billionth of a second or less, enough energy must be

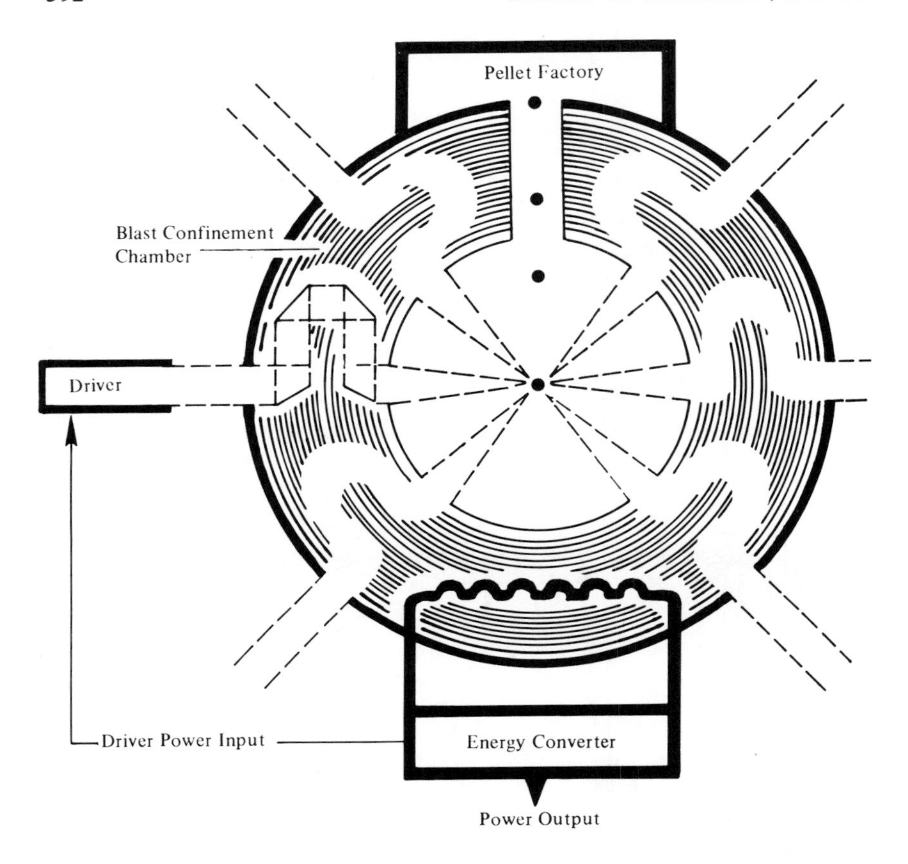

FIGURE 7-2 An inertial confinement scheme for fusion. Fuel pellets are dropped into a vacuum chamber and subjected to focused high-energy beams that cause the pellets' outer layers to evaporate explosively. The resulting forces compress and heat the pellets' cores, generating fusion reactions.

delivered to the pellet to compress and heat the core to a state sufficient for fusion reactions. The performance of available laser systems is far below that required for reactor purposes, although lasers adequate for proposed scientific breakeven experiments (creation of conditions that, in a reactor, could lead to net energy output) are nearing completion.

TIMETABLE FOR DEVELOPMENT AND CRITERIA FOR DEPLOYMENT

Strategies to supply energy in the future—new coal technology, solar power, perhaps local geothermal energy, accelerated and advanced nuclear fission technology—all take long times to develop and put to use. Technologies anticipated for extensive use in the first quarter of the next century would thus have to be fully developed by the end of this century. The appropriate course for further development and resolution of problems must be clear by 1985 or 1990. Advanced fission power systems, including breeder reactors, are being developed around the world in approximate accord with this schedule.

In the competition among potential long-term energy sources, the main criteria for choices among alternatives are likely to be timing of availability, cost and reliability of delivered energy, versatility of application, and environmental and social acceptability. A technology with major liabilities by any of these criteria, compared to the competition, presumably will have to offer major advantages by other criteria if it is to be extensively deployed.

With respect to timing, fusion is far behind the fission breeder reactor, and it is behind at least some forms of solar energy technology.* Three stages of feasibility must be considered in judging fusion's position in this regard.

Scientific feasibility for fusion means achieving, in the laboratory, simultaneous conditions of fuel temperature, density, and confinement time that would lead, if they occurred in a reactor, to output power exceeding the input power.

Technological feasibility means building a device that actually produces a net output in usable form, in essentially continuous operation. Such a device must incorporate sophisticated fuel-handling and energy-conversion equipment not needed to establish scientific feasibility and demonstrate long-term operation of magnets, lasers, vacuum pumping systems, and so on. Central to solving problems of technological feasibility for fusion is finding and testing combinations of materials that can perform adequately in the fusion environment of intense high-energy neutron flux, extreme temperature gradients, intense magnetic fields or laser pulses (or both), and other stresses.

Commercial feasibility means translating technological feasibility into a device that can produce continuous and reliable power under conditions and costs that are attractive to users.

*See statement 7-2, by H. Brooks, Appendix A.

Neither magnetic confinement nor inertial confinement fusion has yet passed even the threshold of scientific feasibility, although both are widely considered likely to do so ultimately; with luck, by the early to middle 1980s. Some work related to technological feasibility is under way in advance of the expected demonstration of scientific feasibility, but the engineering and materials problems are so formidable that it is hard to imagine passing through technological feasibility to commercial feasibility in less than 15 years after that demonstration—that is, before the turn of the century. Substantial effects on electricity budgets, then, could not be expected before 2020.

With respect to the cost of delivered energy, fusion will share with the other long-term sources the characteristic that its costs are dominated by the capital costs of the power plants, with fuel costs a minor factor. Simply on the grounds of technological complexity, it seems likely that the capital costs of fusion power plants—and hence delivered energy costs—will be higher than those of fission breeder reactors. With respect to versatility of application, first-generation fusion reactors (those based on the D-T reaction) are not likely to be much different from fission breeders. Both will produce electricity in large blocks.

Fusion's disadvantage in timing and its likely disadvantage in energy cost or versatility place considerable weight on the magnitude of any advantages it may enjoy over its competitors in environmental and social acceptability. Here it is important to understand that fusion is not one potential technology but many and that these imply a substantial range of characteristics pertinent to environmental hazards, safety, and (perhaps) links to nuclear weaponry. It seems almost certain that even the least attractive of the fusion possibilities will present lower risks than fission breeder reactors in most of these characteristics, but it is quite possible that the margin of superiority would not be judged great enough to offset fusion's disadvantages in time of availability and perhaps cost. Unfortunately, the approaches to fusion that appear to offer the greatest environmental advantages over fission—e.g., tritium inventories and neutron activation so small that the difference in accident risk and waste management problems compared to fission is many orders of magnitude—are the approaches that now appear to be the most difficult to achieve. They tend to entail advanced reactions, advanced materials, and advanced designs. Herein, then, lies a dilemma: Pushing fusion too rapidly toward early commercialization is likely to favor fusion technologies whose environmental advantages over fission are not great enough to be worth the trouble, but delaying fusion too long in pursuit of the more difficult approaches promising theoretically attractive characteristics takes the chance that the fusion option will not be available in time to offer an alternative to fission breeders.*

THE PROGRAM

The program that is moving forward in the face of this dilemma combines vigorous pursuit of demonstrating scientific feasibility, of both magnetic confinement and inertial confinement schemes, with study of reference designs of fusion reactors as a first step toward technological and commercial feasibility. The first reference designs were begun in 1967. These were not intended to serve as blueprints, but as preliminary exercises to ensure that researchers faced every problem associated with reactors designed around any given concept. Reference designs were problem-seeking rather than problem-solving exercises and, as such, registered considerable success. Long lists emerged of the scientific and technical questions demanding attention: more radiation-resistant alloys, different systems for cooling the reactor, and higher energy-handling ability per unit area of reactor wall. As those involved in the elaboration of reference designs turn their attention to these problems, the designs themselves are discarded in favor of better ones. To date, the problems look difficult but not insurmountable. Reference designs have looked progressively more attractive as conceptual solutions for various problems have emerged. New reference designs will appear in due course. The reference design exercises and experimental work proceed more or less continuously. If the fusion program succeeds, the designs for practical fusion reactors may be unlike any envisaged to date.

But will the program succeed? Scientific successes in raising temperatures and improving confinement must be balanced against the accumulating evidence that the technical

One of the formidable difficulties for the nuclear fusion program to overcome is the tendency of groups working on some phase of complicated technology to use up design flexibility in the solution of their problems and leave impossible tasks for others. Technology and engineering cannot simply be ordered to the specifications of plasma physics, nor can the laws of physics and materials science be ordered to the specification of fusion researchers. The program must continuously seek to balance the competing demands of plasma dynamics, materials science, technology, engineering, and power at a reasonable cost.

*See statement 7-3, by H. Brooks, Appendix A.

CONCLUSIONS

• Although the development of nuclear fusion faces considerable uncertainties, it should be pursued and reevaluated in 5 years. By that time, large scientific breakeven experiments in both magnetic and inertial confinement will have been attempted. More realistic engineering designs and guidance for further research on technological obstacles should then emerge naturally.

• Principal attention should be directed first to the problems of pure fusion reactors, before the question of fusion-fission hybrids is considered.

• The immature state of fusion research and development offers the opportunity to given attention to the environmental and safety characteristics in the earliest stages of design. Consideration of these characteristics is so important to decisions on major investments in fusion that the opportunity should not be wasted.

• A small effort should be directed to fuel cycles other than deuterium-tritium. Pure deuterium has a much lower reaction rate but no critical tritium regeneration problem, and it wreaks less structural damage from high-energy neutrons. In the so-called neutronless fuel cycles, all particles and products are electrically charged, and in theory there is no radioactivity. Smaller devices might be built, but the required plasma temperatures are much higher, and the energy balance is probably unfavorable.

• High priority should be given to study and testing of structural materials, and assessments of their availability must be undertaken.

• Research and development in nuclear fusion has enjoyed singularly fruitful international cooperation. This cooperation should be encouraged and extended to speed progress and reduce the cost to each individual country.

NOTES

1. For an overview of fusion's long-term prospects, see J. P. Holdren, "Fusion Energy in Context: Its Fitness for the Long Term," *Science* 200 (1978): 168–180.

2. D. J. Rose and M. Feirtag, "The Prospects for Fusion," *Technology Review,* December 1976, pp. 20–43. See also the report of the Fusion Assessment Resource Group of the Supply and Delivery Panel: National Research Council, *Supporting Paper 3: Controlled Nuclear Fusion: Current Research and Potential Progress,* Committee on Nuclear and Alternative Energy Systems, Supply and Delivery Panel, Fusion Assessment Resource Group (Washington, D.C.: National Academy of Sciences, 1978).

3. Uses have also been proposed for the neutrons and radiation produced by D-T fusion; for example, radiolysis of water to yield hydrogen, or transmutation of radioactive fission products to shorter-lived or stable isotopes.

4. From the report of the Fusion Assessment Resource Group of the Supply and Delivery Panel (see note 2).

8 Geothermal Energy

The United States' vast resource of geothermal heat—thought to come from the radioactive decay of elements in the earth's crust—is certain to become increasingly useful as time goes on and fuel prices go up. At present it provides very little useful energy, and the technical and economic barriers it faces make it unlikely to become one of the nation's main energy sources before the end of this century, if ever. Furthermore, geothermal deposits of the types most useful at present—natural steam and hot water reservoirs—are rather localized, mainly in the western states and often far from potential users. Still, even the small currently useful part of the resource can be important in a world of rising fuel prices and declining supplies. If current research and development are successful in solving some rather intractable technical problems, and if the economics of energy become more favorable, the contribution of the now inaccessible parts of the resource could be a significant factor in reducing oil imports.

Geothermal energy now contributes less than 1/40 of a quad in the United States, all from steam and hot water fields. At The Geysers, near San Francisco, the only geothermal electric power plant in the United States and the largest in the world, with a 565-megawatt (electric) (MWe) capacity in mid-1979 and expanding, exploits one of this country's rare commercial-size steam reservoirs. In addition, about 15 megawatts (thermal) (MWt) of thermal energy from hot water reservoirs are used directly in various places for space heating, low-temperature industrial process heat, and similar uses.

Where it is usable, geothermal energy can be a very economical source of heat for direct use. However, most of the more accessible, higher-grade hot water reservoirs are in parts of the western states remote from industry and population. Because steam can be pipelined efficiently for only a few miles, there is little market for this heat unless it can be converted to electricity, which could be transmitted hundreds of miles with relatively small losses. Power generation from the highest-grade geothermal resources is technically feasible now; rises in the prices of other fuels and advances in geothermal technology could render it economically feasible as well.

In addition to the technical and economic problems of locating and exploiting the different kinds of geothermal resources, development faces a number of institutional constraints. Federal and state leasing policies, for example, often conflict. Leasing itself is slow and costly to bidders. The tax status of geothermal development is unclear; the resources are treated as minerals in some states and as water resources in others. The form of future utility contracts with geothermal steam producers is uncertain. All of these problems will retard development and increase costs unless they are corrected. Their influences are discussed more fully later in this chapter, under the heading "Future Development of the Geothermal Resource."

It is apparent that federal and state governments will largely determine the speed with which the present small geothermal industry can be expanded. By sponsoring research and development in exploration and production techniques, federal funds can provide the technical means. By streamlining leasing procedures, putting the geothermal resource on an equal tax footing with oil and gas, and providing financial assistance, especially in high-risk ventures, federal legislation could greatly speed the day when geothermal energy assumes a competitive place in the U.S. energy market. The timing and extent of these measures will dictate when that day arrives.

GEOTHERMAL RESOURCE TYPES

The geothermal resource is divided, for the purposes of this report, into six categories.

HOT WATER RESERVOIRS

In many places in the United States, especially in the western states, are underground reservoirs of geothermally heated water, some of which are tapped for space heating and the like. Some electricity is generated abroad by such deposits, but because of their higher salinity the hotter geothermal brines, more suitable for this purpose, are very corrosive to generating equipment. This is a very large part of the geothermal resource, though, and the Department of Energy devotes a large proportion of its geothermal funds to investigating economical ways of using it.

NATURAL STEAM RESERVOIRS

Under rare geological conditions the pressures in hydrothermal reservoirs are so low that the water has boiled to steam, as at The Geysers power development. While this form of reservoir is the rarest, it is the easiest and most economical to tap, largely because there is little corrosion problem and little need to deal with the large amounts of brine that must be drawn from and reinjected into a hot water reservoir. It is a very inexpensive source of electricity, but because of its rarity it is not likely ever to contribute more than a few gigawatts of generating capacity to this country.

GEOPRESSURED RESERVOIRS

In some deep sedimentary basins in the United States, notably along the Gulf Coast, are deposits of brine, highly pressurized by the weight of the overlying land. These deposits are mostly at temperatures below 180°C and are assumed to be nearly or completely saturated with natural gas. The temperatures are generally too low and the deposits too deep for economical electricity generation, but the credit for producing gas as a by-product, if it is exploitable along with the heat, offers some potential for space heating and similar direct uses, if the costs of alternatives rise rapidly.

NORMAL GEOTHERMAL GRADIENT AND HOT-DRY-ROCK RESOURCES

The various types of steam and hot water reservoirs are relatively easy to tap, because they supply their own working fluids. However, most of the potentially exploitable geothermal heat is stored in dry rock. Even the normal geothermal gradient (about 30°C/km of depth) provides throughout the world temperatures usable for electric power generation (180°C) at the accessible drilling depth of 5.5 km and temperatures useful for direct heating (80°C) at a depth of 2.2 km. In many places the geothermal gradient is higher than this. Where it is higher than 40°C/km, the resource is called "hot dry rock."

Though these two types are by far the largest in heat content, they cannot yet be exploited. To do so, it will be necessary not only to supply the working fluid by injecting water, but also to find or create in some way a permeable network of channels through which the water can flow to be heated. A number of projects around the world are investigating ways to exploit the resource, but none has reported results that can be applied commercially.

MOLTEN MAGMA

The final and most speculative part of the resource is the heat in molten rocks, or magma. Very few accessible magma bodies are known, and there is no technology for recovering heat from them, but their high temperatures and great heat content make them potentially very important.

ESTIMATED HEAT CONTENTS AND PRODUCIBILITY

Table 8-1 summarizes the total estimated heat contents of each of these six geothermal resource types in the United States, divided by temperature range. These estimates are based on U.S. Geological Survey (USGS) Circular 726,[1] the most authoritative and up-to-date source of data on the geothermal resource of the United States. For the purposes of this study, however, the USGS estimates have been retabulated using somewhat more conservative assumptions. The USGS, for example, includes in its resource base estimates of all of the geothermal heat above 15°C to a depth of 10 km, and assumes that within these limits all heat at temperatures above 90°C could be used for space heating and similar applications and that temperatures above 150°C could be exploited for electricity generation. This study considers 6 km a generally more reasonable maximum depth for geothermal drilling, 80°C the minimum commercially useful temperature for low-temperature heating applications, and 180°C the minimum temperature for electricity generation.

Note that the currently useful part of the resource, hot water and steam reservoirs, is a very small share of the total.

Table 8-2 lists estimates of the amounts of energy actually producible from geothermal reservoirs of the six types.

The uncertainty of the estimates in these tables is obviously very great. Geothermal energy is a largely unexplored resource; from 1970 to 1974 about 130 geothermal wells were drilled, while at least 140,000 oil and gas wells were completed. In fact, the size of the resource can be debated endlessly. Obviously, though, the sum is so large that production is limited not by the size of the accessible resource but by technology, economic considerations, and legal, social, and environmental issues.

TABLE 8-1 Estimated Thermal Energy Contents of Accessible U.S. Geothermal Reservoirs of Several Types (quads)[a]

Reservoir Type	Reservoirs at Temperatures from 80°C to 180°C			Reservoirs at Temperatures Above 180°C			Total Identified and Undiscovered
	Identified	Undiscovered	Total	Identified	Undiscovered	Total	
Hot water	784	3,198	3,982	505	7,530	8,035	12,017
Natural steam	0	0	0	106	73	179	179
Geopressured[b]	30,905	37,086	67,991	2,442	2,930	5,372	73,363
Normal gradient[c]	947,000	—	947,000	306,000	—	306,000	1,253,000
Hot dry rock[d]	52,500	—	52,500	111,000	—	111,000	163,500
Molten magmas	0	0	0	0	3,500	3,500	3,500
TOTAL	1,031,189	40,284	1,071,473	420,053	14,033	434,086	1,505,559

[a]Heat contents above 80°C to 6-km (19,685 ft) depth except where noted. National Parks are not included.
[b]Does not include heat content of reservoir rock or dissolved natural gas, or energy recoverable mechanically from high-pressure fluid. Includes entire heat content of fluid above 50°C, in onshore reservoirs only, to depths of about 6–7 km.
[c]Assumed geothermal gradient, 30°C/km.
[d]Assumed geothermal gradient, 40°C/km.

TABLE 8-2 Estimated Total Potentially Producible Thermal Energy from Accessible U.S. Geothermal Reservoirs of Several Types (quads)[a]

	Heat Contents		
Reservoir Type	80°–180°C	Above 180°C	Total
Hot water	1,991	4,018	6,009
Natural steam	0	45	45
Geopressured[b]	2,244	177	2,421
Normal gradient[c]	9,470	3,060	12,530
Hot dry rock[d]	525	1,110	1,635
Molten magmas[e]	0	35	35
TOTAL	14,230	8,445	22,675

[a]Heat contents above 80°C to 6-km (19,685 ft) depth except where noted.
[b]Heat contents above 50°C to depths of 6–7 km, in onshore reservoirs only, not including the heat content of dissolved natural gas. Economical production technology not yet demonstrated.
[c]Assumed geothermal gradient, 30°C/km. Heat-extraction technology not yet demonstrated.
[d]Assumed geothermal gradient, 40°C/km. Heat-extraction technology not yet demonstrated.
[e]Practical heat-extraction technology not yet developed.

TECHNICAL AND ENVIRONMENTAL CONSIDERATIONS

HOT WATER RESERVOIRS

In this type of underground reservoir the water circulates convectively throughout the reservoir at a nearly uniform temperature, ranging from only slightly above atmospheric temperatures to 350°C or higher. Hydrostatic pressure is generally high enough to keep it from boiling even when greatly superheated, and the water generally remains in the reservoir long enough to become saturated with minerals.

Hotter reservoirs usually are more saline than cooler ones, due to the increased solubility of most minerals at high temperatures. Geothermal waters with total dissolved solids contents ranging from less than 0.1 percent to more than 30 percent are known in the United States. The cooler, less saline waters are used directly in many places for such purposes as space heating. The hotter, more saline waters are used in a few places abroad for generating electricity, but this is not economical in the United States.

Energy Content

The potential for hot water reservoirs in the 80°C–180°C range is based on a retabulation of USGS Circular 726,[2] omitting reservoirs in national parks. This defines a resource base of 784 quadrillion Btu (quads) (Table 8-1). This yields a calculated total heat content of 3982 quads for U.S. hot water reservoirs at temperatures between 80°C and 180°C (Table 8-1). The potentially producible energy above 180°C is calculated to be 505 quads of heat, and the total resource base in the temperature range from 180°C to 850°C is calculated at 8035 quads (Table 8-1). If half this heat can be recovered, 1991 quads of useful heat is potentially producible from U.S. hot water reservoirs with temperatures between 80° and 180°C (Table 8-2), and 4018 quads from hot water reservoirs warmer than 180°C.

At present no electricity is produced in the United States from hot water geothermal sources, and direct use of geothermal hot water in this country amounts to only about 15 MWt.[3] However, both types of use are expected to increase steadily in the United States throughout the next few decades, as they already have in several other countries.

Environmental Considerations

The relatively cooler reservoirs have a major disadvantage in that larger volumes of water must be withdrawn and reinjected to produce a given amount of heat, at a risk of subsidence and aquifer disruption. Given reasonable care, however, nonelectrical uses of such deposits should do little environmental damage because of their small scales.

With hotter, more saline reservoirs (such as those in the Imperial Valley of California) the volumes of fluid that must be extracted and reinjected to produce a given unit of heat are considerably smaller, but accidental spills present hazards of soil salination (already a natural problem in the Imperial Valley) and water pollution. Also, at least some reservoirs produce brine high in the air pollutant hydrogen sulfide. The land-use conflict between geothermal development and agriculture is also serious, as is the possibility of crop damage by geothermal effluents.

NATURAL STEAM RESERVOIRS

Except for variable amounts of noncondensible gases such as carbon dioxide and hydrogen sulfide, natural geothermal steam is usually quite pure and can be piped directly from a well to a turbine generator system. This is the case at The Geysers. Elsewhere, natural steam fields are rare; the USGS has identified another steam field in Yellowstone National Park and suggests the existence of another in Mount Lassen National Park, but because of their locations, these cannot be developed.

Energy Content

From data in USGS Circular 726,[4] the accessible U.S. resource base of natural steam is estimated at 179 quads of heat above 80°C (Table 8-1). If one fourth is assumed potentially producible, this represent 45 quads of useful energy (Table 8-2).

The technologies for producing and using natural steam commercially are well developed. However, there is no reliable method for locating and evaluating new steam reservoirs. Except in Yellowstone Park and in extensions of The Geysers field, no major steam discoveries have been confirmed in recent years. Unless new fields are discovered, all U.S. expansion of natural steam use will be at The Geysers. There, in 1975, the net installed generating capacity was 502 MWe, and plans at the time were to increase capacity to 1238 MWe by 1981, and ultimately to about 2000 MWe. Long delays in licensing and certification have interfered with this schedule, and the rate at which even the proven reserves at The Geysers will be developed is still in question. However, The Geysers field apparently extends far beyond the area so far developed,[5] and Reed and Campbell[6] estimate a maximum potential generating capacity of up to 5000 MWe.

There are good physical, geological, and historical reasons to believe that natural steam fields of commercially exploitable size are rare. It is evident, however, that The Geysers field is not the only such reservoir. We therefore assume here that one or more discoveries will occur, adding a total productive capacity of the order of two thirds that of The Geysers, which appears to be unique in size if not in kind.

Environmental Considerations

Steam reservoirs have one great environmental (and economic) advantage; the only water that must be withdrawn is the actual steam that goes through the turbines. Potential pollutants are, therefore, largely restricted to relatively volatile gases, although some dust and a few rocks come up. There is no need to deal with gargantuan volumes of dirty water and tons of silica scale. (However, The Geysers condensate does contain many environmentally harmful chemicals and must be reinjected.) The relatively small quantities of water involved and the already low pressures in these reservoirs render subsidence hazards small or nonexistent.

GEOPRESSURED RESERVOIRS

Geopressured brine reservoirs are relatively common in deep sedimentary basins. In particular, large areas along the Gulf coasts of Texas and Louisiana are underlain by thick beds of water-filled sandstones and shales. The weight of the overlying sediments produces pressures up to several thousand pounds per square inch above normal hydrostatic pressure.

The water does not circulate deeply, so that it reaches only moderately elevated temperatures, but its volume is so great that the total heat content is thousands of quads. The brine is believed in general to be saturated with natural gas, which could conceivably be an important supplement to the nation's fuel supply. There is no direct evidence that heat or natural gas or both can be extracted economically from geopressured hot water reservoirs, but large-scale field experiments to investigate this have begun in Texas and Louisiana.

Energy Content

Based on this study's retabulation of estimates in USGS Circular 726,[7] the identified resource is estimated to be 30,905 quads in reservoirs at temperatures between 80°C and 180°C, and 2442 quads in the single listed reservoir with a temperature higher than 180°. On the basis simply of the relative land area involved, it is estimated that the accessible geopressured resource base consists of 67,991 quads of heat at temperatures between 80°C and 180°C and 5372 quads at temperatures above 180°C, or a total of 73,363 quads of heat above 80°C (Table 8-1). This is only heat in the reservoir fluid and does not account for a small amount of mechanical energy potentially recoverable by flowing the naturally pressurized water through a turbine, or for the much larger energy content of dissolved natural gas.

If the 3.3 percent recovery factor derived by Papadopulos and his colleagues[8] is representative for this resource, the potentially producible heat from accessible geopressured reservoirs in the United States can be estimated at 2421 quads (Table 8-2). However, relatively rapid decreases in reservoir pressure were observed when geothermal water was tapped by exploratory oil and gas wells in Gulf Coast geopressured areas, leading some reservoir engineers to believe that it may be impossible to maintain the high per-well production rates necessary to economical exploitation. Accordingly, the fraction of the heat that may actually be producible remains uncertain.

There are strong differences of opinion about the technical and economic feasibility of producing commercial power from geopressured reservoirs. There is, however, general agreement on several major issues. First, the resource base (Table 8-1) is large enough to deserve a thorough investigation. On the other hand, the relatively low temperatures and the high cost of drilling the necessary deep holes make it unlikely in the near future to become an economical source of heat for generating electricity. If, however, the geopressured brine is nearly or completely saturated with natural gas and if this can be economically separated, dried, and pipelined, then the credit for natural gas recovery may make this a very economical source of relatively low-grade heat.

It is anticipated that by about 1981 enough information will have been collected to permit an intelligent decision on the feasibility of constructing electrical generating plants using geopressured reservoirs. As much as 25,000 MWe of generating capacity might eventually be installed along the Gulf Coast. It is, of course, also possible that the results of research now beginning will discourage this.

Environmental Considerations

There is little information on the chemical compositions of these brines beyond their salinities (1.5–9.0 percent), but their geological origin as coastal sediments and the presence of large amounts of methane strongly suggest large concentrations of hydrogen sulfide, a harmful air pollutant. A more certain and ultimately even more serious environmental problem is the hazard of subsidence, since the pressure in these reservoirs is due simply to the weight of the overlying land. It seems clear that subsidence problems will limit exploitation. To be sure, full reinjection of the brine might prevent subsidence, but the pumping required would use up all the captured mechanical energy and more.

HOT DRY ROCK

Particularly where the crust is thin or has been disturbed by volcanism or faulting, higher than normal geothermal gradients are often encountered. This offers the possibility of reaching usefully high temperatures with shallower, less expensive holes than would be needed where the geothermal gradient is normal (about 30°C/km of depth). For purposes of discussion, a gradient of 40°C/km is the dividing line between the normal gradient and hot-dry-rock resource types.

The main technical barrier to exploiting this part of the geothermal resource is the lack of a method for extracting heat from deeply buried dry rock. The approach most widely investigated is to use water as a working fluid. However, much of the hot-dry-rock resource is embodied in impermeable rocks, which lack channels through which the water can flow to be heated. In such a case it is necessary to create artificially a large permeable region through which the water can circulate before being withdrawn and used. Bodvarsson and Reistad[9] have discussed the possibility of recovering heat by forced circulation of water through fractured fault zones or open contacts between lava beds or between dikes and the surrounding country rock. Others have investigated the creation of such regions by using conventional and nuclear explosives and by injecting water to flash explosively to steam.[10,11] None has reported commercializable results.

Once a large enough permeable region has been found or created, water must be injected until a steam pressure or water table high enough for extraction is built up. No one knows whether usable per-well flow rates can be achieved in this manner or how much water might have to be invested.

Energy Content

While information on the subject is fragmentary, it is conservatively estimated that 5 percent of the land area of the United States is underlain by rock in which the geothermal gradient is 40°C/km or more. This suggests a resource base of at least 327,000 quads of heat at temperatures above 80°C and depths less than 6 km, of which about 105,000 quads would be in rock at temperatures between 80° and 180°C and about 222,000 quads in rock hotter than 180°C. Most such areas are in sparsely populated parts of the western United States, and it is assumed that, if environmental considerations permit, half this area ultimately may be accessible to geothermal energy development. This yields an estimated resource base of 52,500 quads in rock with initial temperatures between 80°C and 180°C and 111,000 quads in rock hotter than 180°C, for a total of 163,500 quads (Table 8-1).

There is no technical basis for assuming that any energy will be recoverable from this part of the geothermal resource. An assumed recovery factor of only 1 percent, however, would indicate 1635 quads of producible heat at depths less than 6 km, of which 525 quads would come from rock at temperatures between 80°C and 180°C and 1110 quads from rock hotter than 180°C (Table 8-2).

Environmental Considerations

The chief environmental problem with this scheme is that very large volumes of water may have to be invested to bring the reservoir up to producible condition. The amount required will be considerably less if the reservoir is to generate steam rather than hot water. If the rock is naturally permeable, there is no way to be sure of how large a volume of rock would need to be wetted before a usable steam pressure or water table could be attained.

The more important version of the hot-dry-rock technology is likely to be that based on artificially fracturing impermeable hot rock. With hydraulic fracturing, the environmental impacts of development will be comparable to those of native steam or hot water production, depending on how the reservoir is created and operated. There will be some risk of induced seismicity if water is injected into a tight fault in shear-stressed rock. Given reasonable siting care, this risk should be minor. With nuclear explosive fracturing techniques, the main environmental constraint is likely to be seismic.

NORMAL-GRADIENT GEOTHERMAL HEAT

So-called normal-gradient geothermal energy represents most of the nation's geothermal resource base. Exploitation of normal-gradient resources presents in essence the same technical problems as that of hot dry rock. The greater depth at which a given temperature can be reached, however, intensifies the difficulties and increases the potential costs. Development of normal-gradient resources, therefore, will lag behind that of hot dry rock. Presumably, hot-dry-rock extraction techniques could be merely extended to the greater depths necessary when and if it becomes economical to use heat from normal-gradient resources.

Energy Content

The normal-gradient resource is calculated to contain about 3.76 million quads of heat at depths between 2.2 km and 6 km and temperatures between 80°C and 195°C. Because of geological, topographic, and land-use constraints, two thirds of this is inaccessible to geothermal development. This reduces the resource-base estimates to 306,000 quads in rock hotter than 180°C and 947,000 quads at temperatures between 80°C and 180°C (Table 8-1).

In the absence of a demonstrated technology for extracting heat from this deeply buried, relatively low-grade energy source, there is no satisfying basis for estimating the amount of useful heat that might eventually be recovered from it. However, given a recovery factor of 1 percent, it would represent 3060 quads of potentially producible heat (above 80°C) from rock at temperatures above 180°C, and 9470 quads at temperatures between 80°C and 180°C (Table 8-2).

Environmental Considerations

As noted, the technology for exploiting this resource has not been demonstrated, and environmental effects are difficult to estimate. In general, the problems associated with hot-dry-rock exploitation will probably apply.

MOLTEN MAGMA

The extreme case of hot dry rock is magma, or molten lava, which may be found at temperatures higher than 650°C, in pools at the surface or in reservoirs below volcanoes. Aside from a few in national parks, the existence of such bodies and their depths are generally speculative, and practical means of extracting heat from them have yet to be demonstrated. However, there is now some evidence of magma development, and some research is being done.

Energy Content

Because the only known lava pools in the United States are in Hawaii Volcanoes National Park and are thus inaccessible to development, the identified, accessible resource base represented by molten magmas is zero. Some of the Alaskan volcanoes are accessible, and the probability of the existence there of molten magma bodies at drillable depths is high enough that they can be considered an undiscovered resource base. Their heat content above 300°C, to an unspecified depth, is estimated by Smith and Shaw[12] to be 7900 quads. Recalculating this to a reference temperature of 80°C, and assuming that one third of the useful heat exists at depths less than 6 km, yields a resource base estimate of 3500 quads (Table 8-1). If 1 percent of this can eventually be recovered, the product would be 35 quads of useful heat (above 80°C).

Environmental Considerations

Because of the lack of well-defined plans, it is impossible to discuss the potential impacts of magma exploitation. Any proposed scheme will warrant most serious environmental scrutiny before it is allowed to proceed.

PRODUCTION COSTS

Cost figures for geothermal heat mean very little unless they are associated with a use efficiency, which in the temperature regime of most geothermal reservoirs may be very high if the heat is used directly and very low if it is converted to electricity. Btu for Btu, geothermal heat from the highest grade, most accessible reservoirs is competitive with heat from coal at $20/ton, or about $1 per million Btu. In the relatively rare cases in which usable geothermal deposits are close enough to potential users for the heat to be used directly, it is therefore a rather inexpensive source of heat.

However, because of the relatively low temperatures involved, the efficiency with which electricity can be generated from geothermal heat is low. Even the highest-grade geothermal resources, natural steam fields, allow a generating efficiency of only 20 percent, compared to typical efficiencies of about 35 percent for conventional thermal power plants. This means that to be a competitive source of heat for generating electricity, geothermal heat must be at most about half the cost of competing sources, Btu for Btu. The economic success of a geothermal industry therefore depends as much on large increases in the prices of competing fuels as it does on gains in efficiency. Table 8-3 lists cost estimates for electric and nonelectric uses of geothermal resources. (Magma is not included because there is no established or proposed technology for extracting heat from it.)

NATURAL STEAM

When it exists in an accessible reservoir whose volume and permeability are sufficient to guarantee a long enough lifetime at a usable production rate, natural steam is a particularly economical energy source. While many questions about the identification, development, and internal mechanics of natural steam fields remain, the technology for producing and using steam is highly developed at The Geysers and elsewhere. Since it requires no fuel handling, combustion, or smoke- and gas-abatement equipment, a natural steam power plant is relatively simple and inexpensive.

Ownership of The Geysers steam field is distributed among several companies, one of whom manages the currently productive part of the field, produces the steam, and sells it to an electric power company, which generates and sells the electricity. At The Geysers the steam is sold at a price largely unrelated to the actual cost of producing it. The 1975 cost of steam to the power company is reported as 6.89 mills per kilowatt-hour (kWh) of electricity generated, with a "cycle heat rate" (representing the overall thermal conversion efficiency of the plant) of 21,000 to 22,000 Btu/kWh.[13] From this, the calculated cost of heat to the utility in 1975

TABLE 8-3 Representative Heat, Plant, and Generating Costs for Geothermal Energy Systems of Several Types (constant 1975 U.S. dollars)

Reservoir Type	Temperature (°C)	System Costs (dollars per kilowatt of installed capacity)				Heat Cost (dollars per million Btu)	Generating Cost (mills per kilowatt-hour)
		Field Development[a]	Generating Plant	Other	Total		
Hot water (1)	90–125	130–500	none	—	—	0.40–2.50	(Nonelectric)
Hot water (2)	150–270	160–600	400–700	15[b]	600–1300	0.40–1.50	15–45
Natural steam	240	150	255–280	15[b]	420–445	0.35	11–14
Geopressured (1)	80–200	390–450	none	—	—	0.75–2.00[c]	(Nonelectric)
Geopressured (2)	180–200	390–450	440–550	130–260[d]	970–1150	—	20–35[e]
Normal gradient	80–200	300–700	300–600	15[b]	600–1300	—	30–60
Hot dry rock	180–300	230–600	300–600	15[b]	600–1200	—	20–40

[a] Including reinjection system.
[b] Local transmission (20 miles).
[c] Without credit for methane production.
[d] Natural-gas separation and pipeline to market (50 miles).
[e] With credit for methane produced at $2 per thousand cubic feet.

was about $0.32 per million Btu. (This price, however, may not represent the present cost of finding, developing, and producing steam from a new field elsewhere; the 1976 cost of steam at The Geysers was 11.35 mills/kWh or $0.53 per million Btu, and the 1977 cost was then estimated at about 14.5 mills/kWh, or approximately $0.67 per million Btu.)

The capital cost for Generating Unit no. 14, now being built at The Geysers, is $149 per kilowatt (electric) (kWe) (not including the wells and steam-collection system), but estimates for future units are considerably higher.[14] For an 80 percent load factor this gives a delivered cost for power of 9.2 mills/kWh, including the steam cost and the 0.5 mill/kWh paid the field operator to dispose of excess condensate, but not including distribution or customer service costs or general company overhead. Greider[15] estimated the capital cost of a typical generating plant using natural steam at $210/kWe installed, with which were associated capital investments of $148/kWh for field development and $15/kWh for local transmission (20 miles), representing a total investment of $373/kWe in the complete system. (For geothermal developments in which heat is produced and used at the same site, it is usually the capital investment in the complete system that is stated as a "plant cost." This is equivalent to including in the capital cost of a coal-fired power plant its *pro rata* share of the investment in the coal mine and the transportation system furnishing coal to the plant.) Greider[16] estimated that electricity from such a plant should sell for about 10–13 mills/kWh at the plant bus-bar (before transmission). However, to allow for the addition of the hydrogen sulfide abatement or recovery system now required, Greider suggests that the investment in the generating plant should be increased to the range of about $255–$280/kWe (1975 dollars), which would raise the price of power somewhat.

HOT WATER RESERVOIRS

The cost of heat from hot water geothermal systems varies widely with the depth and temperature of the reservoir, the production rates and load factors of the wells, the chemical characteristics of the fluid, and the distance from producer to consumer. For a variety of typical nonelectric uses, Towse[17] gives a range of $0.372–$5.217 per million Btu, including the costs of reinjection wells and of distribution systems where needed. Omitting one obviously uneconomic case (heating a single dwelling) reduces the top of this range to $2.24 per million Btu. In general, the lowest costs are for large-scale, on-site, industrial users with steady loads; the highest are for residential heating, where individual loads are small and intermittent and distribution lines are long. By excluding other relatively small-scale space-heating applications and two cases for which wells were fairly deep and brine concentrations very high, the cost range in

Towse's list is further reduced to $0.372–$1.184 per million Btu, with an average of $0.86.

When geothermal heat from hot water wells is to be used for generating electricity, this broad range of costs is extended further by the fact that, at lower temperatures, more water is needed to deliver a given amount of heat, while the efficiency with which the heat can be used decreases rapidly. For a given generating capacity, large increases are required in the number of both production and injection wells and in either the number or the size of most surface facilities. Thus, for 50-MWe power plants using geothermal water of low salinity at 149°C, Swink and Schultz[18] list estimated total system costs of $1190–$2885/kWe of installed generating capacity, for a wide variety of possible power cycles, and corresponding generating costs of 26.7–64.7 mills/kWh. For water from the same reservoir, but with a lower production rate per well, Bloomster[19] estimates a generating cost of 26 mills at the plant. For 150°C–200°C hot water sources in which the water flashes directly to steam for the turbine, Milora and Tester[20] estimate a total system cost of $650–$1826/kWe and a generating cost of 16.1–43.0 mills/kWh. Substituting a binary cycle plant, in which the hot water is used to heat a second fluid that in turn drives the turbine, would reduce total system cost to $632–$1773/kWe and generating costs to 15.7–41.8 mills/kWh. For a 200°C low-salinity reservoir and an isobutane binary cycle, Bloomster predicts an electric power cost of 16.4 mills/kWh at the plant. Greider,[21] for a typical binary system using water at 204°C, estimates the price of electricity at 16–20 mills/kWh.

Apparently, for generating electricity, the higher temperature hot water systems should be competitive with energy sources of other types because both heat and power plant costs are comparatively low. (The average generating cost for conventional power plants is now about 25 mills/kWh.) However, costs increase rapidly as reservoir temperature or production rate per well decreases and as depth or salinity increases.

GEOPRESSURED RESERVOIRS

With the possible exception of one or two on the Gulf Coast, geopressured reservoirs are in the class of lower temperature hot water reservoirs considered above. However, in general they are at relatively great depths (3–5 km or more), and well costs are correspondingly high. So, therefore, are heat costs. For a large on-site user, Towse[22] estimates a heat cost of $0.77 per million Btu for geopressured water at 82°C from wells about 2.6 km deep, including the cost of reinjection. For a similar user of higher-grade heat, he estimates a heat cost of $1.16–$1.18 per million Btu for water at 121°C from wells 4 km deep, again including reinjection costs.

The Gulf Coast geopressured water is generally believed to be nearly or completely saturated with dissolved natural gas. Besides contributing substantially to the fossil fuel supply of the United States, this gas, credited against the cost of producing the geothermal water, could reduce the cost of heat enough to make geopressured water an economical source of low-grade heat for direct use.

Power generating costs would not be so competitive. A Dow Chemical study reported by Weeden[23] estimates a heat cost of $2.00 per million Btu for geopressured water from a 4.5-km-deep reservoir at 163°C, including reinjection costs. Credit for recovering gas reduces this to $0.63 per million Btu. The study included cost estimates for a 25-MWe steam power plant with methane separators, hydraulic turbines to recover mechanical energy, and two disposal wells for each production well. The estimated total capital investment came to $2420/kWe of capacity. With credit for gas production, estimated generating costs came to 46 mills/kWh, almost twice today's average generating cost for conventional facilities. The report concluded that electricity generation would be economical only with higher temperatures, more efficient conversion techniques, or higher natural gas prices.

Other authorities have produced cost estimates that vary widely according to the water temperature, necessary drilling depth, and conversion technology. Generating costs, for example, have been estimated as low as 21 mills/kWh for certain reservoir characteristics and selected technologies.

HOT DRY ROCK

Since the technology for extracting heat from hot dry rock has not yet been fully developed, cost estimates are necessarily speculative. Most methods proposed for heat production from this source involve drilling an injection hole and a recovery hole to depths comparable to those required to reach geopressured reservoirs, so that in similar formations drilling costs might be similar for the two cases. At least initially, however, hot-dry-rock systems will probably be developed in igneous or metamorphic rock, in which drilling costs should be much higher than in the soft Gulf Coast sediments. No salable by-product such as methane would be expected from such a system, but at least in part this and higher per-foot drilling costs should be countered by the higher temperatures reached at a given depth.

Bodvarsson and Reistad[24] suggest that recovering heat by circulating water through fractured fault zones or open contacts between lava beds or between dikes and the surrounding country rock may be at least marginally economic for producing low-temperature heat where the geothermal gradient is only normal (giving, for example, a rock temperature of 135°C at a depth of 3.5 km). The economic potential is significantly better where the gradient is higher.

For a hydraulically fractured system in granite at a depth of 3.9 km and a temperature of 250°C, Milora and Tester[25] estimate a total capital cost of $625/kWe (1975 dollars) for a 100-MWe binary cycle plant, and a generating cost of 15.6 mills/kWh. In a systematic examination of the effects of geothermal gradients, well depths, flow rates, and rock temperatures, they present the possibility of generating costs less than 10 mills/kWh where gradients are very high and up to about 30 mills where they are normal, in both cases with relatively high flow rates through the system. While their cost estimates for hot dry rock are within the range of similar estimates for natural hot water systems where reinjection is required, they of course assume the success of a technology that has not yet been demonstrated.

NORMAL-GRADIENT SYSTEMS

The possibilities and costs of extracting and using geothermal heat in areas where the geothermal gradient is normal (approximately 30°C/km of depth) have already been treated indirectly, in the discussion of hot dry rock. Costs would be increased, of course, by the need to explore and drill to greater depths and by a small decrease in production rates per well, because of the pressure drop involved in producing fluid through a longer string of casing. With the flexibility in location provided by normal-gradient systems and with the cost decreases expected from improved drilling and energy conversion technology, these deep natural energy supplies may become economic in the future, if a proven technology becomes available.

MOLTEN MAGMA

No estimates for the costs of heat and electric power from molten rock will be possible until at least a primitive heat-extraction technology has been developed. It can only be hoped that the high intensity of the heat in molten magma will compensate for the probably high cost of the sophisticated systems that will be required to recover it.

FUTURE DEVELOPMENT OF THE GEOTHERMAL RESOURCE

Besides the technical and economic obstacles already described, geothermal energy faces a number of financial and institutional barriers. The extent to which these difficulties will be solved in the future, and the costs of doing so, are extremely uncertain. At present it is possible only to identify the major problems and the general development trends that could be expected to follow from their solutions.

INSTITUTIONAL PROBLEMS

In addition to the technical and economic problems of locating and exploiting the different kinds of geothermal resources, development of a geothermal industry faces a number of institutional constraints. Federal leasing policy is an especially important one. One feature of federal law that tends to retard development is the 20,000-acre maximum lease size for any company in any one state. The leasing process itself is long and costly enough to bar small firms and to substantially increase the costs of even the large successful bidders. Furthermore, the time lag between lease application and competitive sales is now about 3–4 years.

Leasing is further complicated by the fact that resource areas are often broken up among federal, state, and private ownership. This means that a given project might be subject to regulation by authorities from federal to local levels, with disparate or conflicting standards.

Another delaying influence is the tax treatment of geothermal development. In general such development is not eligible for many of the tax benefits that oil and gas development receive, even though they all require similar drilling operations. Federal and state courts and executive agencies differ on the question of whether geothermal resources are to be considered minerals, subject to equivalent taxation with oil and gas, or water resources, which are not given special tax treatment but are subject, especially in the West, to complicated state regulations designed to allocate water rights.

Finally, looking ahead to wide commercialization of this resource, there is an institutional mismatch between those who explore for and produce geothermal heat and those who purchase the heat for power generation. Most of the former are oil companies, who have shown a preference for fast-write-off, high-risk opportunities. The latter are utilities, whose long planning horizons bias them in favor of low-risk, regulated, fixed rates of return. This may produce conflicts between producers and users in negotiating contracts for steam.

All of these institutional problems must be dealt with along with the technical and economic ones to which each geothermal resource type is subject. The extent to which federal and state leasing policies, taxations, and regulation can be rationalized will largely determine the effectiveness of current and future research, development, and demonstration in exploration and production technologies. It will also affect the costs of production and therefore the competitiveness of the price at which geothermal energy can be sold.

FUTURE TRENDS IN GEOTHERMAL ENERGY PRODUCTION

The Supply and Delivery Panel[26] constructed the three potential production scenarios shown in Table 8-4. The numbers are of course extremely speculative, but they serve to illustrate the range of possibilities under three levels of effort aimed at reducing technical, economic, and institutional constraints. In the business-as-usual scenario all present constraints remain in effect. The national-commitment scenario assumes that geothermal energy is given a high priority and that most institutional problems are corrected, that financial incentives such as loan guarantees are offered to the industry for high-risk projects, that a strong and consistent demonstration plant program is enacted, and that federal funding greatly improves the technologies for exploration. The enhanced-supply scenario falls between these two extremes.

In summary, the ultimate resource potential of geothermal energy is very difficult to estimate because of the lack of development and assessment of the most abundant resources, and because of the absence of a demonstrated technology for extracting the energy from hot dry rock or geopressured brines. In consequence of this situation the estimates set forth in Table 8-4 indicate the relatively modest total contribution that could be expected from geothermal sources within the time period of this study. The maximum potential realizable by 2010 with a national commitment is much lower than that of nuclear fission, and substantially lower even than what is at least theoretically achievable with a national commitment to all areas of solar development.

CONCLUSIONS ON GEOTHERMAL ENERGY

Sources of geothermal energy are not indefinitely sustainable in the same sense as solar energy. However, their total energy is so large that their potential as an energy source will depend mainly on their economic producibility, not on resource considerations.

At present, the only usable geothermal resources are deposits of hot water or natural steam. In the long-term future, it may be possible to

TABLE 8-4 Estimated Installed Geothermal Energy Capacity for Generating Electricity and for Nonelectrical Uses (megawatts and quads)

	Installed Generating Capacity (megawatts)								Heat Required (quads)		
Year	Hot Water	Natural Steam	Geopressured	Normal Gradient	Hot Dry Rock	Molten Magma	Total	Equivalent (quads)[a]	Electrical[b]	Nonelectrical	Total
Business as Usual											
1990	720	1,800	200	10	200	0	2,930	0.09	0.35	0.02	0.37
2010	7,800	4,500	2,100	710	3,700	60	18,870	0.56	2.26	0.18	2.44
Enhanced Supply											
1990	1,000	2,500	310	10	260	10	4,090	0.12	0.49	0.05	0.54
2010	15,400	5,000	3,100	1,510	6,800	360	32,170	0.96	3.85	0.32	4.17
National Commitment											
1990	2,600	4,000	600	60	1,000	10	8,270	0.25	0.99	0.06	1.05
2010	25,600	6,500	4,200	2,600	21,000	1,000	60,900	1.85	7.28	0.96	8.24

[a] Quads per year assuming continuous operation of full generating capacity.
[b] Quads per year assuming 20 percent overall conversion efficiency and 80 percent load factor.

extract heat from the natural thermal gradient in the earth's crust and from unusually hot rock formations lying close to the earth's crust. As there is no demonstrated technology for using these resources, the cost and the amount of energy that might be producible can be only grossly estimated. The use of dry rock depends on developing a fracture system large enough to be economical as a source of heat. The possibilities of achieving this, and the environmental effects of doing so, are speculative.

The only widespread potential geothermal resource, the natural thermal gradient, is the most speculative in practical exploitability. As an indefinitely sustainable source, it also suffers the inherent disadvantage that the normal heat flux from the inside of the earth is only about 1/1000 of the solar energy flux falling on the same area.

One potentially large source of rather low temperature geothermal energy is the geopressured brines of the Gulf Coast. These brines may also hold very large amounts of dissolved natural gas. If the heat and gas can be exploited simultaneously, this might be an attractive resource. Too little is known about it today: Considerable effort is justified in assessing its potential.

Considered in all of its potential, the geothermal resource represents extremely large amounts of energy. However, for a variety of technical, economic, geographical, and institutional reasons geothermal energy will probably not be a major contributor to the national energy system until well into the twenty-first century, if ever. It may, however, become an important source of inexpensive heat for localized use at relatively small scales. While it warrants serious exploration and continued development, it cannot be considered among the most important of the long-term energy alternatives.

NOTES

1. D. W. White and D. L. Williams, eds., *Assessment of Geothermal Resources of the United States—1975,* U.S. Geological Survey Circular 726 (Washington, D.C.: U.S. Government Printing Office, 1975).
2. *Ibid.*
3. J. H. Howard, ed., *Present Status and Future Prospects for Nonelectrical Uses of Geothermal Resources* (Berkeley, Calif.: Lawrence Livermore Laboratory (UCRL-51926), 1975).
4. White and Williams, eds., *op. cit.*
5. T. C. Urban, W. H. Diment, J. H. Sass, and I. M. Jamieson, "Heat Flow at The Geysers, California, U.S.A.," in *Proceedings, Second United Nations Symposium on the Development and Use of Geothermal Resources,* available from Superintendent of Documents (Washington, D.C.: U.S. Government Printing Office, 1975).
6. M. J. Reed and G. E. Campbell, "Environmental Impact of Development in The Geysers Geothermal Field, U.S.A.," in *Proceedings, Second United Nations Symposium on*

the Development and Use of Geothermal Resources, available from Superintendent of Documents (Washington, D.C.: U.S. Government Printing Office, May 1975).

7. White and Williams, eds., *op. cit.*

8. S. S. Papadopulos, R. H. Wallace, Jr., J. B. Wesselman, and R. E. Taylor, "Assessment of Onshore Geopressured Geothermal Resources in the Northern Gulf of Mexico Basin," in D. E. White and D. L. Williams, eds., *Assessment of Geothermal Resources of the United States—1975,* U.S. Geological Survey Circular 726 (Washington,D.C.: U.S. Government Printing Office, 1975).

9. G. Bodvarsson and G. M. Reistad, "Econometric Analysis of Forced Geoheat Recovery for Low-Temperature Uses in the Pacific Northwest," in *Proceedings, Second United Nations Symposium on the Development and Use of Geothermal Resources,* available from Superintendent of Documents (Washington, D.C.: U.S. Government Printing Office, May 1975).

10. Y. D. Diadkin and Y. M. Pariisky, "Theoretical and Experimental Grounds for Utilization of Dry Hot Rock Geothermal Resources in the Mining Industry," in *Proceedings, Second United Nations Symposium on the Development and Use of Geothermal Resources,* available from Superintendent of Documents (Washington, D.C.: U.S. Government Printing Office, May 1975).

11. M. C. Smith, R. L. Aamodt, R. M. Potter, and D. W. Brown, "Man-Made Geothermal Reservoirs," in *Proceedings, Second United Nations Symposium on the Development and Use of Geothermal Resources,* available from Superintendent of Documents (Washington, D.C.: U.S. Government Printing Office, May 1975).

12. R. L. Smith and H. R. Shaw, "Igneous-Related Geothermal Systems," in D. E. White and D. L. Williams, eds., *Assessment of Geothermal Resources of the United States—1975,* U.S. Geological Survey Circular 726 (Washington, D.C.: U.S. Government Printing Office, 1975).

13. F. J. Dan, D. E. Hersam, S. K. Kho, and L. R. Krumland, "Development of a Typical Generating Unit at The Geysers Geothermal Project—A Case Study," in *Proceedings, Second United Nations Symposium on the Development and Use of Geothermal Resources,* available from Superintendent of Documents (Washington, D.C.: U.S. Government Printing Office, May 1975).

14. Dan, Hersam, Kho, and Krumland, *op. cit.*

15. B. Greider, "Status of Economics and Financing of Geothermal Energy Power Production," in *Proceedings, Second United Nations Symposium on the Development and Use of Geothermal Resources,* available from Superintendent of Documents (Washington, D.C.: U.S. Government Printing Office, May 1975).

16. B. Greider, personal communication, January 1977.

17. D. F. Towse, "Economic Considerations" in J. H. Howard, ed., *Present Status and Future Prospects for Nonelectrical Uses of Geothermal Resources* (Berkeley, Calif.: Lawrence Livermore Laboratory (UCRL-51926), 1975).

18. D. G. Swink and R. J. Schultz, *Conceptual Study for Total Utilization of an Intermediate Temperature Geothermal Resource,* (Idaho Falls, Idaho: Aerojet Nuclear Co. (ANCR- 1260), April 1976).

19. C. H. Bloomster, "An Economic Model for Geothermal Cost Analysis," in *Proceedings, Second United Nations Symposium on the Development and Use of Geothermal Resources,* available from Superintendent of Documents (Washington, D.C.: U.S. Government Printing Office, May 1975).

20. S. L. Milora and J. W. Tester, *Geothermal Energy as a Source of Electric Power* (Cambridge, Mass.: MIT Press, 1976).

21. Greider, "Status of Economics and Financing of Geothermal Energy Power Production," *op. cit.*

22. Towse, *op. cit.*

23. S. L. Weeden, "Geopressured Geothermal Energy—Will It Work?" *Ocean Industry,* June 1976, pp. 119–126.

24. Bodvarsson and Reistad, *op. cit.*

25. Milora and Tester, *op. cit.*

26. National Research Council, *U.S. Energy Supply Prospects to 2010,* Committee on Nuclear and Alternative Energy Systems, Supply and Delivery Panel (Washington, D.C.: National Academy of Sciences, 1979).

9 Risks of Energy Systems

Worldwide concern for the protection of public health and the environment has shown remarkable growth in the past 15 years, evidenced in the United States by the passage of landmark legislation, the creation of the Environmental Protection Agency, and the proliferation of regulations to mitigate, for example, the health and environmental risks of energy systems.

This concern is one of many that must be balanced in formulating energy policy. Furthermore, many aspects of this concern are new: Our knowledge of several important risks, as well as our knowledge of how to control them, is recent and incomplete.

Energy policy must be formulated with the knowledge available. Even were such knowledge greater than it is today, difficult decisions would still have to be made. The risks of various energy systems are of different types that cannot all be reduced to common measures. Judgment will continue to dominate these decisions.

The purpose of this chapter is to review the known risks, to indicate the difficulties of ascertaining some of the most important suspected risks, and to recommend both practical courses of action in the face of uncertainty, and steps to improve judgment with better information. Among the major categories of risk considered are those relating to industrial operations, to atmospheric pollution, to shortage of water supply, and to change in climate. For each of these, we have considered the risks posed by energy systems based on fossil fuels, nuclear fuels, and solar energy.

The chapter takes up the nature of risk and the government's actions to control the risks of energy systems. The risks posed by the major energy systems to health, agriculture, climate and water supply, and ecosystems are described, and wherever possible, compared. After some discussion, the major findings are summed up in nine conclusions. The reader may wish to read these conclusions first.

RISK

In ordinary language, as well as in this report, the word risk is used in two ways: to convey the possibility (probability) of loss or to denote a dangerous element or factor. This chapter examines the risks associated with the three principal groups of energy systems—fossil fuel, nuclear, and solar—particularly in the generation of electricity, as this provides a convenient base for comparison.

Risks have been grouped by origin in the various steps of each energy cycle, including extraction and processing of the energy resource; its transportation and storage; its use in the production of another fuel (liquid fuels from coal, for example), electricity, or power; the disposal of waste, and finally end-use. (In engineering literature, fuel cycle is usually synonymous with energy cycle, but in official regulatory practice,[1] fuel cycle excludes mining, operation of waste disposal sites, and transportation.)

The complete evaluation of risk depends on the nature and amount of the dangerous element or factor (termed "insult" by some environmentalists) and an understanding of how it stresses or interacts with its targets, of how the targets are affected (termed "insult" in medical literature), and of how they react in turn. Such target reactions can then affect other objects or systems.[2]

The comparison of risks is often simplified by consistent comparisons—similar kinds of risks that arise when different energy systems are employed for the same specific purpose, such as the production of a stated amount of electricity. The matching of risks may be difficult. Consider, for example, the number of deaths associated with the production of 1 quadrillion Btu (quad) of electricity in 1 year from oil or uranium. Practically all cancer deaths due to the use of oil for 1 year would occur within the 30 years following, but those from uranium might be projected to occur over thousands of years. Would 30 deaths in 30 years be better, worse, or equal to 30 deaths in 1000 years? The decision calls for judgment, or for specific sociological information that shows how the bunching of deaths leads to more or less damage than spreading them over the years.

ASSESSMENT OF TOTAL RISK

Owing to the many quite different types of risks incurred in the operation of an energy system, we considered it undesirable to force them into the terms of some arbitrary measure to obtain a sum. Instead (where possible), comparisons were made on the basis of individual, similar risks. Since the facts are frequently insufficient for clear-cut quantitative analysis, estimates of cost and benefit may become highly speculative. This is especially true if sociological data are in question, but as explained later in this chapter, it is also a frustrating problem even in dealing with the chemical and toxicological aspects of atmospheric pollution. The evaluation of total damage, based on all contributing factors, must therefore be a matter of judgment.

Additional judgment must be brought to bear on decisions about "*acceptable* levels of risk." In this case, the assessment of cost may have to be considered and should include the consequences of being wrong. Society may (or may not) prefer a larger risk that can be estimated with confidence to one estimated to be smaller, but with great uncertainty.

REGULATION

The federal government attempts to protect public health and the quality of the environment through the Environmental Protection Agency (EPA), the Nuclear Regulatory Commission (NRC), the Occupational Safety and Health Administration (OSHA), and other agencies, based on such acts of Congress and subsequent amendments as the Clean Air Act of 1970 and the Clean Water Act of 1972 (as amended), the Nuclear Regulatory Commission Act of 1975, the Federal Mines Safety and Health Act of 1977, the Surface Mining Control and Regulation Act of 1977, the Energy Supply and Environmental Coordination Act of 1974, and the Resource Conservation and Energy Act of 1976. Some 20 congressional committees deal with this field and compete with one another in producing legislation.[3] At a conservative estimate, close to 90 units of the federal government, most of which function independently of one another, set or enforce standards.[4]

While there is no doubt about the need for regulation, its explosive growth and resulting complexity is bewildering and, on occasion, too costly or perhaps even self-defeating. An example of the complexity may be drawn from the electric utilities industry, where the regulation of risk, though not the only consideration of national policy, is a major factor in determining the 8- to 10-yr lead times for construction of fossil-fueled plants, and the 10- to 12-yr lead times for nuclear power plants.[5] It has been estimated that some 90 permits are now required to open and operate a surface coal mine.[6]*

Administrative and Legal Aspects

Most of the standards governing the risks of concern in this chapter have been established by administrative action under federal legislation. The administrator of the Environmental Protection Agency, for example, lists environmental pollutants and sets standards for their allowable concentrations in, say, drinking water or the atmosphere by a process that usually includes a hearing. After proposed regulations are reviewed by the public and departmental consultants, and a final version adopted, the final action is the promulgation of a regulation with the force of law, which is published in the *Federal Register*.

The enforcement of the regulation is by administrative and court action, by court action in civil proceedings for an injunction and civil penalties, or by court action in criminal proceedings. Where the risks arise from a utility plant or other facility licensed under a quasi-judicial process (such as a nuclear plant licensed by the Nuclear Regulatory Commission), consideration of the elements of risk as applied to that facility is an important aspect of the licensing process.

The attainment and maintenance of ambient air quality at the levels set by the Environmental Protection Agency are principally the responsibility of the states,[7] a responsibility they exercise under state implementation plans that have been approved by the EPA. The plans vary greatly in detail from state to state. Local agencies may also comprise parts of the regulatory process.

The complexity and repercussions of such implementation plans are worthy of more than passing notice. For example, as will be considered under "Emissions and Wastes," the vast majority of urban areas are not in compliance with the ozone standard, and the agency expects that even 10 years from now, some will find compliance very difficult despite envisioned steps to curb automobile emissions and to limit traffic patterns.

Practical Aspects

Should an unattainable standard be relaxed? Under the Clean Air Act, standards are to be reviewed every 5 years and may be revised by the administrator of the Environmental Protection Agency on the basis of that review. The act requires, however, that the standards be based squarely on

*See statement 9-1, by H. Brooks, Appendix A.

scientific criteria for adequate health protection: Neither the cost of achieving such standards, nor even the availability of the requisite technology to achieve them, is germane to such a determination.[8]

Whether governmental regulation extends in such cases beyond practicable or just limits may be questioned. As a case in point, the ozone standard was recently reviewed and relaxed, a decision that aroused criticisms of "not enough" and "too much." The several problems of how to regulate need to be studied, and the ozone case will provide a profitable subject for one such analysis.

Since the enactment of the Clean Air Act of 1970, several major policy steps have been taken that will tend to keep permissible ambient air levels of pollutants low or impede their rise. In the case of the new ozone standard, the administrator has set it with consideration for groups of "particularly sensitive citizens such as bronchial asthmatics and emphysematics who in the normal course of daily activity are exposed to the ambient environment."[9] This requirement will be of particular importance for the determination and interpretation of epidemiological exposure-effect curves (see "Dose-Effect Curves") and poses major problems in determining how much weight should be given to small sensitive subgroups of the population.

Two striking and important changes have been made in the regulation of emissions. Areas of "no significant deterioration" have been designated where only a 10 percent increment over preexisting ambient air levels of pollutants would be permitted.[10] Enforcement here will be complicated by the inability to control extraterritorial emissions that are atmospherically transported to the restricted region over distances that are sometimes as great as many hundreds of miles.

Second, it is now required of new utility plants (among other stationary sources) that the best available control technology (BACT) be employed regardless of ambient air quality levels, emission standards, or the quality of fuel used.

While this may provide benefits, one economic effect will be to lessen the advantage of using low-sulfur coal or oil and to enhance the market for high-sulfur eastern coal. The cost of power plant construction is increased by about 25 percent. (See chapter 4.)

Among the major arguments to support this policy is that of caution. It is not certain that adequate protection is provided by present primary and secondary ambient air quality standards. "Best available control technology" represents a philosophy of playing it safe. Its principal disadvantage is cost. On the other hand, if present standards prove materially inadequate, retrofitting would cost much more, or it could even be impractical.

Although dissatisfaction with various aspects of the government's handling of the regulatory process is often expressed (e.g., of nuclear waste, as described in chapter 5 under "Management of Radioactive Waste"), the tremendous importance of regulation cannot be ignored. Considering the legislative expansion of environmental and public health protection in the last 15 years,[11] it is too much to expect that any society could master the art of such regulation so rapidly. We are still in the period of learning by sequential experience the extent to which our regulatory objectives can be achieved, and how.

There are limitations to the control of risk. As the reduction in risk becomes more refined, the incremental benefit eventually diminishes and the cost rises disproportionately. No amount of regulation can ensure a risk-free society, nor should it be assumed that such a goal is desirable.*

HEALTH

The risks associated with several energy systems are compared here on the basis of the production of electricity. Electricity now accounts for 11 percent of final energy demand and 29 percent of primary energy use, and its use is increasing relative to that of other forms of energy. Comparisons are given for each step in the cycle; thus, some of the results (e.g., mining, transportation) are applicable to other end-uses.

The fossil fuel and nuclear systems are of primary interest. Estimates for the risks of solar, fusion, and geothermal energy cycles are still speculative, although it does appear that many uses and forms of solar energy would be no more hazardous than energy systems now in use and, in some forms, less so.[12]

The Risk and Impact Panel of this study reported health effects relative to the operation of an electric generating plant of 1-GWe (10^9 watts = 1 gigawatt (electric) (GWe)) capacity, at 33 percent efficiency and 75 percent capacity factor for 1 year. Such a "GWe-plant-year" corresponds to a fuel input of about 0.0673 quads, and an electrical output of 0.0225 quads, approximately 1/300 of current national electricity production. The fuel required to operate such a plant for 1 year would be 3 million tons of coal, 12 million barrels of oil, 67 billion ft^3 of natural gas, or 150 tons of uranium oxide (U_3O_8) obtained from 75,000 tons of mined ore (0.2 percent) for a light water reactor operated on today's once-through fuel cycle. (With reprocessing and recycle of uranium, this latter requirement could be reduced to 120 tons of uranium oxide.) It should be noted that a GWe-plant-year of electricity is equal to 75 percent of the GWe-year of electricity used in regulatory procedures. A 1-GWe power station will

*See statement 9-2, by H. Brooks, Appendix A.

serve a regional population of about 1 million, including urban, suburban, industrial, and rural components.

ROUTINE INDUSTRIAL RISKS

The routine risks engendered by industrial activities to supply energy involve fatal and nonfatal outcomes.

Fatal Accidents

The lowest accidental death rates in the generation of electricity are for light water reactors and natural gas systems (0.2 deaths per GWe-plant-year). The rate of accidental deaths for electricity from oil is somewhat higher (0.35) and that for coal is very much the highest (2.6 for surface mining, 4.0 for deep mining).* These rates of accidental death are set out in Table 9-1. To place them in perspective, recall that a GWe-plant serves a population of about 1 million. In 1974, the accidental death rate of the general population was 500 per million, of which 220 deaths were due to motor vehicle accidents, 80 to falls, 30 to burns, and 10 to accidents with firearms.

The higher rate of accidental deaths for electricity from coal is largely due to deep mining and to transportation. Both could be improved. In general, small mines suffer twice as many accidental deaths as large mines, per ton of coal mined. If the standards observed in all coal mines were brought up to the standards of the safest, the accidental death rate could be reduced to perhaps a quarter of its present value. However, future trends are difficult to estimate. Until 5 years ago, the accident rate was declining about 4 percent/yr. Since then, it has risen slightly (per ton).[13] In the future, automation and other technical improvements in deep mining should enhance safety, but the rapid expansion of production, with a less-experienced and younger work force, and a possible shortage of mining engineers may raise accident rates.

The high mortality rate due to coal transportation is an estimate that is not directly based on coal-train-miles. It reflects conditions that are likely to change. For future planning, this estimate could easily be high by a factor of 2 or 3, in our opinion. The matter needs more precise study.

*Statement 9-3, by J. P. Holdren: The coal numbers are dominated, as noted later, by an unrealistic estimate of deaths in coal-train accidents. They shouldn't be cited without a disclaimer.

TABLE 9-1 Accidental Deaths During Routine Operation, by Energy Source (per gigawatt-plant-year)

Energy Source and Quantity Required	Extraction	Processing	Transport	Power Station	Total[a]
Coal (3×10^6 tons)		0.02	2.3[b]	0.01	
Deep	1.7				4.0
Surface	0.3				2.6
Oil, onshore and offshore (12×10^6 barrels)	0.2	0.08	0.05	0.01	0.4
Natural gas (67×10^9 ft^3)	0.16	0.01	0.02	0.01	0.2
Uranium oxide[c] (150 tons from 75,000 tons of ore)	0.2	0.001	0.01	0.01	0.2

[a] Totals do not add due to rounding.
[b] The estimates are not based on coal trains per se, but on the overall rate of train accidents. Furthermore, many accidents with trains are not the fault of cargo nor of the carrier, and the responsibility for them may be incorrectly charged. For meaningful statistics, the matter needs further study. A forthcoming review cites figures based on the exclusive use of unit trains that scale to 0.5 deaths per gigawatt-plant-year, less than one-fourth the entry in the table. Carl W. Gehrs, David S. Shriner, Steven E. Herbes, Harry Perry, and Eli Salmon, "Environmental, Health, and Safety Implications of Increased Coal Utilization," in *Chemistry of Coal Utilization*, tech. ed. M. A. Elliott, chap. II, suppl. vol. 2 (New York: Wiley Interscience, in press).
[c] With reprocessing, the uranium oxide requirement could be reduced to 1.4 tons. Presumably, the mean extraction risk would be reduced proportionately, and the processing risk increased. The net result could be lower total risk.

Source: Data for coal are from MITRE Corporation, Metrek Division, *Accidents and Unscheduled Events Associated with Non-Nuclear Energy Resources and Technology* (Washington, D.C.: MITRE Corporation, (M76-68), December 1976), p. 51, except power-station entry. For oil, natural gas, and power stations, U.S. Council on Environmental Quality, *Energy and the Environment—Electric Power* (Washington, D.C.: U.S. Government Printing Office, 1973). For uranium oxide extraction, U.S. Atomic Energy Commission, *Comparative Risk-Cost-Benefit Study of Alternative Sources of Electrical Energy* (Washington, D.C.: U.S. Atomic Energy Commission (WASH-1224), December 1974). For uranium oxide processing, Nuclear Energy Policy Study Group, Spurgeon M. Keeny, Jr., Chairman, *Nuclear Power: Issues and Choices* (Cambridge, Mass.: Ballinger Publishing Co., 1977), p. 175. For uranium oxide transport, U.S. Atomic Energy Commission, Directorate of Regulatory Standards, *Environmental Survey of Transportation of Radioactive Materials to and from Nuclear Power Plants* (Washington, D.C.: U.S. Atomic Energy Commission (WASH-1238), 1972); and U.S. Nuclear Regulatory Commission, *Final Generic Statement on the Use of Recycle Plutonium in Mixed Oxide Fuel in Light Water Cooled Reactors* (Washington, D.C.: U.S. Nuclear Regulatory Commission (NUREG-0002, or GESMO), 1976).

Nonfatal Accidents and Occupational Disease

Underground coal mining is again the most hazardous energy-cycle activity, as shown in Table 9-2 (15,000 days lost per GWe-plant-year), followed by oil and surface mining (about 3500 days lost), fission and natural gas (1500–2000 days lost). These estimates are less certain than those for low side. The production and use of synthetic fuels from coal (with which we have little recent experience) may be considered to have the same risk as coal from extraction through transport and perhaps about the same as oil in processing.

Underground coal mining is more hazardous than underground mining for other materials: The frequency of injuries is about 50 percent higher and their severity is about 25 percent greater. In the case of oil, extraction accidents account for 60 percent of work days lost. With natural gas, accidents related to drilling were at least 10 times greater than accidents in other steps of the energy cycle (per quad).[14]

Routine accidents in hydroelectric plants in 1972 were about one half as frequent and one tenth as severe as the average for all electric generating plants.

From a medical point of view, underground coal mining adds considerably to the risk of the coal fuel cycle, but it has never been completely and satisfactorily analyzed, owing to the many factors

TABLE 9-2 Accidental Injuries and Workdays Lost During Routine Operations, by Energy Source (per gigawatt-plant-year)[a]

Energy Source	Accidents	Workdays Lost
Coal mining[b]		
Mined underground	112	15,000
Surface mined	41	3,000
Oil	32	3,600
Gas	18	2,000
Nuclear	15	1,500

[a] A permanently disabling accident was credited with 6000 workdays lost, and a temporary disability with 100 workdays lost. The figures are for 1977.

[b] Synthetic liquid fuel from coal might be estimated to have a rate equal to that for coal plus an allowance for the conversion process.

Source: National Research Council, *Risks and Impacts of Alternative Energy Systems*, Committee on Nuclear and Alternative Energy Systems, Risk and Impact Panel (Washington, D.C.: National Academy of Sciences, in preparation), chap. 2.

involved—sociological as well as occupational.[15–18] The principal physical factors are dust, noise, and gases (including engine emissions). The present respirable dust standard is set at 2 mg/m^3 of air. "Respirable dust" signifies particles that are small enough to reach the depths of the lung (bronchioles and alveoli). Larger particles may be eliminated in the larger air passages (nose, trachea, bronchi). Respirable dust is defined in practice as the material trapped by a particular "respirable mass sampler," whose deposition characteristics are assumed to represent the depths of the lung.

The problem may first be seen as coal worker's pneumoconiosis, a reaction of the lung to coal dust, diagnosed by X-ray, that is unimportant in itself, but that may develop into progressive massive pulmonary fibrosis, which seriously impairs pulmonary function. In 1969–1971, the prevalence of pneumoconiosis among miners was 30 percent and of massive pulmonary fibrosis, 2.5 percent. On the basis of meticulous studies abroad, the present dust standards—if enforced—should lower the incidence of simple pneumoconiosis to less than 3 percent, and of massive fibrosis more or less proportionately to 0.25 percent. Although the standard applies to respirable dust, it should be emphasized that dust composed of larger particles can have irritating effects on the larger air passages, and perhaps the gastrointestinal tract, that may be quite important.

"Black lung" is not a medically defined disease but a legally established state of eligibility for benefits, loosely defined by Congress in 1972. The term includes coal worker's pneumoconiosis and progressive massive fibrosis, as well as other diseases affecting the heart and respiratory system that may not necessarily result from coal mining. Measures that reduce the threat of pneumoconiosis-fibrosis types of illness may therefore have a much smaller effect on the whole spectrum of black lung conditions, for which disability payments now total more than $1 billion/yr (a total certain to rise). The larger dusts, noise, and gases (including engine emissions) presumably give rise to significant types of illness outside the pneumoconiosis-fibrosis category. The prevalence of such illnesses should be investigated.

It is essential that the entire problem be studied, with attention to both the medical and sociological factors (and particular attention to such complicating factors as smoking habits). The spectrum of conditions that can be attributed to work in the mines must be clarified to deal with them medically and to establish fair standards for disability compensation that are in line with national policy for all workers.

With respect to cancer,[19] a study of deaths in coal miners only (the working population), excluding disabled and retired men, did not detect an excess risk. In contrast, a study of 533 miners employed in 1937 and followed for 28 years did show an excess of deaths from cancer, particularly for digestive-system cancer. The problem should be reviewed again and prospective studies initiated, especially since recent legislation has led to major improvements in mining conditions. It should not be automatically assumed that the exposures experienced in mining will increase the cancer incidence rate. Significant evidence from the United Kingdom shows that lung cancer is diminished for unknown reasons in underground coal miners.[20]

In the case of uranium miners, a lung cancer mortality rate of about 0.2 per GWe-plant-year can be estimated from the doses and cancer-induction factors given later in this chapter (1400 person-rem $\times$ (2×10^{-4}) cancers per person-rem).*

PUBLIC HEALTH RISKS

Epidemiological Methods

In dealing with the quantitative assessment of risk due to emissions and wastes, the dose-effect curve is used whenever feasible. For the discussions that follow, it is important to consider the advantages and limitations of the method, particularly in the examples detailed below.

Two kinds of toxic agents are generated in energy cycles, artificially radioactive elements (fission products or elements activated by neutron adsorption) whose half-lives may be short or very long, and chemicals. Much more is known (for present purposes) about the mode of action of the ionizing radiations than that of the chemical agents at the low levels of dosage that are of primary concern. It appears that radiation primarily induces late effects (e.g., cancer), whereas the chemicals produce more immediate effects, although they might produce late effects as well.

Ionizing Radiations The systematic study of the Japanese atomic bomb survivors has provided an outstanding example of the knowledge that can be gained from a large-scale epidemiological study, and the time and effort required to achieve it.

Results for leukemia are illustrated by the dose-effect curves in Figure 9-1, for Hiroshima and Nagasaki. The death rate (mean annual rate for the period 1950–1974) is plotted against radiation dose (per person) received in 1945.[21] For the entire study period, there was a crude excess of 67 cases of leukemia in the total population of about 83,000 persons. The curves illustrate the great sensitivity of the epidemiological method at its best, when a specific class of rare disease due to a specific cause is fully

*Statement 9-4, by H. Brooks: This is based on historical data. As in the case of coal mining, it could be reduced much further.

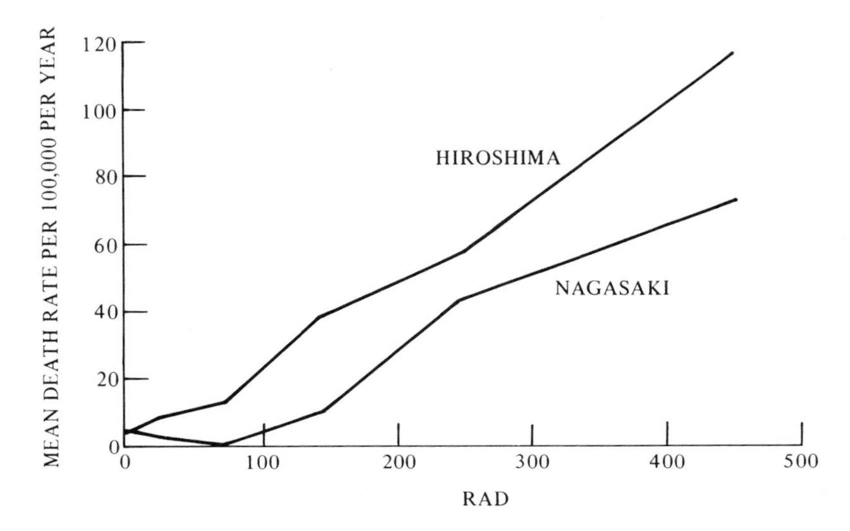

FIGURE 9-1 Mean annual leukemia death rate at Hiroshima and Nagasaki per 100,000 persons at risk, by dose and city from 1950 to 1974. The initial numbers at risk were as follows: for Hiroshima—controls, 43,700; 10–49 rads, 10,800; 50–400 rads or more, 6,000; and for Nagasaki—controls, 15,200; 10–49 rads, 3,800; 50–400 rads or more, 4,100. The number of leukemia cases were as follows: for Hiroshima, 42 control and 68 exposed; and for Nagasaki, 12 control and 22 exposed. Based on the observed control rate for both cities combined, the expected number of leukemias in the exposed population was 23, and the excess, 67. Source: G. W. Beebe, H. Kato, and C. E. Land, *Mortality Experience of Atomic Bomb Survivors, 1950–1974* (Hiroshima, Japan: Radiation Effects Research Foundation (RERF TR 1-77), 1977), p. 24.

investigated over a period of 25 years.[22] If, however, leukemia had not been specified, the 67 extra deaths would have been lost in the variations of the overall gross mortality rate.

In addition to illustrating the importance of a suitable endpoint, the Japanese studies demonstrate the importance of dosimetry. From individual estimates, made years before death, each decedent could be placed in one of five broad dose classes. Dose (rads, rems, or grays) relates to the energy absorbed by the target organs.[23] But even dose may require further specification. The Nagasaki curve lies below that for Hiroshima in Figure 9-1 because the Nagasaki doses involved less than a 2 percent contribution from neutrons, and those at Hiroshima a 19–27 percent contribution. There is reason to believe that the shape of dose-effect curves for neutrons differs from that for gamma rays (and X-rays), as suggested by Figure 9-1, and there is a possibility that neutrons and gamma rays differentially induce different types of leukemia.[24,25] The neutron-weighted curve (Hiroshima) resembles a linear one, whereas the other curve does not.

Sensitive though the Japanese study may be, is it sensitive enough to estimate the risk in the regulatory range of particular interest today, below 5 rads, and especially below 1 rad? This region of dose effect is poorly defined statistically by the two curves in Figure 9-1 owing to the small numbers of leukemia cases below 50 rads. In fact, the two curves are not significantly different statistically. They illustrate the uncertainty of extrapolating from higher, relatively well-established regions of the dose-effect curve to the lowest regions, which are of greatest interest. The problem of allowing for natural background radiation in the very low dose region, where effects even in very large populations have been undetectable, is discussed later in this section of chapter 9 under "Fission."

Finally, the nature of the population at risk must be examined. Is it of uniform sensitivity or does it contain subpopulations whose response in fact determines the overall results? In the case of the Japanese data, age at exposure determined how rapidly leukemia first occurred (Figure 9-2).[26] As a result, early reports concluded that children under 15 were the most sensitive, but after 10 years, disease began to occur in those who had been exposed at 45 years of age or older, and this group can now be seen to be the most affected. The occurrence of other forms of cancer only began some 15 years after exposure (in all age groups) and is still increasing after 30 years.

Chemical Agents The problem is much more complex for chemical agents, since the dose to the target tissue is known only under exceptional circumstances. Even the exposure level (concentration in air or water) may not be known quantitatively, and it is rarely specified for individuals on the basis of where they spend their time, e.g., indoors or outdoors. The parameter of exposure in air pollution studies is almost invariably a measurement taken at some distance from those at risk, for example, from a single monitoring station in a metropolitan area.

Additional confusion may result from the imprecise use of the term "dose." Dose might refer to the ambient exposure, or as in medicine, it might specify the amount of agent entering the body (by mouth or lung) but not necessarily reaching the target organ. (This could be true of radioactive substances that emit alpha and beta particles.) Thus, the agent may be inactivated in the digestive tract or excreted; if absorbed, it may be inactivated by the blood or liver, or it may be excreted by the kidneys. All these mechanisms tend to reduce the effect per unit quantity of agent or even to establish a threshold of exposure below which there is no damage.[27]

However, some innocuous substances are activated in the body. In the

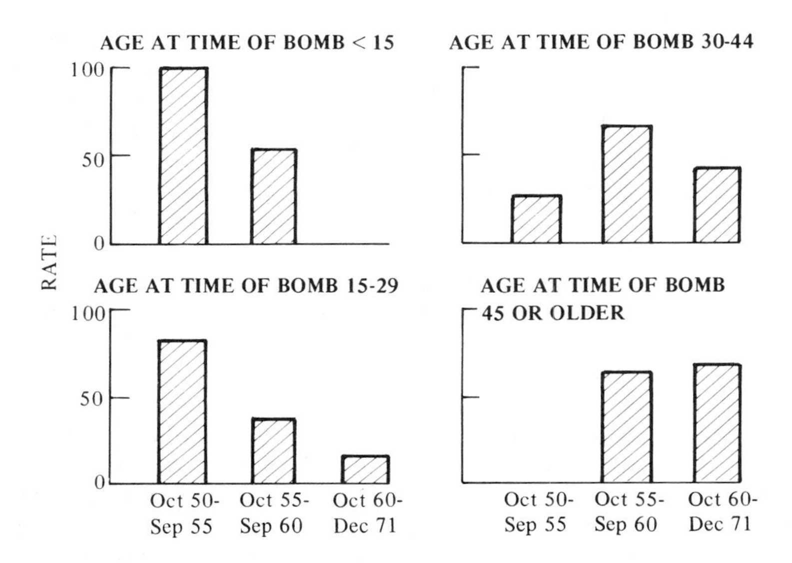

FIGURE 9-2 Leukemia death rate per 100,000 persons during three successive time intervals as dependent on the individual's age at exposure to atomic bomb radiation. The exposure occurred in 1945. The rate is adjusted for sex and city, and for doses of 100 rads or more. Note the much longer latent period in the case of the older cohort. Source: M. Ichimaru, I. Ishimaru, and J. L. Belsky, *Incidence of Leukemia in Atomic Bomb Survivors, by Dose, Years After Exposure, Age, and Type of Leukemia, 1950–1971, Hiroshima and Nagasaki* (Hiroshima, Japan: Radiation Effects Research Foundation (RERF TR 10-76), 1976), p. 14.

case of the respiratory and digestive tracts, and also the skin, direct contact between chemical and epithelial tissue does occur and may enhance the tissue's vulnerability.

Two examples illustrate the complications.[28] First, the accumulation of lead in the human lung appears to be a threshold phenomenon. No increment in tissue level occurs until the atmospheric level rises above 1.35 $\mu g/m^3$. Second, the levels of cadmium in liver, lung, and kidney are largely independent of the atmospheric level but dependent on smoking. To investigate the toxicity of atmospheric cadmium at the usual levels would likely be impossible in a population of unidentified smokers.

Other variables that are known to have major effects on the dose-effect curve in epidemiological studies are past medical history, socioeconomic status, and the weather.

It should be noted that the laboratory demonstration of a toxic agent's entering the cells does not guarantee a toxic epidemiological outcome. First, it has been argued on probabilistic grounds that unless the concentration within the cell exceeds a certain level, the toxic agent is unlikely to reach its intracellular target.[29] Second, although the target is hit, recovery is likely to occur unless the toxic dose is high enough to saturate the cell's recovery system.[30] Third, if a cancer cell is produced, that cell must override the resisting host to produce a tumor. And fourth, the toxic substance may not always be toxic. Selenium, for example, becomes progressively more toxic and eventually lethal at levels of 5–10 ppm (and above) in the diet, but it is an essential element for the good health of many domestic animals at a minimum level of 0.1 ppm.[31] Such factors will affect the shape of the dose-effect curve (the toxic effect of selenium has a threshold) and may complicate its interpretation, especially in the low-dose region.

In the case of man, the epidemiological study of smoking provides the major chemical example of the knowledge that can be gained when large numbers of exposed and cooperative individuals are available and when the level of exposure is relatively high.[32] As with all exposures continuing over a period of years, assessment at one particular time (e.g., at the time of the study inquiry, as is often the case in pollution studies) may have little to do with health at that time or in the future. Thus, in the case of smoking, even after termination of the habit, significant though diminished increments in death rate occur 10 or more years afterward from chronic bronchitis and emphysema, and from pulmonary heart disease.[33]

A cancer dose-effect equation for cigarette smoking has been obtained using data drawn from a 20-yr prospective study of some 34,000 British physicians.[34] The equation was based on the age-standardized incidence of bronchial carcinoma in *those* physicians who began smoking at 16–25 years of age and who had each reported the number of cigarettes smoked per day at a relatively constant rate (but not more than 40/day) from that age onward. No cancers were observed prior to 40 years of age. For the age range 40–79 years, the fitted equation for annual risk of bronchial cancer (per person) was the following.

$$0.27 \times 10^{-12} \times (\text{cigarette/day} + 6)^2 \times (\text{age} - 22.5)^{4.5}$$

Cancer therefore was a nonlinear function of exposure, and its incidence rate depended on intensity of exposure and years of exposure.

Other types of dose-effect curves induced by irritating particulates are seen in asbestosis, byssinosis, and silicosis.[35]

Estimating Health Risks

Adequate dose-effect curves are rarely available for the very low ranges of exposure that are now at issue in the regulatory process, especially when applied to the whole population. Nonetheless, risk must be estimated in many such cases. Some approaches that have been employed to make these estimates are listed below.

1. When the mode of action of the toxic agent is adequately known, as well as the dose-effect curve, and the population at risk is sufficiently well defined, the extrapolation or interpolation of the curve poses no great difficulty (e.g., carbon monoxide).[36] This is rarely the case.

2. Experiments with animals might be used to set upper bounds of permissible doses, provided that additional margins of safety in dosage or exposure have been incorporated in the process.[37,38]

3. Further large-scale epidemiological studies could be undertaken to define the dose-response curve in the low-dose region. The sensitivity of such investigations, however, may not be sufficient to supply the desired information.[39–41] The logistics of such studies must be carefully prepared in advance. Especially important is the accuracy of the dosimetry (compare approach 1, above). It is also important to know if a threshold of response is likely and if there are particularly sensitive groups of individuals whose reactions differ significantly from the mean.

4. High-dose results have been extrapolated on the assumption that effect will continue to follow dose on a curve of the same shape in the unexplored lower-dose region. In the case of radiation protection, this has been done by linear extrapolation.[42] That is to say, if one cancer death results from a dose of 100 rem to each of 100 persons (10^4 person-rem), one death is assumed to follow a dose of 10 rem to each of 1000 persons (10^4 person-rem), or even 0.1 rem to 100,000 persons (10^4 person-rem). In view of the discussion above regarding thresholds (for chemicals, the population dose or exposure would be in person-mg), such a step most likely overestimates the risk (at least in the case of toxic substances). For this reason, decisions based on this assumption are often considered to be conservative.

Although such a conservative decision may appear to be the prudent one, is it in fact? Reduction of exposure to extremely low levels of pollutants may be both costly and troublesome. Would the cost and trouble be warranted, considering that the risk is hypothetical and that the expenditure of equal cost and effort elsewhere would yield tangible benefits (including the reduction of other health risks)? A recent example of the complexity of such choices was the resetting of the ozone standard (discussed in previous and subsequent sections).

Fission

Natural Background and Federal Regulations The natural radiation background to which all of us are exposed—from earth, rocks, outer space, building materials, and food— might be a source of cancer or mutation (Table 9-3).[43] In the United States, the mean annual dose per individual is about 80 millirem (mrem) to the soft tissues, 120 to bone surfaces, and 180 to the lungs. The total-body mean is about 85 mrem. In Denver, the city of highest exposure owing to its elevation, the annual dose averaged over the entire body is about 125 mrem.

Total background includes man-made sources as well as natural background. Medical X-ray exposure is the chief anthropogenic component and averages about 70 mrem/yr in the United States. Fallout adds 3 mrem. Nuclear power now adds less than 0.01 mrem, and for a 300-reactor program, the contribution would not exceed 0.1–0.2 mrem/yr.

The damage from background has been estimated two ways: by the use of factors based on a combination of experience, experiments, and

TABLE 9-3 Average Dose-Equivalent Rates in the United States from Various Sources of Natural Background Radiation (millirem per year)

	Site				
Source	Gonads	Lung	Bone Surfaces	Bone Marrow	Gastro-intestinal Tract
Cosmic radiation[a]	28	28	28	28	28
Cosmogenic radionuclides	0.7	0.7	0.8	0.7	0.7
External terrestrial[b]	26	26	26	26	26
Inhaled radionuclides[c]	—	100[d]	—	—	—
Radionuclides in the body[e]	27	24	60	24	24[f]
TOTAL[g]	80	180	120	80	80

[a] Includes a 10 percent reduction to account for structural shielding.
[b] Includes a 20 percent reduction for shielding by housing and a 20 percent reduction for shielding by the body.
[c] Doses to organs other than lung included in "Radionuclides in the Body."
[d] Local dose-equivalent rate to segmental bronchioles is 450 mrem/yr.
[e] Excluding the cosmogenic contribution shown separately.
[f] This does not include any contribution from radionuclides in the gut contents.
[g] Totals do not add due to rounding. The mean annual whole-body dose for the United States is approximately 85 mrem.

Source: Adapted from National Council on Radiation Protection and Measurements, *Natural Background Radiation in the United States* (Washington, D.C.: National Council on Radiation Protection and Measurements (NCRP Rep. 45), 1975).

judgment, and by field studies. If the cancer factor of the Risk and Impact Panel (2×10^{-4} cancer deaths per person-rem) applies at such low dosage, it might be projected that natural background contributes some 4000 cancer deaths to the annual total of 350,000 deaths each year from cancer in the United States.[44] Since the factor is based on a linear extrapolation from a much higher dose range, it may very likely be an overestimate.

Another approach has been to compare the cancer death rates of geographic regions whose natural backgrounds differ appreciably. Large background differences in India and in Brazil are known, but other uncontrolled variables preclude a sufficiently sensitive analysis.[45] In the United States, cancer mortality by state has been studied against natural background.[46] The analysis indicated that the states with the highest background have the lowest cancer mortality rates. It is interesting to conjecture how much more publicity the study would have received had it shown an increase rather than a decrease in cancer. The investigators attempted to remove the possibly confounding effects of a variety of socioeconomic factors but failed to change the result. The study (as its reporters no doubt realized) fails to meet the rigorous criteria (discussed in "Dose-Effect Curves" of this chapter) that are considered essential when small differences are at stake: The dosimetry was not individualized but was based on a state-wide mean of exposure. The vital statistics were based on death certificates, whose diagnostic bias was not controlled, and the socioeconomic variables were not related to individuals. As it stands, therefore, the study performs the function of again raising an interesting question that would require great effort and sophistication to resolve, assuming that such resolution is feasible.

Federal regulations to limit radiation exposure had their origin in the recommendations proposed by radiologists and medical physicists through their International Commission on Radiological Protection (ICRP) and National Council on Radiation Protection and Measurements (NCRP), long before nuclear power plants were built. These permissible doses have been progressively lowered as knowledge of radiation biology has grown and as the technical ability to limit exposure has improved. The maximum permissible doses (prospective guidance limits) recommended by the NCRP[47] are based on annual cumulative dose.

Occupational exposure:	5 rem to the individual
Population exposure:	0.5 rem for any one individual
	0.17 rem, average for population

Medical exposure and natural background are not counted. The guidance limits for strictly partial-body dosage are higher.[48] Since proper use of the guidance limits involves the ALARA principle (exposure to be *a*s *l*ow *a*s *r*easonably *a*chievable), almost no one receives the maximum under ordinary circumstances, and the average for the exposed population is far below it. However, as standards are driven lower and lower by regulation, protection tends to become more costly, and cost-benefit considerations become important. It may well be that unnecessarily costly environmental restrictions would lead suppliers to use alternative technologies that have greater risks. Any fresh evidence of effects—or lack of effects—in the low-dosage range (relative to the standards) would therefore be of the greatest value.

A recent highly publicized study based on death certificates[49] claimed that cancer mortality associated with employment at an atomic plant was significantly increased in that segment of the working population that had received cumulative doses of less than 10 rem. The doses to the entire population ranged from below 1 rem to above 25 rem (with 75 percent below 5 rem), accumulated over periods up to 20 years and more.

The same statistical data have been reinvestigated by others,[50–52] who employed better methods of analysis, and the whole issue has been reviewed.[53] No increase in leukemia was observed, including myelogenous leukemia, the hallmark of radiation-induced cancer. No increase in cancer from doses below 10 rem was noted. For doses above 10 rem, small excesses of cancer of the pancreas and of multiple myeloma were noted, but they could not be shown to have been caused by occupational exposure to radiation. Further and more refined study is to be undertaken.

In the case of fission products from the present generation of light water reactors, the principal sources of public exposure are radioactive gases and emissions into the atmosphere—radon-222 (^{222}Rn) in uranium mines and mills, and krypton-85 (^{85}Kr), tritium, iodine-129 (^{129}I), and carbon-14 (^{14}C) in other parts of the fuel cycle. Inhalation makes the lung a primary target. The target for iodine is the thyroid gland. Other agents include bone seekers.

In the routine operation of light water reactors, the general population is exposed to radiation by emissions into the atmosphere and by the cooling water discharged into local water (Figure 9-3). Spent fuel is handled separately, for eventual reprocessing or long-term storage. For regulatory purposes, the dose to an individual member of the population is maximized by estimation for a hypothetical individual who spends all his or her time at the plant boundary and obtains all his or her food and water from the immediate area. Such an individual is subjected to airborne radioactive gases, receives external exposure from radioactive particulates deposited

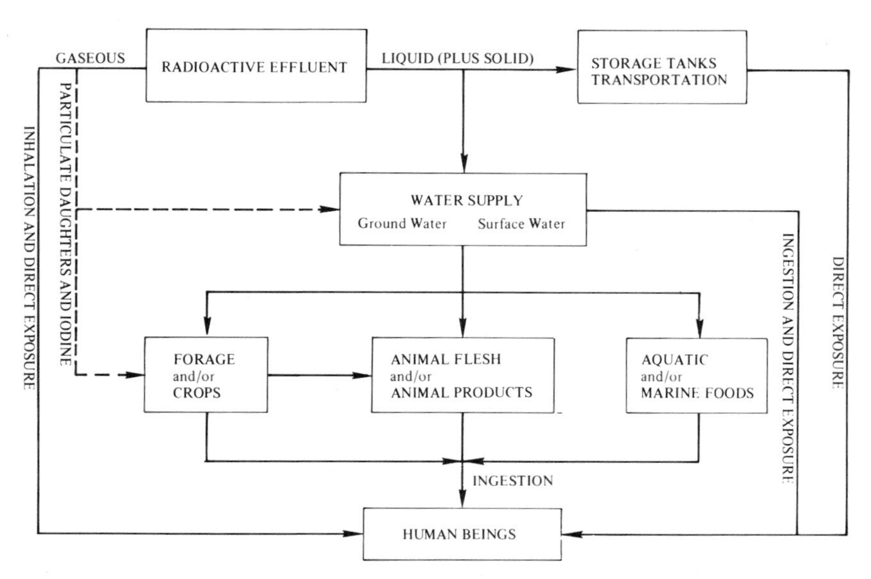

FIGURE 9-3 Generalized exposure pathways to man from the operation of a light water reactor. Source: U.S. Nuclear Regulatory Commission, *Final Generic Environmental Statement on the Use of Recycle Plutonium in Mixed-Oxide Fuel in Light-Water Cooled Reactors* (Washington, D.C.: U.S. Nuclear Regulatory Commission (NUREG-0002, or GESMO), 1976).

on the ground, inhales airborne radionuclides, and drinks water and eats foodstuffs, including fish, that contain radioactivity, as indicated by Figure 9-3.

The radioactivity of these emissions declines with radioactive decay and with dilution or transport by natural factors. Obviously, human exposure will be very sensitive to the action of these factors as well as to lodgement in the food chain and the rate at which radioactive materials are excreted.

For regulatory purposes, ^{85}Kr, tritium, ^{129}I, and ^{14}C are especially important nuclides. (Radon gas is important for mining and milling.) The emissions data in Table 9-4 are associated with the routine operation of a light water reactor for 1 GWe-plant-year.

The federal regulations that control exposure are promulgated by the EPA and the NRC, and the regulations of both agencies are enforced by the NRC. The NRC has decreed that reactor emissions should be as low as reasonably achievable. Current design guides for a light water reactor limit the annual air doses to any member of the public from that plant to 10 millirads (mrad) for gamma radiation; 15 mrem for iodine and particu-

TABLE 9-4 Major Radionuclides in the Emissions of a Light Water Reactor

Radionuclide	Half-life (days)	Radioactivity (curies per gigawatt-year)[a]
Effluent Gases		
Carbon-14	2.1×10^6	9.5
Iodine-131	8.1	0.3
Iodine-133	0.9	1.1
Krypton-85	3,900	290
Tritium	4,480	47
Xenon-133	5.3	3,200
Xenon-135	0.38	1,100
Effluent Liquids		
Cesium-134	750	0.01
Cesium-137	11,000	0.02
Iodine-131	8.1	0.26
Tritium	4,480	43

[a] For conversion to curies per gigawatt-plant-year, multiply by 0.75.

Source: Extracted from U.S. Nuclear Regulatory Commission, *Final Generic Environmental Statement on the Use of Recycle Plutonium in Mixed Oxide Fuel in Light Water Cooled Reactors* (Washington, D.C.: U.S. Nuclear Regulatory Commission (NUREG-0002, or GESMO), 1976).

lates; and for effluents, 3 mrem (whole body) or 10 mrem to any one organ.[54]

The Environmental Protection Agency has declared that by 1983, emissions entering the general environment from the uranium fuel cycle (per GWe-year produced by the fuel cycle) are to be less than 50,000 Ci[55] of ^{85}Kr, 5 mCi (millicuries) of ^{129}I, and 0.5 mCi (combined) of ^{239}Pu and other transuranic radionuclides with half-lives greater than 1 year. (It should be noted that the EPA "uranium fuel cycle" includes neither mining nor waste disposal and that a "GWe-year" is 1.33 times as much electricity as produced in a "GWe-plant-year," which is corrected for load factors.)

The EPA has also directed that beginning in 1980, the dose received by any member of the public from the uranium fuel cycle shall not be more than 25 mrem whole-body, 75 mrem thyroid, and 25 mrem other organs (radon and its daughters excepted).[56]

The EPA takes the position that permissible levels should be as low as regulation can drive them at some practical cost. The specification of what cost is practical, however, is a matter of opinion. One definition by the Nuclear Regulatory Commission for reactor design is that as an interim measure, \$1000 per reduction of 1 person-rem is a favorable cost-benefit ratio.[57] Taking the risk of cancer death as 2×10^{-4} per person-rem, the commission's policy entails a cost of \$5 million per avoided cancer death.

On the other hand, while equally concerned for health, and emphasizing that industrial and other practices should always involve the ALARA principle (as low as reasonably achievable), the National Council on Radiation Protection and Measurements is not convinced that its current permissible doses should be radically changed.[58] These doses are also recommended by the International Commission on Radiological Protection [59] and are generally used today throughout the world. The National Council considers the risk of cancer to be overestimated in the low-dosage range for gamma and beta rays by the process of linear extrapolation from high-dosage and high dose rate experience. It cautions government policy-making agencies against taking such extrapolations as accurate, and as a result, adding heavy margins of safety in setting permissible doses that could become unduly restrictive. To examine the whole matter as fully as possible, the council has set four committees to work. Their reports should begin to appear late in 1979. As noted above, the council was the original proponent of protection standards, and its views cannot be dismissed as those of a biased party.

There is merit in the intent of both positions. The problem is not a matter of choosing between them but of using both to make the best decision possible. A basic difficulty, of course, is that data for the exposure range of interest are not available. Present decisions must rely on conjecture. Government agencies may tend to take a more protective position, in part because they prefer to err on the side of caution. Furthermore, as recently evidenced in the case of ozone, such a position can turn out to be impractical, and it may be reversed. This committee observes that the EPA 25-mrem annual standard for individual members of the general population is equivalent to one quarter of the average natural-background dose for the United States, and is within the range of its regional variation. The annual background dose in Denver, for example, is about 50 mrem higher than that in the Mississippi Valley. The utility of promulgating such a standard is not clear.

Routine Reactor Operations The individual steps of the nuclear energy cycle—from mining to waste disposal—are outlined in Table 9-5, with their individual contributions to occupational and population radiation dosage.[60] Several facts stand out. First, the larger risks to employees are from mining, milling, and reactor operation; to the general population, they are from mining, milling, and reprocessing. For the employees, the

TABLE 9-5 Estimated Radiation Dose Commitments Delivered by Routine Operation of Various Parts of the Uranium Energy Cycle to Existing Populations

	Contributions to Lifetime Dose Commitment (person-rems to whole body, or, in parentheses, to key organs with significantly greater exposure, per gigawatt-year of energy produced)[a]		
Operation	Employees	Domestic Population	Foreign Populations
Mining[b,c]	250 (lung 1370)	600 (bone 1960, kidney 2285)	—
Milling[c]	80 (lung 660, bone 320)	120 (bone 390, kidney 450)	—
Conversion	1 (lung 8, bone 320)	10 (bone 24, kidney 3)	—
Enrichment	0.7 (lung 14, bone 6)	0.02 (gastrointestinal tract 1.6, kidney 0.7, if fuel is recycled)	—
Fuel fabrication	12 (lung 462)	0.6 (bone 10, kidney 1.6)	—
Reactor operation[d]	1240	76 (bone 272, thyroid 195)	52 (bone 250)
Reprocessing[e]	25	360 (bone 890, skin 2200)	240 (bone 750, skin 7900)
Transportation, irradiated fuel storage, and waste management	4	0.2	—
TOTAL[f]			
Without reprocessing	1600 (lung 3800)	800 (kidney 2800, bone 2700)	50 (bone 250)
With reprocessing	1600 (lung 3300)	1000 (kidney 2600, bone 3100)	270 (bone 1000, skin 8300)

[a] The dose commitment delivered to any person by a given release of radioactivity to the environment is the integrated dose he or she will receive for the rest of his or her life from the decay of the released radioactivity, assuming the person remains in the neighborhood and continues to be exposed to radiation from ground, water, and locally produced food. Values given in person-rems equal the number of persons at risk multiplied by the mean dose per person (in rems). (For conversion to curies per gigawatt-(electric)-plant-year, multiply by 0.75.)

[b] The figures overestimate the risk from mining by a factor of 2 (Conyers Herring, personal communication to H. I. Kohn, August 1979).

[c] Figures for mining and milling assume proper management of mill tailings and abandoned mines and allow for the escape of radon gas from open mines and mill tailings before being covered.

[d] Occupational doses per unit of energy produced have varied severalfold, depending on such factors as the age of the reactor and the sophistication of the protective measures employed. Population doses from noble gas effluents of boiling-water reactors have until recently been several times larger than the value shown. Reactors with a thorium-uranium fuel cycle (not yet developed commercially) would produce an additional occupational hazard via gamma radiation from ^{232}U. (See chapter 5.)

[e] Figures assume no retention of the nuclides ^{3}H, ^{14}C, and ^{85}Kr in the body. Population doses would be greatly reduced if any or all of these nuclides were retained. Skin dose is mainly due to ^{85}Kr.

[f] Totals do not add due to rounding.

Source: Taken from National Research Council, *Risks Associated with Nuclear Power: A Critical Review of the Literature, Summary and Synthesis Chapter,* Committee on Science and Public Policy, Committee on Literature Survey of Risks Associated with Nuclear Power (Washington, D.C.: National Academy of Sciences, 1979), p. 40.

total risk is practically the same with or without reprocessing (about 1200 person-rem/GWe-plant-year). In the general population, the total risk is about 750 person-rem with reprocessing and 600 person-rem without. Using a cancer-rate factor of 2×10^{-4} per person-rem, the total cancer risk would be 0.4 per GWe-plant-year, an estimate adopted by the Risk and Impact Panel and about equal to that of the United Nations.[61]

Of the principal emissions from a nuclear plant under routine conditions (see Table 9-4), ^{14}C eventually achieves universal distribution into all living things through the food chain (as illustrated in Figure 9-3). This and its long half-life (5570 years) indicate that its long-term risk should be calculated. Estimating that 7.5 Ci of ^{14}C will be generated per GWe-plant-year, and assuming that the world population will be constant at 4 billion, the worldwide population dose would be 1700 person-rem/GWe-plant-year, equivalent by application of the linear hypothesis to 0.3 cancers per GWe-plant-year.[62]

The number of serious genetic defects per GWe-plant-year has been estimated to be about 0.5 by the Risk and Impact Panel.[63] That panel reported an estimate of 100 cases of all kinds of severe genetic diseases per million person-rem, due to gene mutation and chromosomal abnormalities, and expressed in the first generation after irradiation, with diminishing frequency for the next half-dozen generations. In addition, there is an important class of diseases for which mutation is a partial cause but for which the magnitude of the mutational component is unknown. This includes a variety of congenital abnormalities and constitutional diseases, such as diabetes, cancer, heart disease, and mental retardation. The uncertainty in estimating the radiation-induced incidence of diseases in this group is very great, but as a crude estimate, was taken to be equal to the first. "Severe genetic defects," then, implies conditions or diseases that substantially reduce life expectancy, seriously impair normal physical or mental activity, or require prolonged medical attention.

The panel estimates a risk of 2×10^{-4} per person-rem for severe genetic defects, expressed mostly within the first 5 to 10 generations. The estimates are very uncertain, depending heavily on mouse data and on speculations about the role of mutation in human disease. The estimates could easily err by a factor of 5 in either direction.

For a large domestic nuclear power program of 300 reactors (each of 1-GWe capacity), the projected annual increment in risk (on the basis of the linear hypothesis, and recalling the dangers of extrapolating genetic effects from mice to people) would ultimately be about 100 cancer deaths (added to an annual rate of 340,000) and about the same number of serious genetic defects (added to an annual rate of about 30,000). Foreign populations (*in toto*) would suffer about 5–15 percent of these estimates.

Disposal of Radioactive Waste This subject is discussed at length in chapter 5, and the literature about its risks has recently been reviewed.[64] The principal point is that under routine conditions, radioactive waste contributes only a small fraction of the total risk (dose) of the nuclear energy cycle. Nevertheless, the government's failure to institute a waste disposal program has led to a loss of public confidence: The public now sees waste disposal as a difficult and dangerous undertaking.

Although the numbers of curies to be handled and sequestered is impressively large (as indicated in Table 9-6), many experts agree that their disposal—intelligently managed—will do no more than elevate background in some areas, or give rise to small pockets of higher exposure levels that can be effectively isolated. Table 9-5 estimates less than 4 person-rem/GWe-plant-year as the lifetime dose commitment.

It is important to emphasize that none of the events leading to possible catastrophe in nuclear power reactors can occur in a properly designed radioactive waste repository.[65] The principal dangers are that some of the waste might be carried to the discharge areas of an aquifer, or that the waste-storage area might be accidentally tapped in mining operations. The health hazard in the first case would arise from prolonged low-level exposure to radionuclides that enter drinking water or the food chain. The maximum credible risk in either case is comparable to that of improperly treated tailings from uranium mills.

The greatest health hazard associated with high-level waste disposal is likely to arise in connection with the transportation of the waste to the permanent repository or to a reprocessing plant. For this reason, locating the repository near its satellite reactors or reprocessing plants would be desirable (though not essential).

Under routine operations, the waste originates as follows. Approximately one fourth the charge of nuclear fuel in a light water reactor is replaced each year. The spent fuel elements are now stored in pools near the reactor. At some time in the future, these elements will have to be stored in separate facilities (away from reactor, or AFR pools). For long-term retrievable storage, most would be placed in separate canisters, as damaged fuel elements are now stored.

Between 90 and 99 percent of the actinide toxicity of the fuel would be removed in reprocessing, but the spent fuel from fuel elements containing recycled actinides may have up to 10 times the actinide toxicity of spent fuel removed from reactors loaded with fresh, slightly enriched uranium fuel. Most of this increased toxicity is due to plutonium isotopes plutonium-238 (^{238}Pu) and ^{241}Pu, with half-lives of 88.9 and 14.6 years, respectively) that are not important as long-term disposal risks (i.e., risks that persist for millennia). Table 9-6 sets out the principal fission products in spent fuel and their activities over a lengthy period.

After several hundred years of relatively rapid decline (as illustrated in Figure 9-4), the radioactivity is within a factor of 10 of that in the original ore, and from that point forward, the radioactivity is dominated by the slowly decaying actinides. The shielding requirements for this latter phase are much less demanding than the initial requirements.

The technical problems of waste disposal are not considered major. (These are discussed in detail in chapter 5.) Among the schemes that have been proposed, deep-mined repositories in geologically sound locations seem to offer storage at reasonable cost and acceptable risk.*

The government urgently needs to initiate a program of radioactive waste management, including that of mines and mills. While improved schemes may be developed in the future, waiting for their emergence and demonstration does not seem sensible in view of the practical measures that can be taken now. Rather than searching for a once-and-for-all solution, research should be undertaken to assure that each increment of waste is disposed of by the best technology available. The committee's recommendations on the management of radioactive waste can be found in chapter 5.

Combustion

To control pollution of the atmosphere from combustion, the Environmental Protection Agency establishes and enforces two sets of standards: one set that limits emissions and another that stipulates the ambient air quality to be maintained (or bettered, if possible). Standards for emissions from power plants are detailed in chapter 4. In this chapter, five principal pollutants are considered. For each, one or more national ambient air quality standards have been set that may not be exceeded anywhere. These standards are defined as the amount of pollutant in a cubic meter of air, or as an allowable fraction of the total atmosphere.

There are similarities and differences in the nature of the standards applied to chemical and radioactive pollutants. The emission standards for combustion devices (chapter 4) are analogous to those for radionuclides emitted from reactors. On the other hand, the ambient air quality standards for chemicals relate to possible exposure, whereas the standards for radiation are stated in terms of the *absorbed dose* in the tissue of interest. In this respect, the practice of radiation protection is more sophisticated. The difference is not academic, since the major difficulty in estimating the hazards of the chemical pollutants stems from lack of knowledge of the dosage.

*Statement 9-5, by J. P. Holdren: The decision on acceptability is not this committee's to make.

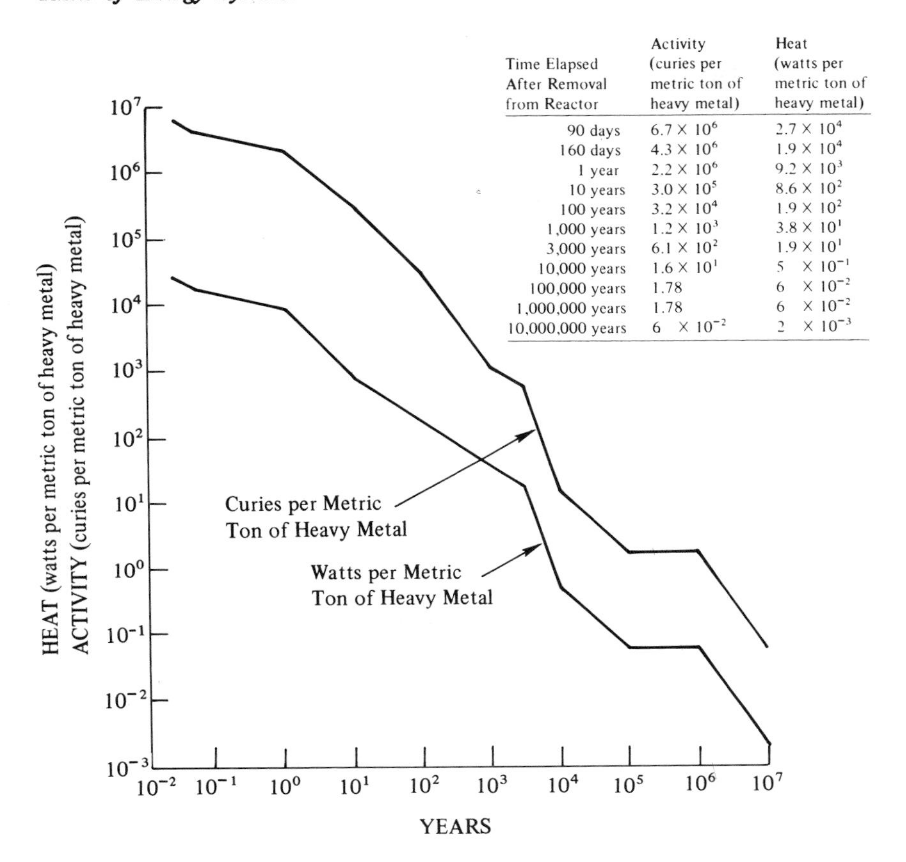

Time Elapsed After Removal from Reactor	Activity (curies per metric ton of heavy metal)	Heat (watts per metric ton of heavy metal)
90 days	6.7×10^6	2.7×10^4
160 days	4.3×10^6	1.9×10^4
1 year	2.2×10^6	9.2×10^3
10 years	3.0×10^5	8.6×10^2
100 years	3.2×10^4	1.9×10^2
1,000 years	1.2×10^3	3.8×10^1
3,000 years	6.1×10^2	1.9×10^1
10,000 years	1.6×10^1	5×10^{-1}
100,000 years	1.78	6×10^{-2}
1,000,000 years	1.78	6×10^{-2}
10,000,000 years	6×10^{-2}	2×10^{-3}

FIGURE 9-4 Decay of radioactivity in spent fuel from a light water reactor: time dependence of activity (curies per metric ton of heavy metal) and of heat production (watts per metric ton of heavy metal). Fuel characteristics: burnup, 25,000 MWd per metric ton of heavy metal; power at shutdown, 35 MWe per metric ton of heavy metal. Source: For activity and power projections up to 3000 years, U.S. Energy Research and Development Administration, *Alternatives for Managing Wastes from Reactors and Post-Fission Operations in the LWR Fuel Cycle* (Washington, D.C.: Energy Research and Development Administration (ERDA-76-43), 1976), Table 2-17; for projections over 3000 years, B. L. Cohen, "The Management of Radioactive Waste: Waste Partitioning as an Alternative" (Paper prepared for the U.S. Nuclear Regulatory Commission, Washington, D.C., 1976).

Regulations and Current Levels of Pollution The standards represent the considered judgment of experts, based on available data from experience with man, plants, and animals, and supplemented by the results of laboratory experimentation. (See "Research on the Health Effects of Air Pollutants," below.) Whether the standards now in force were set at the

TABLE 9-6 Some Fission Products of Biological Interest Associated with a Typical Batch of Spent Fuel Removed from a 1-GWe Boiling-Water Reactor[a]

Radionuclide	Half-life	Radioactivity (curies)
Cesium-134	2.19 years	5.3×10^6
Cesium-137	30 years	2.8×10^6
Carbon-14	5,770 years	0
Iodine-129	17,000,000 years	1
Iodine-131	8.1 days	1.7×10^7
Krypton-85	10.4 years	2.8×10^5
Strontium-90	28 years	1.9×10^6
Tritium	12 years	1.8×10^4
Xenon-133	5.3 days	3.1×10^7
Total of all activity, including actinides		2.7×10^9

[a]These data are for a typical reload of 172 bundles, one-quarter of the total assembly. The bundles contain about 32 metric tons of uranium (enrichment about 2.6 percent). The total inventory of the reactor core is therefore 4 times that indicated in the table, plus an additional allowance for all the actinides. See chapter 5, Table 5-15, for a detailed listing.

Source: Extracted from U.S. Nuclear Regulatory Commission, *Final Generic Environmental Statement of the Use of Recycle Plutonium in Mixed Oxide Fuel in Light Water Cooled Reactors* (Washington, D.C.: U.S. Nuclear Regulatory Commission, (NUREG-0002, or GESMO), 1976), p. IV-C-12.

most efficient levels for all circumstances cannot be decided, since more information is needed in the lower ranges of exposure. Table 9-7 displays data for certain of these pollutants as of 1977. As more experience and knowledge are gained, the standards will be reevaluated as required by law, and possibly improved.

Accepting the standards as they are, the practical comparison of risks from various fuels amounts to a comparison of how readily their emissions can be controlled to comply with the standards. For *stationary* sources, the use of any fossil fuel will lead to the production of nitrogen dioxide from the nitrogen of the atmosphere and, in the case of oil and coal, an increment from their own nitrogen content. Oil and coal contain sulfur, from which sulfur dioxide will be produced. Finally, the combustion of oil and especially coal releases a variety of organic compounds and trace metals, and the combustion of coal releases a small amount of radioactivity. The combustion of coal will therefore be the most costly to control, that of natural gas the least.[66] It should also be noted that the hydrocarbon and nitrogen oxide emissions of the transportation sector are both significant and costly to control. The principal emissions from combustion are set out by source in Table 9-8.

With the trend in recent years toward more stringent control, some diminution in emissions is to be expected, although it may be partially masked by increased consumption of fossil fuel. The total emissions for 1970, 1974, and 1977 (Table 9-8, bottom three lines) show a steady decline in particulates that totals 45 percent, declines of about 5 percent in sulfur oxides and volatile organic compounds (approximately the equivalent of nonmethane hydrocarbons), and increases of 3 percent in carbon monoxide and 10 percent in nitrogen dioxide. The increases in coal and oil consumption during this period were 19 and 20 percent, respectively.

Considering the United States as a whole, a rough idea of how well some of the standards were being met in 1977 can be obtained from an EPA report[67] compiled from the records of state and local monitoring agencies in the National Aerometric Data Bank. Some figures from that report are given below.

- Total suspended particulates (annual geometric mean): Based on 2699 stations, 40 percent exceeded the secondary standard (a guide only) and 17 percent exceeded the primary standard.
- Sulfur dioxide: Of 1355 stations, 2 percent exceeded the primary annual mean; of 2635 stations, 2 percent exceeded the primary 24-hour mean.
- Nitrogen dioxide (annual mean): Of 933 stations, 2 percent exceeded the primary standard.
- Carbon monoxide (8-hour mean): Of 456 stations, 46 percent exceeded the primary standard.
- Ozone (1-hour mean): Of 524 stations, 86 percent exceeded the primary standard.

Clearly, the pollutants can be divided into one class that is fairly well controlled (sulfur and nitrogen oxides), and one that is not (particulates, carbon monoxide, and ozone). The "uncontrolled class" is largely a product of the automobile, reinforced by industrial chemical processes, petroleum refining, oil and gas production, and the use of organic solvents (as indicated in Table 9-8).

The stationary and transportation sources also interact. Both produce emissions (nitrogen dioxide, hydrocarbons) that under the influence of sunlight lead to the formation of ozone and the photochemical oxidants (Table 9-8).

In addition, the mix of emissions from the two sectors can interact at the cellular level to increase or diminish individual toxic reactions, at least in plants (see "Agriculture and Plant Life").

TABLE 9-7 National Ambient Air Quality Standards[a]

			Maximum Permissible Concentration[d]	
Pollutant[b]	Standard[c]	Period	Micrograms per Cubic Meter	Parts per Million by Volume
Particulates (total suspended particulates)	primary	annual	75	
		24 hours	260	
	secondary	annual	60	
		24 hours	150	
Sulfur dioxide (SO_2)	primary	annual	80	0.03
		24 hours	365	0.14
	secondary	3 hours	1,300	0.50
Carbon monoxide (CO)	primary	8 hours	10,000	9.0
		1 hour	40,000	35
Ozone (O_3)[e]	primary and secondary	1 hour (between 9 a.m. and 9 p.m.)	240	0.12
Nitrogen dioxide (NO_2)	primary	annual	100	0.05

[a]The standards are maximum allowable rather than desirable levels. Annual standards are mean levels that are not to be exceeded. Short-term standards are means that may be exceeded not more than once a year. The objective of regulation is to promote keeping ambient air levels below the standards (which must be reviewed by the Environmental Protection Agency every 5 years, and revised if necessary). This table is accurate as of February 8, 1979. Announcements on reviews will be made in the summer of 1979 for nitrogen dioxide and carbon monoxide and in May 1980 for sulfur dioxide and the particulates. The standards were set in 1970 (excepting ozone in 1978), so their 5-year reviews are overdue. A standard was promulgated for lead in 1978 (1.5 $\mu g/m^3$, averaged on a quarterly basis). When originally promulgated, the ambient air quality standards applied throughout the United States regardless of pollution levels. The law has since been broadened to permit EPA to apply standards differentially on the basis of three classes: (1) where air quality should be preserved as it is, well below the standard, (2) where moderate deterioration would accompany controlled growth, and (3) where air quality could be allowed to deteriorate to national standards from present levels.

[b]Federal Reference Methods are specified for the determination of pollutants. Sulfur oxides (SO_x) are calculated as sulfur dioxide (SO_2), and nitrogen oxides (NO_x) as nitrogen dioxide (NO_2). Ozone is no longer an indicator for the class of photochemical oxidants (including peroxyacetylnitrates and aldehydes), but stands only for itself. There is no standard for a class of photochemical oxidants. In addition, hydrocarbons (nonmethane) averaged over a 3-hour period, have a maximum of 160 $\mu g/m^3$ (0.24 ppm), for use as a guide in devising implementation plans to achieve oxidant standards.

[c]The primary standard is for the protection of human health. The secondary standard is to protect "welfare," which in the words of the act "includes, but is not limited to, effects on soils, water, crops, vegetation, man-made materials, animals, wildlife, weather, visibility, and climate, damage to and deterioration of property, and hazards to transportation, as well as effects on economic values and on personal comfort and well being."

TABLE 9-7 *Footnotes Continued*

[d]The standards represent arithmetic averages for the periods specified, except for the annual particulates standard, which represents the geometric mean.
[e]U.S. Environmental Protection Agency, *Title 40—Protection of Environment*, chap. 1, subchap. C, "Air Programs" Fed. Reg. 44:8208–8221. Feb. 8(1979); U.S. Environmental Protection Agency, "Calibration of Ozone Reference Standards" Fed. Reg. 44:8221–8233 Feb. 8(1979).

Source: U.S. Environmental Protection Agency, *National Air Quality Monitoring and Emissions Trends Report, 1977* (Washington, D.C.: U.S. Environmental Protection Agency (EPA-450/2-78-052), 1978).

TABLE 9-8 Nationwide Emissions from Combustion in 1977, with Totals for 1970 and 1974 for Comparison, by Source (millions of metric tons per year)[a]

Source	Particulates	Sulfur Oxides (SO_x)	Nitrogen Oxides (NO_x)	Volatile Organic Compounds (VOC)	Carbon Monoxide (CO)
Transportation	1.1	0.8	9.2	11.5	85.7
Stationary fuel combustion	4.8	22.4	13.0	1.5	1.2
Electric utility	(3.4)	(17.6)	(7.1)	(0.1)	(0.3)
Industrial processes[b]	5.4	4.2	0.7	10.1	8.3
Miscellaneous	1.1	0.1	0.2	5.2	7.5
1977 TOTAL	12.4	27.4	23.1	28.0	102.7
Comparison totals					
1974	17.0	28.4	21.7	28.6	99.7
1970	22.2	29.8	19.6	29.5	102.2

[a]The emission estimates for particulates, sulfur oxides, and nitrogen oxides embrace a broader range of substances than are measured by routine ambient air quality monitoring equipment (see footnote *b*, Table 9-7). VOC are not quite equivalent to nonmethane hydrocarbons, the usual category.
[b]Industrial processes include emissions from chemicals, petroleum refining, metals, mineral products, oil and gas production and marketing, organic solvent use, and other processes.

Source: U.S. Environmental Protection Agency, *National Air Quality, Monitoring, and Emissions Trends Report, 1977* (Washington, D.C.: Environmental Protection Agency (EPA-450/2-78-052), 1978).

Finally, combustion of organic fuels from whatever source produces carbon dioxide, whose direct effect on health is observed only under extreme conditions[68] but whose potential effect on climate is a major concern (discussed under "Global Climate"). Per Btu of energy released, carbon dioxide production stands in the ratio of 1.0 to 0.8 to 0.6 for coal, oil, and natural gas; for coal-derived liquids it is 1.4.[69]

The Clean Air Act as amended requires that the Environmental Protection Agency review the ambient air quality standards every 5 years and change them if the criteria indicate such action. The first such review was completed early in 1979.[70] Ozone, which had been regarded as a parameter of the photochemical oxidants, was redefined as a pollutant, and its permissible level was raised from 0.08 to 0.12 ppm.[71] The agency argued that no significant disadvantage to health or welfare would result from such a change and that the level of smog would be unaffected. As a formal result, it is expected that the 86 percent violation rate recorded in 1977 (and noted above) will be decreased in 1979.

Court actions are being instituted challenging the agency's position, some arguing that the standard should not have been relaxed, others that it was not relaxed enough. A critical survey of the decision-making process for the new ozone standard would enable standard-setters of the future to learn from this experience. The scientific and sociological factors that led to relaxing the ozone standard should be compared to those that led to greatly increased stringency in the radiation standard discussed previously.

Projections The National Energy Plan (NEP) of 1977 called for the use of 13.5 additional quads of coal by utilities and 4.5 quads by industry in 1990. A joint analysis by six national laboratories[72] concluded that, of various factors, the degradation of air quality would be the major constraint in reaching this goal and that its prevention depends critically on siting. The analysis, which concentrated on particulates and sulfur oxides, was based on county-by-county (not within-county) data and assumed that the increase in use would tend to be directly proportional to current use (August 1977). Because such a distribution places increased use in or near many nonattainment or limited areas, approximately 50 percent of the projected industrial use of coal and 25 percent of that projected for utilities was constrained.

These estimates are inflated by the use of county-level analysis that ignores the adjustments possible within counties. Overall, the estimated atmospheric levels of pollutants were about the same on regional and national maps as today's levels, owing to the emission controls required of new plants.

We judge that these results do not rule out doubling the use of coal. They demonstrate the necessity for integrating the several problems of siting (on both a within-county and a regional basis) with other aspects of energy planning.

As indicated in the following section, the epidemiological evidence concerning the health impact of air pollution from coal combustion products is inconclusive. The widely quoted studies indicating severe impact of sulfates resulting from the atmospheric transformation of sulfur dioxide emissions are seriously flawed. On the other hand, there is also no evidence to suggest conclusively that present standards are too tight, or even that there are no health risks at present ambient levels. The present regulatory strategy must be considered as conservative in the light of existing evidence, but probably justified in view of the prospective rapid growth in the use of coal, especially for electric power generation, and the much higher cost of retrofit compared with tight standards on new plants. The case for conservatism is reinforced by the prospect that the largest growth in electric generation is likely to occur during the next 15 years, with a slowdown thereafter, as suggested by the various CONAES scenarios. Moreover, the possibility of a slowdown or even a moratorium on nuclear growth owing to public opposition also argues for holding coal emissions as low as practical in case coal expansion has to take place at an even greater rate than projected in the CONAES estimates.

The one emission whose effect is independent of siting is carbon dioxide. Increasing fossil fuel combustion will increase carbon dioxide emissions and tend to raise the level of carbon dioxide in the atmosphere. This problem is discussed under "Global Climate."

Research on the Health Effects of Air Pollutants The setting of national air quality standards was a landmark in the history of public health protection in this country. The original standards were selected in 1970 by reviewing the available epidemiological and clinical evidence, deciding on a level at which a minimal effect was observed in man, and setting the standard at a level below it.[73–77] For the fossil fuels, the major designated pollutants that emanate from stationary sources are the particulates,[78] sulfur oxides,[79–81] and nitrogen oxides.[82–84] For the mobile sources[85] they are ozone,[86] carbon monoxide,[87] nonmethane hydrocarbons,[88] nitrogen oxides, and as a secondary product, photochemical oxidants.[89] Toxicity was judged by excess morbidity, primarily of the respiratory system, by excess mortality occurring during major fog pollution episodes, or on a daily or other short-term basis associated with fluctuations in the sulfur dioxide and particulate levels over the course of a year or more.

In addition, it had been recognized that coal tar is a classical source of chemical carcinogens. An excess of cancer had been demonstrated as an occupational hazard in certain industrial operations where the workers were heavily exposed to fumes from coal processes[90] (present-day practices would reduce such exposure). These carcinogens may include mutagens. Control of particulates would presumably diminish exposure of the public to the chemical carcinogens and mutagens released by combustion. It is anticipated that the use of synthetic liquids or high-Btu gaseous fuels from coal will be cleaner than burning coal directly.[91]*

The following discussion reviews some trends in health research that bear on the primary standards. The analysis establishing sulfate (at prevailing levels) as an important determinant of mortality has been rejected. There is expanding interest in the many roles played by the particulates. Appreciation has increased for the great complexity of the epidemiology; for example, the difficulty of assessing the risk is magnified (perhaps disproportionately) as the level of pollution diminishes. Quantitatively significant health effects have not been established below the standards; if they exist (and they might), more sophisticated investigation must be used to find them.

Past studies on sulfur dioxide and sulfates were most valuable in showing the complexity of the epidemiological problems. With increased experience, investigators and reviewers have become much more critical of investigational design.[92–97] In fact, two investigators who supplied the classical findings that related fluctuating daily mortality rates to changing ambient levels of sulfur dioxide have independently concluded that their original analyses (considered sophisticated at the time) were not sophisticated enough, and they no longer accept their own conclusions.[98,99] One investigator[100] could show the same associated fluctuations several years later, even though the general level of sulfur dioxide had fallen by a factor of 10. He considers that both mortality and sulfur dioxide levels are fluctuating in response to some other factor. The other investigator[101] has undertaken a study of how the many variables at work, including daily fluctuations in temperature, may interact or otherwise confound the analysis, and of what the statistical and methodological requirements of such investigations might be.

Another study[102] that attempted to establish the mortality risk factor for sulfate, based on intercity comparisons, concluded that per 100,000 persons at risk, 3.25 deaths occurred per microgram of sulfate in a cubic meter of air. The study has been widely quoted, and the coefficient has been used to quantify the hazards of increasing the use of coal.

The study was a retrospective one, based on vital and other public statistics. The investigators had no control over the chemical methods used to estimate ambient air levels of sulfate. The parameter of exposure for any given year was usually obtained thus: Each urban area had one or perhaps two sampling stations; one analysis was performed biweekly; of the 26 analyses per annum available in about two thirds of the urban areas studied, the lowest value of each set was selected as the best parameter of annual exposure for that area.

*Statement 9-6, by H. Brooks: While this may be true, it may not be true for coal-derived fuels substituted for petroleum-derived fuels, especially the heavier liquids.

The study and its mortality-rate factor for sulfate have been criticized[103,104] and rejected.[105–107]* The critical comments point out defects, for example, that result from multicolinearity of the sulfate measure with a number of urban characteristics. When one critic redid the analysis, correcting for defects, the original positive effect for sulfate disappeared.[108,109] A report issued by the National Academy of Sciences[110] concludes that there is insufficient evidence to establish an ambient air quality standard for sulfate in addition to that for sulfur dioxide.

For particulates, one review[111] that synthesized the published statistics of 17 major pollution episodes found an excess of 1 percent in the concurrent mortality rate for each increment of 100 $\mu g/m^3$ of total suspended particulates (TSP). The levels in 16 of the incidents, occurring from 1930 to 1975, were measured or estimated to fall between 500 and 5000 $\mu g/m^3$, well above the present primary standard (Table 9-7).

In time-series studies in single cities (mean annual levels, 130–215 $\mu g/m^3$), the mean annual mortality increment was 0.6 percent per 100 $\mu g/m^3$. For intercity comparisons (cross-sectional studies) based on mean annual rates, an almost significant excess of 6 percent mortality per 100 $\mu g/m^3$ was found.

The positive association between particulates and mortality leaves open the question of "cause of death"—the toxic agent and the pathological process.[112] Particulates could be the surrogate for a large number of combustion products, including sulfates. Population mortality rates serve as indicators of many different factors, social and biological, and thus depend not only on immediate circumstances, but also (and generally much more) on the population's past history. It will be of great importance to determine if such results can be found in the range of exposures at or below the levels achieved under current standards[113] (the data were drawn chiefly from 1950–1972).

The same report[114] also found a positive association for manganese in a few specific locations and concluded that these areas should be restudied when the 1980 census data become available.

The linear relationship between mortality and the level of total suspended particulates in the range above the standard is of major interest and indicates the need for further intensive research (some is already under way). Presumably, a more sensitive and specific endpoint than mortality would be desirable.[115] Improved dosimetry, especially at or below the levels of current standards, would be essential in sorting out the many interrelated problems and factors that apply to atmospheric pollutants in general, as illustrated by the following considerations.[116–118]

*Statement 9-7, by J. P. Holdren: I believe there is still some disagreement among knowledgeable analysts on whether the sulfate studies have been discredited or merely challenged.

1. Probably less than 1 percent of total suspended particulates from an uncontrolled source can reach the lungs, owing to filtration by the upper respiratory tract (the respirable particulates (RP) are less than 2 μm in diameter). The trapped particulates are mostly excreted through the digestive tract.

2. As fly-ash emissions from stationary sources continue to be reduced, the larger particles are removed more and more efficiently, but the smaller respirable particles (less than 1 μm) continue to be emitted. If (as generally assumed) the respirable fraction is the toxic fraction (because it reaches the lungs), the proposed standard of 99 percent reduction in particulate emissions from stationary sources would not yield proportional benefits to health, although tangible gains (cleanliness, visibility) will be realized.

3. The fine particulates include complex organic compounds (including mutagens and carcinogens) and various toxic metals such as arsenic, lead, mercury, and zinc (from trace amounts in coal). During transit, the emitted gases (sulfur dioxide, nitrogen dioxide) form aerosols through condensation and coagulation and may react with other fine particulates,[119] which can lead to increases in particulate size and may change their eventual distribution in the body.

4. The particulates may travel hundreds of miles from their point of origin. The emitted gases that accompany them may be oxidized to sulfates and nitrates (catalyzed in part by transportation pollutants), and other chemical changes may occur, influenced by varying conditions of temperature, humidity, and solar irradiation.[120] Precipitation may bring them down, or their nature may be influenced by substances originating in the territory over which they travel. The qualitative nature of the particulates, therefore, changes in transit and also with place and season. Both increases and decreases in toxicity may be significant.[121]

5. A number of sources, perhaps widely distributed, are responsible for the pollutants observed at any particular place, since pollutants travel far. These independent contributions vary diurnally and seasonally, and are differentially affected by meteorological conditions. The task of estimating the pollution profile for large regions (or metropolitan areas) on the basis of projected energy plans is complicated by the multiplicity and uncertainty of all these factors. The analysis and modeling necessary to the task have been initiated[122–125] and a projection for the National Energy Plan of 1977 has been completed.[126] Such profiles, moreover, must ultimately be related to individual dosage to determine epidemiological effects.

Besides dosimetry, the factor of time must be emphasized in further epidemiological research. Experience has shown that important aspects of the epidemiological studies necessary in this area cannot be hurried to meet the demands of policy,[127] despite the significance of this consideration. Chronic and late effects take time to develop. We know of one study well under way that tends to meet the requirements discussed here.[128] It is a prospective study of comparable populations in six small cities. One such study, however, cannot answer the massive array of questions that confront the policy maker who needs critical examination of the validity of present standards. Much more work will have to be done, and over an extended period of time. We favor conducting much of this work outside government laboratories to ensure its independence, and to provide flexibility to draw upon the country's scientific manpower.

CATASTROPHES

Nuclear power plants, liquefied natural gas, and large dams pose a danger of catastrophic accidents. The greatest potential risk is that of catastrophic accidents in nuclear power plants.*

Gas[129] The principal potential hazard of natural gas is associated with leaks in the pipeline distribution system. In the case of liquefied natural gas (LNG), its cryogenic properties create additional hazards. Contact, for example, will damage human tissue. On vaporization and exposure to oxygen and a source of ignition, the gas will burn and possibly explode. The worst LNG accident on record occurred at a storage facility in Cleveland in 1944: The accident killed 130 people and caused $10 million worth of property damage. It is highly unlikely that an accident of this type could occur today, since the development and use of materials that resist brittle failure at cryogenic temperatures have largely eliminated the cause. Nevertheless, if liquefied natural gas is released in an accident, it may form a vapor cloud that travels several miles before igniting. In 1972, 40 workers were killed by inhalation of the vapors in an emptied tank on Staten Island. Liquefied natural gas is now being shipped from foreign countries in tankships for receipt in ports with large storage facilities. The degree of hazard in shipping and storing LNG is controversial.[130]

Hydroelectric Power[131] In the 40-yr period from 1918 to 1958, 1680 deaths occurred from the failure of five dams: a statistical average of 40 deaths per year. The total number of dam failures was 33. Assuming that the average number of dams in the United States over this period was 1000, a failure rate of about 8×10^{-4} per dam-year can be estimated, and a major disaster rate of 1.3×10^{-4} per dam-year.

*See statement 9-8, by H. Brooks, Appendix A.

From 1959 to 1965, nine major dams of the world failed (of an estimated 7800). The worldwide failure rate was about 2×10^{-4} per dam-year. The probability of failure of a particular dam is difficult to estimate, but can be strongly dependent on geological setting, surface faulting and seismicity, as well as on the type of construction, size, and other factors. Damage will depend on such factors as topography, storage volume, the cause and mode of failure, the size of the population at risk (some thousands to more than 100,000), and on mitigating factors, such as evacuation.

Nuclear Power Reactor accidents are discussed in chapter 5 from a physical and engineering point of view. A major conclusion of that discussion is that the more serious the accident, the less probable its occurrence. Two important implications follow. First, the relatively frequent minor accidents provide opportunities to improve design and lessen overall risk. Second, there is a "dominant risk," defined by maximum cumulative damage, that may be used to characterize and compare particular reactors. The dominant risk is chosen on the basis of a maximum value for the expression: (probability of incident) × (damage per incident).

It should be noted that even large accidents in light water reactors are not analogous to the explosion of an atomic bomb. There is no explosion or release of neutrons; buildings outside the reactor plant are not damaged physically, nor are fires ignited. Property, ground, and water damage are due to radioactive contamination. The danger to human beings is the inhalation or ingestion of radioactive substances or gases, or irradiation from such substances released to the atmosphere or deposited on the ground. (The occupational hazard within the plant could at some time involve exposure to neutrons.)

Three Mile Island In the case of light water reactors, accidents with dire consequences have not occurred in some 400 reactor-years of operation, but an accident that severely damaged the reactor core did occur at the Three Mile Island plant in Pennsylvania on March 28, 1979. It involved a partial failure of the cooling system. The analysis of what happened is being conducted by a presidential commission, whose report is due in October 1979, and other groups.

Of particular interest is the report of the Ad Hoc Population Dose Assessment Group, staffed by technical experts from three government agencies.[132] During the period March 28 to April 7, it was necessary to vent quantities of radioactive gas. These gases constituted an atmospheric risk to approximately 2 million people who reside within a 50-mile radius of the plant. The principal emissions were xenon-133 (^{133}Xe) and, to a lesser degree, ^{131}I, whose half-lives are 5.3 and 8.1 days, respectively. The initial estimate of the total population dose was 3500 person-rem. The mean dose was about 2 mrem (0.002 rem) per person. The highest dose anyone received was below 100 mrem. These estimates are likely to be too high, since they do not allow for the protective effects of evacuation, nor for protective shielding by buildings in the case of persons staying indoors. The amount of radioactivity in milk due to the incident was trivial.

Using the linear dose-response hypothesis, it can be estimated that less than one cancer and less than one serious genetic effect were induced in the entire population at risk (the existing cancer mortality rate for that population is 3400 per year).

At the time of the committee's final deliberations, it appeared that the health risks of this accident were negligible (a single automobile accident during the evacuation would have done more damage), but the doubts and fears raised by the incident could have far-reaching consequences.

WASH-1400 In 1975 the Nuclear Regulatory Commission published the results of a study undertaken to estimate the probability of the great variety of accidents that might occur in the operation of nuclear reactors. The Reactor Safety Study (known as WASH-1400 or the Rasmussen Report)[133,134] employed decision-tree and fault-tree analysis, based on judgment of risk at each step of a specific sequence of events leading to an accident, and tempered by the operating experience that had been accumulated.

The NRC has recently withdrawn its approval of the absolute probabilities of risk of reactor accidents calculated in the report,[135] without commenting whether individually or as a group they are too high or too low. Nonetheless, as the external review group of the NRC has pointed out,[136] WASH-1400 is still the most comprehensive attack on the problem of reactor safety, and its approach, collected materials, discussions, and experience are of great importance in providing a base from which the examination of this complex problem can advance.

Reactor accidents that involve "breach of containment" release radioactive materials into the atmosphere and thus may affect the public. The accident at Three Mile Island did not breach containment; its radioactive emissions were released by the operator in the course of controlling the rising pressure within the reactor during its shutdown (necessitated by insufficient cooling).*

WASH-1400 estimated the frequency of accidents of this severity as 1 in 300 to 1 in 30,000 reactor-years of operation. The range has proved to be

*Statement 9-9, by J. P. Holdren: Actually, part of the release resulted when a design error caused automatic pumping of contaminated water into an unsealed building.

correct, since the accident is the first in about 400 reactor-years of experience.†

Table 9-9 lists three types of progressively more serious accidents and their diminishing probabilities. WASH-1400 is interesting in its attempt to deal with human dosage, which is especially sensitive to meteorological conditions at the time of the accident and the distribution of population around the reactor site.

Taking the 68 reactor sites then in operation or under construction, six typical weather-type sites were selected, each located in one of six geographical regions (eastern river valley, eastern seacoast, southeastern, midwest lakeside, midwest plains, west coast). For each site, a specific surrounding population was constructed over a 50-mile radius (on the basis of the actual populations associated with reactors in that geographical region). Although for each composite population the area distribution is less variable than that for the group of existing populations that it represents, care was taken not to average out the extremes within it. Weather data obtained from six typical sites were employed to predict conditions at the six composite sites. However, the calculations do not allow for variations in wind direction as the plume travels downwind from the reactor, during which it might be dispersed over hundreds of miles. Dosage calculations were made separately for external exposure, inhalation, and ingestion. A correction was made for evacuation of the population from an area within 25 miles downwind of the reactor, which typically would reduce the early health effects by a factor of 3.[137]

Table 9-9 gives the WASH-1400 estimates of damage for (1) core meltdown, (2) core meltdown followed by aboveground breach of containment, and (3) an accident of the type described in (2) followed by adverse conditions of wind, weather, and population density in the path of the radioactive cloud that is released. Although the nominal probabilities given in Table 9-9 are now in question, it is useful to consider them with respect to order of magnitude, and especially relative to one another: 1 in 20,000, 1 in 1 million (dominant risk; see chapter 5, "Reactor Accidents"), and 1 in 1 billion reactor-years, respectively.[138] The reactors are assumed to be 1-GWe plants. Power plant experience in the United States totaled about 400 reactor-years by January 1979, and we are accumulating about 70 reactor-years of experience every year.

Table 9-9 provides estimates of property damage and health effects appearing within 1 year of the accident, and those appearing over a period of years—thyroid nodules (for which there is effective treatment), cancer deaths, and genetic defects. The table illustrates the range of morbidity

†Statement 9-10, by J. P. Holdren: Proof of correctness is far too much to assert for a single data point barely inside the range.

and mortality that might result from different types of accidents and also indicates the tremendous complexity of making such estimates.

The table reveals that the consequences of a single accident may be very large but that such accidents are improbable, and the consequences are small *per year* when averaged over a period of years. Thus, if the chance of an accident is 1 in 1 billion years, a total of 3000 deaths averages only 1 death per 333,000 years. Such averaging, however, does not consider that the social disruption of a catastrophe can be much greater than that associated with a distributed set of small accidents producing the same number of deaths.

It is of interest to compare the hypothetical catastrophic risks of Table 9-9 with those of routine power plant operation. Core meltdown with aboveground breach of containment (the dominant risk) carries a probability of once in a million reactor-years and leads (in round numbers) to less than 10,000 deaths from all causes. Over the same hypothetical period of 1 million reactor-years, the routine operation of the nuclear plant would be associated with 200,000 deaths, that of an oil-fired plant also with 200,000, and that for coal with 2 million (Table 9-1).* If we suppose that the nuclear catastrophe is underrated tenfold, the risks of routine operation and those of catastrophes become approximately equal.

WASH-1400 has been criticized for (*inter alia*) underestimating some hazards owing to the use of median rather than mean estimates,[139–141] and for underestimating the uncertainty of its results. The recent report by the Risk Assessment Review Group to the Nuclear Regulatory Commission[142] and the Ford/Mitre Report[143] both make the latter point. The review group of the NRC points out both overly conservative and inadequately conservative assumptions in the probability estimates, and the group concludes that the uncertainty is seriously understated. Both groups note that reactor experience provided an upper bound at the times of their reports.

AGRICULTURE AND PLANT LIFE

The effects of pollutants on plant life are judged in terms that are quite different from those applied to judge the effects of pollutants on human health. Cancer, for example, is not a problem in the case of crop plants, nor are the late effects of exposure that take years to develop. While genetic effects might be of interest in special cases, the rare occurrence of a mutant in a wild or cultivated population is not important. Major interest

***Statement 9-11, by J. P. Holdren: Table 9-1 includes only the deaths caused by industrial accidents, mostly to workers except in the case of coal trains. The comparison is meaningless.**

TABLE 9-9 WASH-1400 Estimates of Probabilities of and Damage to the Public from Three Types of Major Accidents to Light Water Reactors, with Statistical Rates of U.S. Population for Comparison[a]

Accident		Public Property Damage Excluding Reactor (millions of dollars)	Health Effects Within One Year			First Year's Health Effects Plus Delayed Health Effects			
Type	Probability Per Reactor Year[b]		Deaths	Ill-nesses	Population at Risk Within 25 mi[c]	Cancer Deaths[d]	Thyroid Nodules[d]	Genetic Defects[e]	Population at Risk Within 500 mi[c]
Core meltdown	5×10^{-5}	1	negl.	negl.	negl.	3 (0.1)	3 (0.1)	75 (0.5)	1×10^4
Plus above-ground breach of containment (dominant risk)	1×10^{-6}	1,000	1	300	4,000 (within 5 mi; density 50 per square mile)	5,100 (170)	42,000 (1,400)	3,750 (25)	2×10^6 (density 77 per square mile)
Plus adverse wind, weather, and population density	1×10^{-9}	14,000	3,000	45,000	1×10^6 (within 25 mi; density 500 per square mile)	45,000 (1,500)	240,000 (8,000)	26,000 (170)	10×10^6 (density, includes maximum)[f]
Statistical rates for U.S. population[g]	—	—	9,000	—	1×10^6	1,700	800	800	1×10^6

[a] Among the major factors allowed for in projecting damage are the following: absolute probability of core melt, relative probability of various radioactive release categories after core melt; probability of various types of weather conditions; probability that a particular population density distribution will be exposed. See text. Specific details are given in WASH-1400 (especially Tables 5-4 and 5-5, and in Appendix VI, Section 9). A survey and some additional discussion are given in U.S. Nuclear Regulatory Commission, *Overview of the Reactor Safety Study Consequence Model* (Washington, D.C.: U.S. Nuclear Regulatory Commission (NUREG-0340), 1977).

The ranges of uncertainty in these estimates are under review, and most likely will be increased. The present ranges are as follows: for accident probabilities, one-fifth to five times the stated value; for cancer deaths, one-sixth to six times; for thyroid nodules, one-third to three times; for prompt health effects, one-fourth to four times; and for genetic effects, one-sixth to six times.

Effective evacuation is assumed for 30 percent of the population, partially effective evacuation for 40 percent, and ineffective evacuation for 30 percent. Evacuation for up to 25 miles downwind of the reactor is allowed for.

[b] The Nuclear Regulatory Commission no longer accepts the absolute probabilities given here. They are still of interest, especially relative to one another. See text.

[c] Health effects within one year are seen only in those persons with doses of more than 50 rem. Their number may be quite small, although large numbers may receive very small doses and be liable to delayed effects. Population figures were supplied by Norman Rasmussen.

[d] Assumed to occur from the tenth to the fortieth year after the accident. The statistical mean annual rate during this period is given in parentheses; the actual annual rates would vary considerably. The estimates are based on a sum of an organ by organ survey. The probability of cancer death (per person-rem) depends on three dose-factor components: that for internally deposited radionuclides (5×10^{-4} cancer deaths per person-rem), that for external exposure (1.2×10^{-4}), and that for a "dose-rate and dose factor." (Below 10 rem/day or a total dose of 25 rem, the absorbed dose is progressively less effective; the maximum reduction of 80 percent is at less than 1 rem/day or at a total dose of 10 rem or less.) By and large, this comes to a rule of thumb of 1×10^{-4} cancer deaths per rem (whole body).

[e] Total for all generations. The mean annual rate is given in parentheses. All types of defect are included: 20 percent of the dominant effects and 10 percent of the recessive effects are assumed lost per generation.

[f] Mean population density in the United States is 77 per square mile. In this case, urban areas of great density are included.

[g] The health rates given are for the United States, per million of population per annum.

Source: Compiled from U.S. Nuclear Regulatory Commission, *Reactor Safety Study: An Assessment of Accident Risks in U.S. Commercial Nuclear Power Plants* (Washington, D.C.: U.S. Nuclear Regulatory Commission (WASH-1400), 1975).

centers on the immediate effects of pollutants on plant growth,[144–147] the associated effects in ecosystems, or both.

It was estimated in 1973[148] that about $100 million of crop loss was due to photochemical oxidants and that another $13 million loss was due to sulfur dioxide. Such estimates do not account for all effects on plant life, but do establish that a general problem exists. In heavily polluted areas such as the Los Angeles Basin, it has become necessary to abandon many citrus groves, and elsewhere truck crops, particularly of leafy plants, are increasingly hard to cultivate. Such effects extend to natural ecosystems. In some forested areas under observation, lower dosages of pollutants tend to eliminate trees, then, as duration of exposure lengthens, affect tall shrubs, and later, other plants.

The photochemical oxidants[149] (primarily the products of vehicles) are considered the most important agents, followed by sulfur dioxide.[150,151] Sulfate, which has been considered in epidemiological studies, is not directly toxic to plants (see "Water and Climate").

For sulfur dioxide, there may be a practical threshold of toxicity, owing to the presence of a detoxification system.[152] When the system is saturated, toxicity will occur. The absorption of sulfur dioxide by the plant can be beneficial to growth in soils that are deficient in sulfur.[153] Nitrogen dioxide alone is relatively innocuous but indirectly important through interaction with the photochemical-oxidant system.[154]

Biological interactions between sulfur dioxide and the photochemical oxidants (e.g., ozone) depend on plant species and apparently vary a great deal: They can lead to additive, reduced, or more than additive responses. Much more research is needed to clarify the nature and importance of their combined effects on vegetation.[155]

One of the most striking aspects of the general problem is that plants vary so greatly in their sensitivity.[156] Among trees, the white pine is especially sensitive.[157] Examples among vegetables are pinto beans, potatoes, and lettuce. Most sensitive of all appear to be the mosses and lichens that may show signs of injury after exposure to sulfur dioxide at levels one third those of the human annual ambient air quality standard.[158] The first European Congress on the Influence of Air Pollution on Plants and Animals (1968) recommended that mosses and lichens be employed as biological indicators of pollution because they are easily handled and exhibit a range of sensitivity that greatly exceeds that of most higher plants.

The variation in sensitivity among species indicates clearly that genetic factors are important, and within species, clones of more or less resistant plants have been produced experimentally.[159,160] In Connecticut, the selection, breeding, and production of tobaccos resistant to ozone saved the shade-tobacco industry. On the other hand, the production of spinach declined sharply, in part because the plants available were so susceptible.

Aside from the importance of such work in its own right, the investigations with plants may provide information about the subtle biochemical interactions of pollutants that would be much more difficult to discover initially with animals, but once known or suspected, could be effectively sought.

The interaction between sulfur dioxide and the photochemical oxidants supports the view that under various circumstances, interactions within the biological system negate analysis based on a model of "one agent, one effect." Furthermore, the two (or more) agents are likely in this case to come from two different sources—sulfur dioxide from an electric utilities plant, photochemical oxidants from vehicular traffic. The apportionment of responsibility, therefore, may likewise involve interactions.

WATER AND CLIMATE

This section is concerned with the effects of energy systems on precipitation, climate, and water supply. As dealt with here, these tend to be very large-scale problems, affecting regions of the nation or even the entire world.

ACID PRECIPITATION

The formation of acids in the atmosphere from combustion-generated sulfur dioxide (SO_2) and nitrogen dioxide (NO_2) acidifies rain and snow.[161–163] It was estimated that about 60 percent of the effect 10 years ago was due to acid-sulfate aerosols, and 40 percent to acid-nitrate aerosols. During the past 10 years, however, the nitric acid moiety has become relatively more important, presumably because of the increased use of low-sulfur fuels. Best available control technology (cf. chapter 4) will continue for some years to reduce sulfur oxide emissions relative to those of nitrogen oxides.

The distribution pattern of acid rain, determined by meteorological conditions, may extend for many hundreds of miles from the source, as first dramatically demonstrated in Sweden and Norway, where the effects were attributed to emissions from central Europe and the United Kingdom. In the United States, the Northeast is the focal area, and the pH of precipitation now averages annually about 4 – 4.3, with values as low as 2–3 observed at particular locations during storms. To the west and northwest, the pH gradient rises (to fall eventually in certain restricted areas), and in the desert regions, the pH averages about 7 (neutrality). The

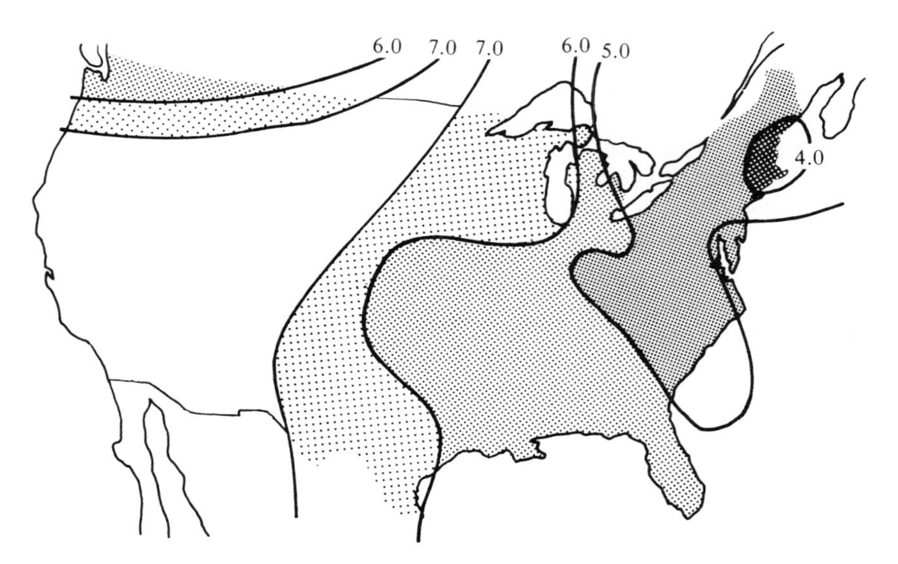

FIGURE 9-5 Acidity of precipitation falling in the United States during June 1966. Source: Gene E. Likens, "Acid Precipitation," *Chemical and Engineering News,* November 22, 1976, 54(48):30.

maps in Figures 9-5 and 9-6 illustrate the increasing spread of the low-pH region over a period of some 15 years (1955–1972). These trends were confirmed in 1978.[164]

The ecological effects are greatest in waters that contain the least dissolved matter—waters that are poorly buffered. Thousands of lakes in southern Norway and Sweden have shown a decline in fish populations, associated with increased acidity of the water, in turn associated with acid precipitation. A similar trend has been reported for the Adirondack Mountain region. Effects on terrestrial systems have been more difficult to isolate unambiguously, perhaps because changes register less quickly. A recent report[165] points out that as a result of acid precipitation, the forest-floor leaching mechanism in a New England coniferous ecosystem has changed from a carbonic-organic acid type to a mineral acid type, which may accelerate leaching and increase the concentrations of dissolved trace metals of potential toxicity. Damage to forests and sport fishing has been estimated at $100 million annually.[166]

Comparing various fuel cycles, it appears that the use of all fossil fuels involves some risk owing to the production of NO_2, and the use of those containing sulfur (coal, oil) present an additional risk.

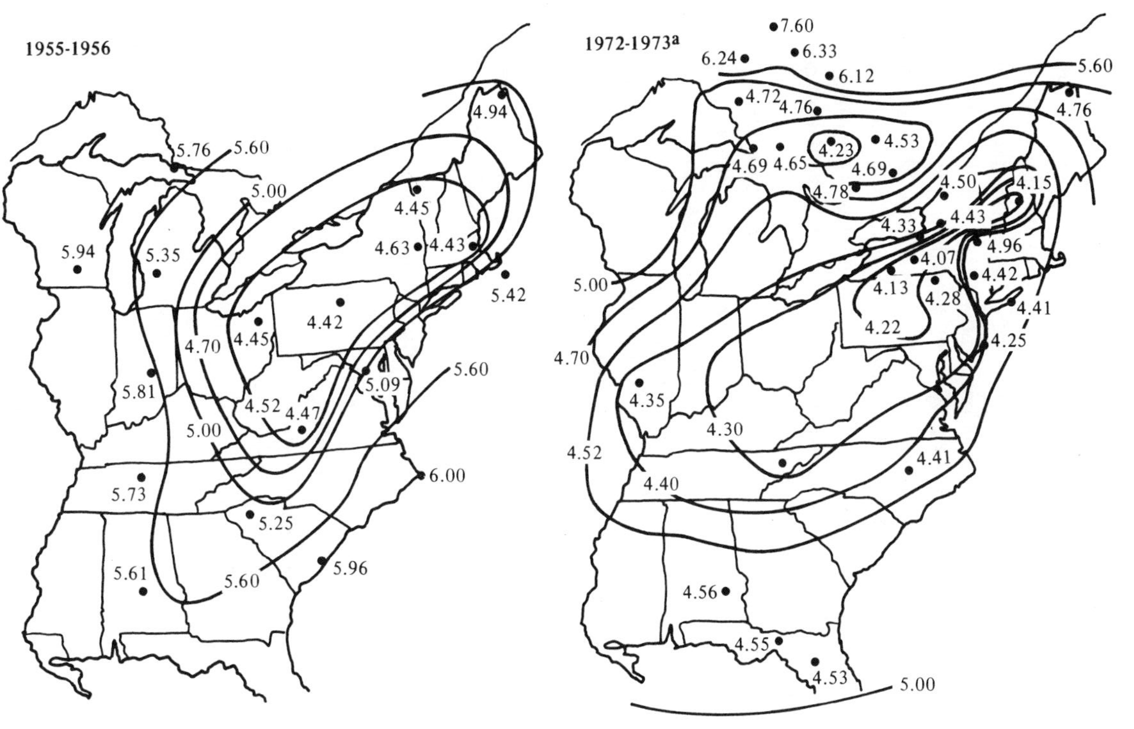

[a]Data from Oak Ridge, Tenn., for 1973-1974; data from Tallahassee, Fla., for 1974-1975; data from Gainesville, Fla., for 1976.

FIGURE 9-6 Acidity of precipitation in the eastern United States in 1955–1956 and 1972–1973. Source: Gene E. Likens, "Acid Precipitation," *Chemical and Engineering News,* November 22, 1976, 54 (48): 31.

GLOBAL CLIMATE[167–171]

Energy systems can influence climate on a large scale by the production of heat and by the production of carbon dioxide, both discharged into the atmosphere.[172] Other less important factors include the discharge of particulate matter and water vapor, and the resulting changes in albedo. From a global point of view, heat production is trivial: It amounts to just 0.01 percent of the sun's input of 150 W/m^2, when averaged over the earth's surface. On a regional or local basis, however, the concentration of industrial and other energy systems may produce sufficient heat to change patterns of wind, precipitation, humidity and cloudiness, and to elevate temperature. While of significant local interest, these effects—which can develop and disappear rapidly since they are reversible—are much less important than those of carbon dioxide. Carbon dioxide production could become the chief factor limiting the use of fossil fuels.

Although quantitatively a minor constituent of the atmosphere (330 parts per million by volume (ppmv) in 1979), carbon dioxide takes part in the control of temperature through the so-called "greenhouse effect." Virtually transparent to visible light, carbon dioxide strongly absorbs certain infrared wavelengths (heat) radiated from the earth's surface, to which other atmospheric gases are transparent. An increase in carbon dioxide in the troposphere, therefore, alters the path of the radiation of heat from earth into space, and thereby elevates the temperature of the lower troposphere—that portion of the atmosphere 5 miles from the earth's surface and below—in which variations in weather are largely determined.

The concentration of carbon dioxide in the atmosphere has been growing in parallel with increasing fossil fuel consumption throughout the world.[173] The future rate of increase will depend on the continuing availability of oil and natural gas and especially on the increasing use of coal. To the extent that coal is substituted for the other two fuels, the problem will be worsened: Coal produces some 200 lb of carbon dioxide per million Btu (89 metric tons per trillion joules), oil produces 80 percent as much, and natural gas about 57 percent. Synthetic liquids from coal produce 140 percent as much.[174] Nuclear power makes no contribution to this problem, nor do solar technologies.*

It is estimated that the atmospheric concentration of carbon dioxide has increased some 10 percent since the beginning of continuous measurements in 1958. At present rates of growth in the consumption of energy (according to one estimate[175]), it could rise from the present level of 330

*See statement 9-12, by H. Brooks, Appendix A.

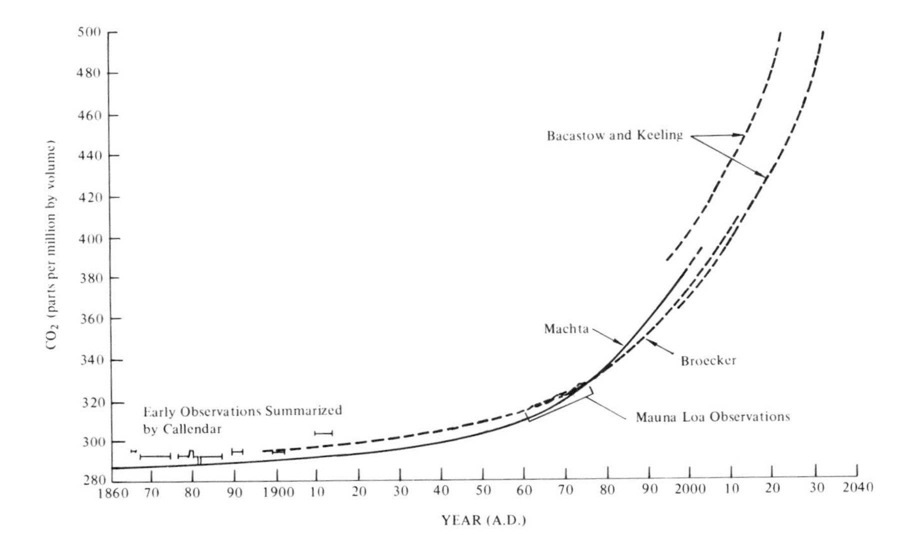

FIGURE 9-7 The record of carbon dioxide concentration from 1860 to 1975, measured at several locations, and some estimates of possible future trends. Source: W. W. Kellogg, *Effects of Human Activities on Global Climate* (Geneva, Switzerland: World Meteorological Organization (Tech. Note 156), 1977).

ppmv to 360–400 by the year 2000, and might double by 2040 (Figure 9-7). Uncertainty about the quantitative role of the biosphere in the overall carbon dioxide cycle adds uncertainty to these estimates,[176] but the Risk and Impact Panel[177] and other experts consider the projected trend correct.

Figure 9-8 illustrates one set of gross estimates for additional increments of atmospheric carbon dioxide and resulting rise in temperature. The estimates indicate the following.

1. For a doubling of atmospheric carbon dioxide concentration there is a 2°C–3°C rise in the average temperature of the lower atmosphere at middle latitudes, and a 7 percent increase in average precipitation.

2. The temperature rise is threefold to fourfold greater in the polar regions in this model.

3. For each 1°C rise in average temperature for the middle latitudes, there might be a 10-day average increase in the growing season. In the higher latitudes (40°C–50°C), the average growing season could be

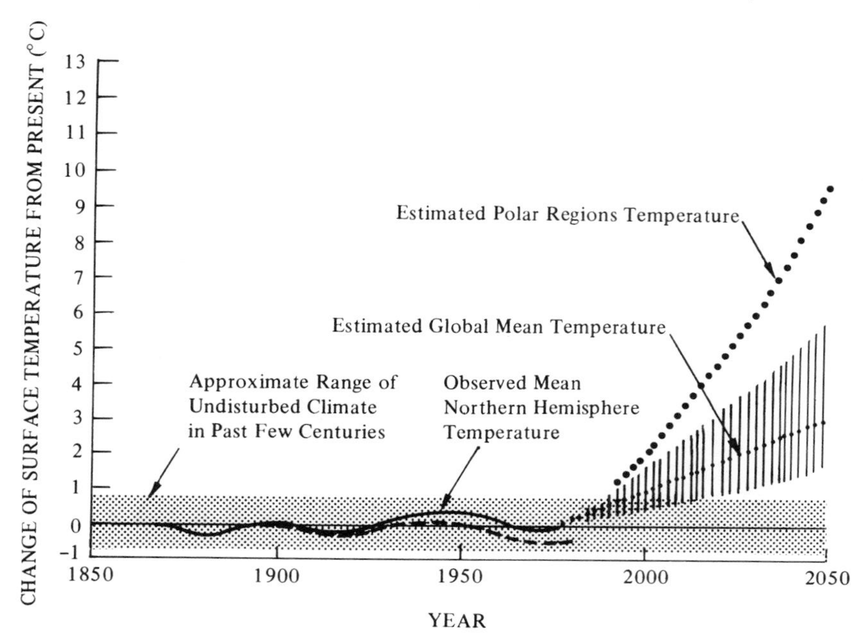

FIGURE 9-8 The mean surface temperature record for the Northern Hemisphere from 1850 to the present (solid line), and one estimate of the course that it might have taken without the addition of anthropogenic carbon dioxide (dashed line). Source: W. W. Kellogg, *Effects of Human Activities on Global Climate* (Geneva, Switzerland: World Meteorological Organization (Tech. Note 156), 1977).

lengthened by 2 or 3 times this amount. (There could be wide local variations from the average.)

4. Extensive and complex changes in precipitation patterns might occur as a result of the diminution in the polar-equator temperature difference, and general enhancement of the hydrological cycle.

A doubling of the carbon dioxide level presumably would have no significant direct effect on human health.[178]

The implications of the changes indicated above are potentially great. Increased temperature in the polar regions would lead to changes in precipitation patterns. Not only would the duration of the seasons be affected in other latitudes, but also humidity, cloudiness, and rainfall, which in turn would affect the extent and location of agricultural lands. Shifts in grazing, agricultural, and forest belts might occur, as well as increases or decreases in their extent. Such regional modulations might be much more important than the rise in temperature alone. Increased temperatures in the polar regions might lead to a "slow" melting of polar ice, in turn leading to a slowly rising water level and changing coastline.

It is important to emphasize, however, that while the trend of such a global picture may be generally correct, it has no degree of certainty today for particular times or places. Present models cannot project when and what will happen at Fargo, North Dakota, or Paris, France. Obviously, the regional effects might be good or bad. But even if the ultimate effects are good, the transitional period will presumably involve significant dislocations and inconvenience at the very least, if not problems that are far worse. Furthermore, the realization that major changes are anticipated is bound to cause distress and political tension.

The problem is a global one that should be analyzed and planned for at an international level. Further study is required to predict when major climatic changes might occur in relation to anticipated carbon dioxide levels, and to determine the influence of changes in carbon dioxide output now, or 15 years from now, on the course of climatic events. The graphs in Figures 9-7 and 9-8 project the *possibility* of perceptible changes by the year 2000, and the *possibility* of significant changes within 50 years. Study is required to predict with greater certainty and precision if and where the changes will occur, to outline their varied distribution, and to indicate more precisely what their consequences will be. Areas of uncertainty as well as certainty in these predictions should be defined as precisely as possible. On the basis of such estimates, society will have to consider building sufficient flexibility into its economic and international organizations to have some chance of adjusting to the changes gradually—in advance of their occurrence, if possible. Such adjustments might well include reduction in dependence on fossil fuels. This is one important reason to increase the diversification in our energy supply system, and to have nonfossil energy sources available for rapid substitution in the future.

WATER SUPPLY

A detailed discussion of the risk that increased energy consumption will induce a shortage of water, or that the limitations in water supply may curtail the use of energy, is given in chapter 4, largely based on the report of the Risk and Impact Panel.[179] Here we note briefly that the water problem may become a major limiting factor of energy availability. The difficulties pertain to mining, to the increased production of electricity, and to the proposed production of synthetic fuels from coal.

Some water is consumed in the routine processes of coal mining. The mining of coal or oil shale can disrupt aquifers and contaminate local drainage systems with acidic wastes.

More dramatic, however, is the issue of land reclamation in the arid western regions containing large deposits of surface coal (the upper Colorado and Missouri hydrological regions). By law, the mined land must be reclaimed, but the report of the Risk and Impact Panel warns that the success of reclamation in the high arid plains of these regions will depend critically on the ability of the soil to redevelop its capacity for aeration, water retention, and biological nutrient regeneration. The amount of water required per year to assist this process, for how many years, and with what promise of success remains an open question. Adequate water supply depends on the particular conditions at specific sites, but in general, the outlook for satisfying additional demands for water from indigenous supplies in these regions is very poor.

On the other hand, large amounts of coal in the eastern production regions lie within the basins of the upper Mississippi, Ohio, and Tennessee Rivers. Even though increased mining and reclamation may indeed be feasible in these and certain western areas, the use of that coal in the production of synthetic fuels or for the production of additional electricity could again raise the question whether local supplies of water are adequate. The generation of electricity consumes 15 times more water than mining the coal it burns.

Suppose the nation required an additional 18 quads of coal per year for the generation of electricity (11.5 quads) and the production of synthetic fuels (6.5 quads) for the next 20 years (a significant amount in national plans). Based on Samuels's criterion, the analysis presented in chapter 4 indicates that it would be desirable (perhaps necessary) to shift a major part of the burden to hydrological regions outside the western states. The adequacy of Samuels's criterion may be questioned: By attempting to guarantee that water flow in the area of interest will not fall below the weekly minimum observed the past 10 years, it may be too restrictive. It may, however, serve as a goal for ecologically sound practices.

An equivalent problem has been studied in greater detail by the six national laboratories that analyzed the water requirements of the President's National Energy Plan of 1977.[180] That plan called for an additional 18 quads of coal—13.5 for electricity and 4.5 for industrial use—and the findings were considered to apply by and large to the plans under the subsequent National Energy Act of 1978. Using a less demanding water shortage criterion (critical surface supply) than that employed by the Risk and Impact Panel, the laboratories' report concludes that such an increase is feasible, provided that particular attention is paid to the many siting problems that will occur.* The problem will not be in the mining of the coal, but in its use.

*Statement 9-13, by H. Brooks: However, the President's Energy Plan does not go beyond 1990 in projecting coal use. Unless energy growth leveled off after that date, there would be a growing problem.

It is clear that regional and interregional as well as local hydrological analysis must become an integral part of national energy planning, both to prevent water-supply failure and especially to obtain optimal use of our hydrological resources. We recommend that the relatively unstudied hydrological regions be examined. Water resources are largely under the control of the states. Two different approaches in law have been used to control them (the Riparian Doctrine and the Appropriation Doctrine). Their utilization in national planning will not be a simple matter. The energy-water problem is, in fact, a part of a much broader one of water as a limiting factor in the activities of society.

ECOSYSTEMS[181]

The following review offers brief summaries of the effects of the principal energy systems on ecosystems, based largely on the report of the Ecosystems Resource Group of the Risk and Impact Panel.[182] The field of study is still young. Its magnitude and extreme diversity are added burdens to investigators who recognize the value of expressing their findings in simple, generally applicable, quantitative terms. The resource group set the following criteria for adverse effects: loss of arable land and of water resources; loss of open space in or near urban areas; intrusion into wilderness areas and loss of beauty; and loss of habitat and loss of wild populations, particularly when leading to extinction of species. It is recognized that much more work should be done to translate qualitative descriptions or observations into quantitative assessments that can be related to energy system activity per unit of time. However, the National Environmental Policy Act explicitly states that unquantified environmental amenities and values must be given appropriate consideration in decision making, as well as economic and technical information.[183]

Decisions on the expanded production and use of energy taken in the face of predicted effects on ecosystems may involve a difficult balancing of values that cannot be made comparable.[184] Threats to ecosystems are generally speculative and subject to a high degree of uncertainty. The causal connections leading from the source to the ultimate (and long-term) consequences are long and complicated, and the consequences themselves are judged differently. To one group, the extinction of a few endangered species may seem a small price to pay for expanded consumer choice in material consumption, but to another group, the benefit of a few more material goods may seem frivolous—hardly worth the destruction of an endangered species.

HYDROELECTRIC POWER[185]

The ecological effects of hydroelectric projects are difficult to quantify and are extremely variable from case to case. If hydroelectric capacity expands in the future, the total ecological damage is likely to increase more than proportionately as the more suitable sites are used. This could be temporarily offset by a major development such as the Canadian James Bay project, but few suitable sites remain in the United States.

Among the adverse ecological consequences of new dam construction are the loss of habitat in the immediate area of the reservoir, subtle effects on the biological productivity of the river below the dam, damage to scenic areas along the wild stretches of the river, damage to the ecological balance of estuaries due to alteration of freshwater flow patterns, accelerated siltation and eutrophication[186] in the artificial lakes behind dams, adverse effects on fish species (such as salmon) that swim up rivers to spawn, and excess evaporation of water from artificial lakes and the resulting increased salinity, particularly in arid regions.

Some of these effects can be reduced or mitigated by proper design measures, but in the opinion of the Ecosystems Resource Group of the Risk and Impact Panel,[187] the ecological damage per unit of energy produced is probably greater for hydroelectricity than for any other energy source.* The social value assigned to these ecological losses varies. There is no question, however, that free-flowing rivers constitute a nonrenewable and rapidly disappearing feature of the American landscape. Their complete destruction would be too high a price for a small contribution to solving the energy problem.

GEOTHERMAL ENERGY[188]

In considering the ecological effects of this source, it is necessary to distinguish existing technologies for sources of steam and hot water from future technologies for hot dry rock or geopressured brines. The existing technologies are limited in their capacity to supply a significant fraction of total energy needs. Their ecological effects are specific to location: Some sources could destroy habitats and release toxic emissions that would affect local flora and fauna. Hydrogen sulfide is a particular problem for some sites.

Hot dry rock may be cleaner, but its exploitation as a source of energy will require methods of fracturing the rock at depth to provide access for water over a sufficiently large volume. In some cases, the procedures used

*See statement 9-14, by H. Brooks, Appendix A.

(such as high explosives) could contaminate groundwater or trigger seismic disturbances. Seismic effects could present significant hazards if large-scale exploitation of the normal thermal gradient in the earth's crust ever becomes feasible.

In the case of geopressured sources, the volume of brine that must be handled is enormous, and its removal could create ecological problems unless a satisfactory technique for reinjection is available: Land subsidence, for example, could be significant.

Too little is known about feasible methods of exploiting geopressured brines or hot dry rock to judge the ecological effects or techniques to avoid them.

SOLAR POWER[189]

The diversity of solar power systems is so great (photothermal, photoelectric, photochemical) that generalizations about the effects of these systems are difficult. The evaluation is complicated in the case of most solar power systems by a complete lack of practical experience. They appear to be no worse as a group than other energy systems, and in many forms, far superior. The caution must be voiced here (as for other alternative sources of energy) that the ecological effects depend on specific processes and siting plans.

In one respect, all solar systems may be expected to be benign. They do not directly contaminate the atmosphere, nor do they dig deep into the earth or strip its surface, particularly in areas that are difficult to reclaim. The centralized stations required for large-scale power production would affect land areas and even the ocean. Thermal stations would require about 6 acres/MWe; placed in the desert for optimal operation, their effect would be amplified by the ecological fragility of the location. On a per-GWe basis, solar central stations would require about 6000 acres, or 15 times the area of a nuclear plant and 10 times the area of a coal plant of comparable capacity (not counting the land required to mine the coal over its life).[190]*

Decentralized rooftop solar collectors represent a minor environmental risk compared to centralized solar collectors. They offer the possibility of generating electricity and space heat, thus using a greater fraction of incident solar energy. The main problems will result from the necessity of removing shade trees. With clustered buildings, this problem can be reduced by sharing the energy among the buildings: Each building does not have to be unshaded throughout the day.

*See statement 9-15, by H. Brooks, Appendix A.

Wind power would require 120,000 acres/GWe (about 20 times the land required for a solar thermal plant), but the land between towers could be used for other purposes, such as agriculture. Access roads would be required for maintenance, as well as transmission lines interconnecting the numerous towers. The environmental effects of these requirements would depend on competing demands for use of the land. Wind power would be most valuable in areas where it could be integrated into a grid supplied by hydroelectric power, which could be used to compensate for the fluctuations in wind. Besides presenting aesthetic problems in many areas, wind-turbine installations may endanger migratory birds and interfere with TV and other communications.

The development of ocean thermal power (OTEC) is almost certain to affect marine ecosystems, particularly if this source of energy is eventually relied upon for a significant fraction of total energy. The mixing of surface and deep waters will bring nutrients to the surface and may also release carbon dioxide to the atmosphere, although in lesser amounts than would be released by producing an equivalent amount of energy through the combustion of fossil fuel. Large-scale deployment of ocean thermal power might ultimately modify ocean currents and temperature distributions, with significant effects on regional or global ecosystems. Very little attention has been given to any of these environmental questions. Most of the effects mentioned, however, would only be of concern with very large-scale deployment and cannot be regarded as major deterrents to the development of OTEC.

Large-scale bioconversion or "biomass" is a popular option advocated by proponents of solar energy. To the extent that biomass is derived from organic wastes or from materials grown on special areas of the ocean or unused lands, the ecological effects would be minimal. This source of energy could supply perhaps 5 quads of total demand by 2010. Bioconversion is estimated to become progressively less desirable above this level as more land is dedicated to the cultivation of energy crops. To obtain an economic harvest and avoid soil depletion, chemical fertilizers and large amounts of water would have to be used. The land requirements would be enormous. At 1 percent average photosynthetic efficiency, 1.5 percent of the land area of the United States—an area about the size of Arkansas—would be required to generate about 10 quads of primary energy. The development of especially hardy species to increase the yields and improve the overall economics of energy from biomass would run some risk of spreading these species where they would be undesirable.

COAL

One of the most disruptive energy sources is coal. Underground mining can affect underground water systems and their associated drainage patterns and lead to subsidence. The underground waters that drain through the mine carry off toxic substances, and the mine itself disturbs the pattern of underground drainage. Strip mining without suitable reclamation is ruinous to surface land, and in some areas the possibility of lasting reclamation must be doubted.

Reclamation does not imply restitution (return to original conditions), but its water requirements may be high in areas where water supply is low. Areas that are relatively flat and possess an abundant water supply with high humidity are the most suitable for reclamation. Much strippable coal exists in areas where it may be effectively impossible to return the land to its prestripped state, even with the abundant use of water transported from elsewhere. If the consumption of coal increases rapidly, pressures will mount to initiate production in areas where reclamation will be relatively ineffective.

Deep mining is less destructive of the surface environment but can lead to subsidence and toxic effects in associated aquatic systems. These latter problems, however, appear susceptible to proper management, especially with improved mining technology.

As discussed in chapter 4, large resources of coal can be neither mined nor stripped but could be recovered by underground gasification. Little is known about the environmental effects of this practice. It could have adverse effects on important groundwater resources and thus indirectly affect ecosystems. The potential problems need to be kept in mind as the technology is developed.

In combustion, coal emits air pollutants that affect plant growth and lead to acid precipitation, which affects freshwater aquatic systems and forests (as explained earlier). The accumulation of carbon dioxide released in all fossil fuel combustion could ultimately result in drastic alteration of both natural and agricultural ecosystems.

The conversion of coal to synthetic fuels (liquids or gas) produces wastes, including contaminated liquids (phenol is a major constituent) that are unsuitable for immediate discharge. Present regulations and experience with coking operations offer assurance that adequate control can be achieved, but this can only be established in practice. The combustion of these fuels should be a relatively clean process (less polluting than combustion of oil or coal), since many impurities are removed in the process of conversion.

OIL

Drilling for oil can lead to accidental discharges that are harmful to local ecosystems. Likewise, spills of oil in transit or from routine operations are a serious hazard that can have devastating effects on marine and freshwater systems. The delayed as well as the prompt effects vary. The refining of oil and the release of pollutants in its combustion may have widely dispersed effects. Pipelines may promote erosion and, when overland, may hinder species migration. That suitable planning can prevent such untoward effects is now being tested by the operation of the Alaska pipeline. The overall ecosystem effects of oil are less serious than those of coal for corresponding energy production levels. Automobile emissions affect not only human health and comfort but also agricultural and plant life systems. (See "Agriculture and Plant Life.")

NATURAL GAS

The extraction and delivery of natural gas can threaten natural habitats. Pipeline leaks, by blanketing with natural gas and thus excluding oxygen, may leave an area barren for several months after the leak has been stopped. Fires and leaks from receiving facilities in marine areas are particularly hard on estuary life. The combustion of natural gas is indirectly damaging to ecosystems through the accompanying oxidation of atmospheric nitrogen. The nitrogen oxides thus formed are an important factor in acid precipitation (discussed under "Water and Climate"). In association with the photochemical-oxidant system or sulfur dioxide, they can be toxic to plant life (as previously noted). In total, however, the effects are much less severe than those of coal.

SHALE OIL AND COAL-DERIVED SYNTHETIC FUELS

The ecological effects of coal-derived fuels are much the same as those for coal itself. On the other hand, the production of oil from shale carries with it the threat of considerably more ecological damage than conventional production of oil. The attractive oil shale resources in the United States are located in limited areas of the western states that are ecologically fragile and that are short of water, of which large quantities would be needed. Solid-waste disposal is a problem. Experimentation has been initiated on conversion of oil shale in situ, a technique that might or might not be less damaging ecologically than retorting in aboveground plants but that may involve serious aquifer disruption.

NUCLEAR POWER

The principal ecological effects of nuclear power result from uranium mining—effects simliar to coal mining, but far less severe (only 3–4 percent as much ore is required today). Recycling uranium and plutonium in light water reactors and lowering enrichment tails could reduce the uranium ore required per gigawatt (electric) by as much as 40 percent and thereby reduce the ecological consequences of uranium mining and also of waste disposal. Light water reactors use about 50 percent more water than fossil-fueled generating plants. If breeders were widely adopted, the remaining ecological effects would be those from thermal pollution, similar to that accompanying the generation of electricity from other sources. The same would tend to be true for advanced converters. If the breeder option is foreclosed but dependence on nuclear power continues, the demand for uranium will lead to the mining of ever lower grade ores, gradually increasing the adverse ecological effects per gigawatt (electric) capacity. Use of the very lowest-grade sources, such as shales, would create environmental disruption comparable to that of strip-mined coal.

DISCUSSION

Before stating the conclusions drawn in this chapter, it may be useful to review some of the problems encountered in formulating them, as well as the concept of risk assessment itself.

LIMITATIONS IN RISK ASSESSMENT

In assessing the risks of energy systems to health, biological systems, and the environment, we have been constrained by the limited data in certain important areas. The major factor here is the early age of this field. While it is true that public health considerations have been in the public view for many years, they have centered on infectious diseases. Our concerns today center on diseases induced by chemicals or radiation. They are more difficult to detect because they are not uniquely associated with their causative agents, and are more difficult to cure because in general they are not so precisely defined and understood.

Two other factors materially increase the difficulty of risk assessment—the no-threshold dose-effect curve and the "late" effect. The latter, exemplified by cancer, may not be observed earlier than 20 years after exposure. The assumption of a no-threshold dose-effect curve is equivalent to the claim that any dose, no matter how small, has a finite chance of damaging someone in the exposed population. Together, both factors create a state of uncertainty in the public mind that leads to the setting of standards at progressively lower levels of dose and smaller probabilities of effect. Epidemiological investigation may not always be able to keep up with the pace of such standard setting, and the validation of a new standard—by scientific study and cost-benefit analysis—may be impossible.

In some cases that deal with very low levels of exposure, it may be asked if such validation could ever be achieved. Nonetheless, epidemiological and other related studies should be conducted, if only to establish the liminal value for risk that can be determined by research. No society can be free of risk, nor is the goal of a risk-free society necessarily worth striving for.

Somewhat analogously, the estimate of risks from rare nuclear accidents may never be precise enough to satisfy those taking a position against nuclear power. Two factors enter here. First, there is the analysis for mechanical failure, such as that reported in great detail in WASH-1400. Second, there is another risk that was recognized in that report but that could not be dealt with so extensively without a great deal more industrial experience—the risk of human error, that management may not be adequate under the stress of unforeseen and previously unexperienced circumstances. It is our impression that this second risk contributed to the damage in the Three Mile Island reactor accident.[191]

Finally, we note the inherent difficulty of comparing one energy system with another, since some of their important risks may be different and thus not strictly comparable. Nor can such differences be reduced to the terms of one common measure. Value judgments must therefore be made in the final determination of the risks society may prefer, from which sources, and at what cost.

For perspective, it may be of interest to note some of the other risks that are current in our society. In 1974, the following annual accident death rates per 100,000 persons at risk applied to the United States: motor vehicle, 22.0; falls, 7.7; drowning, 3.1; burning, 2.9; and firearms, 1.2. A British investigator[192] has summarized United Kingdom experience thus: A one-in-a-million risk of death has been attributed to 400 miles by air, 60 miles by car, three fourths of a cigarette, 1.5 min of rock climbing, 1.5 weeks of typical factory work, and 20 min of being a man 60 years old.

PERCEPTION OF RISK

Society's tolerance of risks that have been lived with is greater than its tolerance of risks associated with new technologies. The personal evaluation of risks tends to be formed when a technology first becomes visible. At the time older technologies were introduced, risks were more readily accepted than they are today. An important question is how subjective evaluations should be taken into account along with the more objective measures, such as fatalities per GWe-plant-year of energy production. If the subjective values are ignored, energy policy may stalemate.

The recent incident at the Three Mile Island reactor bears on this point. From the dosimetry (see "Nuclear Power"), it is clear that damage to public health was negligible, but the public's perception of the dangers of nuclear power (rightly or wrongly) has been greatly heightened. The uncertainty of management and the tenor of the information released in the early days after the accident no doubt played a role in this. It will be of interest to see how public perception is affected by the report of the presidential commission.

A point of difference between the objective quantification of risk and its sociopolitical assessment is the role of "*attitude* toward risk" in the latter. An energy system may be viewed as a hazard by a particular group of people (organized or unorganized) for reasons that may not reflect biological or ecological assessments of the risks. Nuclear power, for example, may be opposed as a symbol of big government, impersonal corporate business, or unrestrained economic growth. Gasoline shortages are viewed by many as tricks on the part of the big oil corporations rather than as a consequence of conditions of supply and demand on the world market.

One conclusion that may be drawn is that adequate information must continually be made easily available to the public, to inform the decision-making process, and to prevent the spread of false conclusions and impressions. We emphasize the continuous nature of the task, since energy policy in general and knowledge of its associated risks in particular will be developing for many years to come.

REGULATION

From the point of view of practical governance, the foregoing considerations are encompassed by the regulatory process, a function that has been evolving rapidly and in a somewhat uncoordinated way (in the areas dealt with here). It should not occasion surprise that this is so. As we have noted, many congressional committees and many units of the executive branch are involved.

Steps should be taken to simplify the regulatory process, but the extent to which such improvements can actually be effected is a matter of conjecture. It would be unrealistic to suppose that the work of a very large number of departments and agencies could be rapidly coordinated and made efficient during a period of rapid expansion in governmental responsibility. The government today is still in the process of learning through its own experience how to regulate and control the risks of energy systems in ways that satisfy many diverse groups.

With respect to the evaluation of risk, it should be recognized as a matter of policy that energy plans require time for their implementation. The assessment of risk therefore can be advantageously made a sequential process. In moving forward, the goal might be set to control overall total risk: As one energy system expands, the risk per unit of product should decline. Ideally, the total risk would not rise and might even fall. Progress toward this goal will also occur whenever the expansion of one system replaces another that is riskier.

SOCIAL AND POLITICAL RISKS[193]

The concerned citizen and policy maker will have to go beyond the scientific and technical comparisons we have made and consider sociological and political risk comparisons as well. An extreme example would be to compare the loss of unique natural beauty with the advantages of local business development. There are many less extreme but important examples. To change the activity of a fuel system, for example, in response to its health risks might be weighed against the risk of reduced employment or of forcing a change in established cultural patterns. Other major comparisons lie completely within the sociopolitical domain, such as the risk of an accelerating oil shortage compared to the economic inflation engendered by increasing the price of oil.

An important aspect of such sociopolitical considerations is the geographical dissociation of risks and benefits. For example, western states might bear more than their share of the adverse consequences of rapidly developed mining (e.g., the boomtown), but the economic benefits would not return to the affected communities in the same proportion. Appalachia has already demonstrated such effects over the years. Likewise, air and water pollution generated at any stage of the fuel cycle are not necessarily borne by the users of the energy in proportion to use.

Traditional cost-benefit analysis, as applied to energy decisions, does not usually include these distributional effects.[194] Extension of the analysis can in principle identify costs and benefits to particular groups or geographical regions, but the balancing of costs to one group against benefits to another, or to the general welfare, is inherently a political judgment. There is a need for some kind of compensation to redress the imbalance in the distributed effects of energy systems.

Finally, we wish to mention four sociopolitical risks that may figure prominently in the deliberation of energy policy and that serve to place other sociopolitical risks in perspective.

1. In the minds of some, the greatest risk associated with the production of nuclear energy is that the associated technology and materials can assist in the proliferation of nuclear weapons. In comparison to nuclear war, all other risks are small. But the closeness of the connection between development and the threat of nuclear hostilities or war is uncertain. As detailed in chapter 5, CONAES is divided on this issue.

2. A related risk is that the protection of nuclear fuel cycles against sabotage and theft will lead to unacceptable security measures, incompatible with our standards for civil liberties.* Some argue that such measures will be necessary in a world of increasing violence and terrorism. Other energy sources, in particular, dams and storage facilities for liquefied natural gas, are subject to sabotage, as are such nonenergy facilities as supplies of drinking water. Nuclear facilities, in principle, may be easier to guard without intrusion into the rest of society by virtue of the small area involved and the limited number of personnel.

3. The third risk is associated with the further large-scale development of national energy systems, whose operation and control become increasingly centralized and increasingly out of the reach of the ordinary citizen. It is argued that numerous self-sufficient energy systems at the household and neighborhood level would be more flexible and responsive to local needs. The development of such systems for the future is certainly desirable, if only to provide for greater diversity of energy alternatives.† Nevertheless, large-scale electrical and gas-distribution systems are working well today. Decentralized systems, to serve their intended purposes, would have to be mass produced, widely distributed, and maintained on a large scale. Without more experience, it is impossible to say how sturdy such decentralized systems could be.

4. The fourth risk is not having enough energy. CONAES has not considered the risks that could be faced by a society in which energy supplies fall short of the citizens' legitimate needs.‡

CONCLUSIONS

1. LIMITS OF RISK CONTROL

The increasing interest in protecting public health and natural resources is of major historical significance, and no doubt will continue to increase in

*See statement 9-16, by B. I. Spinrad and H. Brooks, Appendix A.

†See statement 9-17, by B. I. Spinrad and H. Brooks, Appendix A.

‡Statement 9-18, by J. P. Holdren: The Demand and Conservation Panel found very low growth in energy use compatible with high prosperity. Only *sudden* shortfalls were not considered.

extent and influence. Regulations will always engender controversy. This is especially true today, owing to the tremendous expansion of regulatory activity and to the fact that we are still learning how to manage it. In dealing with these problems, we must recognize that there are practical limits to the refined ascertainment of risk and its control. Energy systems (or other systems) cannot be made risk-free, nor can all improvements be made as a matter of course without significant economic penalty or other adjustment. The public should be made aware of these limitations, and of the possibility that the absolute control of risk may not only be impossible, but undesirable.

2. CONSERVATION

For the most part, conservation is the least risky strategy from the standpoint of direct effects on the environment and public health. (One potential problem is the possibility of indoor air pollution buildup in connection with certain conservation measures in buildings.) The main reason that conservation cannot be the only strategy is that at some level of application, it would give rise to indirect socioeconomic and political effects, mostly through economic adversity, that would predominate over its direct benefits. We cannot be sure where that point is, but all the CONAES technical analyses suggest that it is far from where we are now, possibly at an energy/GNP ratio of about half its present value, given several decades for adjustment. The maximum conservation achievable without adverse socioeconomic effects will likely have health and environmental benefits and therefore should have highest priority in policies to reduce the risks of energy systems.

3. FOSSIL FUELS

Among fossil fuels, natural gas presents the smallest health and environmental risks in both production and consumption, although there is the possibility of serious accidents in the transportation and storage of liquefied natural gas. Oil is next, and coal is much higher in risk. This ranking is likely to persist, although the gap may narrow with improvements in technology. Research is most urgently needed on the health effects of coal combustion by utilities and industry, and on the possible occupational and public health hazards of producing and using synthetic fuels.

We must be prepared for the possibility that adverse health effects, global CO_2 increase and associated climatic change, freshwater supply problems, and ecological considerations will eventually severely restrict continuing expansion of coal use. These problems are likely, though not certain, to become critical at about 3 times current coal output, or less.

4. CARBON DIOXIDE

The accelerating increase in the level of atmospheric carbon dioxide, paralleling the increase in combustion of fossil fuels worldwide, may affect global climate by elevating the mean global temperature. The substitution of coal for oil (or synthetic liquids for oil) and oil for gas will tend to hasten this process, but the combustion of any fossil fuel contributes CO_2 to the atmosphere. Assuming continued growth in the use of fossil fuels, a perceptible change has been projected by one model for the year 2000, and significant change by the year 2030. A serious concern is that climatic changes due to CO_2 would be practically irreversible by the time they were detected. It should be noted that the ultimate effects may be good as well as bad (or mixed), but the transitional period could be disruptive in its effects on agriculture and industry in some regions. Vigorous international efforts should be undertaken to predict the course of carbon dioxide buildup and to determine its climatic and consequent ecological effects. There is a parallel need for planning studies to mitigate possible economic and social disruption, including plans to curtail the use of fossil fuels.

5. NUCLEAR POWER

The routine risks of nuclear power include the induction of cancer and genetic effects by ionizing radiation released throughout the nuclear energy cycle. These risks are very small in comparison to the overall incidence of cancer and genetic effects in the general population, and they could be significantly smaller yet if the most important source of radiation in the nuclear energy cycle—uranium mill tailings—were generally better protected.* There are also risks of severe accidents, whose probabilities have been estimated with a great deal of uncertainty, but whose severities could be comparable to those of large dam failures and liquefied natural gas storage system fires.† There are also risks from the disposal of radioactive waste; these are less than those of the other parts of the nuclear energy cycle, but only if appropriate action is taken to find suitable long-term disposal sites and methods.

It should be clear from the earlier general discussion of risk comparisons that any ranking of the risks of technologies as disparate as coal-fired and nuclear electricity generation is subject to very broad, and in some cases

*Statement 9-19, by H. I. Kohn: This may contradict Table 9-5. Presumably, it assumes that the improper and now illegal practices of the past will be continued.

†See statement 9-20, by H. I. Kohn, Appendix A.

irreducible, uncertainties. However, if one takes all health effects into account (including mining and transportation accidents and the estimated expectations from nuclear accidents), the health effects of coal production and use appear to be a good deal greater than those of the nuclear energy cycle. If in this comparison, though, one takes the most optimistic view of the health effects of coal-derived air pollution and the most pessimistic view of the risk of nuclear accidents, coal might have a small advantage in such a comparison.†

Nuclear power is associated also with risks of nuclear weapons proliferation and terrorism, but the magnitude of these risks (and even whether nuclear power increases or decreases the risks) cannot be assessed in terms of probabilities and consequences.

6. WATER SUPPLY

The supply of water may constrain the continued growth of electrical power, the mining and the conversion of coal to synthetic fuels, the production of oil from shale, and the use of coal by industry. Water scarcity is greater in the West than the East, but affects particular localities in both parts of the country. The projection reported in this study shows that the constraint could indeed be significant when, for example, the amount of additional electricity and synthetic fuels that could be produced from 40 quads of coal is produced nationally. (See chapter 4.) The projection assumes present technology (which could be improved) and does not allow for large-scale use of brackish water or seawater. We urge that regional and interregional hydrological analyses become an integral part of energy production planning, to prevent actual water shortage, and especially to distribute production to make optimal use of our hydrological resources. Such planning will probably be difficult. We note that the water-energy problem is a single manifestation of the broader problem of water as a limiting factor in the growth of society.

7. SOLAR ENERGY

Several solar energy technologies appear very promising from the standpoint of health and environmental risk. Hydroelectric power (classed by convention with solar energy), however, while benign with regard to air pollution, is quite destructive of ecosystems per unit of output. Terrestrial energy farms are also likely to be ecologically destructive if deployed on a scale large enough to provide more than a few percent of total energy

†See statement 9-21, by J. P. Holdren, Appendix A.

needs. (Biomass production at sea could avoid this problem.) For most solar technologies, the main risks are those associated with extracting and processing the requisite large amounts of construction materials.

8. AIR QUALITY STANDARDS AND RESEARCH

The difficult tasks of setting standards for ambient air quality and emissions have been greatly complicated by a lack of precise knowledge of the levels at which epidemiological effects first appear and of the diversity of such effects. Pragmatic decisions must therefore be made in the face of uncertainty—uncertainty magnified by the periodic (and appropriate) review of these standards. Industrialists may claim that the standards are set too low, in order to make them safe regardless of cost and convenience; this inhibits industrial planning and has been a deterrent in the further use of coal. The situation calls for a major research effort into the effects of pollutants, including emissions from mobile sources (nitrogen oxides, ozone, nonmethane hydrocarbons), as well as from the stationary sources (nitrogen oxides, sulfur oxides, particulates) that this chapter considers at length.

The committee recommends that investigation center on the dose-effect curve (or exposure-effect curve) in the region near and below the present ambient air quality standards.

1. The quantitative assessment of exposure, and if possible, of dose per individual, is essential to advance knowledge in this field. One or two centrally located stations are insufficient to monitor an urban area for epidemiological research (they may be sufficient for other purposes). Measurements of indoor and outdoor, residential and occupational exposures are necessary.

2. Mortality is too gross an endpoint to be used alone. Others must be selected for specific types of morbidity and for physiological and biochemical response.

3. While immediate responses are important, late effects in specific individuals may be even more important.

4. The effects of emissions on plants and ecosystems should receive major attention.

5. The magnitude of the several problems to be investigated necessitates undertaking and maintaining long-term studies. Some will take decades to complete.

6. Much of the work cannot be planned from first to last detail, and none of it should be subject to political control. Coupled with the need for an effort adequate in scale to the problems under investigation, there is a

need for flexibility and independence in pursuing the studies that suggests scientists outside the government should conduct much of the work.

7. We should realize that answers will come slowly and decisions will have to be made on an uncertain basis for some time in the future.

9. PUBLIC APPRAISAL OF ENERGY SYSTEMS

There is a need for research that will contribute to better understanding of the factors that determine public perceptions of the health and environmental risks of energy systems, and their acceptance by different subgroups within the public. No strategy for risk reduction in energy systems can be fully acceptable if it does not take into account these public perceptions and judgments, even when they are seen as unfounded by experts.* It is unlikely that the appraisal of risk will ever be able to avoid difficult relative value judgments between different kinds of risks, as well as between risks and economic or other benefits of energy technologies. This is not to say that present methods of risk assessment cannot be improved. Nevertheless, the judgmental factor will continue to predominate in decisions among energy alternatives, and is unlikely ever to be superseded by formal analysis of risks and benefits. This underscores the importance of an informed and open public debate.

NOTES

1. 40 Code of Federal Regulations 190.02 (a), (b), 1978, "Environmental Radiation Protection Standards for Nuclear Power Operations."

2. In addition to the risks associated with the operation of an energy system itself, those associated with construction of power plants and the occupational risks of manufacturing its parts might also be considered (as done in chapter 6). It was recently claimed (H. Inhaber, *Risk of Energy Production* (Ottawa, Ontario: Atomic Energy Control Board (AECB 1119), March 1978); and H. Inhaber, "Risks from Conventional and Unconventional Sources," *Science* 203 (1979): 718–723) that inclusion of these risks brings solar power to a level of risk approximately equal to that of power from coal or oil. The calculations supporting these widely publicized conclusions have been rejected. (See, for example, J. P. Holdren, K. R. Smith, and G. Morris, "Energy: Calculating the Risks (II)," *Science* 204 (1979): 564–568; R. Caputo, "Energy: Calculating the Risks," *Science* 204 (1979): 454; R. Lemberg, "Energy: Calculating the Risks," *Science* 204 (1979): 454, and John P. Holdren *et al., Risk of Renewable Energy Sources: A Critique of the Inhaber Report,* Energy and Resources Group (Berkeley, Calif.: University of California, June 1979).) The inclusion of these risks is worth consideration, but the ramifications might be endless, and ultimately the definition of the risks under investigation would blur. For further discussion of risk and its estimation, see National Research Council, *Risks and Impacts of Alternative Energy Systems,* Committee on

*See statement 9-22, by H. Brooks, D. J. Rose, and B. I. Spinrad, Appendix A.

Nuclear and Alternative Energy Systems, Risk and Impact Panel (Washington, D.C.: National Academy of Sciences, in preparation), chaps. 1, 2, and 3.

3. J. Clarence Davies III and Barbara S. Davies, *The Politics of Pollution,* 2nd ed., *Studies in Contemporary American Politics,* general ed. Richard E. Morgan (Indianapolis, Ind.: Pegasus, 1975).

4. Estimated from inspection of the list of agencies publishing in the *Federal Register* (1979).

5. T. J. Nagel, "Operating a Major Electric Utility Today," *Science* 201 (1978): 985–993.

6. John A. Phinney, consultant to Continental Oil Co., personal communication to H. I. Kohn, September 15, 1978.

7. Public Law 95-95 (August 7, 1977), Clean Air Act Amendments of 1977, Sect. 110.

8. Although the standard is set without reference to technological considerations, penalties or other sanctions resulting from legal proceedings may be withheld if the required technology is unavailable.

9. *Federal Register* 44 (1979): 8202–8233. The preamble to this regulation provides a useful summary of several points raised in this chapter about the Clean Air Act and the problems of regulation in general.

10. Public Law 95-95 (August 7, 1977), Clean Air Act Amendments of 1977, Sects. 160–169, "Prevention of Significant Deterioration of Air Quality."

11. Davies and Davies, *op. cit.*

12. Risk and Impact Panel, *op. cit.*

13. T. Falkie, "Perspectives on Coal Mining, Preparation, and Transportation," in *Actions to Increase the Use of Coal: Today to 1990* (McLean, Va.: Mitre Corp., 1978).

14. Risk and Impact Panel, *op. cit.,* chap. 4.

15. Robert B. Cameron, *An Estimation of the Tangible Costs of Black Lung Disease Related Disability to the Bituminous Coal Mine Operations of Appalachia,* Appalachian Resources Project no. 47 (Knoxville, Tenn.: University of Tennessee, 1976), p. 123.

16. National Research Council, *Mineral Resources and the Environment, Supplementary Report: Coal Workers' Pneumoconiosis,* Commission on Natural Resources, Committee on Mineral Resources and the Environment (Washington, D.C.: National Academy of Sciences, 1975).

17. Office of Technology Assessment, *The Direct Use of Coal: Prospects and Problems of Production and Combustion* (Washington, D.C.: Government Printing Office (052-003-000664-2), 1979).

18. Mitre Corp., *Accidents and Unscheduled Events Associated with Non-Nuclear Energy Resources and Technology,* Metrek Division (McLean, Va.: Mitre Corp. (M76-68), December 1976).

19. Risk and Impact Panel, *op. cit.,* chaps. 4 and 5.

20. F. E. Speizer, "Questionnaire Approaches and Analysis of Epidemiological Data in Organic Dust Lung Diseases," *Annals of the New York Academy of Sciences* 221 (1974): 50–58.

21. G. W. Beebe, H. Kato, and C. E. Land, *Mortality Experience of Atomic Bomb Survivors, 1950–1974* (Hiroshima, Japan: Radiation Effects Research Foundation (RERF TR 1-77), 1977).

22. The analysis can be further refined by considering acute leukemia and chronic granulocytic leukemia separately. T. Ishimaru, M. Otake, and M. Ichimaru, "Dose-Response Relationship of Neutrons and Gamma Rays to Leukemia Incidence Among Atomic Bomb Survivors in Hiroshima and Nagasaki by Type of Leukemia, 1950–1971," *Radiation Research* 77 (1979): 377–394.

23. 1 rad equals 100 ergs of absorbed radiation per gram of tissue. Since different radiations (gamma rays, alpha particles, neutrons) are more or less biologically effective (per rad), the

doses are normalized for comparison to *rem*, the biologically equivalent rad dose of gamma rays.

24. For gamma rays, X-rays, and beta rays (low LET radiations) the following dose-effect relation has frequently proved useful: effect $= aD + bD^2$, where a and b are constants. The initial part of the curve (e.g., up to 50 rads) is practically linear; thereafter, the slope increases constantly. The Nagasaki curve might be of this type. (See Ishimaru, Otake, and Ichimaru, *op. cit.*) A definitive review of the subject is being prepared by the National Council on Radiation Protection (in press, 1979).

25. See note 22.

26. M. Ichimaru, T. Ishimaru, J. L. Belsky, *et al., Incidence of Leukemia in Atomic Bomb Survivors, Hiroshima and Nagasaki, 1959–1971* (Hiroshima, Japan: Radiation Effects Research Foundation (RERF 10-76), 1976).

27. B. D. Dinman, "Non-Concept of 'No-Threshold': Chemicals in the Environment," *Science* 175 (1972): 495–497.

28. D. M. Bernstein, "The Influence of Trace Metals in Disperse Aerosols on the Human Body Burden of Trace Metals" (Thesis submitted to the Department of Environmental Health Sciences, New York University, Graduate School of Arts and Sciences, 1977).

29. Dinman, *op. cit.*

30. R. H. Haynes, "The Influence of Repair Processes on Radiobiological Survival Curves," in *Cell Survival After Low Doses of Radiation,* ed. T. Alper (New York: John Wiley and Sons, 1975). See also, R. F. Kimball, "The Relation of Repair Phenomena to Mutation Induction in Bacteria," *Mutation Res.* 55 (1978): 85–120; and H. I. Kohn, "X-Ray Mutagenesis: Results with the H-Test Compared with Others, and the Importance of Selection and/or Repair," *Genetics* 92 (1979): S63–S66 (supplement).

31. E. J. Underwood, *Trace Elements in Human and Animal Nutrition,* 3rd ed. (New York: Academic Press, 1971).

32. National Cancer Institute, *Epidemiological Study of Cancer and Other Chronic Diseases,* ed. W. Haenszel (Washington, D.C.: Government Printing Office, 1966).

33. R. Doll and R. Peto, "Mortality in Relation to Smoking: 20 Years' Observations on Male British Doctors," *British Medical Journal* 11 (1976): 1525–1536.

34. R. Doll and R. Peto, "Cigarette Smoking and Bronchial Carcinoma: Dose and Time Relationships Among Regular Smokers and Lifelong Non-Smokers," *Journal of Epidemiology and Community Health* 32 (1978): 303–313.

35. For asbestosis see P. E. Enterline, "Pitfalls in Epidemiological Research," *Journal of Occupational Medicine* 18 (1976): 150–156; for cotton dust, J. A. Merchant, *et al.,* "Dose Response Studies in Cotton Textile Workers," *Journal of Occupational Medicine* 15 (1973): 222–230; for silicosis, T. H. Hatch, "Criteria for Hazardous Exposure Limits," *Archives of Environmental Health* 27(4) (1973): 231–235.

36. National Research Council, *Carbon Monoxide,* Assembly of Life Sciences, Committee on Medical and Biological Effects of Environmental Pollutants (Washington, D.C.: National Academy of Sciences, 1977).

37. N. Mantel and M. A. Schneiderman, "Estimating 'Safe' Levels, A Hazardous Undertaking," *Cancer Research* 35 (1975): 1379–1386.

38. N. Mantel, *et al.,* "An Improved Mantel-Bryan Procedure for 'Safety' Testing of Carcinogens," *Cancer Research* 35 (1975): 865–872.

39. Mantel and Schneiderman, *op. cit.*

40. Frederick W. Lipfert, "The Association of Human Mortality with Air Pollution: Statistical Analyses by Region, by Age, and by Cause of Death" (Dissertation submitted to Union Graduate School, Yellow Springs, Ohio, 1978); Frederick W. Lipfert, "Differential Mortality and the Environment," in *Energy Systems and Policy* (in press, 1979); and Frederick W. Lipfert, "Statistical Studies of Mortality and Air Pollution: 1. Multiple

Regression Analysis by Cause of Death; 2. Multiple Regression Analysis Stratified by Age Group," in *Science of the Total Environment* (in press, 1979).

41. Herbert Schimmel and L. Jordan, "The Relation of Air Pollution to Mortality, N. Y. City, 1963–72, II. Refinements in Methodology and Data Analysis," *Bulletin of the New York Academy of Medicine* 54 (1978): 1052–1112.

42. National Research Council, *The Effects on Populations of Exposure to Low Levels of Ionizing Radiation: Report of the Advisory Committee on the Biological Effects on Ionizing Radiations,* Division of Medical Sciences, Committee on the Biological Effects of Ionizing Radiation (Washington, D.C.: National Academy of Sciences, 1974 (first printed 1972)).

43. National Council on Radiation Protection and Measurements, *Natural Background Radiation in the United States* (Washington, D.C.: National Council on Radiation Protection and Measurements (NCRP Report no. 45), 1975).

44. The calculation assumes a mean annual dose per person of 0.1 rem, a population of 200 million, and a factor of 2×10^{-4} cancers per person-rem: $1 \times 10^{-1} \times 2 \times 10^{8} \times 2 \times 10^{-4} = 4 \times 10^{3}$ cancer deaths.

45. United Nations Scientific Committee on the Effects of Atomic Radiation, *Sources and Effects of Ionizing Radiation, 1977,* General Assembly, 32nd Sess. (New York: United Nations (Sales no. 77.IX.1), 1977). The U.N. factor is 2 person-rem to the public per MWe-year, and occupational exposure would double it. The cancer death rate factor is 1×10^{-4} per person-rem.

46. N. A. Frigerio and R. S. Stowe, "Carcinogenic and Genetic Hazard from Background Radiation," in *Biological and Environmental Effects of Low-Level Radiation,* vol. 2 (Vienna, Austria: International Atomic Energy Agency, 1976), pp. 385–393.

47. National Council on Radiation Protection and Measurements, *Review of the Current State of Radiation Protection Philosophy* (Washington, D.C.: National Council on Radiation Protection and Measurements (NCRP Report No. 43), 1975).

48. Separate and much higher annual standards apply to exposure of limited parts of the body, e.g., for occupational exposure, 15 rem to skin, 75 rem to hands, 15 rem to other organs (excepting bone marrow and gonads).

49. T. F. Mancuso, A. Stewart, and G. Kneale, "Radiation Exposures of Hanford Workers Dying from Cancer and Other Causes," *Health Physics* 33 (1977): 369–385.

50. G. B. Hutchison, S. Jablon, and C. E. Land, "Review of Report by Mancuso, Stewart, and Kneale of Radiation Exposure of Hanford Workers," *Health Physics* (in press, 1979).

51. S. Marks, E. S. Gilbert, and B. D. Breitenstein "Cancer Mortality in Hanford Workers," in *Symposium on the Late Effects of Biological Effects of Ionizing Radiation* (Vienna, Austria: International Atomic Energy Agency, in press, 1979).

52. T. C. Anderson, "Radiation Exposures of Hanford Workers: A Critique of the Mancuso, Stewart and Kneale Report," *Health Physics* 35 (1978): 743–750.

53. J. A. Reissland, "An Assessment of the Mancuso Study," National Radiation Protection Board (United Kingdom: Her Majesty's Stationery Office (Report no. NPRB-R79), 1978).

54. 10 Code of Federal Regulations 50, app. I, 1978, "Numerical Guides for Design Objectives...for Radioactive Material in Light-Water-Cooled Nuclear Power Reactor Effluents."

55. A curie (Ci) is the quantity of any radioactive isotope undergoing 3.7×10^{10} disintegrations per second. Note that the radiation emitted may be of any type.

56. 40 Code of Federal Regulations 190.02 (a), (b), *op. cit.*

57. 10 Code of Federal Regulations 50, app. I, *op. cit.*

58. National Council on Radiation Protection and Measurements, *Review of the Current State of Radiation Protection Philosophy, op. cit.*

59. International Commission on Radiological Protection, "Recommendations," *Annals of the ICRP* 1 (1977): 53.

60. National Research Council, *Risks Associated with Nuclear Power: A Critical Review of the Literature,* Committee on Science and Public Policy, Committee on Literature Survey of Risks Associated with Nuclear Power (Washington, D.C.: National Academy of Sciences, 1979).

61. United Nations Scientific Committee on the Effects of Atomic Radiation, *op. cit.*

62. Committee on Science and Public Policy, *op. cit.*

63. Risk and Impact Panel, *op. cit.,* chap. 2.

64. Committee on Science and Public Policy, *op. cit.*

65. *Ibid.*

66. The end-use hazards of natural gas (e.g., of gas stoves and gas heating) have not been evaluated.

67. U.S. Environmental Protection Agency, *National Air Quality, Monitoring, and Emissions Trends Report, 1977* (Research Triangle Park, N.C.: U.S. Environmental Protection Agency (EPA-450/2-78-052), 1978).

68. K. E. Schaefer, "Editorial Summary: Preventive Aspects of Submarine Medicine," *Undersea Biomedical Research* 6 (1979): S-7–S-14 (supplement).

69. G. MacDonald and L. J. Carter, "A Warning on Synfuels, CO_2, and the Weather," *Science* 205 (1979): 376–377.

70. Public Law 91-604 (September 22, 1970), Clean Air Act Amendments of 1970.

71. *Federal Register* 44 (1979): 8202–8233.

72. U.S. Department of Energy, *An Assessment of National Consequences of Increased Coal Utilization, Executive Summary,* vols. 1 and 2, prepared by staff members of the following national laboratories: Argonne, Brookhaven, Lawrence Berkeley, Los Alamos, Oak Ridge, and Pacific Northwest (Washington, D.C.: U.S. Department of Energy (TID-29425), February 1979).

73. U.S. Senate, *Air Quality and Automobile Emission Control,* vol. 1, *Summary Report,* Committee on Public Works. 93rd Cong., 2nd Sess. (Serial no. 93-24), September 1974.

74. B. G. Ferris, Jr., "Health Effects of Exposure to Low Levels of Regulated Air Pollutants," *Air Pollution Control Association Journal* 28 (1978): 482–497.

75. American Lung Association, *Health Effects of Air Pollution,* Medical Section, American Thoracic Society (New York: American Lung Association, 1978).

76. U.S. Senate, Committee on Public Works, *op. cit.,* vol. 2, *Health Effects of Air Pollutants.*

77. National Research Council, *Airborne Particles,* Assembly of Life Sciences, Committee on Biological Effects of Environmental Pollutants, Subcommittee on Airborne Particles (Washington, D.C.: National Academy of Sciences, 1977).

78. *Ibid.*

79. Federal Energy Administration, *A Critical Evaluation of Current Research Regarding Health Criteria for Sulfur Oxides,* technical report prepared by Tabershaw/Cooper Associates, Inc. (Washington, D.C.: Federal Energy Administration, April 11, 1975).

80. J. B. Mudd, "Sulfur Dioxide," in *Responses of Plants to Air Pollutants,* eds. J. Brian Mudd and T. T. Kozlowski (New York: Academic Press, 1975), pp. 9–22.

81. Electric Power Research Institute, Sulfur Oxides: Current Status of Knowledge (Palo Alto, Calif.: Electric Power Research Institute (EPRI EA 316, Project 681-1), 1976).

82. National Research Council, *Nitrogen Oxides,* Assembly of Life Sciences, Division of Medical Sciences, Committee on Medical and Biological Effects of Environmental Pollutants (Washington, D.C.: National Academy of Sciences, 1977).

83. National Research Council, *Nitrates: An Environmental Assessment,* Commission on Natural Resources (Washington, D.C.: National Academy of Sciences, 1978).

84. U.S. Senate, Committee on Public Works, *op. cit.,* vol. 2, pp. 183–315.
85. U.S. Senate, Committee on Public Works, *op. cit.,* vols. 2 and 3.
86. National Research Council, *Ozone and Other Photochemical Oxidants,* Division of Medical Sciences, Committee of Biological Effects of Environmental Pollutants (Washington, D.C.: National Academy of Sciences, 1977).
87. Assembly of Life Sciences, *Carbon Monoxide, op. cit.*
88. National Research Council, *Vapor-Phase Organic Pollutants,* Division of Medical Sciences, Committee on Biological Effects of Environmental Pollutants (Washington, D.C.: National Academy of Sciences, 1976).
89. Division of Medical Sciences, *Ozone and Other Photochemical Oxidants, op. cit.*
90. Risk and Impact Panel, *op. cit.,* chap. 5; and see M. Kawai, H. Amamoto, and K. Harads, "Epidemiologic Study of Occupational Lung Cancer," *Archives of Environmental Health* 14 (1967): 859–864, and R. Doll, et al., "Mortality of Gas Workers—Final Report of a Retrospective Study," *British Journal of Industrial Medicine* 29 (1972): 394–406.
91. C. W. Gehrs, *et al.,* "Environmental Health and Safety Implications of Increased Coal Utilization," in *Chemistry of Coal Utilization,* suppl. vol. 2, tech. ed. M. A. Elliott (New York: Wiley Interscience, in press).
92. Lipfert, "The Association of Human Mortality with Air Pollution," *op. cit.*
93. Electric Power Research Institute, *op. cit.*
94. National Research Council, *Sulfur Oxides,* Assembly of Life Sciences, Board on Toxicology and Environmental Health Hazards, Committee on Sulfur Oxides (Washington, D.C.: National Academy of Sciences, 1978).
95. U.S. House of Representatives, *Staff Report on Joint Hearings on the Conduct of the Environmental Protection Agency's "Community Health and Environmental Surveillance System" (CHESS) Studies,* Committee on Science and Technology and Committee on Interstate and Foreign Commerce, 94th Congress, 2nd Sess., April 9, 1976.
96. Greenfield, Attaway, and Tyler, Inc., *Evaluation of CHESS: New York Asthma Data, 1970–71,* vol. 1 (Palo Alto, Calif.: Electric Power Research Institute (EPRI EA-450), 1977).
97. U.S. Senate, Committee on Public Works, *op. cit.,* vol. 2, pp. 280–291.
98. Schimmel and Jordan, *op. cit.*
99. R. W. Buechley, *SO_2 Levels, 1962–1972, and Perturbations in Mortality,* report for Contract no. 1-ES-2101 (Research Triangle Park, N.C.: National Institute of Environmental Health Sciences, 1976); and "Eleven Years of Daily Deaths in the New York–New Jersey Metropolis" (Paper presented at the 8th International Scientific Meeting, International Epidemiological Association, San Juan, P.R., 1977).
100. *Ibid.*
101. Schimmel and Jordan, *op. cit.*
102. Lester B. Lave and Eugene P. Seskin, *Air Pollution and Human Health* (Baltimore, Md.: The Johns Hopkins University Press, 1977).
103. Schimmel and Jordan, *op. cit.*
104. L. Thibodeau, R. Reed, and Y. M. Bishop, "Air Pollution and Human Health: A Reanalysis," *Environmental Health Perspectives* (in press, 1979).
105. Assembly of Life Sciences, *Sulfur Oxides, op. cit.*
106. Lipfert, "The Association of Human Mortality with Air Pollution," *op. cit.*
107. E. Landau, "NAS Report on Sulfur Oxides: A Critique" (Paper presented at the 72nd Annual Meeting of the Air Pollution Control Association, Cincinnati, Ohio, June 24–29, 1979). See also "The Dangers in Statistics," *The Nation's Health* 8 (1978): 3.
108. Lipfert, "The Association of Human Mortality with Air Pollution," *op. cit.*
109. The statistical sensitivity of the study was such that a difference smaller than 4 percent excess deaths could not have been established at the 0.05 level of significance. This lack of sensitivity is common to studies in this field.

110. Assembly of Life Sciences, *Sulfur Oxides, op. cit.*
111. Lipfert, "The Association of Human Mortality with Air Pollution," *op. cit.*
112. A. E. Martin, "Statistics of Air Pollution," *Proceedings of the Royal Society of Medicine* 57 (1964): 969–975.
113. Schimmel and Jordan, *op. cit.*
114. Lipfert, "The Association of Human Mortality with Air Pollution," *op. cit.*
115. Martin, *op. cit.*
116. Assembly of Life Sciences, *Sulfur Oxides, op. cit.*
117. F. S. Harris, Jr., *Atmospheric Aerosols: A Literature Summary of Their Physical Characteristics and Chemical Composition* (Springfield, Va.: National Technical Information Service (NASA CR-2626), 1976).
118. Landau, *op. cit.*
119. Harris, *op. cit.*
120. *Ibid.*
121. Mary Amdur, "Toxicological Guidelines for Research on Sulfur Oxides and Particulates," in *Proceedings of the Fourth Symposium on Statistics and the Environment* (Washington, D.C.: National Academy of Sciences, 1976), pp. 48–55.
122. J. D. Shannon, *The Argonne Statistical Trajectory Regional Air Pollution Model* (Argonne, Ill.: Argonne National Laboratory (Informal Report ANL/RER-79-1), 1979).
123. B. R. Appel, *et al.,* "Sulfate and Nitrate Data from the California Aerosol Characterization Experiment (ACHEX)," *Environmental Sciences and Technology* 12 (1978): 418–428.
124. G. M. Hidy, P. K. Mueller, V. Deyo, and K. C. Detore, "Design and Implementation of the Sulfate Regional Experiment (SURE)," *in Proceedings of the Symposium on Turbulence, Diffusion and Air Pollution* (Boston, Mass.: American Meteorological Society, 1979), pp. 314–321.
125. Electric Power Research Institute, "SURE Takes to the Air," *EPRI Journal* 3 (1979): 14–17; and J. P. McBride, R. E. Moore, J. P. Witherspoon, and R. E. Blanco, "Radiological Impact of Airborne Effluents of Coal and Nuclear Plants," *Science* 202 (1978): 1045–1050.
126. U.S. Department of Energy, *op. cit.*
127. U.S. House of Representatives, Committee on Science and Technology and Committee on Interstate and Foreign Commerce, *op. cit.*
128. F. E. Speizer, Y. Bishop, and B. G. Ferris, Jr., "An Epidemiological Approach to the Study of the Health Effects of Air Pollution," in *Proceedings of the Fourth Symposium on Statistics and the Environment* (Washington, D.C.: National Academy of Sciences, 1976), pp. 56–68.
129. Risk and Impact Panel, *op. cit.,* chap. 4.
130. *Ibid.*
131. *Ibid.*
132. Ad Hoc Population-Dose Assessment Group, *Population Dose and Health Impact of the Accident at Three Mile Island Nuclear Station* (Washington, D.C.: Government Printing Office, 1979). (A preliminary assessment for the period March 28–April 7, 1979).
133. U.S. Nuclear Regulatory Commission, *Reactor Safety Study,* main report and app. VI (Washington, D.C.: U.S. Nuclear Regulatory Commission (WASH-1400 or NUREG-75-014), 1975).
134. U.S. Nuclear Regulatory Commission, *Overview of the Reactor Safety Study Consequence Model* (Washington, D.C.: U.S. Nuclear Regulatory Commission (NUREG-0340), 1977).
135. U.S. Nuclear Regulatory Commission, *NRC Statement on Risk Assessment and the Reactor Safety Study Report (WASH-1400) in Light of the Risk Assessment Review Group Report* (Washington, D.C.: U.S. Nuclear Regulatory Commission, January 18, 1979).

136. Risk Assessment Review Group, H. W. Lewis, Chairman, *Report of the Risk Assessment Review Group to the U.S. Nuclear Regulatory Commission* (Washington, D.C.: U.S. Nuclear Regulatory Commission (NUREG/CR-0400), 1978).
137. Assuming effective evacuation of 30 percent of the population, partially effective evacuation of 40 percent, and ineffective evacuation of 30 percent.
138. U.S. Nuclear Regulatory Commission, *Reactor Safety Study, op. cit.*
139. While the authors of WASH-1400 argue that median values, which give equal probability of being exceeded or not, may be a better measure of very rare events, the mean value may be a better measure of the probability of accidents involving many events. In WASH-1400, the mean is generally 2–3 times higher than the median, which is within the uncertainty range of about fivefold quoted in the report.
140. Risk Assessment Review Group, *op. cit.*
141. "Report to the American Physical Society by the Study Group on Light-Water Reactor Safety," *Reviews of Modern Physics* 47 (Summer 1975): suppl. no. 1.
142. Risk Assessment Review Group, *op. cit.*
143. Nuclear Energy Policy Study Group, Spurgeon M. Keeny, Jr., Chairman, *Nuclear Power: Issues and Choices* (Cambridge, Mass.: Ballinger Publishing Co., 1977). (Also known as the Ford/Mitre report.)
144. Commission on Natural Resources, *Nitrates: An Environmental Assessment, op. cit.*
145. J. Brian Mudd and T. T. Kozlowski, eds., *Responses of Plants to Air Pollutants* (New York: Academic Press, 1975).
146. Assembly of Life Sciences, *Sulfur Oxides, op. cit.*
147. Division of Medical Sciences, *Ozone and Other Photochemical Oxidants, op. cit.*
148. L. B. Barrett and T. E. Waddell, *Cost of Air Pollution Damage: A Status Report* (Washington,D. C.: Environmental Protection Agency (Publication AP-85), 1973), quoted in *Responses of Plants to Air Pollutants,* eds. J. Brian Mudd and T. T. Kozlowski (New York: Academic Press, 1975), p. 4.
149. Robert L. Heath, "Ozone," in *Responses of Plants to Air Pollutants,* eds. J. Brian Mudd and T. T. Kozlowski (New York: Academic Press, 1975), pp. 23–56.
150. J. Brian Mudd, "Sulfur Dioxide," in *Responses of Plants to Air Pollutants,* eds. J. Brian Mudd and T. T. Kozlowski (New York: Academic Press, 1975), pp. 9–22.
151. Assembly of Life Sciences, *Sulfur Oxides, op. cit.*
152. Mudd, *op. cit.*
153. J. C. Noggle and Herbert C. Jones, *Accumulation of Atmospheric Sulfur by Plants and Sulfur-Supplying Capacity of Soil* (Washington, D.C.: U.S. Environmental Protection Agency (EPA-600/7-79-109), April 1979).
154. Assembly of Life Sciences, *Nitrogen Oxides, op. cit.*
155. Assembly of Life Sciences, *Sulfur Oxides, op. cit.*
156. E. J. Ryder, "Selecting and Breeding Plants for Increased Resistance to Air Pollutants," *Advances in Chemistry Series* 122 (1973): 78–84.
157. Paul R. Miller and Joe R. McBride, "Effects of Air Pollutants on Forests," in *Responses of Plants to Air Pollutants,* eds. J. Brian Mudd and T. T. Kozlowski (New York: Academic Press, 1975), pp. 196–236.
158. Fabius LeBlanc and Chruva N. Rao, "Effects of Air Pollutants on Lichens and Bryophytes," in *Responses of Plants to Air Pollutants,* eds. J. Brian Mudd and T. T. Kozlowski (New York: Academic Press, 1975).
159. Ryder, *op. cit.*
160. Miller and McBride, *op. cit.*
161. Commission on Natural Resources, *Nitrates: An Environmental Assessment, op. cit.*
162. Assembly of Life Sciences, *Sulfur Oxides, op. cit.*

163. G. E. Likens, "Acid Precipitation," *Chemical and Engineering News* 54 (1976): 26–49; and personal communication to H. I. Kohn, October 1978.
164. *Ibid.*
165. Christopher S. Cronan and William A. Reiners, "Forest Floor Leaching: Contributions from Mineral, Organic, and Carbonic Acids in New Hampshire Subalpine Forest," *Science* 200 (1978): 309–311.
166. Commission on Natural Resources, *Nitrates: An Environmental Assessment, op. cit.*
167. National Research Council, *Energy and Climate,* Assembly of Mathematical and Physical Sciences (Washington, D.C.: National Academy of Sciences, 1977).
168. William W. Kellogg, "Review of Mankind's Impact on Global Climate" (Typescript prepared for the Workshop on Multidisciplinary Research Related to the Atmospheric Sciences, National Center for Atmospheric Research, Boulder, Colo., June 21, 1977).
169. C. F. Base, H. E. Goeller, J. S. Olson, and R. M. Rotty, *The Global Carbon Dioxide Problem* (Oak Ridge, Tenn.: Oak Ridge National Laboratory (ORNL-5194), 1976).
170. Risk and Impact Panel, *op. cit.,* chap. 7.
171. J. Williams, ed., *Carbon Dioxide, Climate, and Society* (New York: Pergamon Press, 1978).
172. In addition, dust and nitrogen oxides have the potential to make a quantitative contribution.
173. And possibly also widespread deforestation, although the role of the biosphere is not clear. Reforestation has been recommended: See, for example, G. M. Woodwell, G. J. MacDonald, R. Revelle, and C. D. Keeling, *The Carbon Dioxide Problem: Implications for Policy in the Management of Energy and Other Resources,* report to the Council on Environmental Quality (Washington, D.C.: National Academy of Sciences, July 1979).
174. MacDonald and Carter, *op. cit.*
175. Kellogg, *op. cit.*
176. Williams, *op. cit.*
177. Risk and Impact Panel, *op. cit.,* chap. 7.
178. Schaefer, *op. cit.*
179. J. Harte and M. El-Gasseir, "Energy and Water," *Science* 199 (1978): 623–633; and Risk and Impact Panel, *op. cit.,*chap. 6.
180. U.S. Department of Energy, *op. cit.*
181. National Research Council, *Energy and the Fate of Ecosystems,* Committee on Nuclear and Alternative Energy Systems, Risk and Impact Panel, Ecosystems Resource Group (Washington, D.C.: National Academy of Sciences, in preparation).
182. *Ibid.*
183. Public Law 91-190 (January 1, 1970), National Environmental Policy Act, Sec. 102(B). See also S. F. Singer, "A Quantified Environment," *Science* 203 (1979): 400.
184. H. Brooks, "Environmental Decision Making: Analysis and Values," in *When Values Conflict,* eds. L. H. Tribe, C. Schelling, and J. Voss. (Cambridge, Mass.: Ballinger Publishing Co., 1976), pp. 115–135.
185. Risk and Impact Panel, *Risks and Impacts of Alternative Energy Systems, op. cit.,* chap. 6.
186. Eutrophication refers to the enhancement of the basic nutritional level of lakes and other bodies of water that eventually disrupt the natural ecological relationships between species, permitting "undesirable" ones to outgrow and inhibit the rest, and leading to profound changes in the entire habitat.
187. Risk and Impact Panel, *Risks and Impacts of Alternative Energy Systems, op. cit.,* chap. 6.
188. See chapter 8, "Geothermal Energy," and National Research Council, *Supporting Paper 4: Geothermal Resources and Technology in the United States,* Committee on Nuclear and

Alternative Energy Systems, Supply and Delivery Panel, Geothermal Resource Group (Washington, D.C.: National Academy of Sciences, 1979).

189. See "Solar Energy," Risk and Impact Panel, *Risks and Impacts of Alternative Energy Systems, op. cit.,* chap. 6; and National Research Council, *Supporting Paper 6: Domestic Potential of Solar and Other Renewable Energy Sources,* Committee on Nuclear and Alternative Energy Systems, Supply and Delivery Panel, Solar Resource Group (Washington, D.C.: National Academy of Sciences, 1979).

190. *Ibid.*

191. The President's Commission on the Accident at Three Mile Island, *The Need for Change: The Legacy of TMI* (Washington, D.C.: U.S. Government Printing Office, 1979).

192. Sir Edward E. Pochin, *Why Be Quantitative About Radiation Risk Estimates?,* Lecture no. 2, Lauriston S. Taylor Lecture Series in Radiation Protection and Measurements (Washington, D.C.: National Council on Radiation Protection and Measurements, 1978).

193. Risk and Impact Panel, *Risks and Impacts of Alternative Energy Systems, op. cit.,* chap. 8.

194. National Research Council, *Implications of Environmental Regulations for Energy Production and Consumption,* Commission on Natural Resources, Committee on Energy and the Environment (Washington, D.C.: National Academy of Sciences, 1977).

10 U.S. Energy Policy in the Global Economic Context

The energy upheavals in the winter of 1973–74 made it clear that the United States is strongly affected by energy developments abroad. It is no less obvious that U.S. energy policies, especially in oil and nuclear power, in turn affect other countries. This chapter deals with some international aspects of American energy policy, focusing particularly on the world supply and demand situation.

The chapter opens with an overview of world energy developments before the 1973–74 oil price increase, an overview that brings out the growing dependence on oil and gas. It then reviews the consequences of that price increase. A discussion of the magnitude of global energy resources serves as an introduction to the prospects for energy relations between the United States and the rest of the world.

THE RISE IN OIL AND GAS, 1960–1973

Table 10-1 gives comparable data for production, consumption, and international trade in the principal fuels for the world as a whole and for the world broken down into five regions. Data are shown for 2 years, the earliest year for which comparable data are available and the last year before the increase in oil prices created a new situation; the table also gives the annual percentage growth rate between 1960 and 1973. These growth rates, of course, have not been constant in the past and will not be so in the future.

Between 1960 and 1973 world production and consumption of energy almost doubled, corresponding to an annual growth rate of 5.1 percent. This is virtually identical with the estimated growth rate of world gross national product (GNP) during the period.[1] (It should not be inferred from this that energy consumption and GNP necessarily move proportionately; their relation is discussed in chapter 2.)

In terms of the composition of world energy consumption and production, the most rapidly growing category was nuclear power, of which there was virtually none in 1960. Even in 1973, however, nuclear power supplied less than 1 percent of the world's energy. Apart from this category, whose importance lies in the future, the most rapid growth was in natural gas and petroleum, both of which grew at annual rates of nearly 8 percent. Hydroelectric power, which includes a small amount of geothermal power, increased about 5 percent/yr. World coal production and consumption were virtually stagnant, with an annual growth rate below 1 percent. Reflecting these changes, the share of petroleum in world energy consumption increased from 34 percent in 1960 to 47 percent in 1973, and that of natural gas from 14 percent to 19 percent in the same period. The share of coal dropped from 47 percent to 28 percent, while that of hydroelectric power remained unchanged at 5 percent.

The growth rates of production and consumption of different fuels varied greatly among regions. Only in the United States did consumption grow fairly uniformly, with the share of coal falling and the share of all other fuels rising. For other regions the consumption of petroleum and natural gas rose much faster than that of total energy. The 22 percent growth rate of natural gas in Western Europe is especially remarkable; this was almost entirely at the expense of coal, the consumption of which actually declined. In Western Europe the share of petroleum in total energy consumption rose from 34 percent in 1960 to 61 percent in 1973, while that of natural gas increased from almost nothing to nearly 11 percent; coal dropped from 56 percent to 20 percent.

These drastic changes in consumption patterns were accompanied by no less dramatic changes in the regional patterns of production. In some places these changes complemented one another; for example, natural gas production in Western Europe increased about as much as natural gas consumption there, while coal production fell about as much as coal consumption. In the United States the situation with respect to these two fuels was similar. Neither in the United States nor in Western Europe, however, was there a sufficient increase in petroleum production to match the increase in demand. U.S. crude oil production rose at an average rate of less than 2 percent/yr between 1960 and 1973. (Actually, production reached a peak in 1970 and declined thereafter.) European and Japanese petroleum production remained small, so the sharp rise in demand had to be met entirely from the rest of the world. In the communist countries,

TABLE 10-1 An Overview of World Energy from 1960 to 1973 (quads)[a]

Activity and Fuel Source	World			United States			Western Europe			Other Developed Countries[b]			Communist Countries[c]			Other Developing Countries		
	1960	1973	Average Annual Growth Rate (percent)	1960	1973	Average Annual Growth Rate (percent)	1960	1973	Average Annual Growth Rate (percent)	1960	1973	Average Annual Growth Rate (percent)	1960	1973	Average Annual Growth Rate (percent)	1960	1973	Average Annual Growth Rate (percent)
Consumption																		
All energy sources	131.5	250.4	5.1	44.5	74.7	4.1	26.4	52.2	5.4	8.9	25.1	8.3	39.0	68.2	4.4	12.8	30.2	6.8
Percent of world total	(100)	(100)	—	(33.8)	(29.8)	—	(20.1)	(20.8)	—	(6.8)	(10.0)	—	(29.7)	(27.2)	—	(9.7)	(12.1)	—
Petroleum	45.2	118.5	7.7	18.6	32.3	4.3	8.9	31.7	10.3	3.6	16.2	10.4	6.4	20.0	9.2	7.8	18.3	6.7
Natural gas	18.0	47.8	7.8	14.1	25.3	4.6	0.4	5.6	22.5	0.5	1.9	10.8	2.1	10.8	13.4	0.9	4.2	12.6
Coal	61.4	69.1	0.9	10.1	13.3	2.1	14.8	10.6	−2.5	2.9	3.8	2.1	29.8	35.3	1.3	3.7	6.0	3.7
Hydro[d]	6.9	13.2	5.1	1.5	3.0	5.5	2.3	3.6	3.5	2.0	3.0	3.2	0.8	2.0	7.3	0.4	1.7	11.8
Nuclear	negl.[e]	1.9	n.m.[f]	negl.	0.9	n.m.	negl.	0.7	n.m.	0	0.3	n.m.	0	negl.	n.m.	0	negl.	n.m.

Production[g]																		
All energy sources	132.2	252.1	5.1	41.4	61.7	3.1	17.0	19.7	1.0	6.0	13.9	6.6	40.4	70.4	4.4	27.3	86.5	9.3
Percent of world total	(100)	(100)	—	(31.3)	(24.5)	—	(12.9)	(7.8)	—	(4.5)	(5.5)	—	(30.6)	(27.9)	—	(20.7)	(34.3)	—
Petroleum	46.0	120.8	7.7	14.7	18.8	1.9	0.6	0.8	2.2	1.0	4.7	12.6	7.2	21.6	8.8	22.5	74.8	9.7
Natural gas	18.1	48.4	7.9	14.1	24.8	4.4	0.4	5.3	22.0	0.5	2.9	14.4	2.1	10.5	13.2	0.9	4.8	13.7
Coal	61.1	67.8	0.8	11.1	14.5	2.1	13.6	9.2	−3.0	2.5	2.7	0.6	30.4	36.2	1.4	3.5	5.3	3.2
Net deficit or surplus[g,h]																		
All energy sources	−0.7	−1.7	—	3.0	13.0	11.9	9.4	32.5	10.0	3.0	11.4	10.8	−1.4	−2.2	3.5	−14.5	−56.3	11.0
Petroleum	−0.8	−2.3	—	4.0	13.4	12.2	8.2	30.9	10.7	2.6	11.5	7.6	−0.9	−1.6	4.5	−14.6	−56.4	10.9
Natural gas	−0.1	−0.6	—	negl.	0.5	n.m.	negl.	0.2	n.m.	—	−1.0	n.m.	negl.	0.3	n.m.	negl.	0.6	—
Coal	0.3	1.2	—	−1.0	−1.1	0.7	1.1	1.4	1.8	−0.4	1.0	n.m.	−0.6	−0.9	3.2	0.2	−0.6	n.m.

[a] Details may not add to total due to rounding.
[b] Australia, Canada, Japan, and New Zealand.
[c] Soviet Union, Eastern Europe, People's Republic of China, and North Korea.
[d] Includes geothermal.
[e] Less than 0.1 quad.
[f] Not meaningful.
[g] Includes hydroelectric, geothermal, and nuclear, not shown separately since production was virtually equal to consumption.
[h] Surplus if negative. Equals difference between consumption and production.

Source: Adapted from U.S. Department of the Interior, *Energy Perspectives 2* (Washington, D.C.: U.S. Government Printing Office (Stock No. 024-000-00826-6), 1976), pp. 20–31. In order to make the regions more homogeneous, Canada, Australia, and New Zealand have been merged with Japan to form the "Other Developed" region and taken out of the "Rest of the World," now called "Other Developing." Data on the three transferred countries are taken from Organization for Economic Cooperation and Developments, *Energy Balances in OECD Countries, 1960–1973* (Paris: Organization for Economic Cooperation and Development, 1976).

growth in petroleum consumption was approximately matched by growth in production. Between 1960 and 1973 petroleum production in the "other developing" region[2] grew at an annual rate of nearly 10 percent, from increased output in the Persian Gulf area and from such major new producers as Libya, Nigeria, and Algeria.

It is no accident that the energy deficits of the Western industrial countries were met entirely from increased oil imports and that trade in coal and natural gas remained relatively small. This reflects the low transportation cost of oil. During the 1960s this cost fell further as larger and larger tankers were brought into service. The principal flow of energy products continued to be from the Persian Gulf to Western Europe, but exports to the United States and to Japan had an even larger percentage growth.

Table 10-2 illustrates the development of electricity production from 1960 to 1973. Except in the "other developing" region, electricity grew more rapidly than primary energy, with the highest growth rates occurring in the "other developed" and communist regions. For the world as a whole, electricity consumption grew nearly 50 percent more than primary energy consumption. With the same exception, the share of hydroelectric power dropped everywhere. Nuclear power became significant, particularly in the United States and Western Europe, but the share of fossil fuels (oil and coal) also rose. In 1973 three fourths of the world's electricity was generated from fossil fuels.

Table 10-1 also shows that the growth rate of primary energy consumption varied considerably among the five regions. It was lowest in the United States, and only slightly higher in the communist countries. The highest growth rate was recorded in the "other developed" region, which includes Japan; the "other developing" regions and Western Europe were also well above the world average growth rate. As a result of these disparities, the share of the United States in total world consumption fell from nearly 34 percent to nearly 30 percent, and that of the communist countries from about 30 percent to about 27 percent. Western Europe's share increased slightly, and the shares of the remaining two regions went up considerably. However, the United States remained the world's largest energy user, even compared with the aggregated blocs of countries in Table 10-1.

There was even more variation in the growth rates of energy production. The rate was highest for the "other developing" region, which includes the principal petroleum exporting countries; it was also substantial in the "other developed" region, due mostly to Canada. The other three areas were all below the world average growth rate. In the United States the growth rate of production was 3.1 percent, compared to one for consumption of 4.1 percent. In the communist countries production and consumption grew at the same rate, while in Western Europe there was very little growth in energy production up to 1973. The shares of the five areas in world energy production changed accordingly. In 1960 the United States and the communist countries each produced some 31 percent of the world's energy; by 1973 they accounted together for only 35 percent of the total. The share of the United States dropped below 25 percent and that of Western Europe, already small in 1960, dropped even further.

None of the three noncommunist developed areas were self-sufficient in energy in 1960, and with production growing at a slower pace than consumption their energy deficits increased rapidly: about 12 percent/yr in the United States, 11 percent/yr in the "other developed" countries, and 10 percent/yr in Western Europe. Despite the somewhat lower growth rate, the European deficit remained the largest; 62 percent of European consumption had to be supplied from outside the area in 1973. In percentage terms, the United States was much less dependent on outside sources, 83 percent of consumption being supplied from domestic production in 1973.

Virtually all of the energy deficit of the industrialized Western countries was met by the "other developing" region, which includes OPEC. Energy exports from the latter region grew at an annual rate of 11 percent between 1960 and 1973; in 1973 nearly 65 percent of its production went outside the region, compared to 53 percent in 1960.

THE OIL PRICE RISE OF 1973–1974

The net outcome of the developments just reviewed was that Japan and Western Europe, and to a lesser extent the United States, became heavily dependent on oil from the rest of the world, and in particular from the Middle East. The low price of imported oil and the discovery of large amounts of natural gas provided more competition than the European coal industry, despite considerable government help, could handle. Japan never had large domestic sources of energy, but the decline of coal mining was even steeper there than in Europe. With the possible exception of the United Kingdom, these two areas apparently did not perceive heavy dependence on imported oil as a danger until it was too late.

In the United States, on the contrary, dependence on foreign petroleum had been a matter of official concern since the 1950s, largely under the influence of the domestic petroleum industry, whose high-cost production was threatened by cheaper oil from overseas. The oil import quota program, in force from the late 1950s until 1972, kept the domestic price of crude oil well above the world level. Despite production controls (known as "market demand prorationing") enforced by the major

TABLE 10-2 World Electricity Production from 1960 to 1973[a]

Region	Total[b] (billions of kilowatt-hours)		Growth Rate (percent)	Hydro[c] (percent)		Nuclear (percent)		Fossil (percent)	
	1960	1973		1960	1973	1960	1973	1960	1973
United States	844 (35.3)	1946 (32.2)	6.6	17.8	14.2	0.1	4.3	82.1	81.6
Western Europe	560 (23.4)	1381 (22.9)	7.2	40.2	25.7	0.4	5.1	59.5	69.1
Other developed	257 (10.7)	815 (13.5)	9.3	68.0	36.4	0	3.1	32.0	60.5
Communist	474 (19.8)	1391 (23.0)	8.6	16.0	14.6	0	0.6	84.0	84.8
Other developing	257 (10.7)	510 (8.4)	5.4	23.0	36.9	0	0.6	77.0	62.5
World	2392	6043	7.4	28.6	21.9	0.1	3.1	71.2	75.0

[a] Some percents may not total 100 due to rounding.

[b] Numbers in parentheses indicate percents of world total.

[c] Includes geothermal.

Source: Compiled from U.S. Department of the Interior, *Energy Perspectives 2* (Washington, D.C.: U.S. Government Printing Office (Stock No. 024-000-00826-6), 1976), pp. 171–176, and Organization for Economic Cooperation and Development, *Energy Balances in OECD Countries, 1960–1973* (Paris: Organization for Economic Cooperation and Development, 1976).

producing states, the net effect probably was to accelerate the depletion of domestic petroleum reserves. New discoveries became more difficult to realize; they were overtaken by growing consumption in the early 1960s, so proved reserves peaked in 1968 and then declined. The only recent American discovery of international significance (in northern Alaska) remained unavailable for several years because of opposition to the building of a pipeline. At the same time, environmental concerns about the use of high-sulfur coal for electricity generation further stimulated the demand for imported oil. Other such concerns, expressed through increasingly active and effective citizens' groups, led to abandonments and postponements of hydroelectric power, nuclear power, and offshore oil projects.

These developments made the United States a major factor in the global supply-demand balance. Until the mid-1960s U.S. oil imports were relatively small and were satisfied mostly by nearby areas—Canada and Venezuela. As American demand grew, more had to come from the Persian Gulf area, which was already supplying most of the increasing demand in Western Europe and Japan.

We have already seen that production in the Middle East had expanded rapidly during the 1960s. Large new discoveries and intensified competition among the oil companies had caused crude oil prices to soften in this period. This helped to open up markets for the new output, but it also threatened the level of royalties paid by the oil companies to the countries where reserves were located. To counter this threat, the royalty owners organized themselves in the Organization of Petroleum Exporting Countries (OPEC) and were moderately successful. (One must remember that most of the OPEC countries own little but desert, aside from oil, and it is on the wealth represented by the oil that they must build industrial economies able to survive when the oil runs out.) As demand began to outstrip supply and U.S. oil production approached capacity, OPEC saw its opportunity. Its first achievement along these new lines came in early 1971, when the oil companies agreed to a sizeable increase in royalty rates.

Despite this price increase, oil consumption continued to grow, stimulated by the worldwide inflationary boom that started around 1972. The imbalance came to a head in the fall of 1973, when the fourth Arab-Israeli war was accompanied by an embargo imposed by the Arab oil exporters against certain Western countries, including the United States. How much quantitative effect this embargo had is still unclear, but the psychological effect was unmistakable, reinforced as it was by widespread concern over exhaustion of natural resources. Small quantities of non-Arab oil changed hands at prices as high as $17 per barrel, about seven times the price prevailing before the embargo. Encouraged by this demonstration of the oil importers' vulnerability, OPEC declared a unilateral price increase of about 300 percent, setting the crude price at about $10 per barrel.

It is important to note that the cartelization of the oil market and the resulting price increase were made possible by the sharp increases in the import demand of the United States and, to a lesser extent, other industrial countries, rather than by any worldwide shortage of oil. We shall discuss world reserves later in this chapter.

REACTIONS TO HIGHER OIL PRICES

The sudden quadrupling of oil prices, coming on top of steep price rises in other commodities, had a serious impact on the world economy. Since in the short run energy consumption is not very sensitive to price changes, the consuming countries had little choice but to pay the higher price. Other energy commodities (coal, natural gas, and uranium) also became more expensive. At first energy consumption was held in check not by higher prices, but by a world recession that started in 1974 and reached a trough in 1975.

This recession, the most serious one since the Great Depression of the 1930s, cannot be attributed solely to the increase in oil prices. Industrial production in the developed countries was already losing steam in the summer of 1973, while inflation was advancing to rates not seen since the Korean War. Several important countries had already tightened their monetary and fiscal policies with a view to bringing inflation down, though progress in this direction was not to come until much later. For most countries the oil price increase was of external origin and therefore could not be overcome by domestic economic policies directed at internal inflationary pressures. In fact, the maintenance of restrictive monetary and fiscal policies in the face of rising import prices served to aggravate their depressing effect on income and employment in the oil-importing countries.

To these countries the increase in oil prices presented itself initially as a worsening of the balance of payments. While restrictive monetary and fiscal policies are a standard response to a deterioration in international transactions, they were not effective in the prevailing trade situation because the oil-exporting countries could not immediately increase their imports in line with their higher export revenues. In addition, the tight monetary policy was probably unnecessary because the surpluses of the oil-exporting countries remained in the international banking system, where they were available to finance oil importers' deficits.

Once these facts were recognized and the domestic inflation was reduced to some extent, the principal industrial countries gradually turned their economic policies toward cautious stimulation. The danger of a resurgence of inflation made caution necessary, but in fact most countries were able to bring about a recovery from the 1974–1975 recession while continuing to reduce their rates of inflation.

These developments outside the energy markets are vital to understanding the behavior of energy consumption after the OPEC price increase (Table 10-3). Since comprehensive data are not yet available, this table is limited to oil and electricity and covers only the leading industrial countries. To relate these two main components of energy consumption to overall economic activity, it also gives GNP at constant prices, and for comparison purposes all figures are expressed as index numbers with base 1973.

The main conclusion from Table 10-3 is that oil consumption has *not* continued the rapid growth evident from Table 10-1. Since 1973, oil consumption in the seven countries included has remained stagnant at best, and in some it has fallen considerably. Comparison with the GNP numbers shows that in 1973–1977 oil consumption in each of these countries rose less than GNP, the opposite of what was shown for 1960–1973 in Table 10-1.

Certain differences among the countries are also apparent. The decline in oil consumption relative to GNP was least in the United States and Canada, in both of which there is substantial domestic production. In these countries oil prices were not allowed to rise to the world level for fear of creating large windfall gains to domestic producers; instead the price of domestic oil was kept down by price controls and other devices. This option was not open to the other countries, most of which also raised excise taxes. In the United Kingdom, France, and Germany, moreover, the greater availability of natural gas from the Netherlands and the North Sea provided a substitute for certain oil products.

Recent developments in electricity are less clear than they are in oil. Since fuel accounts for a small part of production costs, electricity prices did not rise as much as primary energy prices. Most of the apparent slowdown in electricity consumption (which is virtually proportional to the production shown in the table) is probably attributable to the behavior of GNP. Demand analyses for electricity show that it responds very slowly to price changes, though the response is large in the long run.

There have also been changes in world energy production since the 1973–1974 price increase, which will be traced here only for oil. As Table 10-4 shows, total world production increased about 7 percent between 1973 and 1977, with most of the increase accounted for by the communist countries. The desire to keep prices high forced the OPEC countries to curtail their production initially, and in 1977 it was about the same as in 1973. Their share of world production fell from 56 percent in 1973 to 52

TABLE 10-3 Oil Consumption, Electricity Production, and Real Gross National Product in the Principal Industrial Countries from 1972 to 1977, in Index Numbers: 1973 Equals 100

Region and Activity	1972	1973	1974	1975	1976	1977	1978
United States							
Oil consumption	95	100	96	94	101	106	109
Electricity production	94	100	100	102	108	113	116
Gross national product	95	100	98	97	103	108	112
Canada							
Oil consumption	95	100	102	100	103	104	107
Electricity production	91	100	106	104	112	120	127
Gross national product	93	100	104	105	110	113	118
France							
Oil consumption	89	100	94	87	94	89	94
Electricity production	94	100	103	102	111	116	122
Gross domestic product[a]	95	100	103	103	109	112	116
Italy							
Oil consumption	95	100	100	97	99	97	102
Electricity production	93	100	102	101	112	114	n.a.[b]
Gross domestic product[a]	94	100	104	101	107	109	112
United Kingdom							
Oil consumption	100	100	93	83	82	85	86
Electricity production	94	100	97	96	98	100	102
Gross domestic product[a]	93	100	98	97	100	102	105
Federal Republic of Germany							
Oil consumption	94	100	89	86	93	92	96
Electricity production	92	100	104	101	112	112	118
Gross national product	95	100	100	99	104	109	110
Japan							
Oil consumption	86	100	97	91	96	100	102
Electricity production	91	100	98	101	109	n.a.[b]	n.a.[b]
Gross national product	91	100	99	101	107	113	120

[a] Real gross domestic product is the gross national product less net factor income from abroad.

[b] Not available.

Source: For oil consumption, Central Intelligence Agency, *International Energy Statistical Review*, National Foreign Assessment Center (Washington, D.C.: Central Intelligence Agency (ER ISER 79-012), Sept. 5, 1979), pp. 14–15. For electricity production, United Nations, *Monthly Bulletin of Statistics*, vol. 33 (no. 8), Aug. 1979, and Energy Information Agency, *Annual Report to Congress 1978*, vol. 2 (Washington, D.C.: U.S. Department of Energy, 1979), pp. 119 and 121. For gross national product and gross domestic product, International Monetary Fund, *International Financial Statistics* (Washington, D.C.: International Monetary Fund, May and Aug. 1978).

percent in 1977. In the U.S. and Canada, production has continued to decline (except for a reversal in 1977 when the Alaskan North Slope came on stream). European output has been rising rapidly as the North Sea discoveries enter into production, but is still relatively small. From 1973 to 1977 there was also a 30 percent increase in non-OPEC production in the developing countries. This suggests that OPEC, like other cartels, will have to contend with outside competitors whose prices are kept high by OPEC but who do not participate in the curtailment of production.[3] However, the non-OPEC "less developed" oil exporters do have a strong interest in the continued existence of the cartel.

After the 1973–1974 price increase the world oil market remained in rough balance until the final months of 1978. During those years the cartel price was adjusted upward in line with U.S. inflation, but not in response to the depreciation of the dollar in terms of other currencies. Many OPEC members had substantial spare capacity.

This balance was upset by the events in Iran, during which that country's oil production fell to almost nothing. Initially the resulting shortfall in world supply was made up by increased output from other members, but soon the spot price of crude oil began moving up as importing countries sought protection against the threat of a shortage. After several months of confusion OPEC decided to raise the cartel price, but was unable to agree on a single figure. The range of oil prices charged by cartel members was put between $18.00 and $22.50 per barrel. Iranian production is well below the pre-1979 level, and there no longer is substantial spare capacity in the other OPEC countries.

The fall of the Shah brought home once more the extreme dependence of the United States and other industrial countries on the Persian Gulf area with its volatile politics. This problem would become especially acute if the Soviet Union were to become a net oil importer, a possibility discussed later in this chapter.

ENERGY RESOURCES AND THEIR DISCOVERY

The world now relies almost entirely on minerals for the production of energy; the only other sources are hydroelectric power and a few renewable fuels such as wood and peat. Until the advent of solar energy and other sustainable sources, the availability of energy minerals is therefore critical to world energy prospects. The number of energy minerals that are of practical importance is small—oil, natural gas, coal, and uranium. There are also materials, such as oil shale and tar sands, from which fuels can be produced synthetically. Other minerals, such as thorium (for nuclear fission) and lithium (for nuclear fusion), may become

TABLE 10-4 World Oil Production from 1973 to 1978 (millions of barrels per day)[a,b]

Region	1973	1974	1975	1976	1977	1978
World	55.8	55.9	53.0	57.3	59.7	60.0
Noncommunist areas	45.8 (82)	45.1 (81)	41.5 (78)	45.0 (78)	46.7 (78)	46.2 (77)
United States	9.2 (17)	8.8 (16)	8.4 (16)	8.1 (14)	8.2 (14)	8.7 (15)
Canada	1.8 (3)	1.7 (3)	1.5 (3)	1.3 (2)	1.3 (2)	1.3 (2)
Western Europe	0.4 (1)	0.4 (1)	0.6 (1)	0.9 (1)	1.3 (2)	1.7 (3)
Other non-OPEC	3.4 (6)	3.6 (6)	4.0 (7)	4.0 (7)	4.5 (8)	4.9 (8)
OPEC	31.0 (56)	30.7 (55)	27.1 (51)	30.7 (54)	31.4 (53)	30.0 (49)
Communist areas	10.0 (18)	10.7 (19)	11.5 (22)	12.3 (22)	13.0 (22)	13.7 (23)

[a]Details may not add due to rounding.
[b]Numbers in parentheses indicate percents of world totals.

Source: Compiled from Central Intelligence Agency, *International Energy Statistical Review*, National Foreign Assessment Center (Washington, D.C.: Central Intelligence Agency, Sept. 5, 1979, issue for 1973, 1976–1978; April 19, 1978, issue for 1975; and Jan. 11, 1979, issue for 1974).

important in the more distant future, but they will not be covered here since technology rather than resource availability constrains the use of these elements in the foreseeable future.

Among these five minerals most attention will be given here to oil and gas, which together account for nearly two thirds of the world's present energy consumption (see Table 10-1), and to uranium. Although coal will continue to be an important source of energy in the future, the known reserves are so large in relation to current and prospective consumption that considerations other than resource availability are likely to be limiting within a wide range of prices. Suffice it to note that on the time scale of a few centuries coal resources are also exhaustible, and that coal will become the source of the world's "petrochemicals." These facts, as well as environmental considerations, are reasons to pursue research at a moderate rate on some of the indefinitely sustainable long-term energy sources, such as solar energy and nuclear fusion. For oil shale and tar sands too, vast resources are known to exist, but much less is known about the cost of producing oil or gas from them; this question is discussed elsewhere in this chapter.

Table 10-5 summarizes measured world recoverable energy reserves in 1974 as reported to the World Energy Conference.[4] The total is in excess of 31,000 quadrillion Btu (quads), which is more than 100 times the present annual world consumption. However, this number is subject to a number of qualifications. The omission of Soviet uranium reserves probably does not make a great deal of difference, but the overstatement of oil reserves from oil shale and tar sands, recognized in a footnote, is more serious, especially since most of these are located in North America.

Perhaps the most important conclusion to be drawn from Table 10-5 is the overwhelming importance of coal reserves, with which the United States is particularly well endowed. It should be borne in mind, however, that some of this coal may be unavailable because of environmental constraints and that the incentive to discover more coal is considerably smaller than for oil, gas, and uranium.

The prospects for increasing domestic coal production are discussed in chapter 4 of this report. Other parts of the world face some of the same difficulties with coal that are faced here, sometimes in more acute forms. In Western Europe, however, new coal mines in Britain are being opened, while in Germany lignite appears to have become more competitive. Western Europe may also be able to obtain more coal from Poland, which has long been a substantial exporter. The large Asian coal reserves appearing in Table 10-5 are mostly in China and are likely to be needed for domestic consumption there.

The reserve position in natural gas is favorable to the extent that gas is a relative newcomer on the world energy scene. Only in the United States

TABLE 10-5 Measured World Recoverable Energy Reserves for 1974 (quads)[a]

Region	Solid Fuels	Crude Oil	Natural Gas	Oil Shale and Tar Sands	Uranium (nonbreeder)[b]	Total
Africa	361.7	526.6	201.7	81.4	198.1	1,369.5
Asia (less U.S.S.R.)	2,608.7	2,212.0	432.6	870.2	3.1	6,126.7
Europe (less U.S.S.R.)	2,446.9	57.1	153.6	117.0	46.4	2,821.0
U.S.S.R.	3,325.5	333.6	577.9	139.0	unknown	4,376.0
North America	5,071.0	301.0	380.6	9,111.0[c]	422.7	15,286.4
South America	49.8	311.5	60.6	23.7	11.9	457.5
Oceania	459.8	9.4	24.9	9.2	99.1	602.2
TOTAL	14,323.3	3,751.2	1,831.8	10,352.1	781.4	31,039.9

[a]Details may not add due to rounding.
[b]Energy content using breeders 60–100 times as great. Thorium resources neglected.
[c]According to the U.S. Department of the Interior, Bureau of Mines, North American tar sands and shale oil reserves may be severely overstated. Development of most of these reserves is not economically feasible at present.

Source: Adapted from World Energy Conference, *Survey of Energy Resources* (New York: U.S. National Committee of the World Energy Conference, 1974), Table IX-2.

and the U.S.S.R. has natural gas been consumed on a large scale for some decades. Europe has switched from coal gas to natural gas only in the last few years; in fact this switch, made possible by large discoveries in and around the North Sea, has been of great help in coping with higher oil prices.

The great oil discoveries of the 1960s in the Middle East and elsewhere were often accompanied by gas discoveries, but the oil has generally been developed first. Even where gas is found in association with oil, it is often flared or reinjected until the necessary facilities for using the gas have been built. Gas transportation over long distances, whether by pipeline or by tanker (after liquefaction) calls for massive investments extending over 10 years or more. Most of these investments are still in the planning stage; by the mid-1980s they may lead to significant international trade in gas. In some cases, such as the gas found in two areas of the Canadian Arctic, proved reserves are not yet large enough to justify the building of pipelines, so these discoveries will become reserves only if either still more is proved or the gas price rises further, assuming environmental and native-claims problems can be overcome. Even greater uncertainty surrounds "unconventional" gas, such as that in coal seams and geopressured brines.

THE DEVELOPMENT OF WORLD OIL RESERVES

The preponderance of oil in world energy consumption calls for a more detailed analysis. Table 10-6 summarizes the history from 1948 to the present. It shows that proved reserves increased nearly tenfold over this period, despite large and growing consumption. Reserves were 68 billion barrels in 1948; 327 billion barrels were produced in the 30 years up to 1978, yet 646 billion barrels remained. This implies that 904 billion barrels were added to proved reserves. Clearly any calculation as to the date when oil will run out is meaningless unless new discoveries are taken into account.

There is no indication in the table that additions to proved reserves are slowing down;[5] the additions in the most recent 10-yr period shown were nearly as large as those in the preceding 20 years taken together. It is true that the largest additions have been concentrated in a few areas, especially in the Middle East, Africa, and the communist countries. Even for the United States, however, the historical picture is less bleak than it is often made out to be; in 1948 U.S. proved reserves were equivalent to about 11 years' current production, and now they are equivalent to about 10 years'. U.S. oil consumption, however, is nearly 3 times that of 30 years ago while production has increased much less. The United States in consequence has changed from a small exporter to a large importer.

TABLE 10-6 Estimated Proven World Oil Reserves from 1948 to 1978 (millions of barrels)

Region	Reserves 1/1/1948	1948–1957 Total		Reserves 1/1/1958	1958–1967 Total		Reserves 1/1/1968	1968–1977 Total		Reserves 1/1/1978
		Production	Gross Additions		Production	Gross Additions		Production	Gross Additions	
World	68,198	47,457	239,899	260,640	94,330	241,243	407,553	185,038	423,333	645,848
Annual average		4,746	23,990		9,433	24,133		18,504	42,333	
Noncommunist areas	61,698	42,730	215,472	234,440	78,281	215,621	371,780	149,702	325,769	547,848
United States	21,488	22,764	31,576	30,300	27,528	28,605	31,377	32,727	30,850	29,500
Other Western Hemisphere	10,210	9,572	23,240	23,878	17,727	28,924	35,075	22,901	34,196	46,370
Western Europe	50	454	1,708	1,304	1,230	1,943	2,017	1,983	26,829	26,863
Middle East	28,550	8,653	149,669	169,566	24,882	104,525	249,209	65,210	182,166	366,166
Africa	100	159	873	814	4,688	46,159	42,285	20,000	36,915	59,200
Asia and Pacific	1,300	1,128	8,406	8,578	2,226	5,464	11,816	6,881	14,814	19,749
Communist areas	6,500	4,728	24,428	26,200	16,049	25,622	35,773	35,335	97,561	98,000

Source: Compiled from American Petroleum Institute, *Basic Petroleum Data Book* (Washington, D.C.: American Petroleum Institute, 1975), and *Oil and Gas Journal*, vols. 46–76, last issue of each volume.

It should be noted that most of the discoveries since World War II occurred in the face of a fall in oil prices (particularly in real terms)—a fall that these discoveries themselves helped bring about. The high level of exploration, in fact, may have been stimulated less by favorable prices than by continuing advances in geological knowledge and improvements in exploration technology. As to geology, Table 10-6 shows that in 1948 half the world's proved reserves were in the Western Hemisphere (chiefly the United States), and most of the rest were in the Middle East. By 1977 the communist countries (chiefly the U.S.S.R.), Africa, and Western Europe had raised their shares in world reserves as one new petroleum province after another was opened up. The share of the Middle East also increased (from 42 percent in 1948 to 57 percent in 1977), but the U.S. share fell from 32 percent to 5 percent.

Progress in technology has been especially marked in offshore exploration; extensive continental shelves are now available for exploration, though many of these have not yet been drilled. The North Sea has been the most spectacular success story, but many other areas (including the entire Atlantic coast from Canada to Argentina) are also considered promising. While technologically feasible, offshore production is generally more expensive than onshore production.

Since Table 10-6 deals only with proved reserves, it is of limited use in analyzing the future. There are many estimates of ultimately recoverable crude oil resources,[6] but they are necessarily speculative and their underlying price assumptions are not always explicit. The most recent expert consensus centers around 2 trillion barrels,[7] about 3 times as much as proved reserves and the equivalent of 35 years' output at current rates.

As pointed out earlier for natural gas, estimates of this type usually do not include production by unconventional methods. They exclude not only oil from oil shale and tar sands, but also the heavy oils found in Venezuela and Canada, of which there may be at least as much as there is of "conventional" oil. How much of these resources can ultimately be recovered depends largely on price. Recovery does not present insuperable problems of technology; it is simply expensive, especially when proper environmental safeguards must be observed. In the case of the Athabascan tar sands in Canada, one plant has been in operation for a number of years and began to achieve profitability at 1976 prices. A larger plant is now producing, and others are under construction or planned.

Economic incentives are also the key to enhanced recovery from ordinary oil reservoirs. At present most of the oil is left in the ground, but if the price were high enough much more could be recovered.[8] Once all these possibilities are taken into account it appears that world oil resources could be produced at present rates until the middle of the twenty-first century.* However, this does not mean that the resource will last that long, since demand will certainly increase for part of that period, until alternative energy sources are introduced, higher prices bring declining consumption, and depletion of the resource lowers production rates.

In the more immediate future there is little need to consider unconventional oil production, for the potential of conventional techniques has by no means been exhausted. The present price level has encouraged intensive exploration, especially outside OPEC areas. It is too early to say what this exploration boom will turn up. In the United States and Canada the number of new oil wells has increased considerably, but most of these are small and might not have been considered commercial before the recent price increase. Elsewhere there have been further successes in the North Sea and substantial discoveries in Mexico, which may again become a large exporter, as it was earlier in this century. Some other non-OPEC developing countries (notably Brazil and India) are making headway in their efforts at self-sufficiency in oil. However, since the Iranian crisis of 1978 many exporting countries have begun to adopt more cautious policies on expanding production, and it is by no means certain that price increases will evoke as much additional production as once hoped. Mexico, for example, has been talking of a limit on oil exports.

Nothing in sight threatens OPEC's dominance in the world oil market; that would take at least another North Sea, and quite possibly two. (The addition of another 10 million barrels of crude oil per day—the equivalent of two North Seas—would cut the demand for OPEC oil to a point where allocation of cartel members' exports would become so difficult that the cartel might break down.)

WORLD URANIUM RESOURCES

Unless public resistance to the growth of nuclear power dictates otherwise, uranium may in due course challenge oil as the leading source of energy. It is not, however, possible to analyze it in as much detail as oil, since data on world uranium resources are even less reliable than those on oil resources. Nevertheless, one recent set of estimates is presented as Table 10-7. It gives reserves at a price of $130/kg of uranium, equivalent to $50/lb of uranium oxide ($U_3O_8$) and fairly close to present market prices; the resources involved may therefore be described as reserves. Most of these reserves are found in the United States,[9] Australia, South Africa, and Canada; there

*See statement 10-1, by E. J. Gornowski, Appendix A.

TABLE 10-7 Estimated World Resources of Uranium Recoverable at Costs up to $130 per Kilogram as of January 1977 (thousands of metric tons)

Region	Reasonably Assured Resources	Estimated Additional Resources	Total
North America	825.0	1709.0	2534.0
Western Europe	389.3	95.4	484.7
Australia, New Zealand, and Japan	303.7	49.0	352.7
Latin America	64.8	66.2	131.0
Middle East and North Africa	32.1	69.6	101.7
Africa south of Sahara	544.0	162.9	706.9
East Asia	3.0	0.4	3.4
South Asia	29.8	23.7	53.5
World (except communist countries)	2191.7	2176.2	4367.9

Source: J. S. Foster, M. F. Duret, G. J. Phillips, J. I. Veeder, W. A. Wolfe, and R. M. Williams, "The Contribution of Nuclear Power to World Energy Supply, 1975–2020," in *World Energy Resources 1985–2020* (Guilford, U.K.: IPC Science and Technology Press, 1978), p. 116.

are also sizeable amounts in France and in a few African nations. Except for Australia, all these countries have had some production in recent years. The Soviet Union and other communist countries are not included in the world totals. The total of 4.4 million tons in Table 10-7 is equivalent to about 150 years of current production, but uranium use will, of course, increase as nuclear power becomes more important.

Australia is the principal source of uncertainty about the world's uranium resources. Not only are many known uranium deposits spread over the Australian continent, suggesting that reserves are larger than the current estimates, but Australia's policy on uranium exports is also a matter of conjecture. The country has no urgent need for nuclear power, being well endowed with coal, hydroelectricity, and natural gas, so domestic demand is not an obstacle to exports. However, Australian minerals policy has generally been one of wariness toward foreigners, mitigated by a desire for development of the outlying regions, where minerals are usually found. At the moment it appears likely that Australia will permit some uranium exports under stringent controls to prevent nuclear proliferation, and only in quantities that are too small to depress the world price seriously. While Canada is an established exporter, its future policies may not be very different from Australia's. However, Canada may also attempt to tie uranium exports to sales of its heavy water reactors. South Africa could be relatively forthcoming with exports but may link their availability to diplomatic support, or at least neutrality, with respect to its domestic policies.

Apart from the estimated reserves shown in the table, it would be useful to know something about resources that may become available at much higher prices. In the United States the Chattanooga shales contain uranium in low concentrations; at a high enough price they would become reserves. Similar low-grade mineralizations occur in other parts of the world, for instance in Sweden. However, if these low-grade areas are mined to feed reactors as inefficient as light water reactors, the amount of shale that has to be moved per unit of power produced is comparable to that moved in the strip mining of coal. The environmental impact is expected to be even more severe than that of strip mining. Uranium can also be obtained as a by-product of other minerals such as phosphate, but the amounts are relatively small.

The present price of uranium offers strong incentives for exploration, which may help clarify the world resource situation. Evaluation of world uranium prospects, however, is greatly complicated by recent reports of an international cartel. The exact significance of these reports is not yet clear, but a cartel (or even the threat of a cartel) could be expected to stimulate importers' interest in U.S. supplies and in such uranium-saving technologies as reprocessing, breeders, and thorium-based reactors.[10] High uranium prices, whether artificial or not, will also encourage exploration in large areas of South America and Asia where the surface has hardly been scratched. These various responses take time, and in the meantime a well-organized cartel could be effective in maintaining, and quite possibly increasing, the world price of uranium.

LONG-RANGE PERSPECTIVES ON WORLD ENERGY FLOWS

The committee has not attempted to produce a set of long-range projections of world energy markets. In addition to the normal difficulties raised by such projections, the uncertainties about fuel reserves and about OPEC's price behavior would call for a bewildering variety of alternative assumptions. Making these world scenarios consistent with the domestic scenarios used in chapter 11 of this report would make the task even more formidable.

In lieu of a formal presentation of alternative global projections, we confine ourselves to a few general remarks on global energy perspectives.[11] The developments described in Tables 10-1, 10-2, and 10-3 will serve as a background to these remarks. To begin with, we assume that the United States undertakes no new energy policy measures beyond those enacted by the end of 1978, and allows existing price controls to expire; the effects of additional U.S. policies on the rest of the world are discussed in the next section.

1. The growth of world energy consumption will slow from the 5.1 percent/yr recorded in 1960–73. This slowdown, which is already evident in Table 10-3, will result from both higher prices and a lower growth rate of world GNP. However, if present patterns of economic growth in the world continue, and if the aspirations of the developing countries for larger shares of economic activity are realized, the average long-term growth rate of energy demand is unlikely to fall much below 3 percent/yr. Even if energy conservation in the United States accomplishes a great deal domestically, it will be more than offset by demand growth in countries at the "takeoff" stage of development. By the year 2010, world energy consumption will probably be 3–4 times as large as it is now. The developing countries will then have a larger share in world energy consumption than they have at present.

2. Electricity demand will probably grow more rapidly than total energy demand for two reasons. First, a large part of electricity cost is due to capital charges, and this will become more true as more capital-intensive forms of electricity generation, particularly nuclear reactors, are introduced. This means that electricity prices are less sensitive to fuel costs. This becomes increasingly true as more advanced reactors are introduced. If primary fuel costs rise more than capital costs, electricity will become cheaper relative to other energy forms. Second, as societies become more affluent they tend to prefer more convenient energy forms, such as electricity or gas, much as they convert more and more grain to animal protein in their food demand. By 2010 world electricity consumption could be 3–5 times as large as at present. If the market is the principal determinant of relative demand, and if there are no noneconomic constraints on the rate at which nuclear capacity can be expanded, then two thirds or more of electricity would probably be supplied by nuclear power, with coal a distant second, consumed mostly in the United States. In our view, expansion of nuclear capacity at so great a rate is unlikely. Also, a breakthrough in solar electric technology, if it came soon enough, could reduce the attractiveness of nuclear power somewhat.

3. In the absence of truly spectacular discoveries elsewhere, the OPEC countries (especially those in the Middle East and Africa) will account for the bulk of the world's oil production in the early part of the twenty-first century. In addition to North America, Europe, and East Asia, even Latin America will by then probably be a large oil importer unless the Venezuelan heavy oils are fully developed. However, North American production, though smaller than at present, will still be substantial.

Cumulative oil production between now and 2010 is likely to exhaust all presently proved reserves of "conventional" oil (Table 10-6). Because of intervening discoveries, however, oil reserves should still be at least as large as they are now, but they will be high-cost reserves, with production costs at least twice those of conventional reserves in the United States.*

4. The Middle East and Africa will become large exporters of natural gas and uranium; U.S. and Canadian uranium will also face a considerable export demand. The degree to which these countries will be willing to satisfy this demand with political conditions acceptable to importers is difficult to foresee.

5. The communist countries are a source of considerable uncertainty. Oil production in the Soviet Union appears to be leveling off, while consumption there and in Eastern Europe continues to grow. The Soviet Union is now a sizeable exporter of oil to the noncommunist world, especially to Western Europe, but may find it difficult to maintain these exports, thus making Western Europe even more dependent on OPEC. Indeed, some analyses suggest that the Soviet Union may have to import oil in the mid-1980s. If so, there would be a considerable impact on the demand for OPEC oil. However, the resulting strain on the Soviet balance of payments may force the regime to adopt stringent conservation measures and to rely more on natural gas, with which the U.S.S.R. is well endowed.

There is even more uncertainty about China's oil resources. In view of China's industrialization plans, which imply rapid growth in energy consumption, it appears unlikely that China will become a major exporter. On the other hand, China can probably not afford large oil imports if domestic discoveries are disappointing. Thus there is little reason to view China as a major factor in the world oil market during the remainder of this century.

6. As oil production gradually falls more firmly under OPEC control, the opportunity for surges in oil prices like those of 1974 and 1979 will increase. Moreover, as OPEC's reserves of low-cost oil are depleted, the incentives to raise prices will increase; this would be true even in the absence of a cartel. The price of uranium, increasing at an accelerating rate as the electric power industry becomes predominantly nuclear, could approach $100/lb of U_3O_8 (in 1972 dollars) by the end of this century if reprocessing is prohibited. Even with reprocessing and recycling of fuel, the uranium price will probably be high enough to make breeder reactors competitive with existing types in some parts of the world, especially in

*Statement 10-2, by E. J. Gornowski: My statement 10-1, Appendix A, also applies here.

Europe. Coal and natural gas will also become considerably more expensive in real terms.

7. Because of their predominance in oil, natural gas, and uranium, the Middle East and Africa will develop an even larger surplus in their energy trades, probably running into the hundreds of billions of 1972 dollars by the turn of the century. The corresponding deficits will be primarily in the industrial countries (except Canada). U.S. invisible items of trade are now quite strong and are supporting the nation's current account. A good part of this flow represents oil company earnings in the world market, which partially offset the high costs of oil imports. In addition, new conservation efforts, new oil finds, and a high propensity to import by OPEC help keep the U.S. external position from deteriorating too much. In the United States the energy trade deficit will be somewhat reduced by the expected growth in exports of coal or uranium if such exports are permitted. If the United States were to limit uranium exports there would be a correspondingly larger demand for U.S. coal. The main reason uranium would normally be preferred by importers is its lower transportation cost.

These projections do not take account of the trade in nuclear power plants and related facilities (and possibly other advanced energy technologies), which may offset a large part of the industrial nations' energy trade deficits but will add to the deficits of many countries that do not produce oil. In the absence of political constraints, worldwide investment in nuclear power between now and 2010 could add up to about $1 trillion (1972 dollars), and much of this will be supplied by North America, Europe, and Japan. Needless to say, other nonenergy exports will also have to expand to cover the growing energy trade deficits of these and the non-OPEC developing countries, and this may present serious problems for international trade. To the extent that economic growth in the industrial countries is slower than in the recent past there will be more political resistance to allowing manufactured imports from developing countries at exactly the time when such imports are most necessary to finance their energy purchases.

CONSEQUENCES OF ACTION ON NATIONAL ENERGY POLICIES

The world energy picture sketched in the previous section is hardly reassuring from either an economic or a political point of view. Let us now consider what difference certain U.S. policies might make.

Conservation in the United States beyond what is induced by higher world oil prices would reduce the growth in world demand for OPEC oil and thus reduce the cartel's power to raise the price and limit production.

The more the conservation effort concentrates on oil (or natural gas in uses where the two are directly substitutable), the greater will be the benefits to the rest of the world, although the magnitude of these benefits should not be exaggerated.

Promotion of domestic energy production, especially of oil and gas and directly substitutable energy forms, would be equivalent to conservation in its external economic effects. For example, even a remote possibility that offshore exploration will turn up large new oil provinces will serve to restrain OPEC. Conversely, a prohibition or retardation of offshore drilling would greatly strengthen the cartel's market power.

Price controls on oil and gas, or other measures shielding domestic consumers from world energy prices, would have effects opposite to that of accelerated conservation and domestic production; they would reinforce the pressure for a higher world oil price. The "entitlements" scheme for oil is especially harmful because it in effect subsidizes imports and penalizes domestic production; this scheme will presumably be phased out with the end of price controls (now anticipated for 1981).

A tariff on imported oil would encourage conservation and domestic output by allowing the domestic price of oil to rise to match the landed price of imported oil (assuming price controls have expired). It would also enable the nation to reduce the monopoly profit that would otherwise go to OPEC. A tariff would be particularly effective if adopted simultaneously by other major oil-importing countries. It might then become a major constraint on the cartel's ability to raise the world price of oil. Import quotas, with competitive bidding for import licenses, would similarly reduce OPEC's power over oil prices.

Abandoning nuclear reprocessing is likely to accelerate the rise of uranium prices. This would increase the incentives for reprocessing in uranium-importing countries. To counter this tendency, the United States (possibly in agreement with Canada and Australia) would have to keep the price of enriched uranium low enough, by subsidies if necessary, to make reprocessing uneconomic. The importing countries would then have to accumulate spent fuel, or possibly send it back to the original sources. Even if European and developing countries' reluctance to depend heavily on North America and Australia for their basic energy needs could be overcome, the combination of declining uranium reserves and increasing stocks of spent fuel will make a policy of subsidizing enriched uranium increasingly expensive in the long run. As the price of virgin uranium rises, the use of secondary materials becomes ever more attractive. If such a policy made a major contribution to preventing nuclear war or large-scale terrorism, the probably high cost to the United States would not be considered prohibitive. However, alternative methods of controlling weapons proliferation (for example, international safeguards programs including international surveillance of reprocessing operations) could be cheaper and more effective, and must be explored.

Abandonment or postponement of the breeder reactor is likely to have effects similar to the avoidance of reprocessing, raising the price of uranium and thus strengthening the interest of other countries in the development of breeders or advanced converter reactors. Under some plausible circumstances, the United States could remain a uranium exporter through the end of this century. Hence, a major delay in the U.S. breeder program, rather than setting an example to others, may accelerate breeder development elsewhere, if only because it would leave less U.S. uranium available for export (or increase U.S. demand for uranium imports). In any case, European work on breeders may be too far along, and too strongly supported by energy projections, to be stopped, despite growing political opposition to nuclear power in many European countries and Japan. To the extent that public distrust of nuclear power in the industrial countries slows its growth, the pressure on uranium supplies will decrease and the above-mentioned problems will be postponed, although the problems of the international oil market will intensify.

A slowdown in the growth of U.S. GNP would help keep down our energy demand and be similar in that respect to the accelerated conservation discussed earlier. However, it would also reduce our demand for nonenergy imports and thus make it more difficult for other countries, especially poor ones, to finance their energy imports. Moreover, slower economic growth, while possibly beneficial from an environmental point of view, would make it more difficult to restore and maintain full employment. Since it would adversely affect investment, it would also retard the turnover of capital stock and thus make it harder to improve energy efficiency.

THE DEVELOPING COUNTRIES AND THE WORLD FINANCIAL SYSTEM

As we have seen, the growing demand for energy in the developing countries will make them increasingly important in the global energy picture. Some of these countries are already considerable importers of oil, and others will become so as their transportation sectors expand. Moreover, the industrialization that is an inescapable aspect of economic development will greatly increase their reliance on electric power, of which they now have very little. (See Table 10-2.) Their agriculture will also shift from animal and human energy to tractors, harvesters, and trucks, and from natural to industrial fertilizers. As personal incomes rise in these countries, they will want better housing with more lighting and appliances, not to mention air conditioning. The more affluent of their citizens will demand motorcycles, automobiles, and air travel. In fact, the total demand for energy in these countries could conceivably rise faster than GNP.[12] Furthermore, we must hope that their economies do grow at reasonable rates, not only in their own interest but also for the sake of global political stability.

No doubt a substantial part of the required energy can be supplied from domestic sources. Oil and gas are found in many developing countries, but most of those with large resources have already joined OPEC. While there does not appear to be much coal in the developing countries, hydroelectricity could be expanded considerably, at ecologically acceptable sites, if financing were available. Sizeable quantities of uranium presumably remain to be discovered in some regions, but uranium (and possibly thorium, of which India has large reserves) is only a small part of the cost of nuclear power.

It is clear therefore that a large part of the energy needed by developing countries will have to be imported. In addition, heavy investments in electric power will be necessary even if the fuel can be obtained inside the country. Electric power, of course, is generally capital intensive, but it will be even more so if oil, gas, and coal are not available so that nuclear and hydroelectric power, or, in the more distant future, solar energy, must be used. In fact, oil is likely to be preempted by transportation uses, and in most developing countries coal would have to be imported from the United States and Australia, the only countries with a large potential for exports. It seems likely, therefore, that the developing countries as a whole will concentrate their investments in nuclear and hydroelectric power, at least by the end of this century, and that they will have to import increasing amounts of oil and uranium.*

This prospect implies further strains in the international financial system, which is already being taxed by the aftermath of the 1973–1974 oil price increase.[13] The developing countries at that time generally had little leeway in their balances of payments for increased oil prices; moreover, the recession in the developed countries induced by the oil price increase had a severe effect on their export earnings. The OPEC countries on the whole did not spend much of their vast new revenue on exports from developing countries. As a result, the non-oil-producing developing countries as a group (with notable exceptions such as India) suddenly found themselves with large trade deficits whose financing continues to preoccupy the international banking community.

The difficulty is not so much that the money is not available; the OPEC

*Statement 10-3 by J. P. Holdren: It is unfortunate that this chapter essentially ignores the potential of renewables other than hydropower, and of geothermal energy, in many developing countries.

surpluses remain in the world banking system and could be invested elsewhere. The problem is rather that the countries with cash surpluses (principally Saudi Arabia, Kuwait, and the United Arab Emirates) have not been willing to lend large amounts directly to the developing countries, although they have made relatively small amounts available to a few selected countries and to international organizations. These countries with surpluses have preferred to invest in short-term assets in the United States and Europe, rather than in long-term investment projects in the developing countries. Consequently, Western banks have had to assume the credit risks of loans to countries whose debt-servicing ability is heavily dependent on continued rapid economic growth. Various international arrangements are now being worked out to diversify these risks. The stakes are high, for without adequate financing the developing countries would have to curtail economic growth, to the detriment of billions of people already close to the subsistence level, and of the international banking system's stability. The developing countries' needs for massive investments in electric power will only magnify their financial problems.

The developed countries, preferably in consultation with the OPEC countries that have cash surpluses, should give high priority to schemes for maintaining a flow of financial resources to poor countries that foster their economic development. This means, among other things, that they should encourage imports from the poor countries even where these imports compete with domestic production. The international institutions active in this field (particularly the International Bank for Reconstruction and Development, the International Development Association, and the regional development banks) need further strengthening. Increased public awareness of the domestic aspects of the energy problem should not lead to neglect of its far-reaching international implications.

NOTES

1. The estimation of world GNP meets with the difficulty that the communist countries do not use this concept; instead, they calculate gross material product, which does not include all services. The United Nations Statistical Yearbook contains an estimate of world GNP excluding services. If services were included, the growth rate would probably be a little higher, though still not very different from the growth rate of energy consumption.

2. This region consists almost entirely of developing noncommunist countries, with South Africa the only significant exception.

3. A table on p. 267 of the *International Petroleum Encyclopedia* (Tulsa, Okla.: Petroleum Publications, 1978) presents a picture for natural gas rather similar to that sketched here for oil. World gas production went up 12 percent from 1973 to 1977, mostly due to rapidly rising output in the U.S.S.R. The U.S. share of world production fell from 48 percent in 1973 to 38 percent in 1977. Contrary to the situation in oil, the OPEC countries are not a major factor in gas.

4. Subsequent reports to the World Energy Conference have not changed these estimates substantially. See *World Energy Resources, 1985–2020* (Guilford, U.K.: IPC Science & Technology Press, 1978).

5. If "probable reserves" are included (as in Exxon Corporation, *World Energy Outlook*, April 1977) additions to reserves are more or less level from 1945 on, with a small dip in the last few years. This somewhat different pattern underscores the importance of timing assumptions in the analysis of resources. Usually reserves are considered "probable" before they are proved. The date at which indicated resources become probable reserves is inevitably somewhat arbitrary.

6. For a summary see Table 3-1 in Workshop on Alternative Energy Strategies, *Energy: Global Prospects 1985–2000* (New York: McGraw-Hill, 1977).

7. See Pierre Desprairies, "Worldwide Petroleum Supply Limits," in *World Energy Resources, 1985–2020* (Guilford, U.K.: IPC Science & Technology Press, 1978), pp. 1–47. The estimate given there for "unconventional petroleum (deep offshore and in the polar zones, enhanced recovery, oil shales, tar sands, synthetic oils)" exploitable toward the end of the twentieth century at a price of $20–$25 per barrel (1976 dollars) is about the same as for conventional petroleum.

8. According to the report cited in note 7, "enhanced recovery can increase the present average recovery of 25–30 percent . . . up to 45–50 percent," p. 15.

9. Table 10-7 does not break down North American reserves by country, but other sources indicate that at least two thirds of them are in the United States.

10. According to Foster *et al.*, "The Contribution of Nuclear Power to World Energy Supply, 1975–2020," in *World Energy Resources, 1985–2020* (Guilford, U.K.: IPC Science & Technology Press, 1978), pp. 126–127, thorium supplies from known sources (primarily as a by-product) are likely to be adequate at present prices.

11. More detail may be found in research inspired by CONAES but not done under its auspices; see H. S. Houthakker and M. Kennedy, "Long Range Energy Prospects," *Journal of Energy Development* 4, no. 1 (Autumn 1978): 1–28.

12. However, this possibility could be offset by the fact that their capital stock will be mostly new, so it can be designed to be efficient at present and prospective energy prices.

13. The most recent manifestation of these strains is the large current account deficit of the United States, which emerged in early 1977 and has had its counterpart in the large surpluses of Germany and Japan. While U.S. oil imports are a major contributor to the deficit, the failure of U.S. exports to grow in real terms is at least as important. The emergence of the deficit was followed by a sharp rise of most European currencies and the yen in terms of dollars, though most of the world's currencies have remained at par with the dollar and the Canadian currency has depreciated. The cheaper U.S. dollar should in due course correct the current account deficit, especially if other industrial countries stimulate their economies from their present recession, thus further encouraging U.S. exports. In the meantime the dollar's depreciation has not made oil imports more expensive because the price of oil is fixed in dollar terms; it has, however, increased the cost of non-oil imports and has thereby aggravated inflation in the United States. Consequently the attempt to keep U.S. oil prices low through price controls, which is one factor in our large oil imports, has led to a larger rise in other prices.

11 Methods and Analysis of Study Projections

CONAES asked several of its panels to develop models of energy and the economy to make plain the interrelations among variables influencing the supply of and demand for the various forms of energy. These models were applied to sets of assumptions about, for example, the growth rate of the economy, changing prices for energy over the next three decades, and consumer response, to picture some plausible states of affairs in the year 2010 and the course of their development. Some of the resulting scenarios are described here to illustrate key interrelations and assumptions.

There is always the danger in presenting models that numerical results will be taken literally. CONAES emphasizes that many uncertain assumptions must be made to construct models, and a great deal must be simplified or left out of consideration. Judgment alone decides whether some factors are important and whether others can be safely neglected, at least in a first approximation. Models cannot predict the future, but simply represent statements contingent on the consequences of assumptions and public policies. Nor can the consequences be regarded as rigorously deduced conclusions from a set of explicitly stated assumptions. Many detailed judgments accompany reason in these cases, judgments about the costs of new technologies, the rate of future resource discoveries, or the likely responses of myriads of producers and consumers to the general political climate and to government regulation. Many of the assumptions themselves are the subjects of wide disagreement among experts. For example, the Demand and Conservation Panel assumed an annual average growth rate for the gross national product (GNP) of 2 percent between 1975 and 2010, and defended this as their assessment of the most probable rate of future economic growth, but most of the economists involved in this study find it implausibly low. All the scenarios are "surprise free" in the sense that they ignore discontinuities such as embargoes, revolutions, natural disasters, international conflicts, and domestic strikes.

The value of models lies in the following.

1. They allow "thought experiments" to be conducted on the likely consequences of specific policies such as supply constraints, energy taxes, mandatory efficiency standards, or price regulation.
2. They allow testing of the sensitivity of outcomes (such as the rate of growth in energy consumption, or the relative consumption of various fuels) to varying input assumptions (such as economic growth rates, prices, population, work-force participation, or life-style preferences).
3. They provide an accounting device that helps ensure internal consistency among the projections.
4. They depict qualitative relations among the various factors affecting energy supply and demand. It is important to caution that the models do not usually prove qualitative statements, but rather illustrate them schematically.

CONAES has employed three different kinds of scenarios, each designed to answer different kinds of questions. The Modeling Resource Group (MRG)[1] employed econometric models to estimate the consequences of various economic and policy assumptions for total energy consumption. These are equilibrium models in which prices are determined endogenously through the interaction of supply and demand schedules for energy resources (using optimization techniques that simulate a competitive market). The MRG investigated the effect on GNP of various policies and levels of energy consumption, modifying supply and demand schedules for such hypothetical possibilities as high or low discovery rates for resources and Btu taxes (i.e., taxes per Btu of primary energy input). The group used econometric models to compute the total net cost to the economy of limitations on various energy supply technologies (e.g., on the expansion of nuclear power, the development of oil shale resources, or the mining of coal). The work of the MRG was largely self-contained, as reported in detail in their report, and has not been used in the other models, although some comparisons between the MRG results and those of the other models are presented in this chapter.

The Demand and Conservation Panel focused on the demand for net energy delivered to the point of consumption (final demand), separated by different energy forms (i.e., electricity, gaseous fuels, liquid fuels, and coal). In the Demand and Conservation Panel's models, energy prices are exogenous and are assumed to increase at various rates between 1975 and

2010. The effects of prices on the final demand for each energy form were estimated by a combination of econometric and technological models, as explained in greater detail below. Different technological models were used for each end-use sector (buildings, transportation, and industry). Specifically, the optimum design for energy-consuming equipment was chosen for each price scenario in such a way that the discounted lifetime cost of each piece of equipment is minimized over time for that particular price assumption, taking into account the normal replacement rate for the equipment. Little or no technological innovation was assumed, other than the application of well-known engineering principles.

Having obtained a set of final energy demands for each form of energy, the Demand and Conservation Panel then estimated the primary fuel requirement needed for conversion to the final fuel form, taking into account conversion efficiency and transportation or transmission losses, as well as processing losses. In the case of synthetic liquids and gases derived from coal, the partition between natural and synthetic fuels was estimated crudely on the basis of judgments by industry consultants about the availability of the synthetics technologies. Initial estimates were corrected in a second iteration by negotiation between the Demand and Conservation Panel and the Supply and Delivery Panel.

It had been hoped that the Supply and Delivery Panel would be able to generate supply curves for each primary fuel, i.e., curves of available supply as a function of price and time. This did not prove feasible. In the opinion of the panel, the political climate for energy resource development is a more influential factor than price in determining investment in energy exploration and development and, thus, future supplies. Price was included in the definition of political climate, but was not the most important factor. The Supply and Delivery Panel expressed the opinion that the energy required for any of the Demand and Conservation Panel's scenarios could be produced for little more than twice the 1975 OPEC price (measured in 1975 dollars), and that much higher prices than this to producers would not bring forth large additional domestic supplies. Not all experts would agree to this assumption. For example, some believe that large unconventional natural gas supplies would be forthcoming at sufficiently high prices.[2]

The Supply and Delivery Panel projected three different "climate" scenarios: business as usual, enhanced supply, and national commitment. The scenario for each primary energy source was defined somewhat differently, according to the source's characteristics. For each scenario, the panel estimated the amount of each form of primary energy likely to be produced in 1990 and in 2010 under the corresponding assumptions.

The next step for CONAES was to try to match the Supply and Delivery Panel's supply projections with the Demand and Conservation Panel's

TABLE 11-1 Scenario Projections Used in the CONAES Study

Scenario	Source	Description
Demand scenarios[a]: A*, A, B, B′, C, D	Demand and Conservation Panel[b]	A, B, C, and D explore the effects of varied schedules of prices for energy at the point of use, from an average quadrupling between 1975 and 2010 (scenario A) to a case (scenario D) in which the average price of energy falls to two thirds of its 1975 value by 2010. Basic assumptions include 2 percent annual average growth in GNP, and population growth to 280 million in the United States in 2010. Scenario A* is a variant of A that takes additional conservation measures into account. Scenario B′ is a variant of B, projecting the effect on energy consumption of a higher annual average rate of growth in GNP (3 percent).
Supply scenarios: Business as usual, enhanced supply, and national commitment	Supply and Delivery Panel[c]	Projections of energy resource and power production under various sets of assumed policy and regulatory conditions. Business-as-usual projections assume continuation without change of the policies and regulations prevailing in 1975; enhanced-supply and national-commitment projections assume policies and regulatory practices to encourage energy resource and power production.
Study scenarios: I_2, I_3, II_2, II_3, III_2, III_3, IV_2, IV_3 (correspondence between study scenarios and demand scenarios: I_2 = A*, II_2 = A, III_2 = B, III_3 = B′, IV_2 = C; scenario D was not used)	Staff of the CONAES study	Based on the demand scenarios; integrations of the projections of demand from the demand scenarios and projections of supply from the supply scenarios. A variant of each price-schedule scenario was projected for 3 percent annual average growth of GNP.
MRG scenarios	Modeling Resource[d] Group	Estimates of the economic costs of limiting or proscribing energy technologies in accordance with various policies.

(TABLE 11-1 *footnotes opposite*)

estimated requirements for final energy. This was accomplished by carrying forward the process initiated by these two panels (and consulting with them as necessary): starting with the scenarios of demand (and a variant for each at 3 percent annual average growth of GNP), estimating the mix of primary fuels likely to meet those requirements (taking into account conversion and transmission losses), then readjusting the requirements to bring the assumed supply policies into line with the political climate likely to accompany the corresponding scenario of demand. For example, the prices and policies leading to projections of greatly moderated demand for energy would most likely correspond to policies that constitute business as usual for supply. The assumptions leading to higher projections of demand would most likely correspond to the conditions for enhanced supply.

Thus, the demand called for by a scenario of low energy consumption was initially matched with a business-as-usual supply scenario. If this did not provide enough energy to meet demand, the enhanced-supply scenario was tried. For the scenarios of high energy consumption, some national-commitment supply scenarios were permitted. The study scenarios do not employ exactly the same fuel mixes as the Demand and Conservation Panel's scenarios to fill the required end-use demands. This results in slightly different primary energy requirements because of the differences in assumed energy-conversion efficiencies. Small differences, usually not more than 10 percent, may be observed between the primary energy inputs (total energy consumption) in the study scenarios and the scenarios computed by the Demand and Conservation Panel.

In the following sections, we describe the assumptions and methods of the Demand and Conservation Panel and the Supply and Delivery Panel, and we present the comparisons of supply and demand incorporated in the study scenarios. This is followed by a separate discussion of the scenarios of the MRG and by a comparison of the results of the MRG with the study

TABLE 11-1 *Footnotes*

[a]Scenario D, a projection of energy consumption in 2010 at prices averaging two-thirds the 1975 prices for energy at the point of consumption, is not taken up in this chapter.

[b]Source: National Research Council, *Alternative Energy Demand Futures to 2010,* Committee on Nuclear and Alternative Energy Systems, Demand and Conservation Panel (Washington, D.C.: National Academy of Sciences, 1979).

[c]Source: National Research Council, *U.S. Energy Supply Prospects to 2010,* Committee on Nuclear and Alternative Energy Systems, Supply and Delivery Panel (Washington, D.C.: National Academy of Sciences, 1979).

[d]Source: National Research Council, *Supporting Paper 2: Energy Modeling for an Uncertain Future,* Committee on Nuclear and Alternative Energy Systems, Synthesis Panel, Modeling Resource Group (Washington, D.C.: National Academy of Sciences, 1978).

scenarios. Table 11-1 summarizes the scenarios discussed in this chapter for ready reference.

ANALYSIS AND SCENARIOS

WORK OF THE DEMAND AND CONSERVATION PANEL[3]

The work of the Demand and Conservation Panel relied primarily on assessments of the technological possibilities for moderating the consumption of energy in the transportation, buildings, and industrial sectors of the economy. The panel projected the extent to which these technological possibilities might be realized under various assumed sets of prices for energy at the point of use. An integrating model for the economy in the year 2010 was used to adjust these sectoral figures for consistency with one another and with final demand. The levels of energy consumption projected by the panel for 2010 range from about half today's per capita consumption to levels twice as high.

In projecting the consumption of energy over the next three decades, the panel chose to fix some demand-shaping variables and to allow others to vary. It was most interested in the effects of changing prices for energy and various policies that would stimulate or discourage energy-conserving practices. Accordingly, the panel fixed the growth rate of GNP (experimenting, however, with a variant), population, and work-force participation, and its scenarios of energy consumption were determined by response to four sets of energy prices. The panel assumed that the decisions of consumers (both industrial and commercial) on the purchase of energy-consuming equipment would be economically rational for each assumed set of prices, and that consumers would seek to minimize the total lifetime cost of such equipment. Policy variables were also simulated in two additional scenarios to test the effects of vigorous conservation accompanied by some voluntary changes in patterns of living, working, and buying (in one case), and a higher rate of economic growth (in the other).

Population Growth

The panel assumed the Series II projection of population growth by the Bureau of the Census in all scenarios. This projection assumes a reversal of the downward trend in fertility (Figure 11-1), resulting in a population of 279 million people in the United States in 2010. The Series III projection, assuming a continued downward trend in fertility, gives a population of 250 million people in 2010. If the Series III projection were realized, the panel estimates that total energy consumption in 2010 would be lower by about 10 percent in all the scenarios. (Series I, II, and III projections by the Bureau of the Census are pictured in Figure 11-2.)

Any assumption of future population growth must be considered arbitrary. The panel points out that the effects of illegal immigration can only be guessed, although they could well be the most significant source of demographic uncertainty.

Work-Force Participation and Other Trends

The panel assumed that the work force in the United States would grow in the direction indicated by prevailing trends: at a lower rate in the future than in the past, and in accordance with recently declining fertility rates and the consequent size of younger-age cohorts, which govern the rate of growth of the labor force. The panel also assumed that the participation of women in the work force would continue to grow. The trend toward shorter working days and fewer working days per year was presumed to extend into the future at about the same rate as in the recent past.

Growth of GNP

Gross national product was used in this study as a measure of economic activity, and in an extended (if not wholly satisfactory) sense as an indicator of national well-being. Over the 30-yr period from 1945 to 1975, GNP grew at an average annual rate of 2.7 percent.* The rate of increase from year to year varied (from 1950 to 1970, the average annual rate was 3.5 percent). The panel assumed that over the next 30 years, the growth of GNP would be rapid in the near term, owing in part to recovery from the 1974 recession and in part to the rapid growth of the labor force in the 1970s stemming from the postwar baby boom. Beginning in the mid-1980s, growth would slow with declining additions to the labor force. The panel selected an average annual growth rate for GNP of 2 percent over the next 30 years.† At this rate, GNP approximately doubles by 2010. One scenario was projected for an annual average growth rate of 3 percent, resulting in a near tripling of GNP (2.8 times the 1975 total) by 2010 (scenario B′). Figure

*Statement 11-1, by R. H. Cannon, Jr.: Over the entire 33-yr period 1946 to present, GNP has followed a 3.4 percent growth curve remarkably closely. During the war years 1940–1945, it was much higher.

†Statement 11-2, by R. H. Cannon, Jr.: Since World War II, GNP has followed a 3.4 percent growth line remarkably closely. Using 2 percent predestines dangerously low energy demand projections.

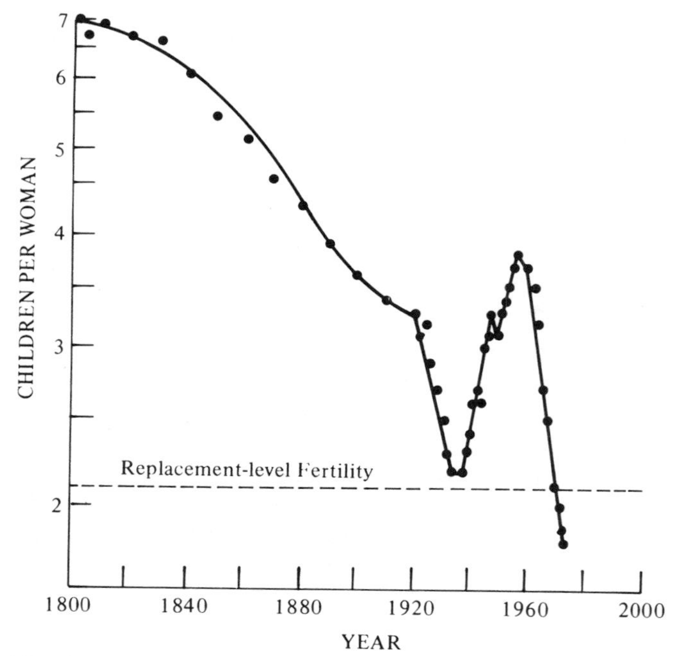

FIGURE 11-1 Total estimated fertility rates in the United States from 1800 to 1976. Source: U.S. Department of Commerce, Bureau of the Census, *Estimates of the Population of the United States and Components of Change: 1940–1976,* Population Estimates and Projections Series P-25, No. 706 (Washington, D.C.: U.S. Government Printing Office, 1977).

11-3 illustrates the two paths, which are approximately linear rather than exponential,‡ and the inset shows the corresponding compound growth rates used by the panel for the subperiods from 1975 to 2010.

The rate of economic growth selected by the panel prompted discussion within the committee and among other participants in the study. Most of the economists express reservations about its likelihood, feeling that a 2 percent average rate of growth would not be consistent with the assumption of full employment. Others point to recent trends of declining growth in productivity and suggest that the growing investment in environmental protection and related areas of health and safety, as well as the shift of employment from the manufacturing to the service sector, will continue to reinforce the trend of declining growth in productivity.

‡Statement 11-3, R. H. Cannon, Jr.: Constant growth-rate curves on this (linear) plot are concave upward. Continuing the rate we've had since 1946 leads to a $5,200 billion GNP in 2010.

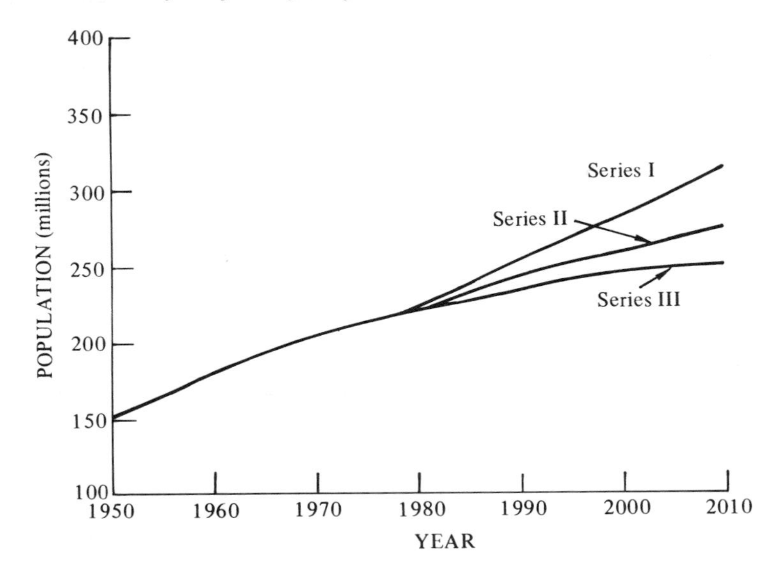

FIGURE 11-2 Estimates and projections of the total population of the United States from 1950 to 2010, showing Bureau of the Census alternative projections from 1975. Source: Adapted from U.S. Department of Commerce, Bureau of the Census, *Estimates of the Population of the United States and Components of Change: 1940–1976,* Population Estimates and Projections Series P-25, No. 706 (Washington, D.C.: U.S. Government Printing Office, 1977).

The Modeling Resource Group used an average annual rate of growth for GNP of 3.2 percent/yr as their base case, but also showed an alternative low value, corresponding to an average of about 2 percent a year, with even greater deceleration.

	D/C Panel	MRG-Low
1975–1980	2.7	3.7
1980–1990	2.3	2.7
1990–2000	1.8	1.2
2000–2010	1.6	0.5

The Modeling Resource Group generated its high, low, and base-case projections for the growth of GNP by projecting the changes that might be expected in three determinants of potential GNP over the period 1975–2010. Those leading to the lower rate of growth are the following.

- Work-force participation declining from 0.73 (its average value from 1950 to 1975) to 0.70 in 2010.
- Unemployment averaging 6 percent.

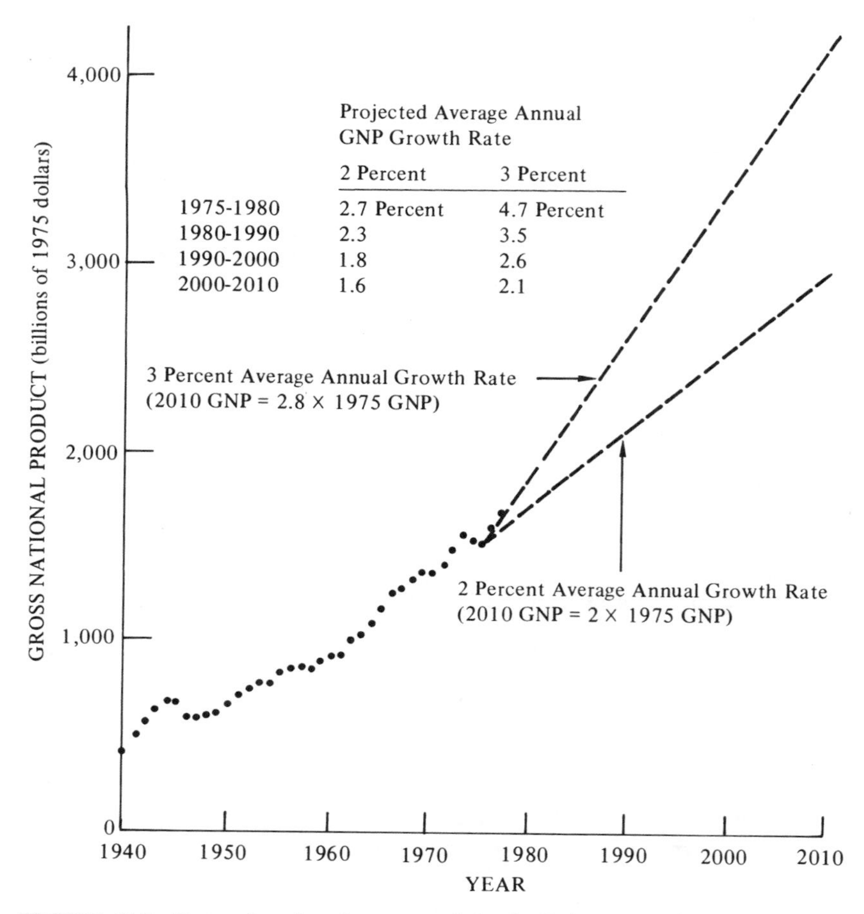

FIGURE 11-3 Past and projected GNP growth in the United States from 1940 to 2010 (billions of 1975 dollars). Source: Adapted from National Research Council, *Alternative Energy Demand Futures to 2010,* Committee on Nuclear and Alternative Energy Systems, Demand and Conservation Panel (Washington, D.C.: National Academy of Sciences, 1979), p. 60.

- Growth in productivity shrinking from 1.57 percent per annum in 1975 to zero by 2010.
- Growth rate of the potential labor force and immigration slowing to 0.2 percent a year after 2010.

Other studies examining the relation of energy consumption and the domestic economy have projected different rates of growth for GNP. The

Energy Policy Project of the Ford Foundation,[4] for example, projected three scenarios to the year 2000 (in 1974). In that study, the zero-energy-growth and technical-fix scenarios assumed that GNP would rise at a rate of 3.5 percent/yr from 1975 to 1985, and at a rate of 3.1 percent/yr from 1985 to 2000. The historical-growth scenario assumed that GNP would rise at a rate of 3.6 percent/yr over the first 10-yr period, and at a rate of 3.3 percent/yr from 1985 to 2000. Exxon Corporation[5] assumed that GNP would grow over the 4-yr period from 1976 to 1980 at an annual rate of 4.2 percent, and from 1980 to 1990 at an annual rate of 3.4 percent, but warns, "A reasonable range of error in estimating long-term economic growth might be perhaps ±0.5 percent per year." The Edison Electric Institute[6] set out three patterns of economic growth to the year 2010—high, moderate, and low. The high-growth case has GNP rising by 4.2 percent/yr; the moderate case, by 3.5–3.7 percent/yr; and the low case, by 2.3 percent/yr. The institute considers the moderate case to be the most likely. (Table 11-34 gives the annual average GNP growth rates projected by CONAES and other energy studies.)

CONAES did not attempt to select a "best" growth rate, but rather estimated the growth of energy consumption for both 2 percent and 3 percent growth rates in GNP from 1975 to 2010. It is important to recognize that several other estimates are higher than this range and would lead to higher energy consumption for a given set of price assumptions. The scenarios of the Demand and Conservation Panel cannot be regarded as bracketing all the possibilities.

Energy Prices

As recapitulated below, the demand scenarios (presented in chapter 2) assume energy prices held constant (scenario C), doubled (B and B′), or quadrupled (A and A*) by 2010. These are average prices of net delivered energy. The panel assumed specific prices for each source of energy by 2010, displayed in Table 11-2, under the categories of these average prices. The relative prices given in Table 11-2 were intended to reflect approximate parity in dollars per million Btu, with adjustments for the relative cleanliness, convenience, and thermodynamic qualities of fuel. Natural gas is thus priced above distillates, and coal below petroleum. Deregulation of prices was assumed in these projections. Unless otherwise specified, the panel's overall assumption was that demand would be met at these prices. (Scenarios A* and A, for example, specify a prohibition against the use of natural gas for industrial boilers.)

The assumed prices listed in Table 11-2 for the year 2010 represent those seen by consumers at the final stage of end-use, expressed in 1975 dollars. The relative increase is the same for all consumers, industrial and residential, and for each end-use. This assumption may overestimate the prices that would be charged for energy consumed in homes and in transportation relative to the prices charged for industrial consumption. It implies that distribution and overhead costs will rise in proportion to primary fuel prices. From 1970 to 1978, in fact, the average costs of primary fuels doubled, but the average prices of delivered energy rose just 30 percent in real dollars.[7] Assuming that overhead, distribution, and capital costs remain constant (in constant dollars) while primary fuel costs increase enough to keep average delivered prices the same, a relative shift in demand would occur from industry to households and transportation and from fluid fuels to electricity. The assumed rises in primary fuel prices would have to be more than double the ratios shown in the table. Some trend of this sort is already indicated in the detailed price assumptions. Prices for electricity (with the largest capital and distribution cost) rise least, while prices for natural gas (with the lowest capital and distribution costs) rise most.

Other Assumptions

Scenarios A, B, B′, and C assume that the structure of the economy will not change markedly over the next three decades. The energy-price/demand extrapolations employed by the panel are consistent with historical data for fuel-price demand elasticities and cross-elasticities, and with regional comparisons. Details are given in the report of the Demand and Conservation Panel.[8]

Scenario A*, a simple variant of scenario A, tests the additional moderation in the growth of energy consumption that might result from some changes in the habits and purchases of consumers and from an accelerated shift in the economy from goods to services (for example, from goods produced to be used once and discarded to goods produced to endure with much more repair and maintenance). Again, this scenario might be criticized on the grounds that high labor costs would make repairs uneconomical. On the other hand, it is possible that advances in information technology and microprocessors could greatly increase the productivity of repair services, as well as making possible better quality control and durability in original manufacture (for example, by replacing low-reliability mechanical devices with electronics of higher reliability). Such developments could shift the optimum balance between initial product cost and repair.

TABLE 11-2 Price Assumptions for Scenarios of Energy Demand[a]

	Oil Prices					
	Distillate No. 2[b]		Utility Residual		Gasoline Before Taxes	
	Dollars per Barrel	Dollars per Million Btu	Dollars per Barrel	Dollars per Million Btu	Dollars per Barrel	Dollars per Million Btu
1975 actual	16.37	2.81	12.40	2.02	19.08	3.64
2010 scenarios						
A, A*	78.58	13.49	59.52	9.70	91.58	17.47
B, B′	39.29	6.74	29.76	4.85	45.79	8.74
C	16.37	2.81	12.40	2.02	19.08	3.64
Ratios of 2010 prices to 1975 prices						
A, A*		4.8		4.8		4.8
B, B′		2.4		2.4		2.4
C		1.0		1.0		1.0

TABLE 11-2 Price Assumptions for Scenarios of Energy Demand (*Continued*)

	Utility Natural Gas Prices (dollars per million Btu)					Utility Coal[c]	
	Utility[d]	Residential	Commercial	Industrial[d]	Consumption-Weighted Average	Dollars per Ton	Dollars per Million Btu
1975 actual	0.75	1.70	1.41	0.99	1.29	17.68	0.81
2010 scenarios							
A, A*	?	19.63	15.89	11.38	14.84	70.52	3.24
B, B′	?	9.82	7.94	5.69	7.42	35.26	1.62
C	?	4.09	3.31	2.37	3.09	17.63	0.81
Ratios of 2010 prices to 1975 prices							
A, A*					11.5		4.0
B, B′					5.7		2.0
C					2.4		1.0

Electricity Prices

	Residential		Commercial		Industrial		Consumption-Weighted Average	
	Cents per Kilowatt-Hour	Dollars per Million Btu	Cents per Kilowatt-Hour	Dollars per Million Btu	Cents per Kilowatt-Hour	Dollars per Million Btu	Cents per Kilowatt-Hour	Dollars per Million Btu
1975 actual	3.11	9.11	3.25	9.52	1.83	5.36	2.70	7.91
2010 scenarios								
A, A*	10.37	30.37	10.83	31.73	6.10	17.87	9.00	26.37
B, B′	6.22	18.22	6.50	19.04	3.66	10.72	5.40	15.82
C	3.11	9.11	3.25	9.52	1.83	5.36	2.70	7.91
Ratios of 2010 prices to 1975 prices								
A, A*								3.33
B, B								2.0
C								1.0

[a] Prices in this table are national averages and are the prices seen by the consumer at the final stage of consumption in 1975 dollars. Prices reflect rough parity on a dollar per million Btu basis, but are adjusted for cleanliness and thermodynamic quality. Price deregulation is implicit. State and federal taxes are not included.

[b] The major form for distillate oil. Diesel oil prices would be expected to grow in similar proportions.

[c] In 1975, electric utilities used 72 percent of domestic bituminous coal tonnage (67 percent of its energy content).

[d] It is doubtful that electric utilities will be burning natural gas in 2010. In 1974, 26 percent of industrial natural gas sales by gas utilities were to electric utilities. Another 3.5 quads of gas were used in electricity generation, but were not purchased from gas utilities.

Source: Adapted from National Research Council, *Alternative Energy Demand Futures to 2010,* Committee on Nuclear and Alternative Energy Systems, Demand and Conservation Panel (Washington, D.C.: National Academy of Sciences, 1979), pp. 8 and 9.

Methodology

The panel investigated the consumption of energy in each of the three principal energy-consuming sectors of the economy—buildings, transportation, and industry—and projected the consumption of energy in these sectors to 2010 under the assumptions of the five scenarios. This sectoral analysis yielded interesting information about energy-efficient technologies and patterns of energy use available for the future, but it did not allow for feedback and other interactions among the sectors. The integrating model was designed to trace the flow of energy through the national economy in 2010, adjusting the energy consumption figures for the three sectors so as to be self-consistent. The energy demand totals for 2010 given here and in chapter 2 were calculated with the aid of the integrating model.

The three sectoral analyses employed different methods, briefly summarized below.

Buildings[9] Residential buildings were defined as those occupied by households, and nonresidential buildings as those occupied by the service sectors of the economy. In the residential sector, energy was assumed to be used for space heating, water heating, air conditioning, the preparation and storage of food, lighting, laundry, and the operation of small appliances. In the nonresidential sector, energy was assumed to be used for heating, cooling, and operations such as cleaning and powering elevators.[10]

An existing engineering-economic simulation model was used to evaluate the effects of rising personal incomes and fuel prices on the cost and use of energy in residential buildings. The model is explained in the report of the Demand and Conservation Panel.[11] The panel used this model to simulate the use of four fuels (gas, oil, electricity, and other) for eight functions (space heating, water heating, refrigeration, food freezing, cooking, air conditioning, lighting, and other) in three types of residential buildings (single family, multifamily, and mobile homes). The fuel consumed for each end-use was estimated in response to changes in stocks of occupied housing units and new residential construction, equipment ownership by fuel and end-use, thermal integrity of new and existing housing units, average unit energy requirements for each type of equipment, and aspects of household behavior reflected in patterns of use.

The economic submodels provided the elasticities that determine the responsiveness of households to changes in economic variables (incomes, fuel prices, and equipment prices). The elasticities were calculated for each of the three major household fuels and each of the eight end uses, each fuel price and income elasticity being separated into two elements—the elasticity of equipment ownership and the elasticity of equipment use. The first gave changes in market shares of equipment ownership in response to changes in fuel prices and incomes; the second gave changes in equipment use with ownership held constant. The submodels also provided equipment-ownership, market-share elasticities with respect to equipment costs. The simulation model was therefore able to estimate consumer responses to changes in operating costs (fuel price times consumption) and changes in capital costs.

The engineering submodels were used to evaluate the variations of purchase prices and energy use with the design of equipment. Detailed submodels were constructed for gas and electric water heaters, refrigerators, and ranges; for the other end-uses, data were combined from various sources to determine relations between energy use and initial cost.

With the simulation model, it was possible to combine the outputs from the various submodels with the initial conditions (from 1970) and the boundary conditions (including policy variables) for the scenario period. The four fuels (i), eight end-uses (k), and three housing types (m) produced 96 fuel-use components ($Q^{i,k,m}$) for each year (t) of the simulation model. The model also provided annual fuel expenditures, equipment costs, and capital costs for improving the thermal integrity of new and existing structures.

Energy use in the nonresidential subsector was projected by a disaggregated model of the commercial demand for energy[12,13] that covered five end-uses (space heating, water heating, cooling, lighting, and other), four fuel types (gas, electricity, oil, and other) and ten commercial subsectors (retail/wholesale, auto repair/garages, office activities, warehouse activities, public administration, education, health care, religious services, hotels/motels, and miscellaneous).

The modeling approach was traditional. The demand for energy, given fuel i, end-use k, and subsector or building type m, was represented simply as

$$Q_{i,k,m} = U_{i,k,m} \times S_{i,k,m}$$

where Q is the energy demand, S the stock of energy-using capital, and U the rate of use.

The stock of equipment was considered fixed over the short term, with only the rate of use changing in response to exogenous factors such as changes in fuel prices. Changes were permitted in the capital stock over the long term in response to rising incomes and the obsolescence of the existing stock.

The energy used in the commercial sector was estimated on the basis of the floor space served. Additions to floor space were calculated by a desired stock estimate (based on population, per capita income, school-age population, etc.), subtracting additions still standing from previous years.

The use of solar energy was estimated separately for the residential and nonresidential subsectors. The share of the capital market that is likely to be absorbed by solar systems in new buildings was approximated by an *ad hoc* relationship suggested by the Solar Resource Group of this study. It was assumed that retrofit of solar installations in existing buildings would make a negligible contribution. (This assumption, made before recent federal legislation providing substantial benefits for retrofit of solar installations, may be conservative. See, for example, "Space Heating and Cooling," in chapter 6.)

Transportation[14] Five categories of passenger transportation and four categories of freight transportation were considered in the analysis of this sector.

Passenger Transportation	Freight Transportation
Automobiles	Truck
Light trucks and vans	Water
Air travel	Air
Mass transportation	Rail
Other	

The model employed features of 10 major models constructed over the past 10 years, with a particular view to evaluating the effect on fuel consumption (under the assumptions of the detailed price scenarios shown in Table 11-2) of changes in the efficiency of fuel consumption by vehicles, load factors, prices for fuel, other operating costs, and capital expenditures for transportation. The effects of public policies were assessed by varying the input parameters.

The model assumes that fuel-price ratios (relative to 1975 in 1975 dollars), load factors, and efficiency ratios will increase linearly over the years ahead, but there is considerable uncertainty about the paths they will actually follow. The paths will probably be S-shaped, but since there is also substantial uncertainty about the endpoints (which are more important than the paths), the linear transition actually assumed is not critical. Other assumptions include the following.

• Automobile travel: Fuel economies of new cars will increase to a plateau by 2000; the average fuel economy of the total fleet will lag behind new-car improvements by 10 years. Gasoline taxes per vehicle-mile (in 1975 dollars) will remain constant on the assumption (based on historical data) that the cost of building and maintaining highways will depend primarily on the vehicle-miles of use. Expenditures per capita on autos (in 1975 dollars) will saturate; auto ownership itself is reaching saturation, and the time spent in auto travel by each person (now about 50 min/day) is not likely to increase greatly. Data from the United States and other affluent countries indicate that there is a saturation effect in auto ownership at high income levels. Auto ownership in the United States is approaching one vehicle per licensed driver, although it is believed that a firmer upper limit in auto ownership is one vehicle per person of driving age (about 71 percent of the population in the United States).

• Light-duty trucks and vans: The assumptions for this mode are essentially similar to those for automobiles, except for a slightly higher gasoline tax per vehicle-mile.

• Air travel: The percentage of passenger transportation dollars spent on air travel will increase with rising incomes and saturation of expenditures on auto travel. The load factor will increase linearly until 2000 and remain constant thereafter. The energy intensity (Btu per passenger-mile) will decrease linearly until 2000 and remain constant thereafter.

• Mass transportation[15] and other passenger modes: All nonfuel operating costs (in 1975 dollars) will remain constant to 2010. For mass transportation, load factors rise in each scenario except B′, as shown below.

Average load factors in 2010 by scenario for mass transportation

(1975)	(17.9 passengers per vehicle)
A*, A	31.0 passengers per vehicle
B	27.0 passengers per vehicle
C	22.0 passengers per vehicle
B′	17.9 passengers per vehicle

The energy intensity of the vehicles used remains unchanged over the period.

• All freight modes: The growth in all modes of freight transport per capita will be proportional to the growth in real GNP per capita. Where load factors and efficiencies change, they change linearly during the 1975–2010 period.

Industry[16] The analysis of energy consumption in the industrial sector concentrated on 14 industries that account for 80 percent of the energy consumed by industry: 9 energy-consuming industries (agriculture, aluminum, cement, chemicals, construction, food, glass, iron and steel, and paper) and 5 energy-producing industries (oil and gas extraction, oil and gas refining, coal mining, synthetic fuels, and electrical generation). The energy these industries are likely to require in the future was estimated by multiplying the projected level of production for each industry in 2010 by its expected energy intensity. The projected growth rates are displayed in Table 11-3. The data available to the panel were inadequate for determination of the energy-price/consumption elasticities within individual industries. The panel identified the technologies available for increasing the efficiency of energy use within each of the 14 industries, and estimated the extent to which these might be applied under the assumptions of each scenario. The technological possibilities are indicated qualitatively in Table 11-4.

In projecting the mix of old and new industrial plants for 2010, the panel assumed the prevailing retirement rate: 2 percent/yr for plants in operation in 1975. A third of today's facilities would thus still be in use in 2010. This assumption may be conservative for the higher-price scenarios, as the energy cost of old equipment might accelerate replacement. In these scenarios, capital not needed for energy supply would be available for accelerated replacement of energy-consuming equipment.

The panel studied the patterns of energy use by the industries named and employed a few additional assumptions to derive preliminary estimates of the fuel mix for industry in 2010. The use of natural gas, for example, was assumed to be increasingly restricted under the assumptions of scenarios A, B, and B′ as a result of higher prices and policies governing scarcer fuels. In these scenarios, the use of natural gas was limited to special applications that justify the use of this high-quality fuel at higher prices (or under the terms of restrictive policies that might be imposed on its use). Since the price of coal is competitive with that of other fuels in the higher-price scenarios, it was assumed that most industrial generation of steam would be fired by coal in scenarios A, B, and B′, and that many existing processes fueled by oil and natural gas would be converted to coal by 2010. Scenario C assumes no restrictions on the availability of natural gas for industrial use.

On the advice of the Supply and Delivery Panel, the Demand and Conservation Panel assumed that solar energy would be economical for some low-temperature industrial applications if the prices of other sources rose appreciably. The panel assumed that the direct use of solar energy would replace that of natural gas to produce low-pressure steam and hot water for agriculture, food processing, and miscellaneous manufacturing processes, accounting for 0.2 quadrillion Btu (quad) of industrial energy consumption in scenario C, for example, and 1.8 quads in scenario A.

Integrating model[17] Since the three sectoral analyses were carried out independently, the panel sought some means to array and correct their results in a model of the economy for 2010. The model used national economic data from the U.S. Department of Commerce[18] for the year 1967 (the most recent at the time the panel conducted the analysis). Corrections were made for data from the sectoral analyses (using 1975 as

TABLE 11-3 Projection of Industrial Growth Rates from 1975 to 2010

Industry	Growth Rate of Production 1960–1972 (percent per year)	Ratio of Production Growth Rate to GNP Growth Rate[a]	Initial Future Growth Rate Approximation[b] (percent per year)		Modified Growth Rates for Final Projections (percent per year)				Reasons[c] for Modification
			For 2 percent per year GNP Growth	For 3 Percent per year GNP Growth	A	B	B	C	
Agriculture	1.6	0.39	0.8	1.2	1.7	1.7	1.7	1.7	1,2,8
Aluminum	6.9	1.68	3.4	5.0	3.2	3.6	5.4	4.0	3,4
Cement	3.6	0.88	1.8	2.6	1.8	1.8	2.6	1.8	
Chemicals	8.4	2.05	4.1	6.1	3.6	3.8	4.8	4.1	3,4,5
Construction[d]	3.8	0.93	1.8	2.8	0.7	0.7	1.0	0.7	1,3,6
Food	3.3	0.80	1.6	2.4	2.2	2.2	3.2	2.2	1,8
Glass	4.3	1.05	2.1	3.1	2.1	2.1	3.1	2.1	
Iron and steel	3.6	0.88	1.8	2.6	1.7	1.7	2.7	1.7	6,7
Paper	5.4	1.32	2.6	3.9	1.9	1.9	2.9	1.9	5,6,7
Other manufacturing and mining activities	4.8	1.17	2.3	3.5	2.2	2.2	3.2	2.2	6,7

[a] The average growth rate in GNP (constant dollars) was 4.1 percent per year during the 1960–1972 period.

[b] Based on the 1960–1972 ratio of production growth to GNP growth.

[c] The reasons are as follows: (1) Growth rate for this industry is relatively independent of the GNP growth rate. (2) Exports will cause a slight increase in the prevailing growth rate. (3) Product is energy intensive. High energy prices will lower the otherwise expected growth in demand. (4) Unique properties of products from this industry will tend to increase demand growth. (5) Market saturation will dampen present growth rates. (6) New technological developments will decrease the otherwise expected growth in demand. (7) Competing products will lower the growth rate of production in the United States. (8) Adjusted for assumed lower population growth rate.

[d] Asphalt paving claims about half the energy used by the construction industry.

Source: Adapted from National Research Council, *Alternative Energy Demand Futures to 2010,* Committee on Nuclear and Alternative Energy Systems, Demand and Conservation Panel (Washington, D.C.: National Academy of Sciences, 1979), p. 100.

TABLE 11-4 Potential for Industrial Energy Conservation for 2010

	Agri-culture	Alu-minum	Cement	Chem-icals	Con-struction[a]	Food	Glass	Iron and Steel	Paper	Other User Industries
Conservation Effect on Consuming Industries[b]										
Basic "housekeeping"		–	–	–		–	–	–	–	–
More recycling		–			–		–		–	
Environmental controls	+		+	+		+		+	+	+
Conversion of gasoline engines to diesel	–									–
Increased yield of product	+							+	+	+
Changing product preference	+			+				+	+	+
New basic process		–	–	–		–		–	–	
More waste heat recovery		–	–	–		–	–	–	–	–

Cogeneration		−	−	−		−		−	−	
Substitute products					−	−				
Feedstock demand				+					+	
Lower-quality raw materials		+		+				+		
Conversion to electricity	+	+				+		+	+	+
Estimated Net Reduction in Energy Intensity by Energy-Consuming Industries[c]										
Scenario A	15	45	40	26[d]	42	34	31	28	36	43
Scenario B	15	37	37	22[d]	35	24	24	24	29	25
Scenario B′	15	37	37	22[d]	35	24	24	24	29	25
Scenario C	5	21	25	16[d]	27	14	18	17	24	15

[a] Asphalt only.
[b] A minus indicates reduction in energy intensity; a plus, increased energy intensity.
[c] Percent reduction or increase in energy use per unit of production.
[d] Does not include chemical feedstocks.

Source: Adapted from National Research Council, *Alternative Energy Demand Futures to 2010,* Committee on Nuclear and Alternative Energy Systems, Demand and Conservation Panel (Washington, D.C.: National Academy of Sciences, 1979), p. 104.

TABLE 11-5 40 Sectors of the Integrating Model

Extraction, Processing, Conversion of Energy	*Production of Goods and Services*
Coal mining	Agriculture
Crude petroleum, natural gas	Mining
Shale oil	Construction
Coal gasification	Food
Coal liquefaction	Paper
Refined petroleum products	Chemicals
Natural gas utilities	Glass products
Coal combined-cycle electricity	Stone and clay products
Fossil fuel electric utilities	Iron and steel
Light water reactors	Nonferrous metals
High-temperature gas-cooled reactors	Intermediate goods
Renewable energy utilities	Rail transport
	Bus transport
Energy Services, End-Uses	Truck transport
Ore-reduction feedstocks	Water transport
Chemical feedstocks	Air transport
Motive power	Wholesale and retail trades
Process heat	Other services
Water heat	Motor vehicles and equipment
Space heat	Consumer goods
Air conditioning	
Miscellaneous uses of electricity	

the base year). The Department of Commerce data were supplemented by specific data on new energy technologies and processes to highlight the end-uses of energy and to express consumption of energy in physical, rather than monetary units, as the price of energy varies with the type of purchaser. A 40-sector model was constructed to characterize the economy. Twelve sectors represent the extraction, processing, and conversion of energy resources, 8 represent energy services or end-uses of energy, and the remaining 20 represent the sectors that produce nonenergy goods and services (see Table 11-5).

The interrelation between these sectors make up a 40 $\times$ 40 matrix of input-output coefficients. A typical element A represents the input required from sector i to produce one unit of output from sector j. The matrix manifests the state of technology by indicating the energy intensity (energy needed per unit of output) for the services and products of each sector. The results of the sectoral analyses were used to define the energy intensities and technologies (described below).

The model presents a simplified picture of energy flow through the economy. By tracing this flow through the sectors, the panel was able to derive self-consistent energy consumption figures for the scenarios.

TABLE 11-6 Demand for Energy Projected by Scenario A* for 2010 (quads)[a]

Energy Form	Industry[b]	Buildings[b]	Trans- portation[b]	Total Demand[b]	Conver- sion Loss[c]	Primary Energy Input	Efficiency (percent)
Coal	9.4	0.6	0.1	10.1	—	10.1	100
Oil	8.0	1.5	10.0	19.5	2.7	22.2	88
Gas	5.8	1.6	0.1	7.5	0.5	8.0	94
Purchased electricity[d]	2.4	2.7	—	5.1	12.6	17.7	29
TOTAL	25.6	6.4	10.2	42.2	15.8	58.0	73

[a]Results are shown to three significant figures to allow display of the small quantities under Transportation.
[b]Demand for energy at point of consumption.
[c]Conversion losses include those incurred in extraction, refining, production, transmission, and distribution.
[d]Includes all energy sources projected for electricity generation—coal, oil, gas, hydroelectric, nuclear, geothermal, and solar.

The panel found it necessary to adopt some accounting conventions for this model. Total primary energy is defined as the total Btu content of the fossil fuels extracted plus the Btu equivalent of energy from hydroelectric, nuclear, solar, and geothermal sources (computed as the heat content of the coal it would take to generate the equivalent amount of electricity). Fossil fuels flow to the energy-producing sectors (for synthetic fuels and generation of electricity) and to the energy-consuming sectors. Tables 11-6 to 11-10 set out the inputs of energy to the energy-consuming sectors and the energy likely to be consumed in producing and distributing these inputs, and Table 11-11 gives figures for 1975. Adding these losses to the inputs flowing into the energy-consuming sectors yields the total primary energy input to the economy.

Each element in the 40 × 40 matrix (A) may change over the next three decades as production technologies change. Specifying all these possible changes would be a laborious task. The most significant changes (for projecting possible paths of reducing energy consumption through conservation techniques) will occur in those aspects of production technologies that have the greatest effect on demand for energy. To identify these aspects, "energy input fractions" ($g_{i,j}$ below)—each the fraction of the total energy intensity of a product from sector j (ϵ_j) that is

TABLE 11-7 Demand for Energy Projected by Scenario A for 2010 (quads)[a]

Energy Form	Industry[b]	Buildings[b]	Transportation[b]	Total Demand[b]	Conversion Loss[c]	Primary Energy Input	Efficiency (percent)
Coal	10.7	—	0.1	10.8		10.8	100
Oil	8.7	2.2	14.0	24.9	3.2	28.1	89
Gas	6.2	2.5	0.1	8.8	0.5	9.3	95
Purchased electricity[d]	2.6	4.8	—	7.4	18.0	25.4	30
TOTAL	28.2	9.5	14.2	51.9	21.7	73.6	70

[a] Results are shown to three significant figures to allow display of the small quantities under Transportation.
[b] Demand for energy at point of consumption.
[c] Conversion losses include those incurred in extraction, refining, production, transmission, and distribution.
[d] Includes all energy sources projected for electricity generation—coal, oil, gas, hydroelectric, nuclear, geothermal, and solar.

TABLE 11-8 Demand for Energy Projected by Scenario B for 2010 (quads)[a]

Energy Form	Industry[b]	Buildings[b]	Transportation[b]	Total Demand[b]	Conversion Loss[c]	Primary Energy Input	Efficiency (percent)
Coal	12.6	—	0.1	12.7	—	12.7	100
Oil	10.0	2.9	19.4	32.3	5.8	38.1	85
Gas	6.9	3.4	0.1	10.4	0.6	11.0	94
Purchased electricity[d]	3.1	6.3	—	9.4	22.7	32.1	29
TOTAL	32.6	12.6	19.6	64.8	29.1	93.9	69

[a] Results are shown to three significant figures to allow display of the small quantities under Transportation.
[b] Demand for energy at point of consumption.
[c] Conversion losses include those incurred in extraction, refining, production, transmission, and distribution.
[d] Includes all energy sources projected for electricity generation—coal, oil, gas, hydroelectric, nuclear, geothermal, and solar.

TABLE 11-9 Demand for Energy Projected by Scenario B′ for 2010 (quads)[a]

Energy Form	Industry[b]	Buildings[b]	Trans-portation[b]	Total Demand[b]	Conver-sion Loss[c]	Primary Energy Input	Efficiency (percent)
Coal	17.5	—	0.2	17.7	—	17.7	100
Oil	14.2	3.7	26.9	44.8	11.8	56.6	79
Gas	9.6	4.6	0.2	14.4	1.4	15.8	91
Purchased elec-tricity[d]	4.4	8.6	0.1	13.1	30.4	43.5	30
TOTAL	45.7	16.9	27.4	90.0	43.6	133.6	67

[a] Results are shown to three significant figures to allow display of the small quantities under Transportation.
[b] Demand for energy at point of consumption.
[c] Conversion losses include those incurred in extraction, refining, production, transmission, and distribution.
[d] Includes all energy sources projected for electricity generation—coal, oil, gas, hydroelectric, nuclear, geothermal, and solar.

TABLE 11-10 Demand for Energy Projected by Scenario C for 2010 (quads)[a]

Energy Form	Industry[b]	Buildings[b]	Trans-portation[b]	Total Demand[b]	Conver-sion Loss[c]	Primary Energy Input	Efficiency (percent)
Coal	9.6	—	0.1	9.7	—	9.7	100
Oil	9.7	4.0	25.9	39.6	9.7	49.3	81
Gas	13.9	7.0	0.1	21.0	4.8	25.8	81
Purchased elec-tricity[d]	5.8	9.3	0.1	15.2	34.8	50.0	30
TOTAL	39.0	20.3	26.2	85.5	49.3	134.8	64

[a] Results are shown to three significant figures to allow display of the small quantities under Transportation.
[b] Demand for energy at point of consumption.
[c] Conversion losses include those incurred in extraction, refining, production, transmission, and distribution.
[d] Includes all energy sources projected for electricity generation—coal, oil, gas, hydroelectric, nuclear, geothermal, and solar.

TABLE 11-11 Demand for Energy in 1975 (quads)[a]

Energy Form	Industry[b]	Buildings[b]	Transportation[b]	Total Demand[b]	Conversion Loss[c]	Primary Energy Input	Efficiency (percent)
Coal	3.8	0.2	—	4.0	—	4.0	100
Oil	5.3	5.4	16.7	27.4	2.5	29.9	92
Gas	8.6	7.6	0.6	16.8	—	16.8	100
Purchased electricity[d]	2.3	3.6	—	5.9	14.2	20.1	29
TOTAL	20.0	16.8	17.3	54.1	16.7	70.8	76

[a] Results are shown to three significant figures to allow display of the small quantities under Transportation.
[b] Demand for energy at point of consumption.
[c] Conversion losses include those incurred in extraction, refining, production, transmission, and distribution.
[d] Includes all energy sources projected for electricity generation—coal, oil, gas, hydroelectric, nuclear, geothermal, and solar.

embodied in the input from sector i (ϵ_j)—were calculated for the base year by the equation

$$g_{i,j} = \frac{\epsilon_i}{\epsilon_j} A_{i,j}$$

and ordered by rank.

Computing the energy input fractions brought several aspects of energy consumption into sharper focus. In the production of automobiles, for example, calculating the energy input fractions revealed that the energy consumed in assembly accounts for only 10 percent of the total energy cost, while more than a third of the total energy cost can be attributed to the energy used in producing steel. Thus, from the perspective of energy conservation, the size, weight, and material composition of automobiles manufactured in 2010 are more important to the energy consumption than the degree to which assembly is mechanized.

Changes in production techniques were defined by multiplying factors, applied term by term to important elements of the A matrix. Energy can be conserved in two ways in the production of goods and services: by reducing energy inputs, or by replacing these inputs with nonenergy inputs. Industry experts on the panel (and others consulted by the panel) estimated the type and extent of conservation that might be practiced as energy prices rise.

The panel selected the technological changes that seemed most plausible and assessed their effects under the conditions of various scenarios. (For a detailed description of this model and a complete set of the energy input fractions used to calculate total demand for energy in 2010, see the report of the Demand and Conservation Panel.)[19]

Demand for Electricity

Each of the resource groups of the Demand and Conservation Panel (buildings, industry, and transportation) was asked to estimate the minimum and maximum amounts of purchased electricity required by each of the basic scenarios (A^* and B' being considered variants). The estimation of demand for electricity, as stressed several times in this report, is difficult and controversial.*

The maximum and minimum figures are set out in Table 11-12. The figures for the maximum amount of purchased electricity in the buildings sector proceed from the assumption that almost all space heating in the high-energy scenarios is provided by electricity. The maximum figures for the industrial sector entail the electrification of many processes and no self-generation of electricity. For the maximum figures in the transportation sector, however, the sector is assumed to continue operating primarily on liquid fuels; major advances in the storage of electrical energy (such as high-efficiency batteries) could radically alter this projection.

The penultimate line of Table 11-12 gives the percentage of total primary energy use claimed by electricity in each projection: a range of one quarter to one half. (See also "Electricity" under "Implications of Study Results: Comparisons of Supply and Demand" in this chapter.)

Minimum Energy-Use Scenario (CLOP)[20]

The Consumption, Location, and Occupational Patterns (CLOP) Resource Group of the Synthesis Panel developed a scenario of energy consumption moderated by shifting attitudes and preferences rather than by rising prices. The group speculated that attitudes emerging in younger generations (for example, thrift in the use of all resources and a preference for environmental quality at the expense of material consumption) might dominate the choices of American society in 2010. Significant changes assumed in the CLOP scenario include major occupational shifts from the employment of individuals in large organizations to self-employment and employment in small groups; increased emphasis on recreational and leisure activities, in forms that consume little material or energy (e.g., cultural activities); and new patterns of settlement that decrease the need

*See statement 11-4, by R. H. Cannon, Jr., Appendix A.

TABLE 11-12 Projections of Maximum and Minimum Demand for Purchased Electricity in 2010 (quads)

	Scenario A[a]		Scenario B		Scenario C		
	Minimum	Maximum	Minimum	Maximum	Minimum	Maximum	1975 (actual)
Buildings	11	22	14	26	21	44	11
Industry	8	17	9	20	18	23	7
Transportation	—	4	—	4	—	4	—
TOTAL	19	43	23	50	39	71	18
Primary energy	74		94		134		
Electricity (percent of primary energy)	26	55	24	53	29	53	28
Average annual electricity growth (percent)[b]	0.2	2.5	0.7	3.0	2.2	4.0	—

[a]A major shift to electricity for ground transportation is not assumed here. However, if about half of transportation were shifted to electricity by 2010 the additional electrical demand, beginning in 1990, would be approximately 5 GWe/yr.
[b]Over the period 1975–2010.

Source: Adapted from National Research Council, *Alternative Energy Demand Futures to 2010*, Committee on Nuclear and Alternative Energy Systems, Demand and Conservation Panel (Washington, D.C.: National Academy of Sciences, 1979), p. 208.

for energy in transportation. This scenario relies heavily on sophisticated technologies to provide high-quality services and conserve exhaustible resources of energy and materials—solar energy, advanced engine designs, extensive telecommunications systems, and rationalized transportation networks.

Energy consumption in the CLOP scenario totals 53 quads in 2010—a value surprisingly close to that projected by the Demand and Conservation Panel in scenario A* (58 quads), and about half the energy consumed today on a per capita basis. The assumptions of this low-energy projection, however, are distinct. Nuclear power has been phased out, and oil and gas imports have been reduced to zero. The demand for liquid and gaseous fuels is about half that of 1975, and no coal is converted to liquid fuels. The accelerated introduction and expansion of solar technologies in the CLOP scenario result in 10 quads from this source by 2010.

The pattern of energy consumption between 1975 to 2010 was not investigated by the CLOP group.

Total Primary Energy Use

Figure 11-4 maps the various paths projected by the Demand and Conservation Panel and shows the endpoints projected by scenario A* and by the CLOP scenario for 2010. These projections were based on the results of the individual sectoral analyses for 1990 and the integrating model for 2010. They indicate how the amount of energy consumed might vary over a wide range—from 53 to 135 quads*—as a result of different prices for energy, rates of growth in GNP, the rate at which available energy-conserving technology is applied, and various public policies. While the projection of lowest growth in the use of energy (A*) assumes some changes in the preferences and life-style choices of energy consumers, these are modest. The projections of the Demand and Conservation Panel depend principally on factors that stimulate (to a greater or lesser degree) the conservation of energy by the application of known technology and practices, rather than depending on fundamental changes in consumer choices or patterns of living and working. This latter pathway to lower energy consumption was investigated qualitatively by the Consumption, Location, and Occupational Patterns Resource Group of the Synthesis Panel, as described in the previous section.

WORK OF THE SUPPLY AND DELIVERY PANEL[21]

Supply and Delivery Panel was originally expected to generate supply curves, i.e., projections of the amount of each energy form or source that could be produced as a function of time and price. However, the panel concluded that production would be so much influenced by nonprice variables that conventional supply curves would be relatively meaningless. Instead, the panel attempted practical assessments of the conditions of resource recovery and the rate of use of new production technologies, measured against the political and economic forces that influence the decisions of energy resource producers, utilities, and heavy industrial consumers of energy. The practical and judgmental nature of their investigation provides frequent counterpoints to the findings of other panels and resource groups. For example, the Demand and Conservation Panel assumed that under the conditions of their scenarios, sufficient capital would be available for investment in producing the energy

*See statement 11-5, by R. H. Cannon, Jr., Appendix A.

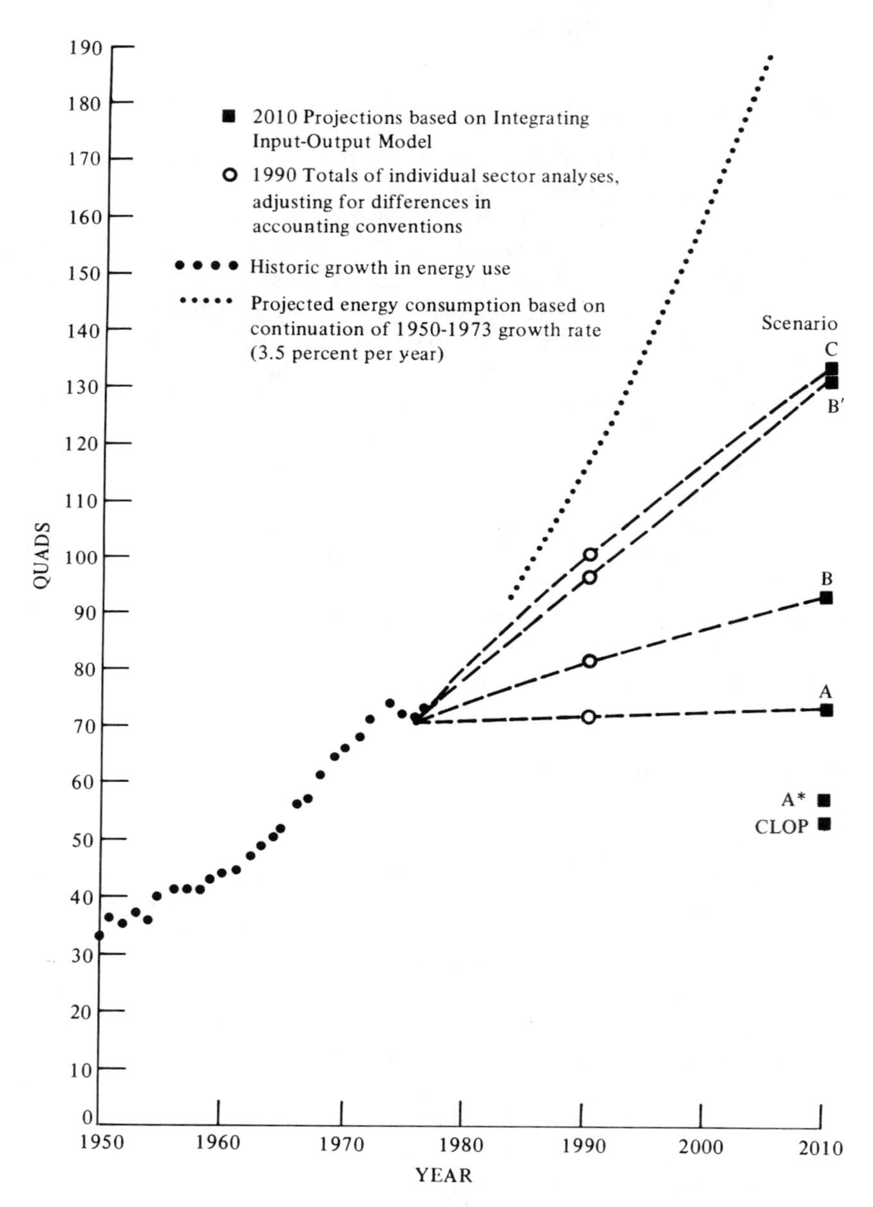

FIGURE 11-4 Projections of total demand for primary energy in the United States to 2010 (quads).

resources needed to meet demand in each scenario. The Supply and Delivery Panel replied bluntly that under prevailing conditions, financiers and potential owners would consider investment in new energy supplies to be unacceptably risky, principally because of uncertainties in government policy, regulatory requirements, and delays in acquiring licenses and permits. Thus, for example, the demands projected in scenario B′ could not be met without significant policy changes or greatly increased petroleum imports.

The committee emphasizes that the major point of both panels' work is that policies enacted to balance energy supply and demand while maintaining satisfactory economic performance must be consistent and sustained. To make the required investments in energy production and end-use capital, consumers and suppliers must be able to read these policies as clear signals encouraging, maintaining, or discouraging various levels of production or end-use. In practice, given policies are likely to be read as different signals by the producers of each energy resource. The Supply and Delivery Panel asked each of its resource groups to estimate the production of its resource or energy form under three sets of conditions and assumptions: business as usual, enhanced supply, and national commitment. The specific conditions assumed for each of these categories varied with the resource. In general, the business-as-usual estimates assumed that existing policies and practices[22] carry over into the future with little change. Enhanced-supply estimates assumed that promising new supply technologies are encouraged by policies effected for energy resource production; for example, relaxing environmental standards, or reducing and expediting required regulatory processes. The national-commitment estimates assumed that the policies and regulatory actions of enhanced supply are pursued with urgency. It must be emphasized that policies of national commitment cannot be effected simultaneously for all supply sources. Giving very high priority to the development of one source necessarily implies lower priority and less stimulus to the development of other sources.[23]

The Supply and Delivery Panel's resource groups reviewed available estimates of reserves and resources, as well as their producibility, and selected those that matched their consensus judgment. In addition, they considered the availability of the various energy supply technologies during the period to 2010 for each of the three sets of assumed conditions. Table 11-13 displays the estimates of each resource group under the three categories of policy.

The specific assumptions made by each group are set out in Table 11-14. The Supply and Delivery Panel emphasizes that neither table should be read as representing self-consistent sets of possibilities. Many of these estimates presume success with technologies that either have not yet been tested or have been demonstrated only on small scales. It is possible that some of the prospective technologies on which these estimates are based cannot be successfully developed with any level of commitment. On the other hand, new supplies from unexpected discoveries (of unconventional gas, for example) or from unexpected technical progress in certain technologies (such as solar photovoltaics or coal liquefaction) could outstrip specific estimates.

Maximum-Solar Scenario

To analyze the extent to which solar energy can substitute for other fuels, a "maximum solar" scenario was developed by the Solar Resource Group of the Supply and Delivery Panel.[24] It represents an estimate of the maximum feasible application of solar energy between now and 2010, from the point of view of technical, rather than economic, feasibility. The group concluded that increased energy prices, within the range considered by the study, would not be sufficient to stimulate widespread use of solar energy by 2010. Heavy subsidies and government mandates would be needed to encourage installation of solar energy systems in that period to the extent described by the maximum-solar scenario.

This scenario assumes that a national policy decision has been made by 1985 to encourage the use of solar energy in all its applications. The government mandates that solar energy be used to supply heat, air conditioning, and hot water to all new buildings, and to supply industrial process heat where technically feasible. A schedule is set in motion for several solar electric central generating stations, and state and local governments are ordered to adopt as rapidly as possible technologies for converting municipal and agricultural wastes to useful energy. These installations might be financed in part from revenues received from taxing nonrenewable energy sources, a practice that would also help make the solar installations more competitive.

Estimates for the maximum potential application of solar energy are set out in Table 11-15. These numbers represent an effective upper bound on solar energy production. Whether this quantity and mix could be effectively used depends on prevailing conditions. The panel made no attempt to estimate the cost of achieving this projection. More details about this scenario and its context in the study are given in chapter 6.

CONAES Projections of Supply Versus Other Projections

For ready comparison, the estimates of the Supply and Delivery Panel for the year 2000 and the year 2010 are compared to the most recent midrange

TABLE 11-13 Domestic Energy Production for Three Sets of Assumed Conditions (quads per year)[a]

Scenario and Energy Source	Year 1977	1990	2000	2010
Business as usual[b]				
Crude oil	19.6	16.0	12.0	6.0
Natural gas	19.4	10.3	7.0	5.0
Oil shale	0	0	0	0
Synthetic liquids[c]	0	(0.3)	(2.3)	(6.1)
Synthetic gas[c]	0	(1.3)	(3.5)	(4.1)
Coal	16.4	25.0	34.0	42.0
Geothermal	0	0.4	0.9	2.4
Solar	0	0	0.1	0.6
Nuclear	2.7	10.0	12.5	15.8
Hydroelectric	2.4	4.0	5.0	5.0
Enhanced supply[b]				
Crude oil	19.6	20.0	18.0	16.0
Natural gas	19.4	15.8	15.0	14.0
Oil shale	0	0.7	1.0	1.5
Synthetic liquids[c]	0	(0.4)	(2.4)	(8.0)
Synthetic gas[c]	0	(1.7)	(3.5)	(4.8)
Coal	16.4	26.6	37.2	49.5
Geothermal	0	0.6	1.6	4.1
Solar	0	1.7	5.9	10.7
Nuclear	2.7	13.0	29.5	41.7
Hydroelectric	2.4	4.1	5.0	5.0
National commitment[b]				
Crude oil	19.6	21.0	20.0	18.0
Natural gas	19.4	18.0	17.0	16.0
Oil shale	0	2.0	2.5	3.0
Synthetic liquids[c]	0	(0.7)	(4.7)	(12.9)
Synthetic gas[c]	0	(1.7)	(4.5)	(7.9)
Coal	16.4	32.5	75.0	100.0
Geothermal	0	2.2	7.8	19.9
Solar	0	3.3	13.1	28.8
Nuclear	2.7	12.0	27.5	42.5
Hydroelectric	2.4	4.1	5.0	5.0

[a]Except for the business-as-usual projections, the entries in this table should not be added to obtain yearly totals; no more than a very few energy sources or technologies could be simultaneously accorded the priorities implied by the enhanced-supply or national-commitment scenarios.

[b]For specific assumptions guiding selection of estimates under this set of conditions, see Table 11-14.

[c]Synthetic fuels are produced from coal and oil shale and are not added in the totals.

Source: Compiled from National Research Council, *U.S. Energy Supply Prospects to 2010,* Committee on Nuclear and Alternative Energy Systems, Supply and Delivery Panel (Washington, D.C.: National Academy of Sciences, 1979).

TABLE 11-14 Assumptions Specific to Energy Resource Estimates

Energy Source	Scenario[a]		
	Business as Usual	Enhanced Supply	National Commitment
Coal	Production limited by difficulties leasing federal land; increasing costs and delays from lack of consistent environmental policies	Increased demand from enactment of consistent environmental policies	Rising worker productivity; availability of more capital; streamlined regulatory policies
Oil and gas	Price controls discourage domestic production; separate permits required for exploration and production in outer continental shelf; delays in leasing; withdrawal of public lands	Accelerated federal offshore leasing; lifting of controls on wellhead prices; streamlined permit processes; improved exploration and production technologies	Relaxation of Clean Air standards; streamlined procedures for environmental impact statements; federal loan guarantees for development and application of new technologies; federal return of withdrawn lands; assignment of priority status to materials and labor for oil exploration and recovery
Coal-based synthetics (oil and gas)	Federal nonrecourse loans for a limited number of new plants: design and construction of such plants takes 6 years (production in seventh year)	Permits for mining, plant construction, and operation acted upon within 12 months: design and construction of such plants takes 5 years	Government agency established to underwrite prices, expedite approval of permits, act to ensure priority assignment of labor, capital, and materials, and so on: design and construction of plants takes 4 years

Solar	No policies enacted to promote solar energy; solar energy finds few applications by 2010	Policies enacted to encourage some technologies (water heating and passive solar design); bioconversion only for municipal wastes; some central generation of electricity	National policy to foster the use of solar energy; federal intervention to accelerate development of demand; application of solar technologies required in new buildings and suitable industrial processes; mandatory conversion of municipal and agricultural wastes
Nuclear	No significant improvement in existing policies and practices; continuing uncertainty about waste disposal and reprocessing; no development work on advanced reactor concepts (except liquid-metal fast breeder reactor (LMFBR)); introduction of LMFBR stalled by unavailability of fuel and lack of sufficient preparation; higher prices for uranium	Major streamlining of regulatory policies to reduce time required for nuclear power plants and fuel cycle facilities; national policy favoring reprocessing and recycling of spent fuels; improved efficiency of light water reactors	Policies enacted to reduce uncertainties and streamline regulatory policies and practices; reprocessing and recycling of spent reactor fuel allowed; disposal of radioactive waste licensed and practiced; breeder development and demonstration accelerated and work sustained on other advanced reactor concepts

[a]Hydroelectric and geothermal projections not made on comparable bases. For geothermal energy, progressively higher projections imply progressively more favorable regulatory policies and industrial patterns, as well as more successful research and development. For hydroelectric generation, growth depends very little on the regulatory and financial climate, because the number of ecologically acceptable sites is limited.

Source: Compiled from National Research Council, *U.S. Energy Supply Prospects to 2010,* Committee on Nuclear and Alternative Energy Systems, Supply and Delivery Panel (Washington, D.C.: National Academy of Sciences, 1979).

projection (Series C) of the Energy Information Administration (EIA)[25] in Table 11-16. The EIA projections were made in 1978.

The EIA projections for both oil and gas, assuming enhanced-recovery techniques, are above even the national-commitment scenario of the Supply and Delivery Panel—about 30 percent above in the case of oil, and 20 percent above in the case of gas. The projection of oil and gas production in the year 2000, including enhanced recovery and all synthetics, totals 62.7 quads in the Series C projections of EIA. For 2010, the production of oil and gas, including all synthetics, totals 67.0 quads in the Series C projection. The Supply and Delivery Panel projects oil and gas production of 50.4 quads in 2000 and 58.3 quads in 2010 under the conditions of the national-commitment scenario. The CONAES projections for fluid fuels appear conservative in this context. Nevertheless, additions to domestic oil reserves in 1978 were only 1.3 billion barrels, compared to production of 3.2 billion barrels. To maintain domestic production of oil through the 1980s would require an average annual finding rate of 4.0 billion barrels.[26]

Several additional points of comparison emerge from inspection of Table 11-16:

- The EIA projections for coal-derived synthetic liquids agree quite well with those of the enhanced-supply scenario for both 2000 and 2010.
- The EIA projections for oil shale are near those of the national-commitment scenario, particularly for 2010.
- If the high- and low-Btu gas projections cast by the EIA are combined, the total corresponds very closely to the national-commitment projections for synthetic gas. Again we see that the CONAES supply projections are on the pessimistic side.
- The EIA nuclear projections lie midway between the business-as-usual and enhanced-supply projections.
- The EIA coal projections lie between the enhanced-supply and national-commitment projections of the Supply and Delivery Panel—closer to enhanced supply.

IMPLICATIONS OF STUDY RESULTS: COMPARISONS OF SUPPLY AND DEMAND

Attempts to draw on the understanding and data emerging from one another's work were important aspects of the work of the several panels and resource groups. In attempting to specify the mix of energy sources appropriate to their scenarios of demand, for example, the Demand and Conservation Panel consulted with members of the Supply and Delivery

TABLE 11-15 Maximum-Solar Scenario

	Contribution (quads)	
Solar Application or Source	1990	2010
Space heat, hot water, and nonresidential air conditioning	0.7	4.4
Municipal wastes	0.8	1.9
Agricultural residues	0.9	3.5
Solar and wind electricity	0.6	12.4
Industrial process heat	0.4	6.6
TOTAL	3.4	28.8

Source: Compiled from National Research Council, *U.S. Energy Supply Prospects to 2010,* Committee on Nuclear and Alternative Energy Systems, Supply and Delivery Panel (Washington, D.C.: National Academy of Sciences, 1979).

Panel. The interaction of supply and demand began with final demand numbers for electricity and for gaseous and liquid fuels. The Supply and Delivery Panel reviewed these demand numbers and rejoined with recommendations modifying the characteristics of demand (asking, for example, "Can some of these demands for liquids be met by electricity?"). The resulting accommodations eventually produced the mix presented in the report of the Demand and Conservation Panel (summarized in Tables 11-6 to 11-10).

The committee staff carried these preliminary attempts further to generate integrated scenarios of supply and demand for the year 2010. These provide numerical examples that illustrate some qualitative conclusions of the study. Although these examples are presented in tables and graphs, they should not be interpreted as predictions, nor can any single scenario be regarded as a preferred path for the next three decades. *There was sufficient disagreement among CONAES members about what was socially desirable and politically feasible to preclude the development of any "most likely" or "most desirable" scenario.* The numerical values were arrived at judgmentally and should not be regarded as the outcome of a complete chain of inference from a formal model and assumptions, although models were used as a partial guide to judgment.

TABLE 11-16 Comparison of Energy-Supply Projections to Midrange Projections of Energy Information Administration (quads)

Energy Source	Supply and Delivery Panel			Energy Information Administration
	Business as Usual	Enhanced Supply	National Commitment	
1990				
Coal	25.0	26.6	32.5	31.2
Oil	16.0	20.0	21.0	23.1
Gas	10.3	15.8	18.0	17.4
Nuclear	10.0	13.0	12.0	9.4
2000				
Coal	34.0	37.2	75.0	46.9
Oil (1)				17.8
Oil (2)	12.0	18.0	20.0	23.2
Shale	0.0	1.0	2.5	1.8
Synthetic liquids	2.3	2.4	4.7	2.6
Gas (1)				15.8
Gas (2)	7.0	15.0	17.0	19.2
High-Btu gas	3.5	3.5	4.5	0.4
Low-Btu gas				4.1
Nuclear	12.5	29.5	27.5	16.9
2010				
Coal	42.0	49.5	100.0	65.2
Oil (1)				13.9
Oil (2)	6.0	16.0	18.0	19.6
Shale	0.0	1.5	3.0	3.6
Synthetic liquids	6.1	8.0	12.9	7.2
Gas (1)				11.6
Gas (2)	5.0	14.0	16.0	15.2
High-Btu gas	4.1	4.8	7.9	1.6
Low-Btu gas				6.8
Nuclear	15.8	41.7	42.5	29.2

Sources: National Research Council, *U.S. Energy Supply Prospects to 2010,* Committee on Nuclear and Alternative Energy Systems, Supply and Delivery Panel (Washington, D.C.: National Academy of Sciences, 1979); and U.S. Department of Energy, Energy Information Administration, *Annual Report to Congress, 1978,* vol. 3, *Forecasts* (Washington, D.C.: U.S. Department of Energy, 1979).

THE STUDY SCENARIOS[27]

To compare the projections of energy demand from the Demand and Conservation Panel with the projections of energy supplies from the Supply and Delivery Panel, a variant of the demand projections was prepared by scaling demand scenarios A*, A, and C for 3 percent annual average growth of GNP, based on the comparison of demand scenarios B and B′. The study scenarios are distinguished by roman numerals (low to high consumption of energy) and subscripts (2 or 3 percent growth of GNP). Study scenario I_2 corresponds to demand scenario A*, for example, and study scenarios III_2 and III_3 correspond to demand scenarios B and B′.

To prepare the projections for comparison with those of the Supply and Delivery Panel, the following changes were made.

1. The demand projections include an 8 percent loss in the distribution of oil. These losses were added to the total figure for demand at the point of consumption to yield total primary demand for oil.
2. The demand projections include a minimum and a maximum figure for electricity purchased by industry, corresponding to maximum and minimum figures (respectively) for the cogeneration and self-generation of electricity by industry (Table 11-12). The study scenarios assume industrial demand for electricity approximately at the midpoint of these two projections. The corresponding decrease in primary fuel use shows up in the demand-supply comparisons principally as lower demand for coal by industry.
3. The figures for substitution of solar energy for direct use of fuels and electricity in buildings and industry, and the figures for the use of "other" sources in industry (such as sawdust burned in paper mills), were taken from the sectoral analyses of the Demand and Conservation Panel and were added to the projections.
4. Following the suggestion of the Industry Resource Group of the Demand and Conservation Panel that gas and oil are readily interchangeable in about 50 percent of industrial applications, the study scenarios assume that industry will use natural gas to fill somewhat less than 50 percent of its projected demand for oil (the percentage varies with scenario, depending on the relative prices and availability of these fuels).

The basic principles guiding the comparison of demand and supply scenarios were (1) to assume with the Demand and Conservation Panel that energy-efficient technology would be adopted over the period 1975–2010 to the extent economically justified under the assumed schedules of price, and (2) to fill scenarios of demand from the Supply and Delivery Panel scenarios that best matched the assumed conditions. For example, the scenarios of low demand were generally compared to business-as-usual scenarios of supply, unless the demand for a particular energy form was greater than business-as-usual supply projections could meet.

Although the assignment of production levels from the Supply and Delivery Panel's scenarios to the Demand and Conservation Panel's scenarios of demand was an exercise of judgment, there is actually less room for maneuver and arbitrariness than might be supposed. Coal and nuclear fuels can be substituted for one another in the generation of electricity, for example, but because of competitive demands for coal to produce synthetic fluids, the growth of coal consumption must be very large if some growth in nuclear power is not also maintained (except in the cases of lowest assumed growth in demand). *A more significant finding of this exercise is that the demand for fluid fuels cannot be met by business-as-usual supply projections, even in the cases of the lowest projected growth in demand, without rising imports.* The scenarios of high demand require energy production levels from the Supply and Delivery Panel's national-commitment scenarios. The complete set of scenarios is displayed in Tables 11-17 to 11-24.

It is convenient for discussion to classify these scenarios in four groups—"low energy consumption" (CLOP, I_2, I_3, II_2), "low-medium" (II_3 and IV_2), "high-medium" (III_3 and IV_2), and "high" (IV_3). Their range in 2010 and the courses of their development are illustrated in Figures 11-5 and 11-6. To indicate their range, this section compares for discussion study scenarios I_2, II_2, III_2, and III_3, and IV_3.

Nuclear Power

The demand for nuclear energy was met by business-as-usual levels of installation and operation in scenarios I_2 and II_2, by enhanced-supply conditions in scenarios III_2 and III_3, and by national-commitment conditions in scenario IV_3. The Supply and Delivery Panel has expressed the opinion that the nuclear industry possesses the physical and organizational capacity to deploy considerably more nuclear power than is required even for scenario IV_3. In the scenarios developed by the Supply and Delivery Panel, the achievable growth rate of installed nuclear capacity increases rapidly after 1990 (as indicated in Table 11-13).

The Demand and Conservation Panel's projections of demand for electricity, on the other hand, show much more rapid growth in demand for electricity before 1990 than after (as indicated in chapter 2). Thus, the study scenarios do not require nearly so much installed nuclear capacity in 2010 as the Supply and Delivery Panel states is possible. In low-consumption scenarios I_2 and II_2, the use of nuclear power actually declines after 1990, following the general trend of demand for electricity. In medium-growth scenario III_2, the use of nuclear power remains fairly constant after 1990, again following the trend of demand for electricity. In high-medium and high-consumption scenarios III_3 and IV_3, nuclear power continues to grow after 1990, but at a rate below the maximum considered possible by the Supply and Delivery Panel.

TABLE 11-17 Fuel Mix Projected by Study Scenario I_2 (quads per year)

Energy Source	1975	1990	2010	Required Supply Conditions[a]
Oil				
Domestic	20	18	11	BAU, ES
Imported	13	5	12	—
Shale	0	0	0	BAU
Gas				
Domestic	19	13	8	BAU, ES
Imported	1	0	0	—
Coal				
Combustion	13	20	15	BAU
Conversion to synthetic liquid	0	0	0	BAU
Conversion to synthetic gas	0	0	0	BAU
Nuclear	2	8	6	BAU
Solar	0	1	6	ES
Other (hydro, geothermal, etc.)	3	5	6	BAU
TOTAL	71	70	64	
TOTAL, liquid fuels[b]	33	23	23	
TOTAL, gaseous fuels[b]	20	13	8	
TOTAL, electricity[c]	20	25	17	

[a]The Supply and Delivery Panel based its estimates of energy source availability on sets of assumptions about regulatory policies and public attitudes, which were judged more likely to determine availability than cost or price. These assumptions are as follows: Business as usual (BAU)—existing attitudes, policies, and practices are extended into the future with little change; integrated, effective energy supply policies are not established and implemented. Enhanced supply (ES)—a well-balanced, comprehensive set of energy supply policies is enacted and aggressively pursued; decision making and regulatory actions are timely and coordinated; and promising new technologies are appropriately supported. National commitment (NC)—the same comprehensive set of energy policies is pursued as in the enhanced-supply case, but more aggressively in specific areas; adequate energy supplies are given the highest priority in allocating national resources; and calculated risks are taken in deploying promising new energy technologies before they are economically practicable.
[b]Includes losses in production and distribution, but not conversion for synthetic fuels derived from coal.
[c]Includes conversion losses.

Electricity

As noted earlier in this chapter, there is considerable uncertainty today in the projection of demand for electricity. If the demand for electricity in 2010 were to fall near the upper end of the range projected by the Demand and Conservation Panel, significantly larger installed nuclear capacity

TABLE 11-18 Fuel Mix Projected by Study Scenario I_3 (quads per year)

Energy Source	1975	1990	2010	Required Supply Conditions[a]
Oil				
Domestic	20	21	18	NC
Imported	13	8	13	—
Shale	0	0	0	BAU
Gas				
Domestic	19	14	11	BAU, ES
Imported	1	1	0	—
Coal				
Combustion	13	27	25	BAU
Conversion to synthetic liquid	0	0	0	BAU
Conversion to synthetic gas	0	0	0	BAU
Nuclear	2	8	6	BAU
Solar	0	1	6	ES
Other (hydro, geothermal, etc.)	3	5	6	BAU
TOTAL	71	85	85	
TOTAL, liquid fuels[b]	33	29	31	
TOTAL, gaseous fuels[b]	20	15	11	
TOTAL, electricity[c]	20	31	23	

[a] The Supply and Delivery Panel based its estimates of energy source availability on sets of assumptions about regulatory policies and public attitudes, which were judged more likely to determine availability than cost or price. These assumptions are as follows: Business as usual (BAU)—existing attitudes, policies, and practices are extended into the future with little change; integrated, effective energy supply policies are not established and implemented. Enhanced supply (ES)—a well-balanced, comprehensive set of energy supply policies is enacted and aggressively pursued; decision making and regulatory actions are timely and coordinated; and promising new technologies are appropriately supported. National commitment (NC)—the same comprehensive set of energy policies is pursued as in the enhanced-supply case, but more aggressively in specific areas; adequate energy supplies are given the highest priority in allocating national resources; and calculated risks are taken in deploying promising new energy technologies before they are economically practicable.

[b] Includes losses in production and distribution, but not conversion for synthetic fuels derived from coal.

[c] Includes conversion losses.

would be required, and nuclear growth rates closer to those projected by the Supply and Delivery Panel would be shown by the scenarios. For example, the demand for electricity in scenario IV_3 requires 71 quads of primary energy input, of which 30 quads (corresponding to 600 gigawatts (electric) (GWe) of installed capacity) is provided by nuclear power—only slightly less than the enhanced-supply scenario value of 35 quads.

TABLE 11-19 Fuel Mix Projected by Study Scenario II_2 (quads per year)

Energy Source	1975	1990	2010	Required Supply Conditions[a]
Oil				
Domestic	20	18	11	BAU, ES
Imported	13	5	10	—
Shale	0	0	1	BAU, ES
Gas				
Domestic	19	13	10	BAU, ES
Imported	1	0	3	—
Coal				
Combustion	13	25	22	BAU
Conversion to synthetic liquid	0	0	6	BAU
Conversion to synthetic gas	0	0	0	BAU
Nuclear	2	8	7	BAU
Solar	0	1	4	ES
Other (hydro, geothermal, etc.)	3	6	9	ES
TOTAL	71	76	83	
TOTAL, liquid fuels[b]	33	23	26	
TOTAL, gaseous fuels[b]	20	13	13	
TOTAL, electricity[c]	20	31	29	

[a]The Supply and Delivery Panel based its estimates of energy source availability on sets of assumptions about regulatory policies and public attitudes, which were judged more likely to determine availability than cost or price. These assumptions are as follows: Business as usual (BAU)—existing attitudes, policies, and practices are extended into the future with little change; integrated, effective energy supply policies are not established and implemented. Enhanced supply (ES)—a well-balanced, comprehensive set of energy supply policies is enacted and aggressively pursued; decision making and regulatory actions are timely and coordinated; and promising new technologies are appropriately supported. National commitment (NC)—the same comprehensive set of energy policies is pursued as in the enhanced-supply case, but more aggressively in specific areas; adequate energy supplies are given the highest priority in allocating national resources; and calculated risks are taken in deploying promising new energy technologies before they are economically practicable.

[b]Includes losses in production and distribution, but not conversion for synthetic fuels derived from coal.

[c]Includes conversion losses.

Scenarios of higher electrical demand, calling for even more nuclear capacity, could be envisaged. The Edison Electric Institute, for example, projects a demand for electrical generation of almost 80 quads in the year 2000.[28] Under these assumed conditions, the national-commitment level of nuclear capacity would be required.

An illustrative example of higher electricity demand can be envisioned

TABLE 11-20 Fuel Mix Projected by Study Scenario II_3 (quads per year)

Energy Source	1975	1990	2010	Required Supply Conditions[a]
Oil				
Domestic	20	21	18	NC
Imported	13	7	11	—
Shale	0	0	2	ES
Gas				
Domestic	19	15	14	ES
Imported	1	0	2	—
Coal				
Combustion	13	31	35	BAU
Conversion to synthetic liquid	0	0	9	BAU, ES
Conversion to synthetic gas	0	0	3	BAU
Nuclear	2	8	8	BAU
Solar	0	1	4	ES
Other (hydro, geothermal, etc.)	3	6	9	ES
TOTAL	71	89	115	
TOTAL, liquid fuels[b]	33	28	37	
TOTAL, gaseous fuels[b]	20	15	18	
TOTAL, electricity[c]	20	36	39	

[a]The Supply and Delivery Panel based its estimates of energy source availability on sets of assumptions about regulatory policies and public attitudes, which were judged more likely to determine availability than cost or price. These assumptions are as follows: Business as usual (BAU)—existing attitudes, policies, and practices are extended into the future with little change; integrated, effective energy supply policies are not established and implemented. Enhanced supply (ES)—a well-balanced, comprehensive set of energy supply policies is enacted and aggressively pursued; decision making and regulatory actions are timely and coordinated; and promising new technologies are appropriately supported. National commitment (NC)—the same comprehensive set of energy policies is pursued as in the enhanced-supply case, but more aggressively in specific areas; adequate energy supplies are given the highest priority in allocating national resources; and calculated risks are taken in deploying promising new energy technologies before they are economically practicable.

[b]Includes losses in production and distribution, but not conversion for synthetic fuels derived from coal.

[c]Includes conversion losses.

by varying the assumptions of the study scenarios. Suppose that half the demand for fluid fuels in the buildings sector were replaced by demand for electrical resistance heat. (The projections used for the base cases in this example are study scenarios that assume 3 percent annual average growth of GNP, as shown in Table 11-25, rather than the scenarios of the Demand and Conservation Panel.) In scenario I_3, the primary demand for liquid

TABLE 11-21 Fuel Mix Projected by Study Scenario III_2 (quads per year)

Energy Source	1975	1990	2010	Required Supply Conditions[a]
Oil				
Domestic	20	20	16	ES
Imported	13	9	7	—
Shale	0	0	1	ES
Gas				
Domestic	19	14	14	ES
Imported	1	0	2	—
Coal				
Combustion	13	24	26	BAU
Conversion to synthetic liquid	0	1	12	ES
Conversion to synthetic gas	0	0	0	BAU
Nuclear	2	11	13	ES
Solar	0	1	3	ES
Other (hydro, geothermal, etc.)	3	5	8	ES
TOTAL	71	85	102	
TOTAL, liquid fuels[b]	33	29	32	
TOTAL, gaseous fuels[b]	20	14	16	
TOTAL, electricity[c]	20	33	37	

[a]The Supply and Delivery Panel based its estimates of energy source availability on sets of assumptions about regulatory policies and public attitudes, which were judged more likely to determine availability than cost or price. These assumptions are as follows: Business as usual (BAU)—existing attitudes, policies, and practices are extended into the future with little change; integrated, effective energy supply policies are not established and implemented. Enhanced supply (ES)—a well-balanced, comprehensive set of energy supply policies is enacted and aggressively pursued; decision making and regulatory actions are timely and coordinated; and promising new technologies are appropriately supported. National commitment (NC)—the same comprehensive set of energy policies is pursued as in the enhanced-supply case, but more aggressively in specific areas; adequate energy supplies are given the highest priority in allocating national resources; and calculated risks are taken in deploying promising new energy technologies before they are economically practicable.

[b]Includes losses in production and distribution, but not conversion for synthetic fuels derived from coal.

[c]Includes conversion losses.

and gaseous fuels in buildings totals 10.5 quads. If the efficiency of building heat averages 60 percent (i.e., 40 percent of the heat is lost up the chimney), then the delivered useful heat totals 0.6 × 10.5 quads, or 6.30 quads. If half this amount were supplied by electrical resistance heat, the requirement would be one half of 6.30 quads, or 3.15 quads of electricity.

TABLE 11-22 Fuel Mix Projected by Study Scenario III_3 (quads per year)

Energy Source	1975	1990	2010	Required Supply Conditions[a]
Oil				
Domestic	20	21	18	NC
Imported	13	16	14	—
Shale	0	0	2	ES
Gas				
Domestic	19	14	14	ES
Imported	1	1	1	—
Coal				
Combustion	13	29	34	ES, NC
Conversion to synthetic liquid	0	2	19	NC
Conversion to synthetic gas	0	1	7	ES
Nuclear	2	11	18	ES
Solar	0	1	3	ES
Other (hydro, geothermal, etc.)	3	5	10	ES
TOTAL	71	101	140	
TOTAL, liquid fuels[b]	33	38	47	
TOTAL, gaseous fuels[b]	20	16	20	
TOTAL, electricity[c]	20	38	48	

[a]The Supply and Delivery Panel based its estimates of energy source availability on sets of assumptions about regulatory policies and public attitudes, which were judged more likely to determine availability than cost or price. These assumptions are as follows: Business as usual (BAU)—existing attitudes, policies, and practices are extended into the future with little change; integrated, effective energy supply policies are not established and implemented. Enhanced supply (ES)—a well-balanced, comprehensive set of energy supply policies is enacted and aggressively pursued; decision making and regulatory actions are timely and coordinated; and promising new technologies are appropriately supported. National commitment (NC)—the same comprehensive set of energy policies is pursued as in the enhanced-supply case, but more aggressively in specific areas; adequate energy supplies are given the highest priority in allocating national resources; and calculated risks are taken in deploying promising new energy technologies before they are economically practicable.

[b]Includes losses in production and distribution, but not conversion for synthetic fuels derived from coal.

[c]Includes conversion losses.

Meeting this demand would require 6.4 quads of waste heat, or a total of 9.5 quads of primary energy equivalent. Thus, the demand for 30.2 quads of primary electricity in 1990 in the buildings sector would rise to 39.7 quads.

For this calculation, it was assumed that the switch to electricity is

TABLE 11-23 Fuel Mix Projected by Study Scenario IV_2 (quads per year)

Energy Source	1975	1990	2010	Required Supply Conditions[a]
Oil				
Domestic	20	21	18	NC
Imported	13	14	14	—
Shale	0	0	2	ES
Gas				
Domestic	19	16	14	ES
Imported	1	0	3	—
Coal				
Combustion	13	25	31	ES
Conversion to synthetic liquid	0	1	19	NC
Conversion to synthetic gas	0	3	7	ES
Nuclear	2	12	20	ES, NC
Solar	0	0	2	BAU, ES
Other (hydro, geothermal, etc.)	3	7	10	ES
TOTAL	71	99	140	
TOTAL, liquid fuels[b]	33	35	47	
TOTAL, gaseous fuels[b]	20	18	22	
TOTAL, electricity[c]	20	39	52	

[a]The Supply and Delivery Panel based its estimates of energy source availability on sets of assumptions about regulatory policies and public attitudes, which were judged more likely to determine availability than cost or price. These assumptions are as follows: Business as usual (BAU)—existing attitudes, policies, and practices are extended into the future with little change; integrated, effective energy supply policies are not established and implemented. Enhanced supply (ES)—a well-balanced, comprehensive set of energy supply policies is enacted and aggressively pursued; decision making and regulatory actions are timely and coordinated; and promising new technologies are appropriately supported. National commitment (NC)—the same comprehensive set of energy policies is pursued as in the enhanced-supply case, but more aggressively in specific areas; adequate energy supplies are given the highest priority in allocating national resources; and calculated risks are taken in deploying promising new energy technologies before they are economically practicable.

[b]Includes losses in production and distribution, but not conversion for synthetic fuels derived from coal.

[c]Includes conversion losses.

accomplished after 1990. The second set of figures in Table 11-25 shows the reduced requirement for fluid fuels. In Table 11-25(A), the annual percent change in demand for electricity is given between the column for 1990 and 2010. The figures in parentheses in Table 11-25(B) represent the fluids requirements for buildings alone in the base case and the higher-

TABLE 11-24 Fuel Mix Projected by Study Scenario IV_3 (quads per year)

Energy Source	1975	1990	2010	Required Supply Conditions[a]
Oil				
Domestic	20	21	18	NC
Imported	13	20	27	—
Shale	0	0	3	NC
Gas				
Domestic	19	18	16	NC
Imported	1	0	6	—
Coal				
Combustion	13	31	42	NC
Conversion to synthetic liquid	0	1	19	NC
Conversion to synthetic gas	0	3	12	NC
Nuclear	2	12	30	NC
Solar	0	0	2	BAU, ES
Other (hydro, geothermal, etc.)	3	7	13	ES
TOTAL	71	113	188	
TOTAL, liquid fuels[b]	33	42	61	
TOTAL, gaseous fuels[b]	20	20	30	
TOTAL, electricity[c]	20	45	71	

[a] The Supply and Delivery Panel based its estimates of energy source availability on sets of assumptions about regulatory policies and public attitudes, which were judged more likely to determine availability than cost or price. These assumptions are as follows: Business as usual (BAU)—existing attitudes, policies, and practices are extended into the future with little change; integrated, effective energy supply policies are not established and implemented. Enhanced supply (ES)—a well-balanced, comprehensive set of energy supply policies is enacted and aggressively pursued; decision making and regulatory actions are timely and coordinated; and promising new technologies are appropriately supported. National commitment (NC)—the same comprehensive set of energy policies is pursued as in the enhanced-supply case, but more aggressively in specific areas; adequate energy supplies are given the highest priority in allocating national resources; and calculated risks are taken in deploying promising new energy technologies before they are economically practicable.
[b] Includes losses in production and distribution, but not conversion for synthetic fuels derived from coal.
[c] Includes conversion losses.

electricity demand case. The last column of Table 11-25(A) represents the total electrical capacity required for each case (based on 20 GWe/quad, which allows for reserve capacity). If the additional electricity were supplied by nuclear power plants, the situation in 2010 would be as shown

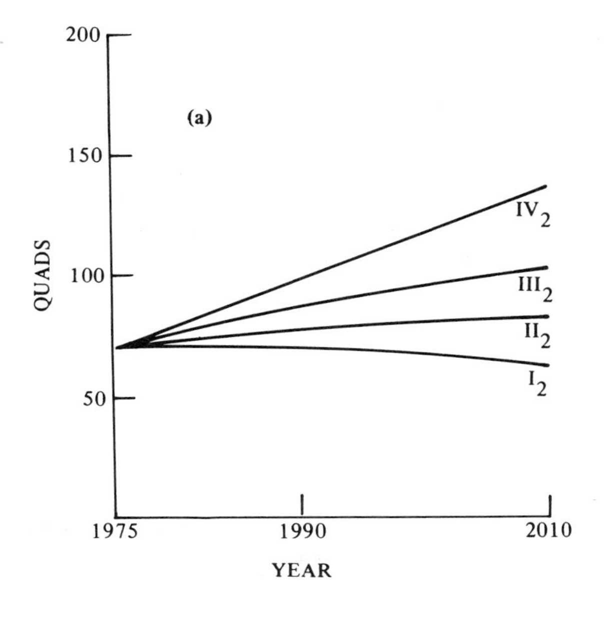

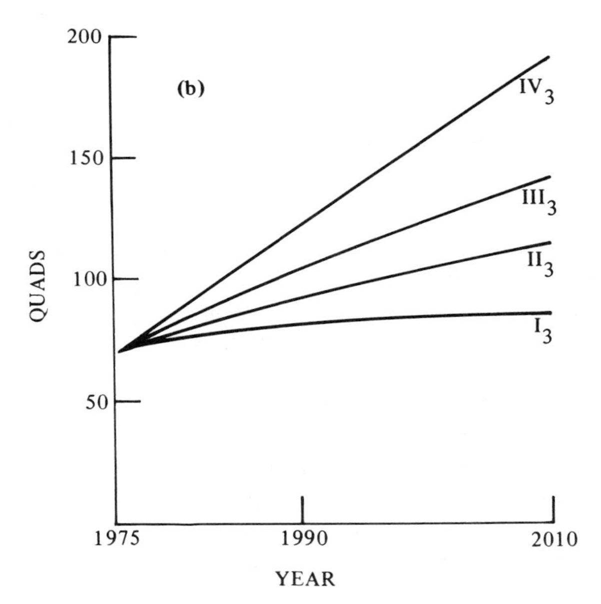

FIGURE 11-5 Projections of total primary energy consumption to 2010 for CONAES study scenarios (quads), with assumed (a) 2 percent GNP growth and (b) 3 percent GNP growth.

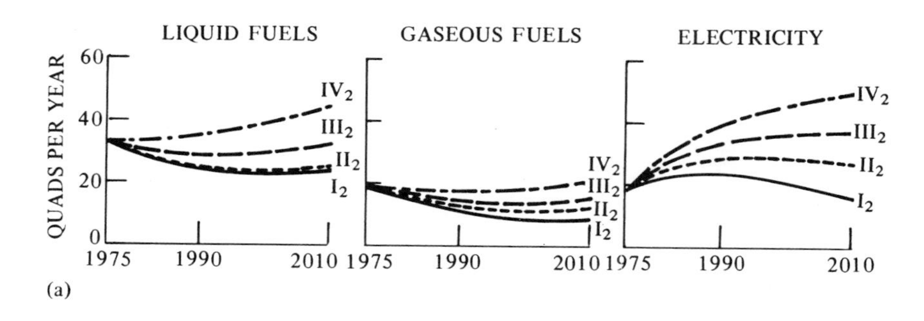

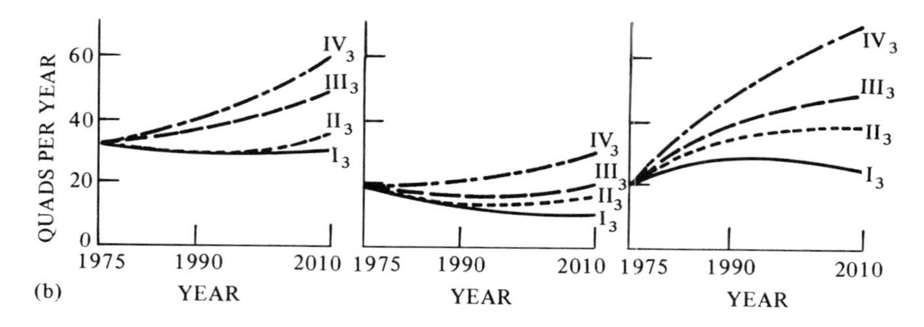

FIGURE 11-6 CONAES study scenario projections of total primary energy consumption for liquid and gaseous fuels and electricity from 1975 to 2010 (quads), with assumed (a) 2 percent GNP growth and (b) 3 percent GNP growth.

in Table 11-26. This table illustrates the sensitivity of outcomes to assumptions about electrification.

A substantial portion of the higher cost of electricity (as compared to liquid and gaseous fuels) is contributed by its capital costs. Recently, for a variety of reasons including regulations, interest rates, and construction labor costs, the capital costs of electrical generating stations have been escalating almost as rapidly as fuel costs. If, however, the capital costs of plants rose no faster than the rate of general inflation in the future, then electricity prices would rise less rapidly than primary fuel costs (which will probably continue to escalate faster than the rate of inflation). Better means of drawing on off-peak power could also contribute to greater use of electricity than indicated by the scenarios presented here. Nevertheless, the probabilities of these several possibilities are not well established.

In attempting to project the growth of demand for electricity, the electrical industry in its 1978 forecast[29] speculated that lower rates of industrial production through 1995, conservation, improved near-term

TABLE 11-25 Demand for Electricity and Fluid Fuels in Study Scenarios If Half the Demand for Fluid Fuels to Supply Building Heat Is Replaced with Electrical Resistance Heat

Scenario	Consumption in 1975 (quads)	Annual Change in Demand for Electricity (percent)	1990 (quads)	Annual Change in Demand for Electricity (percent)	2010 (quads)	Generating Capacity in 2010 (gigawatts)
A. Electricity						
I3						
Base	20	2.74	30	−1.32	23	460
High electricity	—	2.74	30	−0.43	27.5	550
II3						
Base	20	4.37	38	0.13	39	780
High electricity	—	4.37	38	0.96	46	920
III3						
Base	20	4.55	39	1.04	48	960
High electricity	—	4.55	39	1.83	56	1120
IV3						
Base	20	7.11	56	1.60	71	1420
High electricity	—	7.11	56	2.23	87	1740

Scenario	Consumption in 1975 (quads)	1990 (quads)	2010 (quads)	Consumption in Buildings Sector (quads)
B. Total Liquids and Gases				
I3				
Base	53	50	42	(13)
High electricity	—	50	38	(9)
II3				
Base	53	43	55	(11)
High electricity	—	43	48	(4)
III3				
Base	53	57	67	(14)
High electricity	—	57	58	(5)
IV3				
Base	53	60	91	(27)
High electricity	—	60	75	(11)

TABLE 11-26 Maximum Nuclear Contribution, by Study Scenario, in High-Electricity Case Shown in Table 11-25

	Primary Energy Input to Electricity Generation in 2010 (quads)		Generating Capacity in 2010 (gigawatts)	
Scenario	Total	Nuclear	Total	Nuclear
I_3				
Base	23	6	460	120
High electricity	27.5	10.5	550	210
II_3				
Base	39	8	780	160
High electricity	46	15	920	300
III_3				
Base	48	16	960	320
High electricity	56	24	1120	480
IV_3				
Base	71	25	1420	500
High electricity	87	41	1740	820

availability of natural gas (for residential use), and other factors will hold down the rate of growth of electricity demand. The report states,

> We also hear repeated stories of greater electrification of industry, as federal and state governments cut back on industrial use of gas, and industry searches for a dependable supply of energy. This leads us to believe that fuel substitution is in fact taking place, but that it is masked by strong conservation and energy management efforts, which will become evident in the late 1980s. This change, however, is not strong enough to offset the negative elements [lower industrial production, conservation, etc.] discussed above.

One may also question the rapid price increase assumed by the Demand and Conservation Panel for natural gas. While rapid increases in the near term may be anticipated with deregulation, there is some question whether gas should be expected to command as large a premium as assumed (even in the high-price scenarios), representing up to a 12-fold increase over 1975 prices. These prices would not reflect costs of production, and there would likely be a political demand to restrict earnings on natural gas sales. On the other hand, the greater convenience and cleanliness of gas for users would fully warrant a premium over oil.

In the study scenarios, coal was assumed to be the "swing fuel" for electrical generation, filling the demand remaining after the contributions of all other fuels had been summed. The levels of coal-fired electrical generation called for in some of the scenarios, especially by 1990, are rather high. Because of the 6–8 years required to design, construct, and license a coal-fired power plant, some of these levels could be difficult to reach. (The lead time for a nuclear power plant is even longer, by several years.)

Table 11-27 sets out the primary fuel mix for the generation of electricity in the selected study scenarios. In developing these fuel mixes, it was assumed that oil would be reserved for peaking applications, that natural gas would be phased out of the utility sector by 1990, that hydroelectric generation would grow modestly, and that solar and geothermal technologies would make small contributions by 2010. Nuclear contributions were determined from the Supply and Delivery Panel's scenario consistent with the assumed demand policies. If electricity prices increase considerably less than assumed in the Demand and Conservation Panel's scenarios, all these projections may underestimate the substitution of electricity for other fuels.

Liquid and Gaseous Fuels

The supply mixes of liquid and gaseous fuels for the study scenarios are shown in Table 11-28. To obtain an internally consistent supply picture (i.e., essentially equal stresses on the supply systems of oil and gas), it was necessary in scenarios II_2, III_2, and III_3 to shift a significant amount—up to about half—of industrial-sector demand for oil to demand for gas. Without this shift, these scenarios would have called for enhanced-supply or national-commitment conditions for oil supply, but only slightly more than a business-as-usual effort to produce gas. This shift of some oil to gas is consistent with the observation of the Demand and Conservation Panel's Industry Resource Group that gas is a preferred industrial fuel (so long as prices are competitive and supply can be assured), and that gas can be readily substituted for oil in about 50 percent of industrial use.

The numerical values for liquid and gaseous fuels in Table 11-28 include liquid fuels used for electrical generation as well as losses in the production and delivery of domestic liquids and gas. Oil and gas imports were derived as the differences between projected demand and total domestic supply under the supply conditions assumed. It is not suggested that these amounts of imports (particularly oil) will actually be available; rather, the projected needs are based on assumed supply and demand policies. In the cases of lowest growth in demand, enhanced-supply policies (especially in the area of synthetic fuel production) could replace the required imports for a given scenario.

TABLE 11-27 Primary Fuel Mix for Electricity, by Study Scenario (quads per year)[a]

Study Scenario and Energy Source	1975	1990	2010
I_2			
Oil	3	2	1
Gas	3	0	0
Coal	9	11	5
Nuclear	2	8	6
Hydro	3	4	4
Geothermal	0	0	0
Solar	0	0	1
TOTAL	20	25	17
II_2			
Oil	3	2	1
Gas	3	0	0
Coal	9	17	13
Nuclear	2	8	7
Hydro	3	4	4
Geothermal	0	0	2
Solar	0	0	1
TOTAL	20	31	29
III_2			
Oil	3	2	2
Gas	3	0	0
Coal	9	15	16
Nuclear	2	11	13
Hydro	3	4	4
Geothermal	0	1	1
Solar	0	0	1
TOTAL	20	33	37
III_3			
Oil	3	3	3
Gas	3	0	0
Coal	9	19	20
Nuclear	2	11	18
Hydro	3	4	4
Geothermal	0	1	3
Solar	0	0	1
TOTAL	20	38	48
IV_3			
Oil	3	5	3
Gas	3	0	0
Coal	9	23	29
Nuclear	2	12	30
Hydro	3	4	4
Geothermal	0	1	4
Solar	0	0	1
TOTAL	20	45	71

[a]Some totals may not add due to rounding.

DISCUSSION OF STUDY SCENARIOS

The study scenarios cover a wide range of potential energy supply and demand patterns (Tables 11-29 to 11-33). The totals for primary energy consumption vary widely—from 64 to 188 quads in 2010. The scenarios are even more varied in the individual components of supply and demand (as illustrated in Figures 11-7 to 11-9). The highest-consumption scenarios, for example, include contributions from light water reactors with extensive reprocessing, and from fast breeder reactors by 2010, while the lowest-consumption scenarios need only conventional light water reactors with a once-through fuel cycle and relatively small amounts of coal for electricity (Figure 11-9). Similarly, the higher-consumption scenarios entail heavy promotion of synthetic fuel production, while the low-growth scenarios require little or no synthetic fuel (Figure 11-7). The study scenarios show solar energy contributing only a small fraction to total energy consumed, with the exception of scenario I_2 (about 13 percent of primary energy). A special low-energy-growth scenario developed by the Consumption, Location, and Occupational Patterns Resource Group (described previously) shows a more significant solar contribution, about 25 percent.

Characteristics of the Scenarios

High-consumption scenario IV_3 (188 quads in 2010) is characterized by great expansion of nuclear power (including the introduction of breeder reactor technology as early as possible), and by aggressive exploitation of oil shale resources. This expansion would require rapid changes from current policy to encourage the development of energy resources—tax concessions, guarantees for energy investments, and expeditious processing of regulatory matters. Maintaining energy prices at essentially the levels prevailing in 1975 is unlikely to attract the capital to finance immediate expansion on all fronts. In addition, resources would have to be shifted into the development of technology to abate pollution, to protect public health and the environment, and to avert significant damage. Serious social dislocations could accompany the rapid regional shifts in economic activity implied by these scenarios.

The two medium-consumption study scenarios, III_2 and III_3 (102 quads and 140 quads), represent varying degrees of effort to conserve energy, spurred by policy and higher prices for energy, but they do not imply significant shifts in the mix of final purchases made by consumers. The low-consumption scenarios, CLOP and I_2 (58 and 64 quads), assume changes in regional policies and in individual choices aimed at lowering energy demand. The four-times-greater prices for energy and one-third-less consumption of energy (compared to 1975) that characterize scenario

TABLE 11-28 Liquid and Gaseous Fuel Supply Mix, by Study Scenario (quads per year)

	Annual Production[a]		
Study Scenario and Energy Source	1975	1990	2010
I_2			
Domestic oil	20	18	11
Imported oil	13	5	12
Synthetic oil	0	0	0
Shale oil	0	0	0
TOTAL liquids	33	23	23
Domestic natural gas	19	13	8
Imported natural gas	1	0	0
Synthetic gas	0	0	0
TOTAL gases	20	13	8
II_2			
Domestic oil	20	18	11
Imported oil	13	5	10
Synthetic oil	0	0	4
Shale oil	0	0	1
TOTAL liquids	33	23	26
Domestic natural gas	19	13	10
Imported natural gas	1	0	3
Synthetic gas	0	0	0
TOTAL gases	20	13	13
III_2			
Domestic oil	20	20	16
Imported oil	13	9	7
Synthetic oil	0	0	8
Shale oil	0	0	1
TOTAL liquids	33	29	32
Domestic natural gas	19	14	14
Imported natural gas	1	0	2
Synthetic gas	0	0	0
TOTAL gases	20	14	16
III_3			
Domestic oil	20	21	18
Imported oil	13	16	14
Synthetic oil	0	1	13
Shale oil	0	0	2
TOTAL liquids	33	38	47

TABLE 11-28 (*continued*)

Study Scenario and Energy Source	Annual Production[a] 1975	1990	2010
Domestic natural gas	19	14	14
Imported natural gas	1	1	1
Synthetic gas	0	1	5
TOTAL gases	20	16	20
IV_3			
Domestic oil	20	21	18
Imported oil	13	20	27
Synthetic oil	0	1	13
Shale oil	0	0	3
TOTAL liquids	33	42	61
Domestic natural gas	19	18	16
Imported natural gas	1	0	6
Synthetic gas	0	2	8
TOTAL gases	20	20	30

[a]Figures include losses in production and distribution, but do not include conversion losses for synthetic fuels derived from coal.

I_2 in 2010 would make both time and money available to deal with a lightened burden of environmental and public health consequences, but these apparent gains could be offset by the difficulties of achieving national consensus for the necessary policies and their effectuation.

Scenario II_2 is driven primarily by economic changes. Its low rate of growth in energy consumption results from high energy prices (a fourfold increase) and low GNP growth (2 percent annually). However, these price levels would be considerably above production costs and would require special energy taxes, with the resulting revenue returned to the economy (e.g., through the reduction of other forms of taxation).

The midrange scenarios encompass a variety of possible energy prices (or policies) and paths for GNP. In general, the lower part of this range, scenario II_3 (115 quads), presents fewer supply problems because of the slower rate of supply-system expansion (e.g., less need to change procedures for licensing and siting facilities). On the other hand, political pressures for low energy prices could present difficulties, and, of course, the economic growth assumptions are not directly subject to policy control in our models, since they depend primarily on changes in productivity.

TABLE 11-29 Projected Energy Supply and Demand in 2010 for Study Scenario I_2 (quads per year)[a]

	Demand by Energy Source	Demand by Energy-Consuming Sector			
		Industry	Buildings	Transportation	Total[b]
Liquid fuels[b]		8.7	1.6	10.8	21.1 (0.8)
Domestic oil	10.2 (0.8)				
Imported oil	12				
Shale oil	0				
Coal synthetics	0				
SUBTOTAL[c]	22.2 (0.8)				
Gaseous fuels[b]		5.8	1.6	0.1	7.5 (0.6)
Domestic gas	7.5 (0.6)				
Imported gas	0				
Coal synthetics	0				
Direct coal combustion		9.4	0.6	0.1	10.1
Solar		2.0	2.9	0	4.9
Other		2.0	0	0	2.0
Purchased electricity[b,d]		2.7	3.0	0	5.7 (11.3)
Liquid fuels	0.3 (0.8)				
Coal	2.0 (3.3)				
Uranium in light water reactors	1.8 (3.8)				
Uranium in fast breeder reactors	0				
Hydro	1.3 (2.8)				
Geothermal	0				
Solar	0.3 (0.6)				
TOTAL, delivered energy and losses					51.3 (12.7)
TOTAL, primary energy use					64.0

[a] Assumed 2 percent average annual gross national product growth, and energy prices at 4 times 1975 prices in real dollars.
[b] Figures in parentheses indicate the losses in production and conversion of fuels included in sectoral figures.
[c] Liquid fuels subtotal includes liquids used directly in the buildings, industrial, and transportation sectors, and liquids used to generate electricity.
[d] Losses in the transmission and distribution of electricity are included with purchased electricity in the values that are not in parentheses.

Comparison with Projections of Other Energy Studies

The study scenarios are set out against scenarios of demand from other energy studies in Table 11-34. (Note that most of the projections from

TABLE 11-30 Projected Energy Supply and Demand in 2010 for Study Scenario II_2 (quads per year)[a]

	Demand by Energy Source	Demand by Energy-Consuming Sector			
		Industry	Buildings	Transportation	Total[b]
Liquid fuels[b]		6.5	2.4	15.1	24.0 (2.8)
Domestic oil	10.0 (0.8)				
Imported oil	10.2				
Shale oil	0.9				
Coal synthetics	4.0 (2.0)				
SUBTOTAL[c]	25.1 (2.8)				
Gaseous fuels[b]		9.5	2.5	0.1	12.1 (0.7)
Domestic gas	9.1 (0.7)				
Imported gas	3.0				
Coal synthetics	0				
Direct coal combustion		8.5	0	0.1	8.6
Solar		1.8	1.6	0	3.4
Other		2.7	0	0	2.7
Purchased electricity[b,d]		4.6	5.4	0	10.0 (18.7)
Liquid fuels	0.3 (0.8)				
Coal	5.0 (8.2)				
Uranium in light water reactors	2.3 (4.8)				
Uranium in fast breeder reactors	0				
Hydro	1.4 (3.0)				
Geothermal	0.5 (1.3)				
Solar	0.3 (0.6)				
TOTAL, delivered energy and losses					60.8 (22.2)
TOTAL, primary energy use					83.0

[a] Assumed 2 percent average annual gross national product growth, and energy prices at 4 times 1975 prices in real dollars.
[b] Figures in parentheses indicate the losses in production and conversion of fuels included in sectoral totals.
[c] Liquid fuels subtotal includes liquids used directly in the buildings, industrial, and transportation sectors, and liquids used to generate electricity.
[d] Losses in the transmission and distribution of electricity are included with purchased electricity in the values that are not in parentheses.

other energy studies are for the year 2000, while those of CONAES are for the year 2010. A uniform basis for comparison is the column "Annual Average Rate of Growth, Energy Consumption.") The purposes of the studies, as well as their assumptions, techniques, and accounting conven-

TABLE 11-31 Projected Energy Supply and Demand in 2010 for Study Scenario III_2 (quads per year)[a]

	Demand by Energy Source	Demand by Energy-Consuming Sector			Total[b]
		Industry	Buildings	Transportation	
Liquid fuels[b]		4.8	3.1	21.0	28.9 (5.1)
Domestic oil	14.5 (1.2)				
Imported oil	7.0				
Shale oil	1.0				
Coal synthetics	8.1 (3.9)				
SUBTOTAL[c]	30.6 (5.1)				
Gaseous fuels[b]		11.5	3.4	0.1	15.0 (1.0)
Domestic gas	13.0 (1.0)				
Imported gas	2.0				
Coal synthetics	0				
Direct coal combustion		10.0	0	0.1	10.1
Solar		1.1	1.4	0	2.5
Other		2.7	0	0	2.7
Purchased electricity[b,d]		5.2	6.9	0	12.1 (24.6)
Liquid fuels	0.5 (1.2)				
Coal	5.5 (10.2)				
Uranium in light water reactors	4.2 (8.9)				
Uranium in fast breeder reactors	— —				
Hydro	1.4 (3.0)				
Geothermal	0.2 (0.7)				
Solar	0.3 (0.6)				
TOTAL, delivered energy and losses					71.3 (30.7)
TOTAL, primary energy use					102.2

[a] Assumed 2 percent average annual gross national product growth, and energy prices at 4 times 1975 prices in real dollars.
[b] Figures in parentheses indicate the losses in production and conversion of fuels included in sectoral totals.
[c] Liquid fuels subtotal includes liquids used directly in the buildings, industrial, and transportation sectors, and liquids used to generate electricity.
[d] Losses in the transmission and distribution of electricity are included with purchased electricity in the values that are not in parentheses.

tions, vary greatly. All these energy studies (as well as CONAES) faced the difficult problem of casting projections from slippery ground. At the time most of the projections were attempted, the equilibrium of energy and the economy had been disrupted by new prices for energy (following the oil

TABLE 11-32 Projected Energy Supply and Demand in 2010 for Study Scenario III_3 (quads per year)[a]

	Demand by Energy Source	Demand by Energy-Consuming Sector			
		Industry	Buildings	Transportation	Total[b]
Liquid fuels[b]		10.0	4.0	29.0	43.0 (7.4)
Domestic oil	16.5 (1.3)				
Imported oil	14.0				
Shale oil	2.2				
Coal synthetics	13.0 (6.1)				
SUBTOTAL[c]	45.7 (7.4)				
Gaseous fuels[b]		13.9	4.6	0.2	18.7 (3.2)
Domestic gas	13.0 (1.0)				
Imported gas	1.0				
Coal synthetics	4.7 (2.2)				
Direct coal combustion		14.0	0	0.2	14.2
Solar		1.1	1.4	0	2.5
Other		3.5	0	0	3.5
Purchased electricity[b,d]		6.5	9.5	0.1	16.1 (31.7)
Liquid fuels	0.8 (1.9)				
Coal	7.2 (12.3)				
Uranium in light water reactors	5.7 (11.8)				
Uranium in fast breeder reactors	0				
Hydro	1.4 (3.0)				
Geothermal	0.7 (2.1)				
Solar	0.3 (0.6)				
TOTAL, delivered energy and losses					98.0 (42.3)
TOTAL, primary energy use					140.3

[a] Assumed 2 percent average annual gross national product growth, and energy prices at 4 times 1975 prices in real dollars.
[b] Figures in parentheses represent the losses in production and conversion of fuels included in sectoral totals.
[c] Liquid fuels subtotal includes liquids used directly in the buildings, industrial, and transportation sectors, and liquids used to generate electricity.
[d] Losses in the transmission and distribution of electricity are included with purchased electricity in the values that are not in parentheses.

embargo of 1973). A decade or more may pass before these relations approach a new equilibrium. In addition, the recession of 1974–1975 induced short- and long-term changes in the economy that cannot easily be distinguished. While these differences and common difficulties warn

TABLE 11-33 Projected Energy Supply and Demand in 2010 for Study Scenario IV_2 (quads per year)[a]

	Demand by Energy Source	Demand by Energy-Consuming Sector			
		Industry	Buildings	Transportation	Total[b]
Liquid fuels[b]		11.9	6.6	38.6	57.1 (7.3)
Domestic oil	16.5 (1.3)				
Imported oil	27.0				
Shale oil	3.0				
Coal synthetics	13.0 (6.0)				
SUBTOTAL[c]	59.5 (7.3)				
Gaseous fuels[b]		19.4	9.8	0.2	29.4 (5.2)
Domestic gas	14.8 (1.2)				
Imported gas	6.4				
Coal synthetics	8.2 (4.0)				
Direct coal combustion		13.0	0	0.2	13.2
Solar		0.8	0.3	0	1.1
Other		4.5	0	0	4.5
Purchased electricity[b,d]		9.3	14.4	0.2	23.9 (47.4)
Liquid fuels	1.0 (2.4)				
Coal	10.7 (18.2)				
Uranium in light water reactors	7.6 (16.0)				
Uranium in fast breeder reactors	2.0 (4.2)				
Hydro	1.4 (3.0)				
Geothermal	1.0 (3.0)				
Solar	0.3 (0.6)				
TOTAL, delivered energy and losses					129.2 (59.9)
TOTAL, primary energy use					188.1

[a]Assumed 2 percent average annual gross national product growth, and energy prices at 4 times 1975 prices in real dollars.
[b]Figures in parentheses indicate the losses in production and conversion of fuels included in sectoral totals.
[c]Liquid fuels subtotal includes liquids used directly in the buildings, industrial, and transportation sectors, and liquids used to generate electricity.
[d]Losses in the transmission and distribution of electricity are included with purchased electricity in the values that are not in parentheses.

against any but qualitative interpretations and comparisons, they also point up the need for continuing research into the factors that influence demand for energy, and into the several relations between energy and the

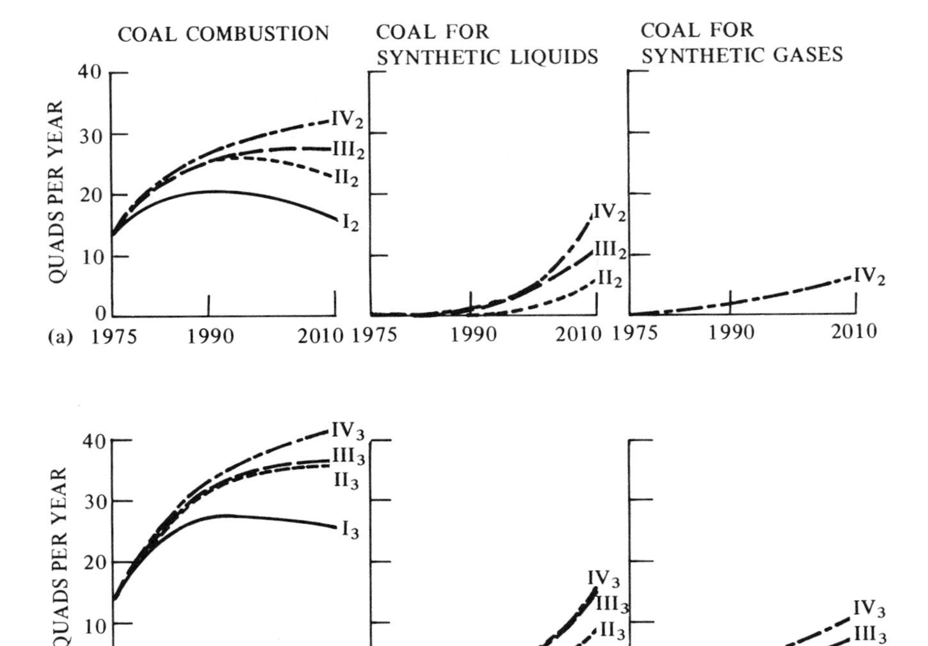

FIGURE 11-7 CONAES study scenario projections of coal consumption for combustion, synthetic liquids, and synthetic gases from 1975 to 2010 (quads), with assumed (a) 2 percent GNP growth and (b) 3 percent GNP growth.

economy (as detailed in chapter 2 under "Econometric Studies . . . " and "Conclusions and Recommendations").

WORK OF THE MODELING RESOURCE GROUP

The Modeling Resource Group of the Synthesis Panel sought to compare the economic benefits and costs of various energy technologies that might be applied to meet the nation's demand for energy over the next three decades.[30] In consultation with the Risk and Impact Panel, the group found that the estimation of risks to life and health presented by various energy technologies is itself uncertain, and for some technologies, unknown. The MRG decided against attempting to reduce the estimation of these risks to a common aggregate measure (such as dollars). It reasoned that first approximations could be made for the economic costs of

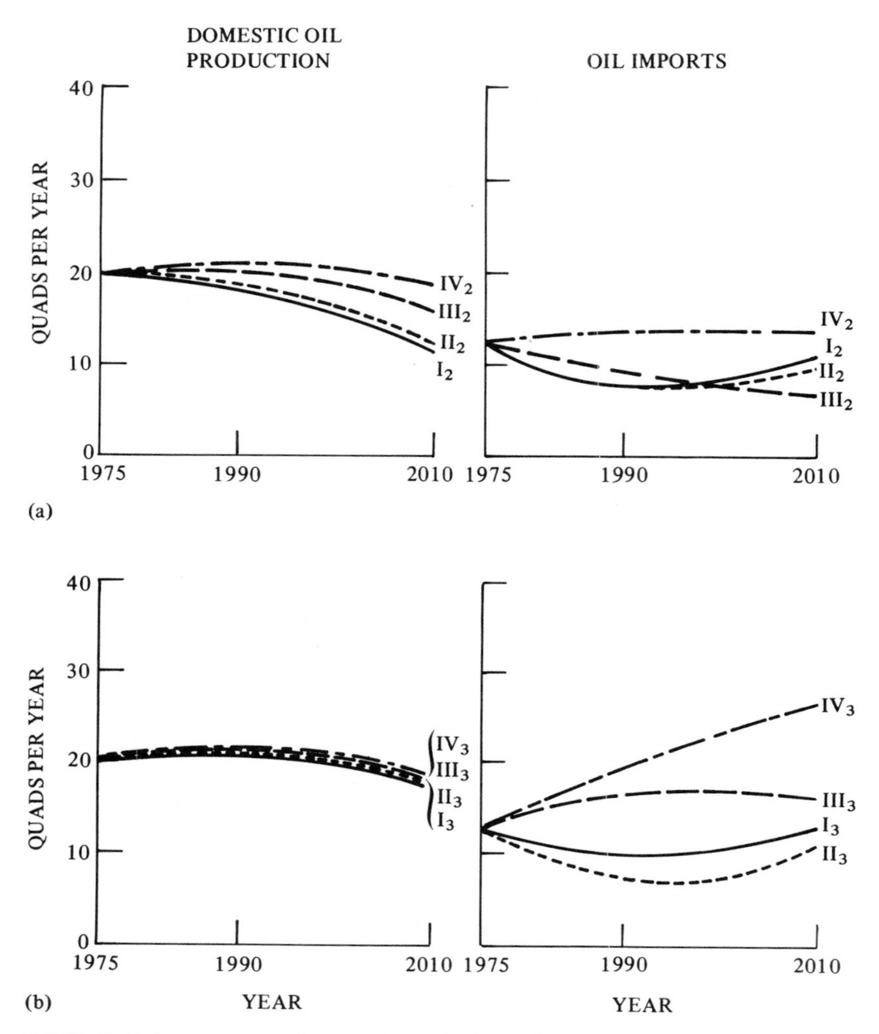

FIGURE 11-8 CONAES study scenario projections of oil consumption from domestic production and imports from 1975 to 2010 (quads), with assumed (a) 2 percent GNP growth and (b) 3 percent GNP growth.

technologies known to present some risk to human health and the environment by assuming that the use of these technologies is limited to some upper bound, and then calculating the effect on the gross national product. The loss of GNP resulting from these limitations or proscriptions represents the price of protecting the environment and human health to whatever degree is achieved with the particular assumed limits.

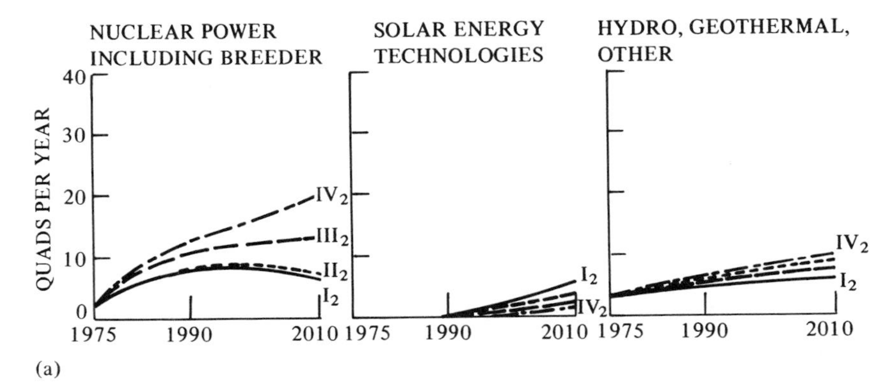

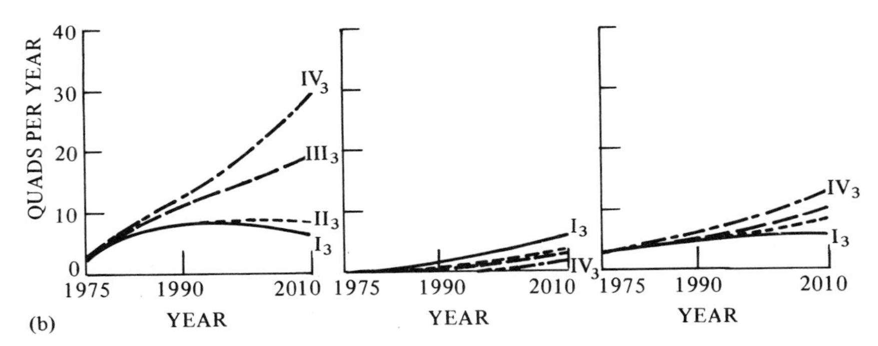

FIGURE 11-9 CONAES study scenario projections of consumption of nuclear power (including breeder); solar energy; and hydroelectric, geothermal, and other energy sources from 1975 to 2010 (quads), with assumed (a) 2 percent GNP growth and (b) 3 percent GNP growth.

Working with six large models of the domestic economy and the demand for energy, the Modeling Resource Group formalized the terms of its task as estimating the effects for outcome variables of changing certain policy variables and realization variables (as illustrated in Figure 11-10). The realization variables selected were the following.

- The growth rate of GNP to the year 2010 unbounded by limits or proscriptions against specific energy technologies.
- Levels of capital cost for existing and future energy technologies.
- The availability of oil, gas, and uranium at various costs of extraction.
- The long-term price and income elasticities of demand for various forms of energy at the point of end-use.

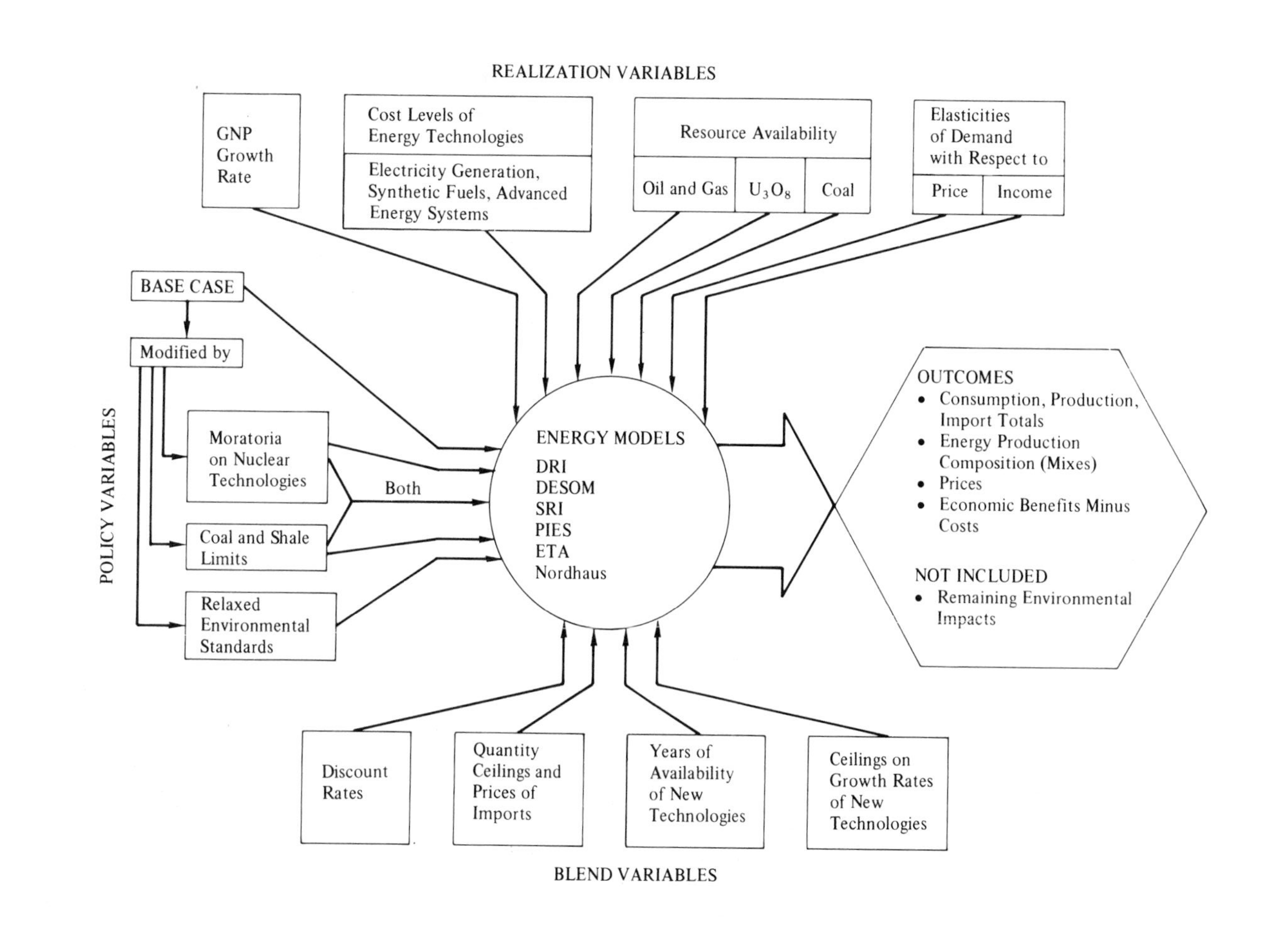
REALIZATION VARIABLES
GNP Growth Rate
Cost Levels of Energy Technologies
Electricity Generation, Synthetic Fuels, Advanced Energy Systems
Resource Availability
Oil and Gas
U_3O_8
Coal
Elasticities of Demand with Respect to
Price
Income
POLICY VARIABLES
BASE CASE
Modified by
Moratoria on Nuclear Technologies
Coal and Shale Limits
Relaxed Environmental Standards
Both
ENERGY MODELS
DRI
DESOM
SRI
PIES
ETA
Nordhaus
OUTCOMES
• Consumption, Production, Import Totals
• Energy Production Composition (Mixes)
• Prices
• Economic Benefits Minus Costs
NOT INCLUDED
• Remaining Environmental Impacts
Discount Rates
Quantity Ceilings and Prices of Imports
Years of Availability of New Technologies
Ceilings on Growth Rates of New Technologies
BLEND VARIABLES

FIGURE 11-10 Compressed view of driving variables, energy models, and outcomes of the Modeling Resource Group's work. The models vary in the detail and composition of the information inputs they require and the information outputs they produce. *DRI* (Data Resources, Inc., energy model) projects final demand for energy, in four principal fuels (coal, electricity, petroleum, and natural gas), and its consumption in 13 geographical regions of the United States by each of four consuming sectors (commercial, industrial, residential, and transportation). *DESOM* (Dynamic Energy Systems Optimization Model) assumes that demand for energy is given and determines combination of energy resources and technologies that meet this demand at least cost over time. *SRI* (Stanford Research Institute model) assumes that end-use demands for energy are functions of prices and simultaneously balances demands and supplies of energy for various points of an energy system (elaborated from data for 20 supply regions and 9 census regions in the United States), at particular times in a 50-yr period. *PIES* (Federal Energy Administration Project Independence Evaluation System), using a reference energy system, solves for the minimum-cost mix of fuels and energy technologies to meet sets of final demands. It integrates separate dynamic models of energy supply and demand to describe market equilibrium in 1980, 1985, and 1990. *ETA* and *Nordhaus* (Energy Technology Assessment model and Nordhaus simulation model) are similar to *PIES*, but incorporate price-sensitive energy demands, allow for consideration of a broader array of energy technologies, and project patterns of energy use further into the future. Source: Adapted from National Research Council, *Supporting Paper 2: Energy Modeling for an Uncertain Future,* Committee on Nuclear and Alternative Energy Systems, Synthesis Panel, Modeling Resource Group (Washington, D.C.: National Academy of Sciences, 1978), p. 9 and Appendix A.

TABLE 11-34 Energy Demand Projections by Various Studies

Energy Study	Period of Projection	Key Assumptions: Population at End of Period (millions)	Key Assumptions: Average Annual Rate of Growth, GNP (percent)	Key Assumptions: Average Annual Rate of Growth, Energy Consumption (percent)	Total Energy Consumption in Final Year of Projection (quads)
Ford Foundation[a]	1975–2000				
Historical		265	3.02	3.4	186.7
Technical fix		265	2.91	1.9	124.0
Zero energy growth		265	2.92	1.1	100.0
Edison Electric Institute[b]	1975–2000				
High		286	4.2	3.8	179
Moderate		265	3.7	3.2	155
Low		251	2.3	1.6	105
Exxon[c]	1977–1990	—[d]	3.6	2.3	108
Bureau of Mines[e]	1974–2000	264	3.7	3.1	163.4
EPRI[f]	1975–2000				
Baseline[g]		281	3.4	3.37	159
High electricity[h]		281	3.4	4.21	196.1
Conservation		281	3.4	2.97	145.6
Five times prices		281	3.4	1.98	114.3
CONAES	1975–2010				
I_2		279	2.0	−0.29	64
II_2		279	2.0	0.45	83
III_2		279	2.0	1.04	102
IV_2		279	2.0	1.95	140
I_3		279	3.0	0.52	85
II_3		279	3.0	1.38	115
III_3		279	3.0	1.95	140
IV_3		279	3.0	2.82	188

[a] Source: Ford Foundation, Energy Policy Project, *A Time to Choose: America's Energy Future* (Cambridge, Mass.: Ballinger Publishing Co., 1974).
[b] Source: Edison Electric Institute, *Economic Growth in the Future*, Committee on Economic Growth, Pricing and Energy Use (New York: Edison Electric Institute, 1976).
[c] Source: Exxon Company, U.S.A., "Energy Outlook: 1978–1990," May 1978. (Available from Public Affairs Department, P.O. Box 2180, Houston, Tex. 77001.)
[d] Not specified.
[e] Source: U.S. Bureau of Mines, *United States Energy Through the Year 2000*, rev. ed. (Washington, D.C.: U.S. Government Printing Office, 1975).
[f] Source: Electric Power Research Institute, *Demand '77: EPRI Annual Energy Forecasts and Consumption Model* (Palo Alto, Calif.: EPRI (EA-621-SR), 1978).
[g] With restrictions on the availability of natural gas.
[h] With no restrictions on the availability of natural gas.

The policy variables represent limits or proscriptions placed on the use of energy technologies.

- A moratorium on new construction of nuclear power plants.[31]
- Limits on the annual production of coal and shale oil.

In addition, the group considered "blend" variables that share the characteristics of both policy and realization variables.

- The discount rates used to compare present values of future costs and benefits.
- The price of oil imports, ceilings on imports, etc.

Growth Rate of Real Gross National Product

The standing of this variable as a purported measure of aggregate welfare of the population has suffered over time, and some attempts have been made to define better measures of economic welfare,[32] whether by adding imputations for the services of the consumer's capital, for the personal value of leisure, and for the value added by work done in households, or by subtracting the cost of certain negative consequences (or the cost of dealing with them), such as police services and environmental degradation.

For the MRG's purpose of identifying a measure of aggregate economic activity that can serve as a driving variable for demand for energy, the GNP seemed as good as any other available measure of aggregate welfare. The principal question, bearing equally on all such measures, is whether the causation flows only one way, from GNP (say) to energy use, or also the other way, in the sense that direct interventions to curtail energy use would in turn have a negative effect on GNP, especially at low levels of energy availability.

Accordingly, GNP growth was classified as a realization variable, assumed to be determined over the long term by a combination of three factors largely independent of energy policy decisions: (1) population growth, (2) technological change, as reflected in labor productivity, and (3) work-force participation. The projections of GNP growth given in Table 11-35 were made by estimating potential GNP on the basis of future trends for each of these variables, assuming no more than a floor level of labor unemployment.

Over the short term, aggregate demand—not potential GNP—shapes the growth of GNP. For this reason, all estimates of GNP growth provide for higher growth over the 1975–1985 period than in subsequent periods, to account for further recovery from the recession of 1974–1975. Recessions

TABLE 11-35 Projected Growth Rate of Real Gross National Product (percent per annum)[a]

	Projections		
Period	Low	Base Case	High
1975–1980	3.7	5.70	7.0
1980–1990	2.7	3.28	4.8
1990–2000	1.2	2.84	4.0
2000–2010	0.5	2.48	4.0
Beyond 2010	0.2	1.80	3.5

[a]Calculated by projecting separately three determinants of potential GNP. (*Population* increases by the Bureau of Census Series II projections (see Figure 11-2). *Labor force participation,* or ratio of actual to potential labor force (the latter equals the male and female population between 18 and 64 years of age), rises from its 1970–1975 value of 0.73 to 0.83 by 2010. *Productivity,* or output per worker, continues to grow from 1975 to 2000 at an average rate of 1.8 percent and grows at an average rate of 1.5 percent thereafter.) Actual GNP is assumed to complete its recovery to the level of potential GNP by 1985, and to remain equal to potential GNP after that point, at levels that ensure an average unemployment rate of 4.8 percent from 1985 through 2010.

Source: Adapted from National Research Council, *Supporting Paper 2: Energy Modeling for an Uncertain Future,* Committee on Nuclear and Alternative Energy Systems, Synthesis Panel, Modeling Resource Group (Washington, D.C.: National Academy of Sciences, 1978), p. 13.

may, of course, recur in the future, but the Modeling Resource Group assumed GNP to remain at its potential level from 1985 on, to avoid underestimating the investments needed to meet future energy demand.

Besides the assumptions listed, the MRG's assumption of an average annual growth rate for GNP of 3.2 percent is based on projections of public and private decisions that influence the composition of GNP, that is, allocations between investment and consumption. Should reduced energy consumption affect the amount of savings allocated to capital formation, then the growth of GNP to 2010 would follow a lower path. The MRG assumed (from empirical and conceptual considerations) that capital scarcity for investment in energy supply is unlikely.

Capital investment in the energy supply sector of the economy represented 2.64 percent of GNP in 1974. Projecting this capital investment for their various scenarios with one of the six models, the MRG noted a similar pattern. Energy-sector investment (as a percentage of GNP) drops over the near term, but returns to 1974 levels by 2010.

Feedback from Energy Use to GNP

In all MRG scenarios, GNP was treated as the principal driving variable influencing energy consumption, but not in turn influenced by it. At the same time, various scenarios considered policies that reduce energy supply, and hence consumption, directly rather than via GNP. The MRG examined the effects of policies that curtail energy supply below levels that would otherwise have prevailed—curtailments imposed for reasons of environmental protection, energy independence, or other national policy objectives. In those cases, the question naturally arose, Does the diminished use of energy also reduce GNP below what it would otherwise have been?

Virtually everyone will agree that such a reverse effect must exist. The real question is its magnitude under various circumstances. One consideration that has an important bearing on this reverse effect is discussed in chapter 2: whether the curtailment of supply is abrupt or gradual, and if abrupt, whether it is foreseen.

The models used by the Modeling Resource Group assume that the government pursues and successfully institutes a full-employment policy by maintaining aggregate demand at a level sufficient to consume whatever is produced. Thus, feedback from the contraction of energy consumption would occur only through reduction in available goods and services. If the shift to lower energy growth is a gradual one—proceeding at a rate no higher than that implicit in the normal retirement cycle of plants and equipment and in the normal turnover of labor—then capital funds and labor released from the production of energy and energy-intensive goods can find employment in other sectors.

The assumptions made about rates of capital turnover were slightly different in the six models, as detailed in the report of the MRG.[33] The theoretical adjustments postulated may be inhibited by obstacles to the mobility of labor and capital. However, it appears that the adjustments necessary to accommodate long-term energy constraints or higher prices would be small compared to the adjustments required by normal technological change and differential rates of productivity growth in the economy.

Changes in energy inputs to the economy influence the pattern of capital investment over time and thus have an additional feedback effect on GNP. Reductions in energy consumption lead to changes in the rate of return on capital that alter rates of saving, investment, and the use of capital. Over time, these effects may lead to significant cumulative changes in the total capital stock, and thus in the productive capacity and output of the economy. This feedback effect can be illustrated by allowing capital to adjust to maintain a constant rate of return, rather than assuming capital and labor inputs as a constant fraction of GNP. If the elasticity of substitution of capital for energy is 0.3, then a 50 percent reduction in energy inputs would lower GNP just 4 percent* if the capital input fraction is held constant, but would lower GNP 11 percent if capital is allowed to adjust to maintain a constant rate of return. The greater the reduction postulated, the greater the difference in effect on GNP between the two assumptions, as illustrated in Figure 11-11.[34]

The critical parameter that describes the quantitative effect of all energy-saving substitutions taken together, and thereby determines the feedback from energy use to GNP, is the long-term price elasticity of demand for energy. In a complete model, a matrix of price elasticities would be used to express the change in aggregate demand for each fuel in terms of the price change for each fuel. In most of the work presented here, however, this complex of effects was represented by a single price elasticity representing the ratio of the percentage change in aggregate demand for all (price-weighted) forms of primary energy to the percentage change in the average (consumption-weighted) price of primary energy. The test of the validity of this gross price elasticity would be how well the simple aggregate model, with a single primary energy, can be made to simulate the behavior of a more complex model with many fuels and many economic sectors.

Estimates of the Feedback from Energy Consumption to Real Income

Three of the models employed by the Modeling Resource Group (DESOM, ETA, and Nordhaus) minimize the discounted economic cost of meeting a set of demands for energy over a long period (subject to technological constraints and a limited range of consumer and producer behavior). In the DESOM model, the path of demand for energy is given *a priori* (the aggregate price elasticity of demand for primary energy is zero). In the ETA and Nordhaus models, the path of demand is obtained by maximizing the discounted sum of economic benefits to the consumer and subtracting the discounted sum of costs incurred by the producer. The optimization features of these three models enabled the MRG to estimate the economic costs (excluding those of research and development) of limiting or proscribing energy technologies in accordance with various policies. The results are displayed in Table 11-36.

Scenarios 2–6 represent alternative policies that restrict the amount of energy supplied to the economy of the United States by limiting the use of one or two energy technologies. For DESOM, the scenario entries represent the increase (over the base case) in the minimal discounted sum of year-by-

*Statement 11-6, by R. H. Cannon, Jr.: Or $176 billion. Our current costs for foreign oil, for example, are about one fourth of this and are driving a harrowing inflation.

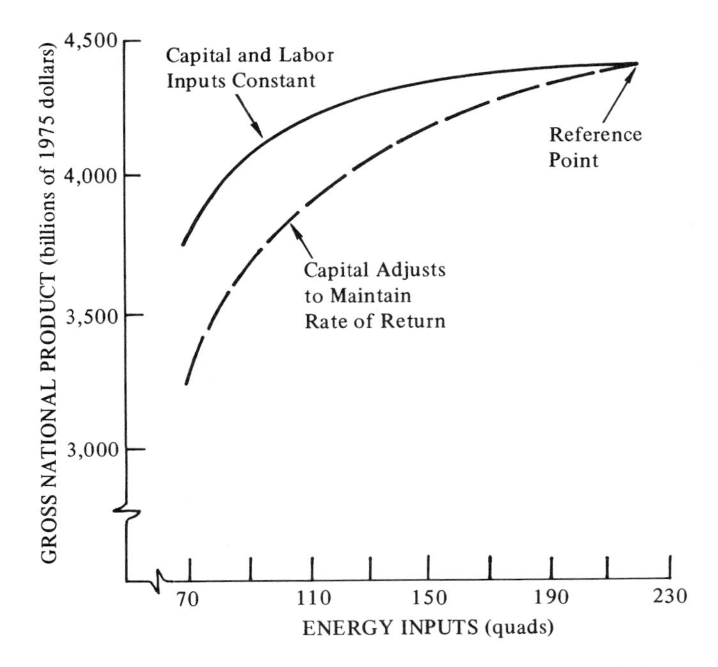

FIGURE 11-11 Effect on the economy of energy scarcity in 2010, under two alternative assumptions about capital. Source: Energy Modeling Forum, *Energy and the Economy* (Stanford, Calif.: Stanford University, Institute for Energy Studies, September 1977), p. 16.

year, constant-dollar costs to achieve the *a priori* path of demand for energy under each policy alternative. For ETA and Nordhaus, the scenario entries represent the decrease (below the base case) in the maximal discounted sum of year-by-year benefits minus costs of the paths of energy consumption in an open, competitive energy market (simulated by the same maximization). These costs (DESOM) or losses of benefits minus costs (ETA and Nordhaus) cut into the real income that might be spent for nonenergy goods and services.

The Modeling Resource Group assumed that the long-term effects of gradual and foreseeable restrictions on the supply of energy could be estimated if the percentage of total capital and labor put to work is independent of energy supply restrictions. Under that assumption, Table 11-36 can be read as discounted sums of precisely the year-by-year implications for real income of restricted energy supplies (without crediting the gains for the environment or public health). Figure 11-12 depicts the results for the ETA and Nordhaus models. The ratio set out on

TABLE 11-36 Estimated Differences in Net Economic Benefits from Six Technology Mixes and Net Economic Costs of Five Alternative Policies to Reduce Environmental Impacts (billions of 1975 dollars)

	Shortfall Below Base Case of Benefits Minus Costs[a]		
Policy Alternatives	DESOM[b] (Costs Only)	ETA[b]	Nordhaus[b]
1. Base case	(0)	0	0
2. Moratorium on all advanced converters and fast breeder reactor[c]	(43)	8	2
3. Moratorium on all nuclear technologies	(105)	46	136
4. Coal and shale limits	(914)	159	64
5. Moratorium on all advanced converters and fast breeder reactor, and coal and shale limits	(1012)	181	72
6. Nuclear moratorium and coal and shale limits	(2325)	358	457

[a] In all policy scenarios, total benefits and costs are the sums of year-by-year benefits and costs, discounted to 1975 at 6 percent per annum. DESOM computes only discounted costs, through 2025, ETA computes discounted benefits and costs through 2050, and Nordhaus computes them through 2060. For each year, benefits estimate the value to the consumer of total amounts of energy consumed, on an incremental basis. For further explanations, see National Research Council, *Supporting Paper 2: Energy Modeling for an Uncertain Future*, Committee on Nuclear and Alternative Energy Systems, Synthesis Panel, Modeling Resource Group (Washington, D.C.: National Academy of Sciences, 1978), sect. III.8.
[b] The main features of the models used are described in the caption for Figure 11-10.
[c] For ETA and SRI, this policy includes a moratorium on light water reactors with plutonium recycle; for the other models it does not.

the horizontal axis is the "total energy consumption (in primary energy equivalents) for 2010 projected by scenario" to the "total energy consumption (in primary energy equivalents) for 2010 projected by the base case." The ratio on the vertical axis is the "discounted sum of year-by-year levels of GNP minus the discounted sum of year-by-year losses in income given by scenario to the discounted sum of year-by-year levels of GNP projected by MRG." *The Modeling Resource Group concluded that the feedback effect from restrictions on energy supplies to GNP is small. Apart from two points, the feedback effect is at most 2 percent of GNP, even with energy supplies restricted to 50 percent of their levels in the base case.**

There are some differences in this result among the models, and the

*See statement 11-7, by R. H. Cannon, Jr., Appendix A.

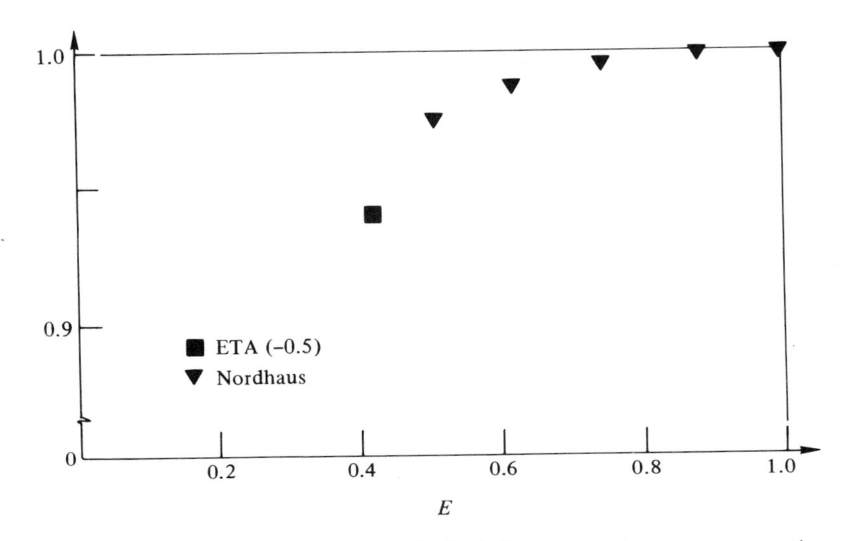

FIGURE 11-12 Estimates of the long-term feedback from aggregate energy consumption on undiscounted GNP for 2010. Nordhaus points represent policies of successive curtailments of the growth rate of energy use. Y' on vertical axis is the ratio of undiscounted GNP in 2010 to that for the Modeling Resource Group's base case. Source: National Research Council, *Supporting Paper 2: Energy Modeling for an Uncertain Future,* Committee on Nuclear and Alternative Energy Systems, Synthesis Panel, Modeling Resource Group (Washington, D.C.: National Academy of Sciences, 1978), p. 109.

Modeling Resource Group concluded that these variations could most adequately be explained by the value each assumes (explicitly or implicitly) for the price elasticity of demand (see Table 11-37).[35]*

Table 11-38 gives the results of assuming a "conservation tax" on energy in the ETA model that holds consumption to a constant level of 70 quads throughout the 1975–2010 period. For the Nordhaus model, as illustrated in Figure 11-12, successively more stringent limits are placed on the growth of energy consumption to achieve zero growth. Optimization determines the fuel mix at any given time in the period.

Any reduction in energy consumption that can be brought about without adding to the price reduces the conservation tax necessary to balance supply and demand, but the effects of various nonprice policies cannot be estimated from historical data. It must also be emphasized that the tax proceeds are assumed to be plowed back into the economy. If the tax were imposed, for example, by OPEC, this assumption would be violated to the extent that OPEC revenues were not offset by increased

*See statement 11-8, by R. H. Cannon, Jr., Appendix A.

TABLE 11-37 Estimated Price and Income Elasticities of Demand for Aggregate Energy in Three Models

Energy Models[a]	Elasticity of Demand for Aggregate Energy with Respect to: Price	Income
DESOM	small[b]	0.75
ETA (price elasticity, −0.25)	−0.25	1
ETA (price elasticity, −0.5)	−0.50	1
Nordhaus	−0.40[c,d]	0.90[d]

[a]The main features of the models used are described in the caption for Figure 11-10.
[b]Since DESOM does not incorporate energy price responses by end-use consumers, its price elasticity reflects only the adjustment (small in absolute value) of the process mix between primary extraction and end use.
[c]Made comparable to ETA price elasticities.
[d]Average of elasticities measured at historical 1970–1972 prices and at 2010 prices of the Nordhaus model base-case projection.

Source: Adapted from National Research Council, *Supporting Paper 2: Energy Modeling for an Uncertain Future,* Committee on Nuclear and Alternative Energy Systems, Synthesis Panel, Modeling Resource Group (Washington, D.C.: National Academy of Sciences, 1978), p. 110.

imports from the United States, and to this extent, the effect on GNP would be greater—up to the fraction of total GNP that constitutes payments for primary energy, probably not more than 5 percent.

The MRG results on the size of feedback effects confirm earlier results by other investigators and add further insight into the effect of the price elasticity of demand for energy. The first feedback study based on an econometric model (to the year 2000) was made by Hudson and Jorgenson[36] and was presented by the same authors[37] in greater detail for the years 1980 and 1985. Using a conservation tax (referred to by the authors as a "Btu tax"), this latter effort calculated that a tax of $0.50 per million Btu in 1980, compressing total energy input by 7.8 percent, decreases GNP by only 0.42 percent, all relative to the no-tax base case.

A few differences between the assumptions underlying these estimates and those made in the applications of the ETA and Nordhaus models should be noted. The Hudson-Jorgenson (H-J) model is an equilibrium model covering the entire economy (with four nonenergy sectors and five energy sectors). This aspect of the H-J model allows a more detailed

TABLE 11-38 Estimates of the Long-Term Feedback from Aggregate Energy Consumption on Cumulative Discounted Real Gross National Product[a]

Upper Limits on Growth Rate of Energy Consumption (percent per annum)	Implied Conservation Tax in 2010 (1975 dollars per million Btu)	Nordhaus		ETA: Price Elasticity Equal to −0.25		ETA: Price Elasticity Equal to −0.50	
		$\overline{E}^b$	$\overline{Y}^b$	$\overline{E}^b$	$\overline{Y}^b$	$\overline{E}^b$	$\overline{Y}^b$
None	0	1.000	1.000	1.000	1.000	1.000	1.000
1.5	0.34	0.893	0.999	—	—	—	—
1.0	1.02	0.751	0.997	—	—	—	—
0.5	1.98	0.631	0.995	—	—	—	—
0.0	3.19	0.531	0.991	—	—	—	—
Upper bound of 70 quads	[c]	—	—	0.419	0.711	0.419	0.982

[a] Total discounted GNP, $35 trillion for Nordhaus, $40.4 trillion for ETA (1975 dollars), calculated over the horizon of the model.
[b] $\overline{E}$ is the ratio of aggregate energy consumption in 2010 in the indicated scenario to that of the base case. $\overline{Y}$ is the ratio of cumulative discounted GNP in the indicated scenario to that of the base case.
[c] Price elasticity, −0.50 only; tax for electricity, 126 mills/kwh; tax for oil and gas, 8.9 dollars per million Btu.

Source: Adapted from National Research Council, *Supporting Paper 2: Energy Modeling for an Uncertain Future,* Committee on Nuclear and Alternative Energy Systems, Synthesis Panel, Modeling Resource Group (Washington, D.C.: National Academy of Sciences, 1978), p. 109.

tracing of the effects on energy-consuming industries of a tax-induced reduction in energy use.

In comparing the conservation-tax rates of the H-J, ETA, and Nordhaus models, it should be kept in mind that in the H-J model the tax is levied on the Btu content of energy as it leaves the energy sector for use by other sectors. In the ETA and Nordhaus analyses, the tax is implicitly levied on the use of primary energy or energy equivalent. This can be read as a levy on Btu content at the point of entry into the energy-producing and energy-conversion sectors. The principal difference is that in the Modeling Resource Group's analyses, the implicit tax on electricity is relatively much higher than in the H-J model owing to the low primary or secondary conversion efficiency of electrical generation.

In summary, the Modeling Resource Group considered only two types of policies: restricting the use of one or two energy technologies, and imposing a blanket tax on all forms of energy. It should be emphasized again that a tax on energy as such, whether on primary or ready-to-use forms, is not a suitable device for balancing the economic costs of curtailing energy use against the environmental benefits or other policy gains. Practical tax proposals to accomplish ends such as these would have to be tailored with care. The CONAES study has not investigated this issue. Qualitatively, the results would be similar to those shown for the blanket Btu or conservation tax, but would lead to a different mix of primary energy sources.

DISCUSSION

One of the puzzles emerging from the models and analyses of this study is the difference in shape of the curves projecting energy consumption over the next three decades. Those of the Demand and Conservation Panel (reflected in the curves of the study scenarios) show a considerable degree of saturation. They rise rapidly in the early part of the period 1975–2000 and level off late in the period. The curves of the Modeling Resource Group tend to be more nearly uniform over the same period.

The Demand and Conservation Panel's results were computed from a model based on prices of net delivered energy, while those of the Modeling Resource Group were computed on the basis of the price elasticity of demand for primary energy. As the price of primary energy rises, it constitutes an increasing proportion of the price of secondary energy. A model that assumes a constant price elasticity of demand for primary energy will correspond to a model in which the price elasticity of demand for secondary energy falls as prices rise. The model of the Demand and Conservation Panel, on the other hand, corresponds to a model that assumes a constant price elasticity of demand for secondary energy. This implies that the curves of the panel's projections should bend more than those of the Modeling Resource Group with rising prices, and that the difference between the two will be more marked with greater assumed increases in the total price of energy.

It is difficult to make a confident choice between the two price elasticities, particularly as to which yields more realistic results.

It is important to note a factor that could make the lower-energy-growth scenarios easier to achieve than implied here. The energy conservation shown in all the scenarios is achievable by the application of *known* technology or of technological principles that have already been demonstrated. It does not incorporate contributions to energy efficiency from major technological innovation. Given a favorable political and economic climate for innovations in energy efficiency, substantial opportunities exist to develop and market ingenious new energy-conserving technologies. In not allowing for human ingenuity, the scenarios may *understate* the actual potential for moderating the consumption of energy. The ingenious use of information technologies (including microprocessors) to direct and control energy more selectively is still in its infancy and may have more potential than can now be envisaged. Such a favorable development could offset any shortfall from the conservation estimated in the scenarios, especially if energetically promoted by a combination of aggressive private marketing and highly supportive public policies.

NOTES

1. National Research Council, *Supporting Paper 2: Energy Modeling for an Uncertain Future,* Committee on Nuclear and Alternative Energy Systems, Synthesis Panel, Modeling Resource Group (Washington, D.C.: National Academy of Sciences, 1978).
2. See, for example, American Gas Association, *Gas Supply Review* 5 (1977); National Research Council, *Supporting Paper 4: Geothermal Resources and Technology in the United States,* Committee on Nuclear and Alternative Energy Systems, Supply and Delivery Panel, Geothermal Resource Group (Washington, D.C.: National Academy of Sciences, 1979) and Gordon J. MacDonald, *The Future of Natural Gas* (McLean, Va.: Mitre Corp., 1979).
3. See also chapter 2 of this report, and for a detailed account see National Research Council, *Alternative Energy Demand Futures to 2010,* Committee on Nuclear and Alternative Energy Systems, Demand and Conservation Panel (Washington, D.C.: National Academy of Sciences, 1979).
4. Ford Foundation, Energy Policy Project, *A Time to Choose: America's Energy Future* (Cambridge, Mass.: Ballinger Publishing Co., 1974).
5. Exxon Corporation, "World Energy Outlook," in *Exxon Background Series,* Public Affairs Department (New York: Exxon Corporation, April 1978), p. 7.
6. Edison Electric Institute, *Economic Growth in the Future,* executive summary, Committee on Economic Growth, Pricing and Energy Use (New York: Edison Electric Institute, February 1976), p. 13.

7. Institute for Energy Studies, *Energy and the Economy,* Energy Modeling Forum Report no. 1, vols. 1 and 2 (Stanford, Calif.: Stanford University, 1977).
8. Demand and Conservation Panel, *op. cit.*
9. *Ibid.,* chap. 3.
10. E. Hirst, W. Lin, and J. Cope, *An Engineering-Economic Model of Residential Energy Use,* (Oak Ridge, Tenn.: Oak Ridge National Laboratory (ORNL/TM-5470), July 1976); and E. Hirst *et al., An Improved Engineering-Economic Model of Residential Energy Use,* (Oak Ridge, Tenn.: Oak Ridge National Laboratory (ORNL/CON-8), April 1977).
11. Demand and Conservation Panel, *op. cit.*, chap. 3.
12. J. R. Jackson and W. S. Johnson, *Commercial Energy Use: A Disaggregation by Fuel, Building Type, and End Use* (Oak Ridge, Tenn.: Oak Ridge National Laboratory (ORNL/CON-14), February 1978).
13. J. R. Jackson, S. M. Cohn, J. Cope, and W. S. Johnson, *The Commercial Demand for Energy: A Disaggregated Approach* (Oak Ridge, Tenn.: Oak Ridge National Laboratory (ORNL/CON-15), April 1978).
14. Demand and Conservation Panel, *op. cit.*, chap. 5.
15. For the sectoral analysis, "mass transportation" was assumed to include school buses, local and intercity buses, subways, and elevated railways.
16. Demand and Conservation Panel, *op. cit.*, chap. 4.
17. *Ibid.*, chap. 6 and app. A; see also C. W. Bullard and R. A. Herendeen, *Energy Impact of Consumption Decisions* (University of Illinois at Urbana: Center for Advanced Computation (CAC Document no. 135), October 1974), reprinted in *Proceedings of the Institute of Electrical and Electronics Engineers* 63 (March 1975): 484 – 493.
18. The Department of Commerce issues these data at irregular intervals.
19. Demand and Conservation Panel, *op. cit.*, chap. 6 and app. A.
20. See the report of the Consumption, Location, and Occupational Patterns Resource Group of the Synthesis Panel.
21. National Research Council, *U.S. Energy Supply Prospects to 2010,* Committee on Nuclear and Alternative Energy Systems, Supply and Delivery Panel (Washington, D.C.: National Academy of Sciences, 1979).
22. Since the work of the Supply and Delivery Panel was mostly done in 1976, "existing policies" generally refer to those of that period. The effects of policy changes since that date (though not expected to be large) are generally favorable to slightly enhanced supplies.
23. The Supply and Delivery Panel did not assess the quantitative trade-offs among supply sources implicit in this warning.
24. National Research Council, *Supporting Paper 6: Domestic Potential of Solar and Other Renewable Energy Sources,* Committee on Nuclear and Alternative Energy Systems, Supply and Delivery Panel, Solar Resource Group (Washington, D.C.: National Academy of Sciences, 1979).
25. U.S. Department of Energy, *Annual Report to Congress, 1978,* vol. 3, Energy Information Administration (Washington, D.C.: U.S. Department of Energy, 1979).
26. *Conservation Foundation Letter,* July 1979, pp. 2–3.
27. The study scenarios were developed by J. M. Hollander and R. Silberglitt.
28. Edison Electric Institute, *op. cit.*
29. "29th Annual Electrical Industry Forecast," *Electrical World,* September 15, 1978, pp. 68–69.
30. The Modeling Resource Group undertook two tasks. The task not described here is a systematic examination of the economic desirability of government-funded research and development of various energy technologies. See Modeling Resource Group, *Supporting Paper 2, op. cit.*

31. As defined by the Modeling Resource Group, "nuclear moratorium" is construed as allowing the operation of existing light water reactors and the completion of those under construction. The MRG estimates that this policy would result in no new additional nuclear capacity after 1983, but would permit 70 GWe of nuclear capacity by that year. The nuclear power plants would be retired after about 30 years of operation.

32. See the appendix to chap. 2, "A Word about GNP," and W. Nordhaus and J. Tobin, "Is Growth Obsolete?" in *Economic Growth* (New York: Columbia University Press, 1972).

33. Modeling Resource Group, *Supporting Paper 2, op. cit.*

34. Energy Modeling Forum, *Energy and the Economy,* Institute for Energy Studies (Stanford, Calif.: Stanford University, September 1977).

35. Demand for aggregate energy, measured in primary energy equivalents.

36. Ford Foundation, *op. cit.*, app. F.

37. "U.S. Energy Policy and Economic Growth, 1975–2000," *The Bell Journal of Economics and Management Science* 2 (Autumn 1974): 461–514, especially section 6.

APPENDIX A

Individual Statements by CONAES Members

GENERAL COMMENTS

KENNETH E. BOULDING

I am glad to accept the report as the product of 4 years of very hard work on an extremely intractable problem, in regard to which there are unusually wide but legitimate divergences of opinion. It may be that the most significant conclusion of this report is its constant emphasis on the profound uncertainties that beset even the most crucial aspects of this problem. This is cold comfort to the decision makers, whose position indeed is not to be envied. With all the evidence and wisdom that the scientific community can muster, we are forced to admit that our areas of ignorance in this subject are very large. Under these circumstances, the best advice to the decision maker is to avoid delusions of certainty and to put a high premium on decisions today that allow a wide range of decisions tomorrow. If there is any conclusion to this report of practical significance, it is that there is a strong case for having our eggs in as many baskets as possible and that we should avoid foreclosing any line of development too soon.

I am prepared to accept the report, therefore, as one of a long series of interim statements, each of which provides the basis for the next. There are a number of aspects of the report that make me uneasy, none of which are wholly indefensible, but which should provide material for further discussion and study. I wish the report had stressed more the divergence of the feeling on the committee instead of trying to concentrate on areas of agreement, for these divergences are a very important part of the picture. They do come out in the concluding comments by members of the committee, and these should be taken not as a sign of failure but as a sign of the immense difficulty and complexity of the problem and as pointers toward further work.

I would raise one or two questions about the general assumptions underlying the study that might lead to misleading conclusions if they were not brought up. One is the assumption implicit in many of the models that the GNP of the United States will grow by at least 2 percent over the next few decades, though there is some recognition that this rate of growth will eventually slow down. It seems to me at least possible that rates of economic growth will be much less than this. It is very dangerous to extrapolate the rates of growth for a very complex aggregate like the GNP.

Two factors suggest that the GNP of the United States may have a much slower rate of growth in the future than it has had in the past few decades. One is that as technological development proceeds, those industries that are subject to productivity increase usually decline in relative importance as the economy moves into industries and occupations where the increase in productivity is very difficult, like education, the arts, medicine, and government. A good deal of the increase in GNP in the last 30 years came out of the extraordinary increase in productivity in agriculture, which released some 30 million people from agriculture to produce other things. This process is now reaching its conclusion, even in agriculture, and agriculture is now such a small proportion of the labor force that even a substantial increase in productivity would not release very many people. Productivity in manufacturing was not rising very rapidly even in the last 30 years, and has now virtually slowed to a stop. Increase in knowledge and improvements in technology will, of course, continue to take place and will in part offset the increasing costs of energy and materials. But whether this offset will be sufficient to prevent virtually stationary or even declining GNP per capita is a real question. If we add to this the change in the political climate in the last 10 or 15 years, which has imposed great uncertainties on private enterprise and has created an increasing unwillingness to take risks, particularly those imposed by somewhat uncertain regulations, the chances of actual decline in GNP per capita seem much greater than the economics profession is willing to recognize.

It may be, of course, that this gloomy prophecy will be falsified as many similar prophecies have been in the past by unexpected advances in knowledge and technology. There are indeed a few possible technical developments that would transform the whole energy and economic picture of the world in a relatively short time. The development of a cheap and portable battery for storing large quantities of electricity might transform the whole energy picture. It would make electricity a real substitute for fuels, which now it is not.

Something indeed which the report implies, but does not perhaps emphasize enough, is the great heterogeneity of the energy problem. Energy sources are very heterogeneous, and energy uses are even more so; the structure and storage of different forms of energy may be much more important in the total picture than the number of "quads." We must not be misled by the physical homogeneity of the measurement of energy in terms of ergs or Btu's, for energy is only significant socially when, where, and in the form that it is wanted by human beings. These times, places, and forms are extremely diverse, and are frequently not substitutable one for another.

I detect in the report a slight prejudice in favor of nuclear power and a certain prejudice against biomass and energy farms, though great pains have been taken to present all points of view fairly and even to try to reach a compromise statement, which like most compromise statements will not satisfy any of the contending parties. The main thing I have learned myself in the course of this study is to recognize the enormous uncertainties that are involved, even in such a basic question of future policy as to whether the net advantages of coal are greater than those of various forms of nuclear energy. We do not really know whether carbon dioxide is worse than plutonium as a hazard to the human race. Professor Holdren's claim that the report underestimates the long-run hazards of nuclear power seems to me to have some weight and broadens still further the spectrum of uncertainty. It is also true, however, as Professor Cannon has suggested, that the report does not deal with the social and human costs of severe energy shortages, which could be very large. The balancing of these costs and benefits in the light of the enormous uncertainties is an unenviable task, and one hopes that the sense of uncertainty will at least modify the heat of the inevitable controversies.

A curious general characteristic of the report that reflects, however, almost all discussions of this subject is that the "risk" always involves costs or negative goods, whereas benefits are often implicitly assumed to be certain. Under these circumstances, risk of the loss of benefits can easily be grossly underestimated, and this can distort the whole judgment in regard to the net benefits of different policies. If overestimation, or over-visibility, of the real costs of different forms of energy leads to a loss of the benefits—often invisible and taken for granted—we may find ourselves in very bad shape. While this is certainly recognized, there is something in the rhetoric of the discussion of this problem that takes benefits for granted and puts all the emphasis on uncertain costs. The environmental and antinuclear movements are particularly subject to this danger, which should be pointed out to them.

The staff of the project is much to be commended for trying to put together the various scenarios that emerged from the Demand and Conservation and the Supply and Delivery panels, respectively, as well as the modeling groups. The value in these scenarios, however, is that they point to a very wide range of possible futures, which have at least some degree of probability and stress the need, again, for highly flexible policies, constantly subject to revision as new data come in from experience.

A methodological point is that in model building and scenario construction there is everything to be said for doing the simplest possible models first, then elaborating the models with successive introduction of complexities. The Modeling Research Group model is a very good case in point. An extremely simple model would have come to much the same basic conclusion that the elaborate model came to, and it would be much more comprehensible to the ordinary reader. The conclusion, for instance, that a substantial rise in the real cost of energy could take place with only a relatively small impact on the GNP will surprise many readers, and it will not be clear to them how this emerges out of the elaborate model, simply because the elaborate model cannot be understood, even by the quite sophisticated reader, who has not actually participated in its construction.

A very simple model, however, will illustrate the point. What might be called the "energy industry" now represents something like 7 percent of the GNP. If the real price of energy quadruples, as it might do by the early twenty-first century, this will rise perhaps to 15 or 20 percent. The proportion itself will not quadruple because the rise in the price of energy will offer very large incentives for conservation and for technical changes that economize energy in both consumption and production. Even if energy goes to 20 percent of GNP, however, this still leaves 80 percent for other things instead of 93 percent. Quite small changes, therefore, in improvement in productivity of other parts of the economy would offset the increase in the real cost of energy, so it is not surprising that unless the overall impact in the rise of energy prices is very small indeed, both in terms of elasticity of demand and of the impact on technology both of production and of utilization, the conclusion that the impact on GNP will not be very large is quite reasonable, although there are circumstances in which the technological changes will fail to come through, and the effect would be much larger and much more deleterious than the above very simple model would suggest.

All model building involves assumptions about constancy of parameters of the system. In social systems, however, parameters are not constant, which is why model building must always be treated as productive of significant but rather dubious evidence, and certainly never as productive of truth. The great uncertainties here are in the area of the future of human knowledge, know-how, and skill. There is a nonexistence theorem about prediction in this area, in the sense that if we could predict what we are going to know at some time in the future, we would not have to wait, for we would know it now. It is not surprising, therefore, that the great technical changes have never been anticipated, neither the development of oil and gas, nor the automobile, nor the computer.

In preparing for the future, therefore, it is very important to have a wide range of options and to think in advance about how we are going to react to the worst cases as well as the best. The report does not quite do this. There is an underlying assumption throughout, for instance, that we will solve the problem of the development of large quantities of usable energy from constantly renewable sources, say, by 2010. Suppose, however, that in the next 50, 100, or 200 years we do not solve this problem; what then? It can hardly be doubted that there will be a deeply traumatic experience for the human race, which could well result in a catastrophe for which there is no historical parallel.

It is a fundamental principle that we cannot discover what is not there. For nearly 100 years, for instance, there have been very high payoffs for the discovery of a cheap, light, and capacious battery for storing electricity on a large scale; we have completely failed to solve this problem. It is very hard to prove that something is impossible, but this failure at least suggests that the problem is difficult. The trouble with all permanent or long-lasting sources of energy, like the sun or the earth's internal heat, is that they are extremely diffuse and the cost of concentrating their energy may therefore be very high. Or with a bit of luck, it may not; we cannot be sure. To face a winding down of the extraordinary explosion of economic development that followed the rise of science and the discovery of fossil fuels would require extraordinary courage and sense of community on the part of the human race, which we could develop perhaps only under conditions of high perception of extreme challenge. I hope this may never have to take place, but it seems to me we cannot rule it out of our scenarios altogether.

I myself feel I have learned a great deal from being on this committee. I became aware of the enormous complexity of the problem in a way I had not been before, and many of my views changed quite radically as a result. If the report can have the effect of questioning established positions on all sides of the great controversies, it will have been well worth doing.

A very crucial problem that underlies the report, particularly important to the scientific community, is where are the areas in which research in pursuit of further knowledge and skill is most likely to pay off in the next generation? Whether this question could have been directly addressed in a separate chapter is a matter of debate, but one hopes that this is the question that will be most frequently asked by those who read the report.

I would personally like to thank the staff of the committee for their patience and persistence in an extraordinarily difficult task. The rewards of anonymous writing are meager, and this report is testimony to the devotion and self-sacrifice of those who have written it.

HARVEY BROOKS

I concur with the above statement.

BERNARD I. SPINRAD

Definition of the Energy Problem: I wish it had been stated more forcefully in the report that we have, not an "energy" problem, but an *oil* problem. By focusing on the primacy of the oil problem, it becomes clearer what we have to do:

- Use less oil
- Get more oil

We could get eloquent about all the reasons that this is *the* problem. The impact of our oil use on a seller's market is the most demanding of them. Industrializing countries of the third world are impacted by our continued consumption of the most appropriate fuel for them at this stage of their developments and by our reluctance to reduce our consumption by fuel substitutions or resource substitutions.

Energy conservation is a fine thing. I am all for it. Yet, it will do us little good unless it is applied in such a way as to conserve oil. For example, if it takes more energy to substitute coal for oil, we should nevertheless do it, and do it quickly. The same thing holds for substituting electricity, made from coal or nuclear power, for oil.

In this context, nuclear power offers, at present, an excellent combination of economy, environmental blandness, and low health effects. It therefore deserves support for its properly regulated expansion.

Similarly, it would be very nice to get, if we can, economical energy from solar sources. However, it would not be worth a great economic "front-end" payment (a payment that could be used for other things, and is rarely recovered) unless this energy replaces oil use.

There are problems, of course, with all substitutions. Some are environmental, some are political. These problems can be resolved by proper use of technology, good management in the public interest, and the replacement of adversary-type political confrontations by honest attempts to understand problems and come up with solutions. I believe that CONAES has made such an attempt, and that is why I endorse the report.

CHAPTER 1 STATEMENTS

1-1 HARVEY BROOKS

Natural gas from "unevaluated and unconventional sources" could also contribute to arresting this decline. Compare Tables 22 and 25 of the CONAES Supply and Delivery Panel report (National Research Council, *U.S. Energy Supply Prospects to 2010,* Committee on Nuclear and Alternative Energy Systems, Supply and Delivery Panel (Washington, D.C.: National Academy of Sciences, 1979), pp. 81, 84). For the enhanced-supply case, which probably corresponds to the new conditions as of 1979, as much as 6 quads could be added to the gas supply by 2010, keeping production approximately level with 1975. This does not include highly speculative abiogenic sources, which have received recent press attention but are too conjectural to be included in quantitative estimates.

1-2 JOHN P. HOLDREN

The assertion that coal and nuclear fission are the "only readily available domestic energy sources that could even in principle reverse the decline in domestic energy production over the next three decades" rests on a judgment I do not share. This judgment is that the obstacles to significant penetration of the energy mix by renewable energy sources in this period are more fundamental and less tractable than the obstacles in the way of expanded use of coal and nuclear fission. The obstacles for the renewables are technical and economic—extensive penetration between 1990 and 2010 would require some technical breakthroughs yielding large cost reductions early in the period, or willingness to spend significantly more for renewable energy supplies than we have been spending for conventional ones. The obstacles hindering coal and nuclear are different—they are environmental and sociopolitical more than technical and economic—but they are neither less real nor more easily circumvented than the liabilities of the renewables. The choices are increased flirtation with CO_2-induced climatic change, other potentially excruciating environmental costs of coal, and nuclear debacles (those arising from malevolence as well as from miscalculation and mismanagement), on the one hand, and the probability of considerably higher energy prices (for renewables), on the other. The notion that society should prefer the former to the latter may be the majority view of this committee, but that position should be recognized as a value judgment that does not deserve to be paraded as the "only" possible outcome.

1-6 EDWARD J. GORNOWSKI

I and others had repeatedly challenged the Modeling Resource Group on the composition of the "consumer market basket." The answers given were broad, with no indication of a real understanding of the specific implications of the model output. Specifically the answers were: "The goods would be less energy intensive"; "we'll have a lot more oboe players"; "consumers will own more pairs of shoes." I submit that although the report carefully qualifies the results of the work of the Modeling Resource Group, chapter 1 places too much emphasis on the results. The models in my opinion are not sophisticated enough to extrapolate so far into the future. Further, the models address only the situation in the terminal years. There has been no real attempt to explain that perhaps serious economic problems could occur during intervening years, and what might be done about these.

1-8 HENDRIK S. HOUTHAKKER AND HARVEY BROOKS

A 2 percent growth rate would be consistent with full employment only if output per man-hour grew much more slowly than in the post–World War II period.

1-10 EDWARD J. GORNOWSKI

This statement needs to be repeated perhaps more often in the preceding discussion. The models really provide mechanisms for carrying out the calculations necessary to indicate the implications of certain types of assumptions. For instance, when one assumes a "high energy tax" that will be "pumped back" into the economy and a high elasticity factor, it is almost obvious that the impact of decreasing energy demand on the GNP will be small. This in no way suggests that I do not agree that the energy/GNP ratio will decrease. It is simply a question of how much, and although a low energy growth may provide an adequate standard of living, a higher energy growth may well provide an even better one.

1-13 JOHN P. HOLDREN

It is misleading not to note that pure price increases were consciously used in the work of the Modeling Resource Group as a *surrogate* for the nearly infinite variety of combinations of increased prices and conservation-inducing policies that might be used in real life in place of price alone. There was no consensus in the committee as to the relative role that price increases and policies not related to price should have in the promotion of more efficient energy use. The issue was finessed by letting price represent the combined intensity of price and nonprice conservation pressures, mainly because the effects of price could be rather easily captured by the economic models at hand, while the effects of policies that could substitute for price increases could not be. Thus the seeming unreality of the price increases the text associates with the lowest-growth energy futures should not lead the reader to reject these futures as implausible; they could come about at lower price levels than stated, with the help of suitable policies. (This clarification also disposes of the otherwise natural objection that such low-energy futures at such high energy prices are inconsistent economically because of the enormous energy supplies that would be forthcoming at these prices.)

1-14 EDWARD J. GORNOWSKI

The reserves of oil shale in this country are vast and should be included with coal and nuclear. The report as written presupposes that oil shale extraction will not be developed to any significant extent. It seems possible to me that we will find ways to exploit the huge reserves.

1-21 HARVEY BROOKS

There is some indication that the potential for unconventional gas sources, such as Devonian shales, coal seams, and geopressured brines, may have been underestimated, especially in relation to prospects for ultimate decontrol of gas prices and proposed tax benefits for unconventional gas sources. Some recent estimates (such as those of the Electric Power Research Institute) of natural gas production by 2000 have run as high as 30 quads. Since gas is substitutable for fuels derived from petroleum in a large number of applications, new sources of gas may represent the most likely favorable future development to offset the forecast rise in demand for oil imports. Hence energetic further exploration and assessment of this possibility are warranted. I would be inclined to give this even higher priority than the suggested pioneer plants for oil shale and coal-derived liquids. However, like other possible favorable developments in supply, enhanced natural gas supplies cannot be counted upon in prudent planning in comparison with fully developed technologies such as coal and nuclear electric power generation, which are already commercially proven.

1-29 HENDRIK S. HOUTHAKKER, EDWARD J. GORNOWSKI, AND LUDWIG F. LISCHER

Who decides what are "unnecessarily high rates of growth in electricity demand"? The development of electricity has been a major contributor to our economic performance and is likely to be equally important for the presently less developed countries.

1-30 LUDWIG F. LISCHER AND EDWARD J. GORNOWSKI

The counter view is not stated. That view holds that even with a moderate growth in demand for electricity after 1990, the development of the LMFBR is not only desirable but necessary. The LMFBR is further along in development than any other advanced reactor. A prudent basis for planning energy policy, it seems to us, should not rely on completely achieving all the goals of conservation and extreme optimism on uranium resources. History tells us that future events rarely, if ever, turn out as planned. Therefore, proceeding at a reasonably expeditious pace with the LMFBR is a necessity if as a nation we wish to have this resource available to us on a commercial basis by the end of this century. At that time it may turn out to be a vital necessity; and if not, at the worst, it provides a reasonable-cost insurance.

In our view, the nuclear industry will not undertake commercialization of advanced converters because at best the converter is an interim solution (good for perhaps 20–30 years) and neither the suppliers nor the users will believe that there is sufficient incentive to bring it to commercial status.

Without reprocessing, the growth of nuclear power will be slow, at best. No manufacturer could afford the high development costs to bring an interim nuclear reactor system to licensable status. If reprocessing is permitted, any advanced converter would have to compete with the fast breeder reactor. In that event, the breeder is the obvious choice.

No mention is made of the Clinch River breeder reactor (CRBR), yet this was the primary issue leading to the formation of CONAES. To proceed with the orderly development of the LMFBR, the construction of the CRBR is of vital importance. The United States has operated successfully the 20-MWe EBR-II at Idaho Falls for over 15 years. CRBR represents the next logical step (350–400 MWe) in scaling up plant size. It is essential to construct and operate a unit of this size prior to proceeding with commercial designs on the order of 1000 MWe. At the present stage of development, one learns little from more paper and analytical work as proposed by some. The direct scale-up from 20 MWe to 1000 MWe is simply too large a step for prudent engineering and design. CRBR is not an outmoded plant; its design has been continually updated, and it has flexibility for accommodating a variety of nuclear fuel and core designs.

(Harvey Brooks: I subscribe to the views expressed in the first two paragraphs of the above statement.)

(Henry I. Kohn: I agree with the general approach of the above statement.)

1-32 HARVEY BROOKS

This is not a very likely example, since nuclear power is only useful as base load. It would be plausible only if the load curve were considerably leveled, e.g., due to the widespread use of electric cars with batteries charged on off-peak power, or the production of hydrogen by off-peak power, or by an inexpensive energy storage system. Some progress, however, is being made in the development of fuel that is more resistant to thermal cycling and hence suitable for use in reactors operating in a load-following mode.

1-36 JOHN P. HOLDREN

Tailings piles, under present practices, are the largest source of ultimate human radiation exposure from the routine operation of nuclear power. If the linear hypothesis about radiation damage is correct, the million-year burden of extra cancer deaths produced by these tailings, although undetectable against the background of cancers from other causes, could amount to a total that almost certainly would be deemed unacceptable if it had to be borne by the present-generation users of the electricity. This situation poses an ethical dilemma that is not made less troublesome by the possibility that other energy sources also produce health costs that are spread over millennia (e.g., toxic effects of trace metals mobilized by burning coal) but that cannot yet be estimated quantitatively. In these circumstances, I am unconvinced that one should "solve" the problem of alpha-emitting wastes from elsewhere in the fuel cycle by making the tailings problem worse by even an iota. If the tailings problem itself were actually solved today, in the form of the existence of a scheme that manifestly would reduce the ultimate human exposure from this source by, say, a factor of 1000, I would feel differently about putting other alpha wastes in the same basket.

1-40 BERNARD I. SPINRAD, HARVEY BROOKS, AND DAVID J. ROSE

This statement, made as a catalog of fears popular among nuclear opponents, is correct. Nevertheless, the fears themselves are neither peculiar to nuclear power nor necessarily commensurate with physical realities. A similar criticism can be made of the sentence at the start of the next paragraph, stating what many supporters of nuclear power believe. Such myths must be discarded and replaced by usable information if reasonable resolution of the outstanding nuclear issues is to be made.

1-42 LUDWIG F. LISCHER, HARVEY BROOKS, AND DAVID J. ROSE

The section "Public Appraisal of Nuclear Power" appears to say that public appraisal dominates (or should dominate) the role of nuclear power in the future. Public perceptions are important. But if they are based on erroneous or distorted information, then there is a role for government and other institutions to correct those perceptions by providing facts in an understandable manner.

1-43 BERNARD I. SPINRAD, HARVEY BROOKS, AND LUDWIG F. LISCHER

Since nothing is certain, this statement cannot be disputed. However, the technical grounds for selecting the LMFBR as the breeder of choice were strong when that decision was made originally, and no intervening technical developments have ensued that would negate the decision. Unless alternative breeders are, in fact, developed toward commercialization first, nothing in the future is likely to change that decision, either.

1-46 BERNARD I. SPINRAD

This argument rests, in my opinion, on double counting of social costs. The massive controls and restrictions on nuclear power have forced internalization of not merely its own intrinsic social costs, but also the social costs that have been artificially loaded onto it by politicized opposition. The social costs of coal seem to be going the same way. If solar power cannot make the grade economically, given this favorable handicap, it doesn't deserve to be further stimulated.

1-47 LUDWIG F. LISCHER

"Additional technical developments," "best technology," "final choice," and "lowest risk" are terms used in several places in this chapter—each time with the implication that nothing can be done now because we do not know what is best or lowest in risk or what additional technical developments will bring. This seems to me a negative outlook. If one waits until everything is known, then nothing is ever accomplished. At some point, one can only learn more by doing rather than by further studying.

1-49 LUDWIG F. LISCHER

Technically the preceding statements on risks are correct, but they do little to help public understanding. All risks are relative, and unless we make comparisons (even if they are less than 100 percent correct), we do little to assist people. For example, to say that the maximum calculated dose received by an individual was 80 mrem (as in the case of Three Mile Island) is less than helpful unless one adds that if one moves from Chicago to Denver he will receive an increased dose of 80 mrem/yr because the natural background radiation in Denver is about that much higher than in Chicago. If public perceptions are important (as stated elsewhere in this chapter), then surely comparisons of risks in an understandable manner are pertinent to energy policy.

1-50 BERNARD I. SPINRAD, HARVEY BROOKS, LUDWIG F. LISCHER, AND DAVID J. ROSE

We cannot concur with this policy, and we think that the stated reason is fallacious. It is important to compare energy-related risks with nonenergy risks that are accepted, to gain perspective.

Overemphasis, often to the extent of single-minded concentration on risk reduction from energy sources—often, of particular energy options—diverts attention from more serious problems. We have:

- Risks of war
- Risks of poverty
- Risks of disease
- Risks of crime
- Risks of "normal" accidents

All of these risks are major, and the rather low risks of properly controlled use of coal and the very low risks of properly controlled use of nuclear energy pale by comparison.

We enter into the "how safe is safe enough?" controversy here. It is unpopular to attack the problem objectively, so the graceful cop-out of calling it a social and political issue was used by CONAES. Yet, how can people make social and political decisions that are valid in the absence of contextual information and evaluation?

Another area where objective thinking needs to be done is the issue of immediate risks versus delayed risks. We would be far more willing to take a risk, such as exposure to a carcinogen, that might lead to morbidity 20 years from now than we are to take a risk of equal probability that would lead to similar harm immediately. (Such a risk might be in the class of letting hunters practice their hobby within half a mile of residences.) Yet, much of the literature, and virtually all of the recent press coverage of risks, concentrates on how much more scary delayed risks are. Is this a real psychological fact, or a learned response? If it is a learned one, shouldn't it be unlearned?

1-51 LUDWIG F. LISCHER AND HENRY I. KOHN

It is interesting to note that both the Clean Air Act and the Water Quality Act essentially prohibit cost-benefit reasoning in standard setting and implementation. (See chapter 9, "Regulation".)

1-52 HARVEY BROOKS AND DAVID J. ROSE

In our opinion the section on emissions gives a misleadingly optimistic impression of the health risks associated with air pollution from the burning of coal, especially in comparison with the risks of nuclear power. Admittedly the epidemiological studies that have so far been conducted are of questionable validity; see chapter 9 for a detailed assessment of these. However, because we cannot quantify adverse health effects with the same confidence as in the case of ionizing radiation, it would be wrong to conclude that the risks of coal are not substantially greater than those of nuclear power with high probability.

1-53 JOHN P. HOLDREN

It is not enough to note that the use of more reasonable uncertainty bounds alone would make the expected number of fatalities from nuclear accidents larger by a factor of 10 or more than the median value stated in WASH-1400. The size of the risk at the upper end of the uncertainty range—its value if the pessimists are right—is also relevant to the public's comparison of the liabilities of this technology with the liabilities of alternative ways to get electricity. (That a large uncertainty is itself a liability, above and beyond the liability associated with the "best-estimate" or "expected" consequences, is a well-established principle in benefit-cost analysis.) WASH-1400's own estimate of the upper limit is higher than its median value of 0.024 deaths per reactor-year by a factor of about 15 (3 in consequences and 5 in probability). But the prestigious Ford/MITRE study (*Nuclear Power: Issues and Choices* (Cambridge, Mass.: Ballinger Publishing Co., 1977), p. 179) found that "the WASH-1400 probability estimate could be low, under extremely pessimistic assumptions, by a factor of as much as 500" and that the expected number of cancers for a given accident "could be several times higher" than in WASH-1400, based on the dose-response modeling alone (leaving out, for example, uncertainties in the dispersion model). The product of these Ford/MITRE "upper limits" on probability and consequences implies an upper limit risk 1500–3000 times the WASH-1400 median value, or 36–72 cancer deaths per reactor-year. This result puts the upper-limit health risk of nuclear power well above the upper-limit health risk from burning coal in new power plants. (See my dissenting view on the coal-nuclear comparison, statement 1-59, Appendix A.)

1-55 LUDWIG F. LISCHER

This statement is in conflict with Sandia and NRC studies that conclude that sabotage could cause embarrassment and public apprehension, but actual harm to the public is extremely unlikely.

1-56 HENDRIK S. HOUTHAKKER

While hydroelectricity destroys old ecosystems, it creates new ones that are not necessarily less valuable. Moreover, hydroelectricity is benign in environmental respects other than damage to ecosystems.

1-59 JOHN P. HOLDREN

The situation is even more ambiguous than the text suggests because it is not actually possible to do what is implied by the words, "if one takes all health effects into account." Specifically, the statement that coal's health effects "appear to be a good deal greater" than nuclear's requires either that one ignore the million-year accumulation of excess cancer deaths plausibly attributable to uranium-mill tailings if the linear hypothesis is accepted (an excess that could amount to 30–400 deaths per GWe-year of electricity, according to the Academy's own recent report, *Risks Associated with Nuclear Power: A Critical Review of the Literature,* Committee on Science and Public Policy, Committee on Literature Survey of Risks Associated with Nuclear Power (Washington, D.C.: National Academy of Sciences, April 1979)), or that one assume without any quantitative support that the health effects of today's coal use over the next million years will be as large or larger. Suppose this troublesome issue, which is completely unresolvable at present, is neglected. Suppose one also neglects genetic illness from nontailings routine emissions, for which there is an uncertainty range spanning at least a factor of 20 on the nuclear side (extending on the high end to consequences about equal to those of the excess cancers) and for which no quantitative estimates at all are available on the coal side. Suppose one neglects, further, the health effects of emissions of oxides of nitrogen, hydrocarbons, and trace metals from coal combustion, which are generally presumed (but not proved) to be smaller than those of the sulfur oxides and generalized particulates that existing dose-response relations include. In the restricted comparison that remains after all these troublesome factors are excluded, the widespread view that coal comes out "a good deal" worse than nuclear rests on various combinations of the following three errors: (1) attribution to coal of practices in mine health and safety, power plant siting, and pollution control that are illegal or inconceivable (and usually both) in the new, large facilities relevant to comparisons with nuclear power; (2) failure to note that the "excess deaths" attributed to air pollution from coal typically deprive the victims of far less life expectancy than the cancers attributed to nuclear power (the ratio is almost certainly more than 10:1); (3) refusal to take seriously the upper end of the range of responsible opinion on conceivable nuclear accident risk, while taking completely seriously the upper-limit estimates of excess deaths from air pollution. When these errors are avoided, the "best estimates" of the years of life lost per GWe-year of electricity from coal and nuclear differ by an amount small compared to the uncertainties associated with each.

1-60 HENRY I. KOHN AND HARVEY BROOKS

The dangers of nuclear power are primarily contingent on the probability of major accidents (or sabotage) that release radioactivity and thus endanger employees and the general public. These probabilities are under discussion (they are largely hypothetical projections) and have large ranges above and below their best estimates (median, mean, or otherwise). The risks in the coal energy cycle, on the other hand, arise from different steps (mining, transportation, routine emissions from power plants), all of which are susceptible to materially better control in the future than the overall average at present. Given these variables, the reader may appreciate the complexity underlying any brief, simple statement.

1-61 HARVEY BROOKS, DAVID J. ROSE, AND BERNARD I. SPINRAD

Although we agree with the statement, we fear it might be interpreted as implying that government planning should accept the most irrational appraisals of risk put forward by politically active minorities.

1-67 HENRY I. KOHN AND LUDWIG F. LISCHER

We object to some of the general attitudes that appear to underlie this paragraph. As one major example, the paragraph fails to mention its beneficiaries as major determinants of schemes for their own benefit. As another, we note that there are problems that cannot be fruitfully solved with money. The subject is politically more complex and philosophically more sophisticated than this paragraph indicates.

1-71 LUDWIG F. LISCHER AND HARVEY BROOKS

But if we place great reliance on CONAES's study projections (which are in reality ranges of possibilities), we may well turn out to be wrong, with far-reaching consequences.

1-74 HARVEY BROOKS

By this risk is meant the probability times fatal consequences, both delayed and prompt, integrated over the full spectrum of possible accidents. If the probability distribution of each event is assumed to be log normal, then the mean is much larger than the median, and hence the mean fatality rate is sensitive to the width of the error bar attributed to the accident risk calculations. The events that make the greatest contribution to the mean risk are those that lead to small incremental exposures over background to relatively large populations; hence the mean risk is also sensitive to whether the linear dose-response hypothesis is adopted.

CHAPTER 2

2-1 ROBERT H. CANNON, JR.

The energy/GNP ratio of the United States started to fall in 1923 (Figure 2-2). Six years later came the Great Depression, which continued until World War II started in 1941. Is this significant?

2-2 JOHN P. HOLDREN

It is most unfortunate that this document nowhere gives adequate attention to the role of population growth as a primary variable driving the growth of GNP and energy use. Over the last 100 years, population growth has been roughly equal in importance to increasing energy use per head in producing the growth of total U.S. energy demand. In the most recent part of this period, population growth has been relatively slow and growth of energy use per head relatively rapid, but population's contribution is still far from negligible. The difference between energy requirements in 2010 depending on whether U.S. population growth in the intervening period is "medium" or "low" is significant even if expressed simply in quads. (See chapter 11.) Given, however, that quads "on the margin" will be supplied from the most expensive and perhaps environmentally disruptive sources, the importance of the incremental contribution of extra population growth will be greater than that suggested by simple addition.

If population growth makes energy problems less tractable for the United States, then of course, its contribution is even less welcome in developing countries where providing adequately for people now alive is already a formidable problem. And, for the world as a whole, the chances of providing enough energy at tolerable economic and environmental costs will be far greater if the population stabilizes at, say, 6 billion people than if it stabilizes at 12 billion. For the United States and for the world, the question of how best to discourage or otherwise limit population growth is a thorny one, for which it is widely admitted there are no easy answers. The question's thorniness and political sensitivity, however, are matched by its importance to energy futures and to practically every other ingredient of the human predicament. The issue will not be made to go away—indeed it will be made much worse—by ignoring it, which sadly is essentially what CONAES did.

2-3 HARVEY BROOKS

This is especially true for long-lasting equipment when future prices are considered. Even if the consumer made a rational trade-off between first cost and lifetime energy cost as estimated from today's prices, he would be unlikely to fully anticipate future price increases in his calculations. In my view it is the necessity of anticipating future prices that provides an important justification for mandatory standards. An equally important consideration is the effect of reduction in aggregate energy demand on world energy prices, which is also not taken into account in the calculations of the individual consumer.

2-4 EDWARD J. GORNOWSKI

While generally supporting market forces as an allocator of resources, this chapter states in several places and for different situations that government regulations and mandates are often required. In special situations, mandates might be appropriate. However, in general, mandates are not as sound as market forces since they obscure the cost of reaching a target and are not adaptable to changing circumstances.

2-8 HARVEY BROOKS

That personal transportation would be reduced by denser living patterns is fairly clear. It is not so obvious for freight, however, since this would involve reclustering of manufacturing activities in city centers, which is contrary to all recent trends, and probably implies regression with respect to environmental standards and manufacturing efficiencies. Some gains might be achieved through clustering of closely interrelated industries in industrial parks, which might also result in energy savings through the sharing of cogeneration facilities. However, without much more detailed analysis of possible future industrial location patterns, and their detailed energy and transportation requirements, it is dangerous to accept overfacile generalizations based on intuitive impressions.

2-11 BERNARD I. SPINRAD

This statement of the potential of selling heat from utility stations is unduly pessimistic. There have been a number of cooperative projects on "energy centers" that include both centralized electricity and steam production. One example, by Consumers Power of Michigan and the Dow Chemical Company, for a dual-purpose nuclear station, seems to have failed more because of licensing delay than for any other reason. Others involving oil and chemical complexes on the Gulf Coast are moving along.

2-13 HARVEY BROOKS

This discussion of cogeneration does not make sufficiently clear the complex trade-offs between market penetration of cogeneration and increased demands for fluid fuels. Cogeneration with coal or on-site coal-derived fluids is not practical today, and in the future would be considerably more expensive than cogeneration with direct use of fluid fuels, especially natural gas. Only where cogeneration replaces purchased electricity supplied from oil- or gas-fired central generating stations would there be a net saving of fluid fuels. The most economical cogeneration installations, and hence those with the fastest potential market penetration, would probably be those using natural gas. From an import savings standpoint they would be most desirable in cases where they displace centrally generated electricity produced with residual oil or distillates. Incentives and regulations to facilitate and encourage cogeneration should be adjusted to take into account the degree of displacement of imported oil offered by each project. However, to the extent that natural gas can displace oil in other applications where coal is not an option, this must be considered also in evaluating cogeneration projects.

2-14 HARVEY BROOKS

The assumption also neglects the effect of political climate surrounding various types of energy systems apart from their relative costs (for example, the political resistance to nuclear plants and the favorable political climate for solar energy). History teaches us, though, that these political factors can be quite volatile in response to various external events such as the Indian nuclear explosion, the accident at Three Mile Island, or an oil embargo.

2-15 HARVEY BROOKS

In my opinion, CONAES did not sufficiently face up to this issue. The price increases assumed in order to reach the lowest energy growth projections are very large even in comparison with the increases that have occurred in the period 1972–1979. They will very probably require taxation on various forms of energy which will raise consumer prices substantially above the market prices, even taking into account the effect of OPEC actions. The political resistance to much more modest price increases that has frustrated the implementation of a national energy policy since 1974 is indicative of the practical difficulty of carrying out the policies necessary to achieve large reductions in the energy/GNP ratio. An alternative would be to use general tax revenues to subsidize conservation investments in all sectors, but concentrating especially on low-income consumers, nonprofit institutions, and small businesses. Because of the effects of U.S. aggregate demand on OPEC prices (and hence on all domestic energy prices if a free market is allowed to operate domestically), the investment of general tax revenues in this manner might have substantial benefits on equity through restraining the rate of increase of world prices. This is true because the source of revenue used for the conservation subsidies is more progressive in relation to income than the impact of higher energy prices. A strategy of conservation investment financed by progressive taxes would be politically more popular than energy taxes to restrain demand growth. However, it would probably be much more complicated to administer, since it would involve millions of governmental decisions to evaluate millions of individual consumer investments and specific conservation technologies. In practice, it may tend to freeze in specific, and ultimately obsolescent, technologies and discourage innovation that would yield much greater conservation in the long run.

2-17 EDWARD J. GORNOWSKI

Conservation must play a major role in improving the future energy situation in the United States. However, emphasis should also be placed on encouraging the development and production of this nation's domestic oil and gas. This will provide a balanced, two-pronged attack to the U.S. energy problem.

CHAPTER 3

3-1 HARVEY BROOKS

There is some indication that the potential for unconventional gas sources, such as Devonian shales, coal seams, and geopressured brines, may have been underestimated, especially in relation to prospects for ultimate decontrol of gas prices and proposed tax benefits for unconventional gas sources. Some recent estimates of natural gas production by 2000 (such as those of the Electric Power Research Institute) have run as high as 30 quads. Since gas is substitutable for fuels derived from petroleum in a large number of applications, new sources of gas may represent the most likely favorable future development to offset the forecast rise in demand for oil imports. Hence energetic further exploration and assessment of this possibility are warranted. I would be inclined to give this even higher priority than the suggested pioneer plants for oil shale and coal-derived liquids. However, like other possible favorable developments in supply, enhanced natural gas supplies cannot be counted upon in prudent planning in comparison with fully developed technologies such as coal and nuclear electric power generation, which are already commercially proven.

CHAPTER 4

4-1 HARVEY BROOKS

The financing of expansion of coal transport capacity will be critically dependent upon long-term supply contracts and hence on stable expectations as to environmentally acceptable types of coal.

CHAPTER 5

5-1 EDWARD J. GORNOWSKI

Other nations do not view the matter of the contribution of nuclear power to nuclear weapons proliferation and nuclear war as being as serious a problem as the United States does. Another aspect of this matter which is not mentioned is the contribution that nuclear power can make in decreasing the energy shortage and thus reducing international tension.

5-2 EDWARD J. GORNOWSKI

A number of comments in the introduction reflect more the discussion in the media than the results of an independent and objective study of nuclear power, particularly in the areas of safety and protection against diversion of nuclear materials.

5-3 LUDWIG F. LISCHER

That assumption carries the risk of future electricity shortages. There are experts who believe that the fraction of 30 percent in 1978 will grow to 50 percent by the year 2000.

5-5 EDWARD J. GORNOWSKI

The mill tailing problem is real and the past cannot be undone, but the new regulations will not permit a repeat of the past mishandling, and the report should recognize this intent.

5-8 EDWARD J. GORNOWSKI

The energy resource benefits of the nuclear option, including the plutonium breeder, appear to outweigh any plausible risks of proliferation and diversion and could justify significant investment in upgrading safeguards.

5-11 LUDWIG F. LISCHER

There is no mention in this paragraph or elsewhere in the summary recommendations of the Clinch River breeder reactor. Although it is briefly discussed further on in the chapter, and as stated the committee was divided on the subject, I believe it important enough to mention in the summary. A commercially available LMFBR may well be a necessity by the year 2000. In order for the United States to proceed with the orderly and timely development of the LMFBR, the construction of the Clinch River breeder reactor is of vital importance as part of a prudent and responsible national energy policy. See my statement 5-18, Appendix A.

5-13 JOHN P. HOLDREN

If there is *no* level of compensation and persuasion at which *any* state will host a repository voluntarily, then one is no longer speaking of a modest technical/economic burden to be tallied up on nuclear power's ledger under "waste disposal," but of a large political cost. I am wary of "solutions" that require the imposition of unwanted burdens, concrete or psychological, on large minorities in the name of the common good.

5-14 LUDWIG F. LISCHER

This appears to be an unrealistic date. Not even a commitment made today for a standard LWR could be in service before 1992 under today's licensing procedures. See my statement 5-17, Appendix A.

5-16 EDWARD J. GORNOWSKI

This section on isotope separation is out of date and could be considerably updated. For example, the date given for full-capacity operation of the new gas-centrifuge plant should be corrected to 1993 (from 1988).

5-17 LUDWIG F. LISCHER

No mention is made of the *likelihood* of developing a converter reactor economy. In my view, the nuclear industry (and very likely the federal government) will not undertake the development and commercialization of advanced converters because at best the converter is an interim solution (good for perhaps 20–30 years), and neither the suppliers nor the users will believe that there is sufficient incentive to bring it to commercial status. Without reprocessing, the growth of nuclear power will be slow at best. No manufacturer could afford the high development costs to bring an interim nuclear reactor system to licensable status. If reprocessing is permitted, any advanced converter would have to compete with the fast breeder reactor. In that event, the breeder is the obvious choice.

5-18 LUDWIG F. LISCHER

The Clinch River breeder reactor is not an inappropriate facility. Its design (now 75 percent complete) after 3 years of NRC licensing review, before that was stopped, has been continually updated. It has flexibility for accommodating a variety of nuclear fuel cycles and core designs that might come from the International Nuclear Fuel Cycle Evaluation (INFCE). The United States has operated successfully the 20-MWe EBR-II at Idaho Falls for more than 15 years. CRBR represents the next logical step (350–400 MWe) in scaling up plant size. It is essential to construct and operate a unit of this size prior to proceeding with commercial designs on the order of 1000 MWe. At the present stage of development, one learns little from more paper and analytical work, as proposed by some. The direct scale-up from 20 MWe to 1000 MWe is simply too large a step for prudent engineering and design.

5-24 LUDWIG F. LISCHER

Studies made by the Nuclear Safety Analysis Center using the Three Mile Island sequence of events indicate that even if the core had melted through the reactor vessel (which it did not), it could not have melted through the concrete below because of the water in the containment. Calculations show that quenching and cooling would be effective and the containment would not be breached.

5-27 JOHN P. HOLDREN

I believe that this and the several preceding paragraphs, while giving a superficial impression of balance, in fact lean consistently toward greater optimism about the neutrality of WASH-1400 than is warranted. Unreviewed criticisms that purport to show excessive conservatism in WASH-1400 are described as "documented," while errors in the opposite direction that have been confirmed by many independent analysts—notably the treatment of common-mode failures and the use of median values in place of means for the computation of actuarial risk—are couched in conditionals. In the case of common-mode failures, the reader might well suppose that the Risk and Impact Panel could not even decide on the *direction* of the effect, which I believe is incorrect. The fact is that the two clearest errors in WASH-1400—the treatment of common modes and the use of medians for means—both lead to underestimations of risk.

The Nuclear Energy Policy Study Group's report (*Nuclear Power: Issues and Choices* (Cambridge, Mass.: Ballinger Publishing Co., 1977), hereinafter NPIC) shows, moreover, that WASH-1400's "central estimate" for excess cancer deaths from a given number of person-rem delivered at low doses was about 3 times lower than the lower limit given in the 1972 report of the Committee on Biological Effects of Ionizing Radiation (BEIR) of the National Academy of Sciences and 30 percent lower than the lower limit given in the 1976 report of the National Council on Radiation Protection and Measurements (NPIC, p. 168). Had WASH-1400's central estimate for this dose-response relation corresponded to the central estimates of the BEIR or National Council on Radiation Protection studies, the actuarial risk from reactor accidents would have been 3–5 times higher from this change alone.

Finally, NPIC's statement that WASH-1400 could be as much as a factor of 500 low refers explicitly to probability only, not to actuarial risk (NPIC, p. 179). If the possibility that the consequence estimates are low is taken into account at the same time, the conclusion is that the actuarial risk is unlikely to be greater than about 3000 times the WASH-1400 "central estimate" of 0.024 latent cancer deaths per reactor year. This "upper limit" should be seen to be well above the "upper limits" for coal usually cited, if one took into account that coal's excess deaths from aggravated respiratory illness deprive the victims of far less life expectancy than do nuclear power's cancers.

5-32 LUDWIG F. LISCHER

But to clear up some of the misinformation, calculations show that the activity of the waste in curies compares to the curies in the total amount of original ore related to the fuel from which the waste came.

CHAPTER 6

6-1 BERNARD I. SPINRAD

However, this liability limitation corresponds to a very small financial subsidy, because accidents for which liability would exceed the maximum that is insured are, at worst, extremely unlikely. At best, given normal industrial learning about safety practices, they will not occur at all.

6-2 HARVEY BROOKS

It is worth emphasizing that the program required to reach the President's goal is a drastic one, involving as it does federal mandating of the installation of many technologies that are far from being economical. I doubt whether this would be accepted politically without heavy subsidies, and even then probably only in the wake of a very severe energy crisis brought on by another embargo or a complete nuclear moratorium.

6-3 HARVEY BROOKS

Solar heating and cooling would be much more nearly competitive if fossil fuels and electricity were priced at their actual replacement costs rather than their average cost.

6-4 BERNARD I. SPINRAD

Crediting solar energy with savings obtained by combining energy-conserving building practices with passive solar building heating (or with active solar heating, for that matter) considerably overstates the contribution of solar energy per se. We have already, in chapter 2, indicated the degree to which conservation can contribute to alleviating energy requirements. The inclusion of conservation effects here counts this contribution a second time.

CHAPTER 7

7-2 HARVEY BROOKS

This statement seems unduly optimistic for fusion; with respect to solar energy probably only photochemical methods of solar fuel production and satellite solar power are less advanced than fusion.

7-3 HARVEY BROOKS

It seems very unlikely to me that fusion reactors could be economically superior to solar photovoltaics as electricity generators by the date of 2020 at which fusion might first reach large-scale use. Only if there are unexpected positive breakthroughs for fusion and unexpected economic difficulties with photovoltaics does the fusion option look like a reasonable hedge. Thus I think this chapter, while realistic as to fusion prospects, is too optimistic in the context of the likely prospects for competing alternatives. The principal justification for fusion research may be the discovery of some entirely new application of the technologies that we do not now foresee. In this perspective the most attractive aspect of fusion research is that it stretches the state of the art in so many different areas of advanced technology, and thus has a high potential for generating important by-product technologies useful in other areas of application.

CHAPTER 9

9-1 HARVEY BROOKS

The preceding discussion seems to imply, but does not say explicitly, that the multiplicity of legislative mandates and regulatory agencies is unnecessary. I wish the committee had been willing to state this more forcefully. I believe that equal or greater safety and environmental quality could be achieved with better rationalized legislation and rule setting; yet most attempts in this direction have fallen down. Part of the problem seems to stem from a tacit assumption that all risks lie on the side of the introduction of technology, and that therefore we have to erect multiple barriers in the way of development so that through pluralistic regulation the assurance against unidentified risk is increased. I believe this confidence in multiple hurdles is misplaced.

9-2 HARVEY BROOKS

Eventually the effort to reduce risks below a certain level is likely to introduce other risks, usually more subtle and unforeseeable than the original risk. See S. Black, F. Niehaus, and D. Simpson, *How Safe Is Too Safe?* (Salzburg, Austria: International Institute for Applied Systems Analysis (WP-79-68), June 1979).

9-8 HARVEY BROOKS

I cannot agree with this statement, at least not without qualifications. Worst-case calculations for large dams indicate that they can cause fatalities comparable to those resulting from "worst-case" nuclear accidents, and the number of immediate fatalities is probably greater in the case of dams. Furthermore, the possibility of learning from small accidents to increase safety is much less for dams than for nuclear reactors.

9-12 HARVEY BROOKS

The harvesting of wood on a large scale would reduce the average amount of carbon stored as biomass in forests and would thus contribute some CO_2 to the atmosphere. OTEC would also release CO_2 to the atmosphere from the deep oceans. However, per unit of energy produced, these effects would probably be less than a third those from fossil fuels.

9-14 HARVEY BROOKS

However, hydroelectricity may be attractive on other grounds. It generates no air pollution and has a low accident rate, although accidents in the construction of dams, as well as the threat of catastrophic dam failure, are significant risks.

9-15 HARVEY BROOKS

Depending on the source of coal, if one includes the land area necessary for mining enough for the full lifetime of a plant, solar and coal-fired electricity are comparable.

9-16 BERNARD I. SPINRAD AND HARVEY BROOKS

Please refer to the discussion of this topic in chapter 5, where it is pointed out that the civil liberties argument is an unprovable allegation. We see no grounds for taking it as a basis for policy.

9-17 BERNARD I. SPINRAD AND HARVEY BROOKS

We would replace "if only" by "but only" in this sentence. This is because centralized energy systems serve population centers more efficiently than decentralized ones do, and serve interregional equity better.

9-20 HENRY I. KOHN

Note the *non*comparability of some of the risks in this comparison. For nonradiation systems, there are property losses and immediate injury. For radiation accidents, in addition and usually more importantly, there are cancer and genetic effects.

9-21 JOHN P. HOLDREN

The situation is even more ambiguous than the text suggests because it is not actually possible to do what is implied by the words, "if one takes all health effects into account." Specifically, the statement that coal's health effects "appear to be a good deal greater" than nuclear's requires either that one ignore the million-year accumulation of excess cancer deaths plausibly attributable to uranium-mill tailings if the linear hypothesis is accepted (an excess that could amount to 30–400 deaths per GWe-year of electricity, according to the Academy's own recent report, *Risks Associated with Nuclear Power: A Critical Review of the Literature,* Committee on

Science and Public Policy, Committee on Literature Survey of Risks Associated with Nuclear Power (Washington, D.C.: National Academy of Sciences, April 1979)), or that one assume without any quantitative support that the health effects of today's coal use over the next million years will be as large or larger. Suppose this troublesome issue, which is completely unresolvable at present, is neglected. Suppose one also neglects genetic illness from nontailings routine emissions, for which there is an uncertainty range spanning at least a factor of 20 on the nuclear side (extending on the high end to consequences about equal to those of the excess cancers) and for which no quantitative estimates at all are available on the coal side. Suppose one neglects, further, the health effects of emissions of oxides of nitrogen, hydrocarbons, and trace metals from coal combustion, which are generally presumed (but not proved) to be smaller than those of the sulfur oxides and generalized particulates that existing dose-response relations include. In the restricted comparison that remains after all these troublesome factors are excluded, the widespread view that coal comes out "a good deal" worse than nuclear rests on various combinations of the following three errors: (1) attribution to coal of practices in mine health and safety, power plant siting, and pollution control that are illegal or inconceivable (and usually both) in the new, large facilities relevant to comparisons with nuclear power; (2) failure to note that the "excess deaths" attributed to air pollution from coal typically deprive the victims of far less life expectancy than the cancers attributed to nuclear power (the ratio is almost certainly more than 10:1); (3) refusal to take seriously the upper end of the range of responsible opinion on conceivable nuclear accident risk, while taking completely seriously the upper-limit estimates of excess deaths from air pollution. When these errors are avoided, the "best estimates" of the years of life lost per GWe-year of electricity from coal and nuclear differ by an amount small compared to the uncertainties associated with each.

9-22 HARVEY BROOKS, DAVID J. ROSE, AND BERNARD I. SPINRAD

Although we agree with the statement, we fear it might be interpreted as implying that government planning should accept the most irrational appraisals of risk put forward by politically active minorities.

CHAPTER 10

10-1 EDWARD J. GORNOWSKI

Generally speaking, this chapter appears too optimistic on the future prospects for oil and gas discoveries, both in the United States and worldwide. The chapter holds out hope that oil production will be maintained approximately constant through 2010 in the United States and through 2050 worldwide. It relies primarily on USGS circular 725 for ultimate U.S. oil reserves. This circular was developed in 1975, is out of date, and is generally considered to be optimistic based on reappraisals now in progress.

CHAPTER 11

11-4 ROBERT H. CANNON, JR.

Assumptions of 2 percent GNP growth (compared to everyone else's 2.9–3.7 percent) and of electricity prices rising as rapidly as fuel prices make all the projections of Table 11-12 very low.

11-5 ROBERT H. CANNON, JR.

Assuming 3.4 percent GNP growth would make the 2010 quad values in Figure 11-4 roughly as follows: scenario A, 125; scenario B, 160; scenario C, 230; scenario D, 270.

11-7 ROBERT H. CANNON, JR.

I believe it terribly dangerous to extend our experience with a very small range of economic quantities to predict what will happen far beyond that range.

11-8 ROBERT H. CANNON, JR.

Price elasticity is the local slope of a highly nonlinear cause-effect curve. To estimate that slope for a range far beyond experience is highly speculative.

APPENDIX B
Glossary of Technical Terms

ACCELERATOR (particle accelerator): A device for imparting large kinetic energy to electrically charged elementary particles such as electrons, protons, deuterons, and helium ions through the application of electrical and/or magnetic forces. Common types of particle accelerators are direct voltage accelerators, cyclotrons, betatrons, and linear accelerators.

ACTINIDES: A group name for the series of radioactive elements from element 89 (actinium) through element 103 (lawrencium). The series includes uranium and all the man-made transuranic elements.

BINARY CYCLE: An energy recovery system based on the transfer of heat from one fluid (e.g., hot brine from a geothermal well) to a second fluid (e.g., pure water or an organic liquid) from which the heat is ultimately extracted for use.

BIOCONVERSION: The conversion of organic wastes into methane (equivalent to natural gas) through the action of microorganisms.

BLANKET: A layer of fertile material such as uranium-238 or thorium-232 that is placed around the core of a fission or fusion reactor. Its major function is to produce fissile isotopes from fertile blanket material.

BOILING-WATER REACTOR (BWR): A light water reactor that employs a direct cycle; the water coolant that passes through the reactor is converted to high-pressure steam that flows directly through the turbines.

BREEDER REACTOR: A nuclear reactor that produces more fissile material than it consumes. In fast breeder reactors, high-energy (fast) neutrons produce most of the fissions, while in thermal breeder reactors, fissions are principally caused by low-energy (thermal) neutrons.

BREEDING RATIO: The ratio of the number of fissionable atoms produced in a breeder reactor to the number of fissionable atoms consumed in the reactor. The "breeding gain" is the breeding ratio minus 1.

Btu (British thermal unit): The amount of energy necessary to raise the temperature of one pound of water by one degree Fahrenheit, from 39.2 to 40.2 degrees Fahrenheit.

CAPACITY FACTOR: The ratio of the amount of product (e.g., electrical energy or geothermal brine) actually produced by a given unit per unit of time to its maximum production rate over that period. Also called "load factor."

COGENERATION: The generation of electricity with direct use of the residual heat for industrial process heat or for space heating.

COMBINED CYCLE: A combination of a steam turbine and a gas turbine in an electrical generating plant, with the gas-turbine exhaust heat used in raising steam for the steam turbine.

CONVERSION RATIO: The ratio of the number of atoms of new fissionable material produced in a converter reactor to the number of atoms of fissionable fuel consumed. See "breeding ratio."

CONVERTER REACTOR: A reactor that produces some fissionable material, but less than it consumes. In some usages, a reactor that produces a fissionable material different from the fuel burned, regardless of the ratio. In both usages the process is known as conversion.

CURIE: A measure of intensity of the radioactivity of a substance; i.e., the number of unstable nuclei that are undergoing transformation in the process of radioactive decay. One curie equals the disintegration of 3.7×10^{10} nuclei per second, which is approximately the rate of decay of one gram of radium.

DEPLETION ALLOWANCE: A tax credit based on the permanent reduction in value of a depletable resource that results from removing or using some part of it.

DRY HOT ROCK (geothermal): See "hot dry rock."

ELASTICITIES OF DEMAND: The arithmetic relations used by economists in quantifying the change in demand for a commodity in response to a change in another economic quantity. In this report, elasticities of demand in terms of price and income are especially important. These elasticities are calculated as the ratio of the percentage change in demand to the percentage change in price or income that evokes it.

FERTILE MATERIAL: A material, not itself fissionable by thermal neutrons, that can be converted into a fissile material by irradiation in a reactor. There are two basic fertile materials, uranium-238 and thorium-232. When these materials capture neutrons, they are partially converted into plutonium-239 and uranium-233, respectively.

FLASHING: The rapid change in state from a liquid to a vapor without visible boiling, resulting usually from a sudden reduction in the pressure maintained on a hot liquid.

FLUIDIZED BED: A body of finely divided particles kept separated and partially supported by gases blown through or evolved within the mass, so that the mixture flows much like a liquid.

FLY ASH: Fine solid particles of noncombustible ash entrained in the flue gases arising from the combustion of carbonaceous fuels. The particles of ash may be accompanied by combustible unburned fuel particles.

FUEL CELL: A device that produces electrical energy directly from the controlled electrochemical oxidation of fuel. It does not contain an intermediate heat cycle, as do most other electrical generation techniques.

FUEL CYCLE: The various processing, manufacturing, and transportation steps involved in producing fuel for a nuclear reactor, and in processing fuel discharged from the reactor. The uranium fuel cycle includes uranium mining and milling, conversion to uranium hexafluoride (UF_6), isotopic enrichment, fuel fabrication, reprocessing, recycling of recovered fissile isotopes, and disposal of radioactive wastes.

GAS-CENTRIFUGE PROCESS: A method of isotopic separation in which heavy gaseous atoms or molecules are separated from light atoms or molecules by centrifugal force.

GASEOUS DIFFUSION: A process used to enrich uranium in the isotope uranium-235. Uranium in the form of a gas, uranium hexafluoride (UF_6), is forced through a thin porous barrier. Since the lighter gas molecules containing uranium-235 move at a higher velocity than the heavy molecules containing uranium-238, the lighter molecules pass through the barrier more frequently than do the heavy ones, producing a slight enrichment in the lighter isotope. Many stages in series are required to produce material enriched sufficiently for use in a light water reactor.

GEOPRESSURED RESERVOIR (geothermal): A hydrothermal reservoir in which the pore fluid is confined under pressure significantly greater than normal hydrostatic pressure, developed principally by the weight of overlying rocks and sediments. Also called "overpressured" and "geopressurized" reservoirs.

GEOTHERMAL GRADIENT: The rate at which the temperature of the earth increases with depth below its surface. This varies widely from place to place, but the average or "normal" geothermal gradient is about 30 degrees Celsius per kilometer of depth (16.5 degrees Fahrenheit per thousand feet).

GROSS DOMESTIC PRODUCT (GDP): Gross national product minus net factor payments abroad (such as income from foreign investments and wages paid to foreign workers). GDP is preferable to gross national product (q.v.) as a measure for international comparisons.

GROSS NATIONAL PRODUCT (GNP): The total market value of the goods and services produced in a national economy, during a given year, for final consumption, capital investment, and governmental use. (Note that GNP does not include the value of intermediate goods and services sold to producers and used in the production process itself.) See "gross domestic product."

HEAVY WATER: Water containing significantly more than the natural proportion (1 in 6500) of heavy hydrogen (deuterium) atoms to ordinary hydrogen atoms. Heavy water is used as a moderator in certain reactors because it slows down neutrons effectively and also has a low cross section for absorption of neutrons.

HIGH-LEVEL WASTE: A by-product of the operation of nuclear reactors that includes a variety of aqueous wastes from fuel reprocessing and their solidified derivatives, such as alkaline aqueous waste, calcine, crystallized salts, insoluble precipitates, salts of cesium and strontium extracts, and coating wastes from chemical decladding of fuel elements.

HIGH-TEMPERATURE GAS-COOLED REACTOR (HTGR): A graphite-moderated, helium-cooled advanced reactor that utilizes the thorium fuel cycle. The initial core is fueled with a mixture of fully enriched uranium-235 and thorium. When operated in the recycle mode, the reactor is refueled with a mixture of uranium-233 (produced from thorium) with the balance of the fissile material provided from an external source of fully enriched uranium-235.

HOT DRY ROCK (geothermal): Naturally heated but unmelted rock sufficiently low in either permeability or pore-fluid content that wells drilled into it do not yield either hot water or steam at commercially useful rates. To be compared with "hydrothermal reservoir."

HYDROTHERMAL RESERVOIR: A body of porous, permeable rock, gravel, or soil containing natural steam or naturally heated water at a temperature significantly above the average temperature at the earth's surface.

ISOTOPE: One of two or more atoms with the same atomic number (i.e., the same chemical element) but with different atomic weights. Isotopes usually have very nearly the same chemical properties but somewhat different physical properties.

KEROGEN: A solid, largely insoluble organic material, occurring in oil shale, which yields oil when it is heated in the absence of oxygen.

LIGHT WATER REACTOR (LWR): A nuclear reactor that uses ordinary water as both a moderator and a coolant and utilizes slightly enriched uranium-235 fuel. There are two commercial light water reactor types—the boiling-water reactor (BWR) and the pressurized-water reactor (PWR).

LIQUEFIED NATURAL GAS (LNG): Natural gas cooled to −259 degrees Fahrenheit so that it forms a liquid at approximately atmospheric pressure. As natural gas becomes liquid, it reduces in volume nearly 600-fold, thus allowing economical storage and making long-distance transportation economically feasible. Natural gas in its liquid state must be regasified and introduced to the consumer at the same pressure as other natural gas. The cooling process does not alter the gas chemically, and the regasified LNG is indistinguishable from other natural gases of the same composition.

LIQUEFIED PETROLEUM GAS (LPG): A gas containing certain specific hydrocarbons that are gaseous under normal atmospheric conditions but that can be liquefied under moderate pressure at normal temperatures. Propane and butane are the principal examples.

LOAD FACTOR: Capacity factor (q.v.).

LOW-LEVEL WASTE: Generally a solid by-product of special nuclear materials production, utilization, and research and development. Examples of solid low-level waste are discarded equipment and materials, filters from gaseous waste cleanup, ion-exchange resins from liquid waste cleanup, liquid wastes that have been converted to solid form by techniques such as mixing with cement, and miscellaneous trash. Low-level liquid waste is generally decontaminated and released under controlled conditions.

MILLING (uranium processing): A process in the uranium fuel cycle in which ore that contains only about 0.2 percent uranium oxide (U_3O_8) is concentrated into a compound called yellowcake, which contains 80–90 percent uranium oxide.

MODERATOR: A material, such as ordinary water, heavy water, or graphite, used in a reactor to slow down high-velocity neutrons, thus increasing the likelihood of further fission.

NUCLEAR WASTE: The radioactive products formed by fission and other nuclear processes in a reactor. Most nuclear waste is initially in the form of spent fuel. If this material is reprocessed, new categories of waste result: high-level, transuranic, and low-level wastes (as well as others).

PARTICULATES: Microscopic pieces of solids that emanate from a range of sources and are the most widespread of all substances that are usually considered air pollutants. Those between 1 and 10 microns are most numerous in the atmosphere, stemming from mechanical processes and including industrial dusts, ash, etc.

PLUTONIUM: A heavy, radioactive man-made metallic element with atomic number 94, created by absorption of neutrons in uranium-238. Its most important isotope is plutonium-239, which is fissionable.

PRESSURIZED-WATER REACTOR (PWR): A light water moderated and cooled reactor that employs an indirect cycle; the cooling water that passes through the reactor is kept under high pressure to keep it from boiling, but it heats water in a secondary loop that produces steam that drives the turbine.

PRIMARY CONTAINMENT: An enclosure that surrounds a nuclear reactor and associated equipment for the purpose of minimizing the release of radioactive material in the event of a serious malfunction in the operation of the reactor.

PYROLYSIS: Decomposition of materials through the application of heat with insufficient oxygen for complete oxidation.

QUAD: A quantity of energy equal to 10^{15} British thermal units.

REACTOR CORE: The central portion of a nuclear reactor, containing the fuel elements and the control rods.

REPROCESSING: A generic term for the chemical and mechanical processes applied to fuel elements discharged from a nuclear reactor; the purpose is to recover fissile materials such as plutonium-239, uranium-235, and uranium-233 and to isolate the fission products.

RESERVES: Resources that are known in location, quantity, and quality and that are economically recoverable using currently available technologies.

RESOURCE (energy): That part of the resource base believed to be recoverable using only current or near-current technology, without regard to the cost of actually recovering it. To be distinguished from both "resource base" and "reserves" (q.v.).

RESOURCE BASE (energy): The total quantity of energy or of any given energy-producing or energy-related material that is estimated to exist in or on the earth or in its atmosphere, independent of quality, location, or the engineering or economic feasibility of recovering it.

SCRUBBER: An air pollution control device that uses a liquid spray for removing pollutants such as sulfur dioxide or particulate matter from a gas stream by absorption or chemical reaction.

SECONDARY RECOVERY: Methods of obtaining oil and gas by the augmentation of reservoir energy, often by the injection of air, gas, or water into a production formation. (See "tertiary recovery.")

SOLAR CONSTANT: The solar radiation falling on a unit area at the outer limits of the earth's atmosphere.

SPECTRAL-SHIFT REACTOR: A reactor in which a mixture of light water and heavy water is used as the moderator and coolant. The ratio of light to heavy water is varied to change (shift) the energy spectrum of the neutrons in the reactor core. Since the probability of neutron capture varies with neutron velocity, a measure of reactor control is thus obtained.

SYNTHESIS GAS: A fuel gas containing primarily carbon monoxide and hydrogen; it can be used after careful removal of impurities, particularly sulfur compounds, for conversion to methane (high-Btu gas), methanol, liquid hydrocarbons, and a wide variety of other organic compounds.

TAILINGS: Waste material from a separation process. Commonly the finely divided waste from a mineral separation operation.

TAILS: Contraction of "tailings" (q.v.).

TAILS (OR TAILINGS) ASSAY: The percentage of valuable material that remains unrecovered in the tailings of a separation process.

TAR SANDS: Hydrocarbon-bearing deposits distinguished from more conventional oil and gas reservoirs by the high viscosity of the hydrocarbon, which is not recoverable in its natural state through a well by ordinary production methods.

TERTIARY RECOVERY: Use of heat and methods other than air, gas, or water injection to augment oil recovery (presumably occurring after secondary recovery).

THORIUM: A radioactive element of atomic number 90; naturally occurring thorium has one main isotope—thorium-232. The absorption of a neutron by a thorium atom can result in the creation of the fissile material uranium-233.

THROWAWAY FUEL CYCLE: A fuel cycle in which the spent fuel discharged from the reactor is not reprocessed to recover residual plutonium and uranium values.

TRANSURANIC ELEMENTS: Radioactive nuclides generated as fission products from the fissioning of nuclear fuel during reactor operation and as induced activity from the capture of neutrons in fuel cladding, reactor structures, and reactor coolant.

URANIUM: A radioactive element of atomic number 92. Naturally occurring uranium is a mixture of 99.28 percent uranium-238, 0.71 percent uranium-235, and 0.0058 percent uranium-234. Uranium-235 is a fissile material and is the primary fuel of light water reactors. When bombarded with slow or fast neutrons, it will undergo fission. Uranium-238 is a fertile material that is transmuted to plutonium-239 upon the absorption of a neutron.

URANIUM HEXAFLUORIDE (UF_6): A compound of uranium, which is used in gaseous form in the enrichment of uranium isotopes.

YELLOWCAKE: A uranium concentrate that results from the milling (concentrating) of uranium ore. It typically contains 80–90 percent uranium oxide (U_3O_8).

APPENDIX C

Resource Groups, Consultants, and Contributors

Each CONAES panel organized a number of resource groups to focus on particular matters. The members of these groups are listed in this appendix. Affiliations of the resource group chairmen, who served also as panel members, can be found in the front matter of this report.

DEMAND AND CONSERVATION PANEL

BUILDINGS RESOURCE GROUP

Roger S. Carlsmith (Chairman), Eric A. Hirst, David A. Pilati

ECONOMICS RESOURCE GROUP

L. Duane Chapman (Chairman), Ernst Berndt, Timothy D. Mount, Timothy T. Tyrrell

INDUSTRY RESOURCE GROUP

Macauley Whiting (Chairman), Richard W. Barnes, Charles A. Berg, Robert T. Clark, William H. Davis, Thomas Gross, Herbert F. Kraemer, T. L. Nabors, Thomas O'Connor, James A. Palmer, E. P. Scheu, Allen C. Sheldon, Robert G. Uhler

INTEGRATION RESOURCE GROUP

Clark Bullard (Chairman), Ernst Berndt, L. Duane Chapman, Joel Darmstadter, Lee Schipper, Robert G. Uhler

TRANSPORTATION RESOURCE GROUP

R. Eugene Goodson (Chairman), Norman M. Bradburn, Stanley Feder, Victor Ferkiss, Kenneth Friedman, David Hawkins, John Hemphill, Donald Igo, Leon N. Moses, Jeffrey L. Staley

RISK AND IMPACT PANEL

CLIMATOLOGICAL EFFECTS RESOURCE GROUP

Stephen Schneider (Chairman), Robert Charlson, William Kellogg, Ralph Rotty, Richard Temkin, J. Dana Thompson

ECOSYSTEM IMPACTS RESOURCE GROUP

John Harte (Chairman), Beverly Berger, F. H. Bormann, Daniel Botkin, Robert Curry, Mark Davidson, Mohamed M. El-Gassier, Kathy Fletcher, Donald Grether, Joel Hedgpeth, Andrew Jacobs, Alan Jassby, Laura B. King, Estella Leopold, Harold Malde, Fred Nichols, Robert Risebrough, Edmund Schofield, O. C. Taylor, Oleh Weres

EMISSIONS AND PATHWAYS RESOURCE GROUP

David Okrent (Chairman), Robert Avery, Leslie Burris, Jr., Stephen Gage, Joseph Hendrie, William Higinbotham, William Kastenberg, Terry Lash, Harold Lewis, Thomas Moss, Donald Schweitzer, Theodore B. Taylor

HEALTH EFFECTS RESOURCE GROUP

Carl M. Shy (Chairman), Kenneth Bridbord, James Crow, Carter Denniston, Frederick de Serres, John Finklea, Herschel Griffin, Leonard Hamilton, John Storer, Niel Wald

SOCIOPOLITICAL EFFECTS RESOURCE GROUP

David Sills (Chairman), Dean E. Abrahamson, H. Paul Friesema, Walter Isard, Samuel Klausner, Todd La Porte, Denton E. Morrison, Dorothy

Nelkin, Marvin Olsen, Elizabeth Peelle, Paul Slovic, Roger Van Zele, C. P. Wolf

SUPPLY AND DELIVERY PANEL

BREEDER REACTOR RESOURCE GROUP

Milton Levenson (Chairman), Simcha Golan (Deputy), Joseph R. Dietrich, George W. Hardigg, Herbert G. MacPherson, Roger S. Palmer, Robert H. Simon, William E. Unger

COAL CONVERSION RESOURCE GROUP

Eric H. Reichl (Chairman), Martin A. Elliott, J. A. Phinney, Howard Siegal

ELECTRICITY CONVERSION RESOURCE GROUP

Herman M. Dieckamp (Chairman)

ENERGY RESOURCES RESOURCE GROUP

James Boyd (Chairman). *Coal:* Robert Stefanko (Chairman), William Bellano, Robert W. Erwin, Earl T. Hays, Thomas C. Kryzer, William N. Poundstone, Michael Trbovich. *Lithium:* Thomas L. Kesler (Chairman), Keith Evans, Ihor Kunasz, James Vine. *Oil and Gas:* Charles J. Mankin (Chairman), Kenneth Crandall, Thomas M. Garland, Ralph W. Garrett, Jr., Richard J. Gonzalez, Michael T. Halbouty, Richard L. Jodry. *Oil Shale:* James H. Gary (Chairman), John Dew, John R. Donnell, John Hopkins, Robert McClements, Jr., Richard D. Schwendinger. *Uranium:* Leon T. Silver (Chairman), Jack Grynberg, Joseph B. Rosenbaum, David S. Robertson, Arnold J. Silverman

FUSION ASSESSMENT RESOURCE GROUP

Peter L. Auer (Chairman), Keith Brueckner, Sol J. Buchsbaum, William Gough, Henry Hurwitz, Gerald Kulcinsky, Michael Lotker, Marshall N. Rosenbluth, Donald Steiner

GAS DISTRIBUTION RESOURCE GROUP

Derek Gregory (Chairman), Henry R. Linden, Harold J. Mall, William C. McDonnell

GEOTHERMAL RESOURCE GROUP

Morton C. Smith (Chairman), James H. Barkman, Allen G. Blair, Clarence H. Bloomster, Myron H. Dorfman, Harry W. Falk, Jr., Donald F. X. Finn, Bob Greider, John H. Howard, Richard L. Jodry, J. F. Kunze, A. W. Laughlin, L. J. P. Muffler, Carel Otte, John C. Rowley, Jefferson W. Tester

NUCLEAR CONVERTERS AND FUEL CYCLE RESOURCE GROUP

John W. Landis (Chairman), Peter W. Camp, Eugene Critoph, Andres de la Garza, Ralph W. Deuster, Albert J. Goodjohn, Robert M. Jefferson, William M. Pardue, John A. Patterson, Alfred M. Perry, Louis H. Roddis, Jr., John J. Taylor, Richard C. Vogel, Alan K. Williams, Warren F. Witzig

OUTPUT RESOURCE GROUP

Donald G. Allen (Chairman), Bruce C. Netschert, Harry Perry, A. C. Stanojev

PETROLEUM RESOURCE GROUP

Allen E. Bryson and Theodore Eck (Co-Chairmen). *Domestic Refining and Facilities:* Ben C. Ball, Jr., Allen E. Bryson, Robert C. Gunness, Harold L. Hoffman, L. R. Roberts, C. Davis Wrigley. *Imports:* Theodore Eck, William A. Johnson, Stuart E. Watterson, Jr.

SOLAR ENERGY RESOURCE GROUP

Melvin K. Simmons (Chairman), Bruce Anderson, Piet Bos, Sheldon Butt, William Dickinson, Elmer L. Gaden, Joseph C. Grosskreutz, Ronal W. Larson, Abrahim Lavi, Aden Meinel, Marshal Merriam, Michael J. Mulcahy, Rosalie Ruegg, Roger Schmidt, Joseph L. Shay, William A. Thomas, Martin Wolf

SYNTHESIS PANEL

CONSUMPTION, LOCATION, AND OCCUPATIONAL PATTERNS RESOURCE GROUP

Laura Nader (Chairperson), Peter Benenson, Norman Bradburn, Paul Craig, Lester Lave, David Pilati, Lee Schipper, Sam Schurr

CURRENT DECISION-MAKING RESOURCE GROUP

David Cohen (Chairman), Charles O. Jones, Lester Lave, David Masselli, Jon M. Veigel, James Walker

MODELING RESOURCE GROUP

Tjalling C. Koopmans (Chairman), Edward G. Cazalet, Kenneth C. Hoffman, William Hogan, Lester Lave, Alan S. Manne, William D. Nordhaus, Milton F. Searl, Philip K. Verleger, Jr., David O. Wood

INTERPANEL RESOURCE GROUPS

INTERNATIONAL RESOURCE GROUP

Hendrik S. Houthakker (Chairman), Kenneth E. Boulding, Harvey Brooks, W. Kenneth Davis, Michael Kennedy, Lester Lave, David Rose, David Sills, Bernard I. Spinrad

SCENARIO RESOURCE GROUP

Lester Lave (Chairman), Donald G. Allen, Norman Bradburn, Clark W. Bullard, Carter Denniston, John H. Gibbons, Joseph Hendrie, John W. Landis, Alan S. Manne, Lee Schipper, Philip K. Verleger, Jr., Cathy J. Witty

INSTITUTIONS RESOURCE GROUP

Lester Lave (Chairman), Donald G. Allen, Sidney Bernsen, John H. Gibbons, Charles O. Jones, Samuel Klausner, Denton Morrison, Harry Perry, Jon M. Veigel, James Walker, C. P. Wolf

CONSULTANTS AND CONTRIBUTORS

CONAES

Leonard S. Cottrell III, Paul F. Donovan, Hans L. Hamester, James E. Just, Stephen Rattien

DEMAND AND CONSERVATION PANEL

J. Daniel Khazzoom, Vasily Kouskoulas

RISK AND IMPACT PANEL

Virginia T. Bemis, Charles F. Cortese, Frederick L. Frankena, Joseph J. Galin, Bernard Jones, Peter D. Miller, Arthur B. Shostak

SUPPLY AND DELIVERY PANEL

Nicholas P. Biederman, Richard O. Bright, William C. Chambers, David F. Cope, R. Bruce Foster, Hal E. Goeller, Jon B. Pangborn, Ray O. Sandberg, Harold E. Shaw, David A. Tillman

SYNTHESIS PANEL

David H. Davis, Mary Foster, Clarence J. Glacken, Nelson Graburn, DeVerle P. Harris, Robert E. Litan, Robert L. Olson, Roger W. Sant, Richard Schoen, James L. Sweeney, Lester D. Taylor, Cathy J. Witty

APPENDIX D

Publications of the CONAES Study

PANEL REPORTS

National Research Council, *Alternative Energy Demand Futures to 2010,* Committee on Nuclear and Alternative Energy Systems, Demand and Conservation Panel (Washington, D.C.: National Academy of Sciences, 1979).

National Research Council, *Risks and Impacts of Alternative Energy Systems,* Committee on Nuclear and Alternative Energy Systems, Risk and Impact Panel (Washington, D.C.: National Academy of Sciences, in preparation).

National Research Council, *U.S. Energy Supply Prospects to 2010,* Committee on Nuclear and Alternative Energy Systems, Supply and Delivery Panel (Washington, D.C.: National Academy of Sciences, 1979).

RESOURCE GROUP REPORTS

National Research Council, *Supporting Paper 1: Problems of U.S. Uranium Resources and Supply to the Year 2010,* Committee on Nuclear and Alternative Energy Systems, Supply and Delivery Panel, Uranium Resource Group (Washington, D.C.: National Academy of Sciences, 1978).

National Research Council, *Supporting Paper 2: Energy Modeling for an Uncertain Future,* Committee on Nuclear and Alternative Energy Systems, Synthesis Panel, Modeling Resource Group (Washington, D.C.: National Academy of Sciences, 1978).

National Research Council, *Supporting Paper 3: Controlled Nuclear Fusion: Current Research and Potential Progress,* Committee on Nuclear and Alternative Energy Systems, Supply and Delivery Panel, Fusion Assessment Resource Group (Washington, D.C.: National Academy of Sciences, 1978).

National Research Council, *Supporting Paper 4: Geothermal Resources and Technology in the United States,* Committee on Nuclear and Alternative Energy Systems, Supply and Delivery Panel, Geothermal Resource Group (Washington, D.C.: National Academy of Sciences, 1979).
National Research Council, *Supporting Paper 5: Sociopolitical Effects of Energy Use and Policy,* Committee on Nuclear and Alternative Energy Systems, Risk and Impact Panel, Reports to the Sociopolitical Effects Resource Group (Washington, D.C.: National Academy of Sciences, 1979).
National Research Council, *Supporting Paper 6: Domestic Potential of Solar and Other Renewable Energy Sources,* Committee on Nuclear and Alternative Energy Systems, Supply and Delivery Panel, Solar Resource Group (Washington, D.C.: National Academy of Sciences, 1979).
National Research Council, *Supporting Paper 7: Energy Choices in a Democratic Society,* Committee on Nuclear and Alternative Energy Systems, Synthesis Panel, Consumption, Location, and Occupational Patterns Resource Group (Washington, D.C.: National Academy of Sciences, in preparation).
National Research Council, *Supporting Paper 8: Energy and the Fate of Ecosystems,* Committee on Nuclear and Alternative Energy Systems, Risk and Impact Panel, Ecosystem Impacts Resource Group (Washington, D.C.: National Academy of Sciences, in preparation).

Index